Random Vibrations

Random Vibrations: Theory and Applications investigates methods and theories involved in random vibration analyses of linear and nonlinear systems, as well as in predicting random vibration induced failures.

This book is a lucid and well-paced introduction to random vibrations, superbly motivated and illustrated through a wealth of convincing applications in various engineering fields. The strong points of the book are its coverage of weakly stationary and ergodic random processes, spectral analysis of random processes, mode displacement superposition method, equivalent linearization technique for nonlinear random vibrations and an updated definition of rain flow cycle for fatigue analysis. Particularly appealing features of the book are its numerous examples and end-of-chapter exercises. This book offers a clear guide to the formulations and mathematical properties of random vibration analysis techniques, with an emphasis on practical applications rather than mathematical development for its own sake. However, some important mathematical formulas have been explicitly deduced in a detailed manner so that readers can go through the material in this book very smoothly and efficiently.

This book is intended for upper undergraduate and graduate students who are interested in learning advanced techniques for performing random vibration analysis, researchers and scientists investigating linear and nonlinear systems under random external excitations, and aeronautical/civil/mechanical/structural/ocean engineers involved in the design and manufacture of real-world stochastically excited engineering systems.

Random Vibrations
Theory and Applications

Yingguang Wang
School of Ocean and Civil Engineering
Shanghai Jiao Tong University
Shanghai, P. R. China

CRC Press
Taylor & Francis Group
Boca Raton London New York

CRC Press is an imprint of the
Taylor & Francis Group, an **informa** business

Designed cover image: Yingguang Wang

First edition published 2025
by CRC Press
2385 NW Executive Center Drive, Suite 320, Boca Raton FL 33431

and by CRC Press
4 Park Square, Milton Park, Abingdon, Oxon, OX14 4RN

CRC Press is an imprint of Taylor & Francis Group, LLC

ISBN: 978-1-032-55639-0 (hbk)
ISBN: 978-1-032-55640-6 (pbk)
ISBN: 978-1-003-43150-3 (ebk)

DOI: 10.1201/9781003431503

Typeset in Times
by KnowledgeWorks Global Ltd.

Contents

Preface

In the field of engineering design, the evaluation of the safety and reliability of the designed system is considered one of the most important tasks. Due to the fact that many of the factors that have a substantial impact on the final system design cannot be determined in a definitive manner, this task is typically difficult to accomplish. When it comes to the design of an ocean-going ship, for instance, there is often a great deal of uncertainty regarding the sea waves and the ocean currents along the navigation route, as well as the wind loads acting on the vessel. For aerospace, civil, mechanical and ocean engineers, the fundamental goal of studying the random vibration of engineering systems is to assess and understand the uncertainty of these dynamic loadings and their impact on the safety and reliability of the engineered system.

Historically, the dynamic loadings or excitations from sea waves, ocean currents, wind pressure and earthquake motion have been sometimes dealt with in a generally crude manner by selecting either a maximal design load versus time function or the average of several significant design load time functions. Both methods are fundamentally deficient because, although they occasionally offer reasonable (albeit conservative) indications for future loading conditions, they are unable to provide the design engineer with a quantitative and precise answer to the question of system safety and reliability in a rational manner.

Random vibration of engineering systems is an engineering subject that specifically addresses such complicated uncertainty problems in three fundamental steps. The first step is to characterize the random external excitations in the most realistic way by using a statistical approach. The second step involves an excitation-response analysis of the engineering system and finally, a safety and reliability assessment based upon various failure criteria is carried out in the third step. Within the framework of this approach, the uncertainty of the entire engineering system design problem is explored and synthesized at its very source, which is the random external excitation that is under consideration. The following two steps do not contain any additional uncertainty, with the exception of the assumptions that are always present in the system dynamic models (mass, stiffness and damping) and failure criteria (yielding and fatigue). The outcome of this kind of random vibration analysis is a description of the safety and reliability of the engineering system in probabilistic terms. This is highly beneficial for the purposes of engineering system design and planning.

This book is prepared with the aforementioned engineering motivation in mind and has been developed during more than 21 years of teaching and research experience in the School of Ocean and Civil Engineering at Shanghai Jiao Tong University (SJTU). In addition, some of the material in the book was also written based on the author's decade-long experience in the ship design industry before joining the faculty at SJTU in 2003. Due to the fact that the course requires a background in engineering dynamics, advanced engineering mathematics, as well as probability, statistics and random processes, it is primarily appropriate for upper-undergraduate and graduate students who are pursuing degrees in aerospace, civil, mechanical and ocean engineering. For practicing design engineers, this book provides an understanding of the basic concepts and some applications of random vibration of engineering systems to enable them to solve real-world practical problems and to stay updated with the latest research in the field. Throughout the book, the focus has been placed on understanding the engineering relevance involved rather than on rigorous mathematical analysis. However, some important mathematical formulas have been explicitly deduced in a detailed manner so that the prospective reader can go through the material in this book very smoothly and efficiently. For those who find the subject particularly interesting and helpful, this book can serve as a jumping-off point for either more advanced statistical theory study or more complicated engineering and research problems within their respective professions.

This book is structured, in short, as follows: There are a total of ten chapters in the text. Chapter 1 contains a brief introduction to the subject of random vibration. Chapters 2–4 introduce the probabilistic theories of random variables and random processes. The majority of these presentations begin with very basic definitions with no presumption of technical background beyond the senior level in aerospace, civil, mechanical and ocean engineering and are very concise and explicit. Chapters 5–7 present the excitation-response solutions for single-degree-of-freedom linear systems, two-degree-of-freedom linear systems and multi-degree-of-freedom linear systems, respectively. In Chapter 8, a concise introduction is provided to the challenging problem of random vibration of nonlinear systems. Some numerical simulation methods for random vibration analysis are elucidated in Chapter 9. Finally, Chapter 10 deals with the central issue of system or structural safety in relation to first-excursion and fatigue-type failures.

The scope of my presentation is limited by the allowable length of the book and my intention to make a solid presentation of the fundamentals. Therefore, I have deliberately not included material on the random vibration of continuous systems and the tools normally associated with nonlinear random vibrations, such as bi-spectral analysis, Markov processes or the Fokker–Planck equation. Likewise, I have not included the material on nonstationary and non-Gaussian random processes. While these topics hold significance, it is crucial to note that there is no universally applicable approach for solving these problems, and it would be impractical to attempt to cover all the possible solution methods. As a result, some educators might wish to supplement this textbook with more advanced, specialized material as appropriate.

This textbook is advised for a regular two-semester academic year of study, with Chapters 1–6 being assigned to the first semester. The book contains an appropriate number of examples and homework problems to facilitate the student's learning process. To facilitate a more extensive study, references are provided.

It is a joy to acknowledge the assistance I have received during the preparation of the manuscript of this book from some of my colleagues and graduate students here at the School of Ocean and Civil Engineering of SJTU. My work in this research area, some of which have been incorporated into this book, has been largely supported by the State Key Laboratory of Ocean Engineering of SJTU over many years. This continuing financial support is gratefully acknowledged.

Yingguang Wang
School of Ocean and Civil Engineering
Shanghai Jiao Tong University
Shanghai, PR China

Author

Yingguang Wang is a faculty member of the School of Ocean and Civil Engineering, Shanghai Jiao Tong University. He earned a graduate degree at the University of Washington, Seattle, USA, in 2001. He worked at a leading American engineering company in Philadelphia as a structural design engineer before joining the faculty at Shanghai Jiao Tong University in 2003.

During his academic career, Dr. Wang has published 76 peer-reviewed academic papers in leading international journals as the first author (and most of these papers are solely authored by Dr. Wang). He is a member of the American Society of Mechanical Engineers and the American Society of Civil Engineers. He has also served as the Associate Editor of the SNAME (Society of Naval Architects and Marine Engineers) *Journal of Ship Production and Design.* From 2010 to present, Dr. Wang has also been a committee member of the SNAME Ship Design Committee, SD-3 Stability Panel.

In 2018, Dr. Wang won the prestigious Albert Nelson Marquis Lifetime Achievement Award for having demonstrated unwavering excellence in the field of engineering. He was a recipient of an Outstanding Reviewer Award bestowed by the *Journal of Ocean Engineering* in 2017 and a recipient of an Outstanding Reviewer Award bestowed by the *Journal of Ocean Engineering and Science* in 2018. Dr. Wang was inducted into Sigma Xi, the Scientific Research Honor Society, in 2023.

1 Introduction

In engineering design, the importance of the methods that can forecast the vibration response of mechanical and structural systems to fluctuating external forces has significantly increased over the last 70 years. The ongoing trend toward lowering the weight and cost of engineering structures and increasing the power-to-weight ratio of machines, engines and vehicles has brought the need to forecast vibration response levels to the forefront in many areas of engineering. Vibration amplitudes that are large are almost invariably undesirable for a number of reasons, including the high noise levels that accompany them, the increased wear and fretting of joints and, often most importantly, the likelihood of component failure that could lead to catastrophic system failure. A failure of this kind may take place at the time when the levels of stress surpass the safe working limits or as a consequence of the accumulation of damage through metal fatigue.

For many years, engineers have been primarily interested in periodic vibration, which refers to situations in which the excitation and response of the system are straightforward periodic functions of time. This type of vibration is typically caused by the lack of perfect balance in rotating or reciprocating machinery and can be transmitted to neighboring systems through support structures and foundations. Following Lord Rayleigh's extensive pioneering work, a comprehensive body of classical theory is now available for the analysis of this type of vibration. This theory is particularly well developed for linear systems and is available in a number of standard textbooks.

Nevertheless, around the middle of the 1950s, a new kind of vibration problem emerged in the aircraft industry, and the classical deterministic approaches were unable to tackle it. It was observed that aircraft fuselage panels, which were located in close proximity to jet engines, were experiencing such high levels of vibration response as a result of acoustic excitation from the jet exhausts. As a result, fatigue cracks could develop and spread quite rapidly. Research studies have shown that the vibration response of these panels was exceedingly complicated, which is a reflection of the extremely complicated nature of the spatial and temporal fluctuations of pressure on the panel surfaces. This type of excitation and response was not only non-periodic and of a highly irregular nature but also lacked repeatability. Therefore, two sequential experiments, carried out under the same settings, would result in vibration amplitude response time histories that were markedly different from one another in terms of the specifics, despite the fact that they had the same overall behavior, on "average". It became apparent that attempting to solve such a problem using the conventional "deterministic" approach was not practical. It was necessary to adopt a new, "statistical" or more precisely "probabilistic" approach wherein the excitation and response were

DOI: 10.1201/9781003431503-1

described in terms of statistical parameters, such as the mean square of the vibration amplitudes. Similar problems emerged at approximately the same time in the design of aircraft to withstand buffeting caused by atmospheric turbulence and in the evaluation of the reliability of payloads in rocket-propelled vehicles. In each of these applications, it was discovered that the vibration response was so erratic and complicated that a probabilistic approach, as opposed to a deterministic one, proved to be significantly more fruitful.

In the probabilistic approach to vibration, which was later developed by structural engineers, both the excitation and the response are modeled as "stochastic" or "random" processes. These processes can be specified in terms of a relatively small number of statistical parameters and functions, such as the power spectral density function. Such processes can be seen as an infinite "ensemble" of possible sample functions, also known as sample realizations.

The realm of engineering vibrations has been considerably enlarged since the introduction of probabilistic methods. Historically, the vibration engineer would often make use of a deterministic analysis in order to meet the requirements of the design. In such an analysis, he (or she) would presume to have a complete knowledge of the dynamic properties of a system, as well as its initial state and the precise temporal history of its excitation. As an illustration, take into consideration a most basic scenario, which depicts a vibration system that is idealized as a spring-mass-damper that is excited by an external force acting on the mass, as shown in Figure 1.1.

If we make the assumption that the value of each component (i.e., m, k, c and $f(t)$) in the vibration system shown in Figure 1.1 is known precisely, then it is possible to compute the motion of the mass (i.e., $y(t)$) exactly by using a deterministic analysis. In a more rational probabilistic analysis, we acknowledge the possibility of having a lack of knowledge regarding one or more of the components in the

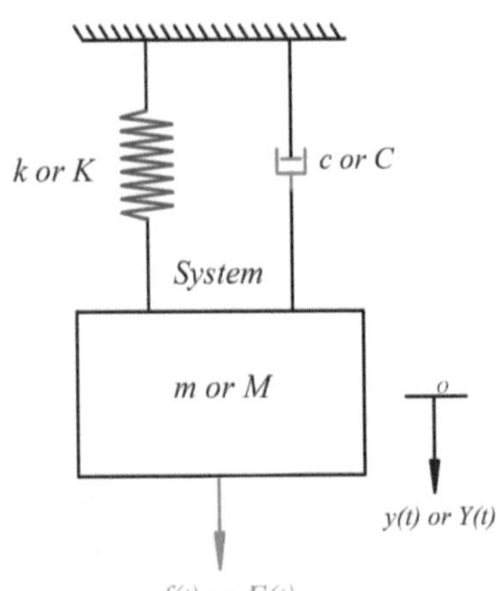

FIGURE 1.1 A spring-mass-damper system. Lowercase letters denote deterministic quantities; capital letters denote random quantities.

vibration system shown in Figure 1.1. For the purpose of this discussion, suppose there is an uncertainty on the spring rate (stiffness coefficient) k. In this book, an uncertain quantity will be referred to as being random. In the event that we select and carefully calibrate one particular spring, we will undoubtedly discover that it possesses a definite extension-contraction rate. Thus, as far as one particular spring is concerned, the probabilistic viewpoint is not appropriate. Let's say, on the other hand, that we are interested in any such springs that are manufactured from a certain process. Prior to selecting a single sample from a batch and conducting tests on it, it is not possible to predict the exact spring rate of this single sample. This is due to the fact that not all springs so manufactured are identical. By examining this example, we can see that the adjective "random" implies a collection of samples, none of which has been separated from the others, and that once an individual sample has been taken out from the collection, it is deterministic. Similarly, when we say that the forcing function is random, we indicate that the analysis is for a collection of excitation forces, each one of which is generally different from the others, but any one of which can be encountered during the actual service of the system. It is evident that the displacement $Y(t)$ must also be random when one or more of the elements in the vibration system shown in Figure 1.1 are random. However, it is feasible to deduce the random nature of the displacement from the random nature of the spring rate, the damping rate, the excitation force and so forth.

In light of the fact that no material is perfectly homogeneous, no bar is perfectly uniform, no rivet perfectly fits a hole, etc., it is evident that the probabilistic viewpoint is more realistic, although in those situations where the uncertainties involved are relatively minor and do not constitute the primary concern, the more straightforward deterministic approach might very well be satisfactory. For instance, by implementing adequate specifications and supervisions, the fabrication of an engineering structure can be controlled to the extent that any uncertainty about the engineering structure's dynamic properties can be overlooked. The uncertainty about the excitation is typically higher, although, occasionally, it is possible to make a wise choice of a suitable representative load or the most severe load, which can then be used for a deterministic analysis, leading to a useful design although perhaps not the optimal design. However, the primary incentive for the adaptation of probabilistic approaches in vibration analyses has been the presence of random external excitations.

In the field of aerospace engineering, noise from jet or rocket engines and gusts are two examples of representative sources of excitations that ought to be dealt with using probabilistic methods. The efflux of a jet or a rocket engine is known as a turbulent flow. It is common knowledge that turbulence is a phenomenon that is characterized by flow instability and that it is accompanied by a complex energy exchange that has not yet been completely comprehended. A small percentage of the energy is transformed into acoustic pressure, and this causes the structural components of the flight vehicle that are located nearby to become excited. Both the malfunction of electronic equipment mounted on the structure and the fatigue failure of the neighboring structure, which is now commonly referred to as acoustic fatigue, are the undesired consequences that can occur as a result of this type of excitation. When an airplane enters into a region that is characterized by gusty winds, the fraction of the total lift that is characterized by irregular fluctuations has the potential to generate higher stresses in the wing structure than those that are caused by the steady portion of the lift. In the case of particular wing configurations, it has been discovered that such fluctuating stresses may end up being the most important factor to take into consideration while designing the structure. If we were to take records of the jet noise pressure at a specific position on the fuselage of an airplane, we believe that we would discover that the records are highly erratic and that each record is distinct from the others. The gust velocity records exhibit the same degree of unpredictability and a lack of resemblance. The same unpredictability and a lack of resemblance to each other are also characteristics of the records of ocean wave elevations measured by a group of wave gauges installed at a specific sea site. In Figure 1.2, a selection of some such records is illustrated in a side-by-side configuration. It is apparent that this type of physical phenomenon must be considered to be random.

There is no shortage of examples of excitations that are essentially random and that act upon non-aerospace structures. Wind pressures, earthquakes and blasts

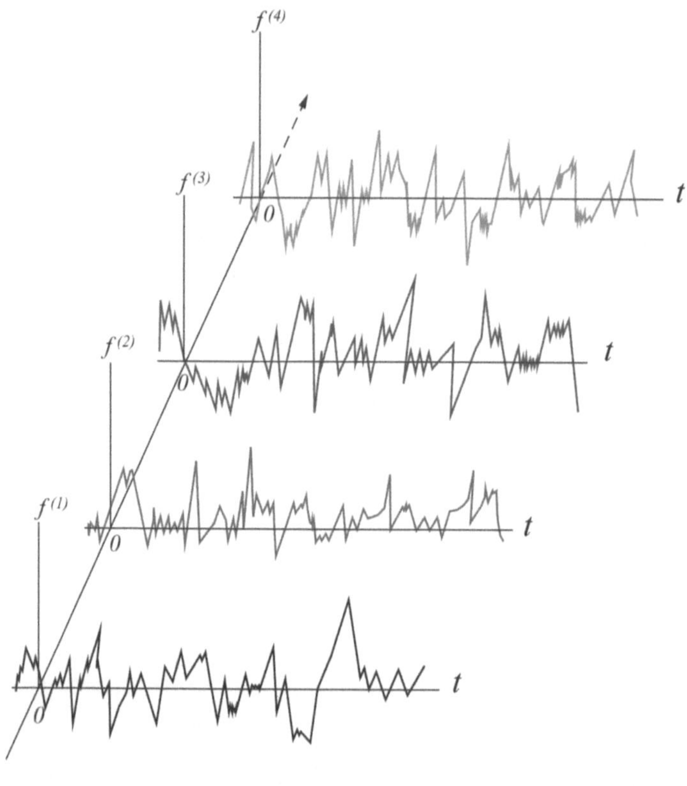

FIGURE 1.2 Records that are typical of a random phenomenon.

that act upon architectural structures are included in this general category. The excitation that a ship hull experiences in an irregular sea is similar in nature to the excitation that an airplane experiences in a gusty atmosphere. It is important to note that the probabilistic techniques in the area of engineering vibrations are increasingly being accepted by government purchasing and certifying agencies as being more appropriate for the analysis of a product's structural integrity.

In spite of the fact that structural and mechanical applications have provided the immediate incentive for developing the probabilistic theory of engineering vibrations on a broad basis, it is important to note that structural and mechanical engineers have inherited a significant amount of knowledge from the early work of physicists on the subject of Brownian motion. The first academic paper treating Brownian motion as a random process was published by Albert Einstein in 1905. It is interesting to note that Einstein's problem can be viewed as a special case of the spring-mass-damper system, in which the spring rate and the mass of the system are both negligibly small. Later on, the work of Einstein was expanded upon by a great number of other individuals, the most notable of whom were George Eugene Uhlenbeck and Leonard Salomon Ornstein. Nevertheless, from the perspective of Brownian motion, one is generally more interested in the behavior of closed systems. By "closed" we mean that either the simple mass particle or the complex continuum, such as a bar, is surrounded by a fluid medium that is in a state of so-called statistical equilibrium. The primary objective is to investigate how the Brownian motion of the mass particle or the bar tends to statistical equilibrium. Because, for such cases, the excitation force and the dissipation force are both provided by the fluid medium, the two types of force are inter-related. In contrast, structural and mechanical engineers are primarily concerned with excitations and dissipations that are fundamentally independent of one another. The contributions of electrical engineers, particularly those working in the domains of communications theory, control theory and information theory have been of great assistance to structural and mechanical engineers as well. These groups have been making substantial use of probability theory since the early 1930s, and they have developed techniques that are helpful to structural and mechanical engineers. In addition, the extensive body of literature on the theory of turbulence is noteworthy. It was during the process of developing this turbulence theory that the important concept of correlation in random phenomena evolved. Last but not least, it is imperative that we should not overlook the contribution made by the mathematicians who advanced the development of the common tool known as probability theory and put it on a rigorous foundation.

Despite the fact that there are healthy cross-fertilizations between various disciplines of study, certain characteristics and challenges are unique in some engineering vibration problems. Therefore, it is necessary to have a unified approach in order to discuss and tackle such problems. An effort has been made to ensure that this objective is accomplished in the following chapters in this book. Approximately 30% of the volume is devoted to acquainting the reader with specific parts of the probability and random process theory, while the other 60% of the volume is devoted to applications.

The discussion of probability and random process theory is restricted to what is required for our purposes; yet, it adheres to the fundamental viewpoint of the contemporary textbooks on this theory. This is done so that the reader can easily resort to such works for more background information. The exposition in the applications portion of this volume is divided according to tradition into single-degree-of-freedom linear systems, two-degree-of-freedom linear systems, multi-degree-of-freedom linear systems, nonlinear systems and numerical simulation methods. In all cases, the dynamic characteristics of a vibration system are assumed to be deterministic, and the external excitation is assumed to be random. The goal is to determine the probability law that governs the behavior of the random response of the system or, in various orders of completeness, statistical properties possessed by the response. Nevertheless, the ultimate objective from an engineering perspective is to be able to make statements about the reliability of an engineered structure or system in the face of random excitations. Therefore, the final chapter of this book is devoted to a discussion of random vibration-induced failures.

2 Random Variables and Related Probability Theory

2.1 INTRODUCTION

It is neither possible nor necessary to provide a comprehensive explanation of modern probability theory here. In spite of this, Chapters 2 through 4 in this book will outline certain elements of this theory that are necessary for our examination of the probabilistic theory of random vibrations. This is done both to ensure that the information in this book is self-contained and to facilitate the development of a common language for later discussions.

2.2 EXPERIMENT, SAMPLE SPACE AND EVENT

2.2.1 SAMPLE SPACE

An outcome of a single observation of a random phenomenon will hereafter be referred to as an *experiment*. In this book, the word "random phenomenon" refers to occurrences that, when observed again under the same conditions, do not nearly produce identical outcomes. The occurrence of outcomes does not follow a predictable pattern; instead, there is a statistical pattern where the frequency of occurrence of the outcome can be measured. In other words, the event's relative frequency varies, but the magnitude of variation generally reduces as the number of observations increases. As a result, the frequency eventually stabilizes at a specific value.

The result of an *experiment* with a random phenomenon is, of course, unable to be predicted in advance. Nevertheless, we are always able to determine the set whose components include each and every conceivable result of an *experiment*. The term "sample space" refers to this particular set in relation to the random phenomenon.

A sample space represents the set of all possible outcomes of an experiment. A sample space can be represented in various ways, depending on the nature of the experiment. One common representation is in the form of an exhaustive list, where each possible outcome is listed. For example, if a coin is flipped, the sample space for this experiment would be {heads, tails}. Another representation is a mathematical set, where each element represents an outcome. For instance, if an *experiment* involves rolling a fair six-sided die, the sample space would be the set $\{1, 2, 3, 4, 5, 6\}$. A further representation is an interval; for example, the sample space of measuring the length of a machined part from a batch of parts is the set of all positive real numbers, specifically the open interval $(0, +\infty)$. Each constituent

DOI: 10.1201/9781003431503-2

within a sample space that represents a certain outcome is referred to as a sample point. The sample space is represented by the symbol Ω, whereas each individual outcome is denoted by the symbol ω.

There are three distinct categories of sample spaces. A finite sample space consists of a limited number of sample points, such as {heads, tails} and {1, 2, 3, 4, 5, 6}. In a countably infinite sample space, there exists an endless number of sample points, such as the set of all integers, represented as $\{n: n = \text{integer}\}$. Ultimately, an uncountable sample space consists of an infinite number of sample points, such as the open interval $(0, +\infty)$ and the closed interval $[c, d]$.

In summary, a sample space represents the full set of possible outcomes of a random experiment. It provides a systematic and organized approach to analyzing probability and is an essential tool for studying randomness and uncertainty.

2.2.2 Event

Subsequently, we present the notion of an event. Assume that the result of an *experiment* of rolling a die is 3. Multiple events have occurred, including: (1) The occurrence of number 3, (2) the occurrence of an odd number, (3) the occurrence of a number bigger than 1, (4) the occurrence of a number smaller than 4, (5) the occurrence of a number equal to or smaller than 6 and so on. It should be noted that events (2) through (5) can also happen when the outcome is not 3. Given that the sample space encompasses all possible outcomes, it follows that every event is a subset of the sample space. An event that consists of only one sample point is referred to as an elementary event, while occurrences that contain more than one sample point are known as compound events. A certain event is defined as an event that includes all the items in the sample space. Therefore, event (5) is a certain event. An impossible event, in contrast to a certain event, does not have any sample points. In this particular case, it is impossible for the event to have a number smaller than 1. Considering both a certain event and an event that cannot occur (an empty set) as subsets of the sample space aligns with the common approach in set theory of treating a set as a subset of itself and the empty set as a subset of any set.

2.3 AXIOMS IN PROBABILITY THEORY

Probability theory is a fundamental branch of mathematics that deals with quantifying the likelihood of a particular event or outcome. It is based on a set of fundamental axioms that provide a framework for understanding and analyzing uncertainty. These axioms provide the foundation for the development of various probability concepts and theorems.

2.3.1 The First Axiom of Probability

The first axiom of probability states that for any event or outcome, there is a number between 0 and 1 that represents the probability or chance of that event occurring. This number is called the probability of the event. In other words, the first axiom is expressed in the following manner: Consider a random event E, and let $P(E)$ represents the probability of the event E occurring, indicating the conditions for E being true. Then, $P(E)$ is a numerical value ranging from 0 to 1, i.e.,

$$0 \leq P(E) \leq 1 \tag{2.1}$$

2.3.2 The Second Axiom of Probability

The second axiom of probability states that for a certain event Ω, $P(\Omega)$ is equal to 1, i.e.,

$$P(\Omega) = 1 \tag{2.2}$$

2.3.3 The Third Axiom of Probability

The third axiom of probability states that for any mutually exclusive and exhaustive event $E_1, E_2, \ldots, E_n$, the probability of the union of these events is equal to the sum of the probabilities of the individual events. This axiom allows for the addition and subtraction of probabilities, providing a unified framework for analyzing and manipulating probabilities:

$$P\left(\bigcup_{i=1}^{n} E_i\right) = \sum_{i=1}^{n} P(E_i) \tag{2.3}$$

Please note that, in this book, the union of two events A and B is denoted as $A \cup B$ and represents the set of outcomes that belong to A or B or both. In this book, the intersection of two events A and B is denoted as $A \cap B$ and represents the set of outcomes that belong to both A and B. Events in a given experiment are considered mutually exclusive if the occurrence of one event precludes the occurrence of other events. The complement of event A is written as $\bar{A}$ and is the outcomes of Ω which do not belong to A.

2.4 SOME THEOREMS IN PROBABILITY THEORY

2.4.1 The Theorem of the Complementary Event

Let us first give an example for clarifying the definition of the complementary event. For example, consider flipping a fair coin. The event "heads" is the complementary event of "tails". The theorem of the complementary event is an important concept in probability theory that states the relationship between the probability of an event occurring and the probability of its complement not occurring.

The theorem of the complementary event states that the probability of one event occurring is equal to the probability of its complement not occurring. This theorem can be stated mathematically as follows:

$$P(\bar{E}) = 1 - P(E) \tag{2.4}$$

2.4.2 The Theorem of the Total Event

Generally, the intersection of two events E_1 and E_2 is not an empty set, and the probability of their union is equal to the sum of the probabilities corresponding to $P(E_1$ and not $E_2)$, $P(E_2$ and not $E_1)$ and $P(E_1$ and $E_2)$. This is expressed as:

$$P(E_1 \cup E_2) = P(E_1) + P(E_2) - P(E_1 \cap E_2) \tag{2.5}$$

2.5 SOME DEFINITIONS IN PROBABILITY THEORY

2.5.1 The Definition of Conditional Probability

Conditional probability is a mathematical concept that measures the probability of an event occurring, given that another event has already occurred. It takes into account the relationship between two events and allows for the determination of how likely a desired outcome is, given a specific condition. The formula for conditional probability is as follows:

$$P(E_2|E_1) = \frac{P(E_1 \cap E_2)}{P(E_1)} \tag{2.6}$$

If the value of $P(E_1)$ is equal to zero, then the conditional probability is considered undefined.

2.5.2 The Definition of Independence in Probability

Independence is an important concept in probability theory. It refers to a situation where the outcomes of one event do not affect the probability of another event occurring. In other words, the occurrence of one event does not influence the probability of another event. Event E_2 is considered independent of event E_1 if

$$P(E_2|E_1) = P(E_2) \tag{2.7}$$

Consider a simple example to illustrate the concept of independence. Imagine flipping a fair coin twice. The outcomes of the second flip, say, heads or tails, are independent of the first flip. Therefore, the probability of getting heads on the first flip does not affect the probability of getting heads on the second flip.

If E_2 is independent of the event E_1, then it follows that the probability of $E_1 \cap E_2$ is the product of $P(E_1)$ and $P(E_2)$:

$$P(E_2 \cap E_1) = P(E_2) \bullet P(E_1) \tag{2.8}$$

2.6 RANDOM VARIABLES AND THEIR DISTRIBUTION CHARACTERISTICS

A straightforward number is frequently used to summarize the results of a randomized experiment. In some of the examples of random experiments that we have looked at, the sample space has been a description of the possible outcomes. There are circumstances in which descriptions of the outcomes are sufficient, but there are other circumstances in which it is helpful to associate a number with each outcome in the sample space. Because we do not know how the experiment will turn out in advance, we do not know what the final value of our variable will be. Because of this, the variable that associates a number with the outcome of a random experiment is referred to as a random variable. We herewith provide a formal definition of the random variable in the field of probability. A **random variable** is a function that assigns a real number to each outcome in the sample space of a random experiment.

Random variables can be classified into two main types: Discrete and continuous. A **discrete random variable** takes only a finite set of possible values, such as integers. Examples of discrete random variables include the number of heads when flipping a coin, the number of successes in a series of Bernoulli trials or the number of units in an inventory. A **continuous random variable** can take on any value within a certain range, such as the real numbers. Examples of continuous random variables include the height of a person, the weight of an object or the temperature in a room.

In this book, the distinction between a random variable and a real number is made via the usage of notation. An uppercase letter, such as the letter X, is used to indicate a random variable. Following the completion of an experiment, the value of the random variable that was measured is represented by a letter in lowercase form, such as $x = 70$ kg.

2.6.1 Probability Distribution Functions and Probability Densities

The cumulative distribution function (CDF) is a function that maps the probability of an event occurring to the corresponding value on the [0, 1] interval. It represents the cumulative probability up to a specific value, where 0 represents an event that never happens and 1 represents an event that always happens. Mathematically, the CDF (or the probability distribution function) $F(x)$ of a random variable X is calculated as follows:

$$F(x) = Pr\{X \leqslant x\} \tag{2.9}$$

One of the fundamental properties of the CDF is that it must be monotonic. This means that the function must increase or remain constant as the value of the input parameter increases. The CDF of a random variable is bounded above by 1. This means that no matter how large the value of the random variable becomes, the probability that it is less than or equal to any specific value approaches 1 as that value approaches positive infinity.

If a random variable is discrete, then the most straightforward method is to state the probabilities for the random variable to take on these discrete values, that is, specify

$$p_X(x) = Pr(X = x) \quad x = a_1, a_2, \ldots, a_n \tag{2.10}$$

where n can have either a finite or infinite value. The notation $p_X(x)$ indicates that this function is a point function, as $a_1, a_2, \ldots, a_n$ represent specific points on the real line. It is known as the probability function of the random variable X, with the argument x being known as the state variable or the range variable of X. It is important to differentiate between a random variable and its state variable. The latter is an arbitrary conceivable value that the former may emerge to be.

According to the definition provided by equation (2.9), a random variable, whether discrete or continuous, can be characterized by the probability of it being less than or equal to a state variable. The probability distribution function of the random variable X, represented by $F_X(x)$, is related to the probability function as follows in the case of discrete X:

$$F_X(x) = Pr(X \le x) = \sum_{x \le x_i} p_X(x_i) \tag{2.11}$$

In this equation, $p_X(x_i)$ is called the probability mass function. When plotted against the state variable x, the discrete random variable has the appearance of a staircase; however, the staircase may or may not have an infinite number of steps depending on the circumstances.

It is important to note that a continuous random variable is associated with a sample space that cannot be counted, and the probability of it taking on a particular value is typically zero. Probability functions are thus not applicable for describing continuous random variables; however, probability distribution functions are appropriate for this purpose. Dealing with the derivative of the probability distribution function, known as the probability density function (PDF) ($f(x')$), is more convenient. A random variable X is classified as continuous when, for each possible value x in the sample space, there exists a non-negative function denoted by $f(x')$ such that the CDF, denoted by $F(x)$, can be defined as follows:

$$F(x) = \int_{-\infty}^{x} f(x')\,dx' \tag{2.12}$$

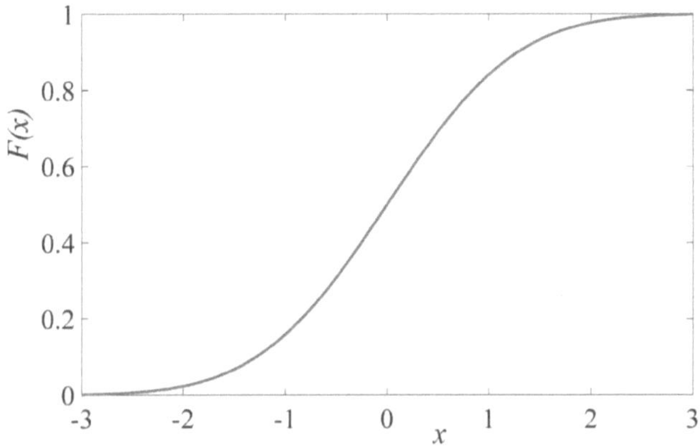

FIGURE 2.1 A cumulative distribution function of a Gaussian random variable.

Properties of the PDF of a continuous-type random variable include the following:

1. Shape: The shape of the PDF reflects the distribution of the underlying random variable. It can exhibit various shapes, such as uniform, bell-shaped (Gaussian) or skewed.
2. Support: The support of the PDF represents the range of values that the random variable can take on. The PDF is always defined within the support.
3. Continuity: The PDF should be continuous for all values of x in the support. It should not contain gaps or jumps.
4. Positivity: The PDF should be non-negative, meaning that it should always yield a value greater than zero.
5. Sum to 1: The PDF should sum up to 1, indicating that the probability mass is spread across the entire range of values.

Figure 2.1 shows a CDF of a continuous Gaussian random variable in the support range [–3, 3]. Figure 2.2 shows the corresponding PDF of the continuous Gaussian random variable in Figure 2.1.

2.6.2 Probability Distribution Quantiles, Median and Mode

A quantile is a statistical measure that divides a dataset into equal-sized intervals and indicates the proportion of observations that fall into each interval. It provides us with a better understanding of the distribution of a dataset by allowing us to identify specific values or ranges that correspond to certain percentages or proportions of the data.

We call the value of a random variable the quantile of order p, denoted by x_p, if $Pr\{X < x_p\} \le p$ and $Pr\{X \ge x_p\} \ge 1 - p$, where $0 < p < 1$. In particular,

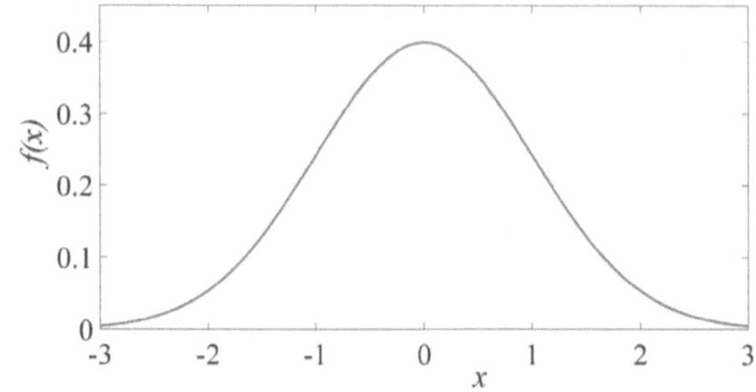

FIGURE 2.2 The corresponding PDF of a Gaussian random variable.

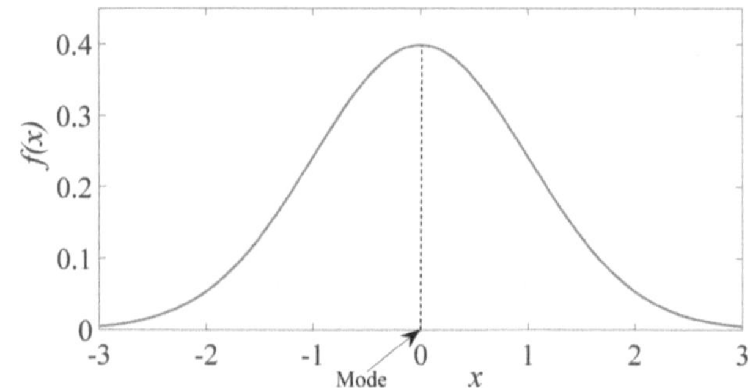

FIGURE 2.3 The definition of the mode of distribution.

we designate x_p the median of the probability distribution for $p = 1/2$. For $p = 1/4$, x_p is called the lower quartile. For $p = 3/4$, x_p is called the upper quartile. The distance between the lower quartile and upper quartile is called the interquartile range, and it covers 50% of the probability distribution.

The mode of a distribution refers to the value at which the PDF or probability mass function reaches its peak, assuming such a value exists for the random variable X (see Figure 2.3). The mode can be defined as the value of the random variable that has the highest probability of occurring in random experiments. The mode of a continuous random variable can be determined by finding the point at which the derivative of the PDF equals zero.

2.6.3 Jointly Distributed Random Variables

The previously stated concept of random variables and probability distribution functions can be expanded to encompass multiple random variables. To be concise, we will focus our discussion on the scenario involving two random variables. For instance, let's examine the statistical properties of the excursion (X) and period (Y) of surface waves in the ocean, as shown in Figure 2.4. The magnitudes of X and Y fluctuate in each cycle, making them a pair of random variables.

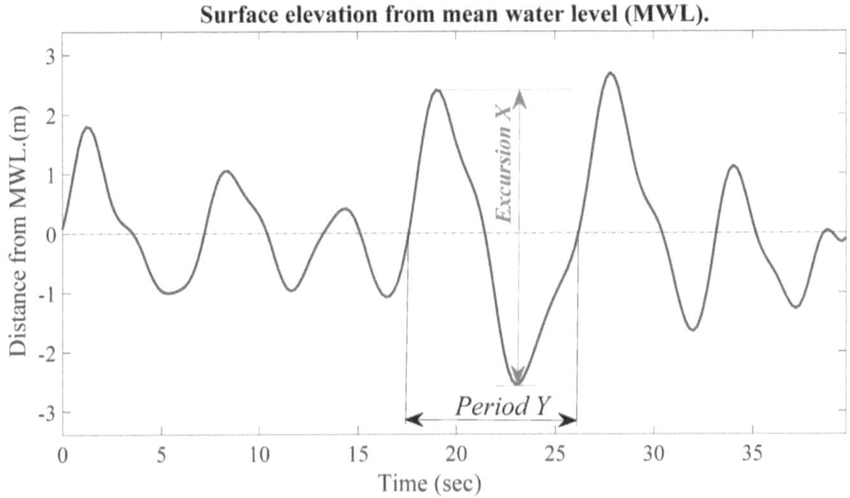

FIGURE 2.4 An example of two random variables.

It's possible that the individual random variables X and Y are statistically independent, but it's also possible that they have some kind of functional connection. In many cases, it is necessary to obtain the statistical characteristics of both X and Y at the same time. In other cases, it is necessary to determine the probability of X given that Y has a particular value that has been stated. If one has knowledge of the joint probability distribution of the two random variables X and Y, then one can use this information to solve the problems at hand.

The upcoming discussion will focus on two random variables. However, the concept and formulas can be readily applied to multidimensional random variables, which are commonly referred to as a random vector.

The joint behavior of two random variables X and Y is characterized by their joint probability distribution function, which is defined as:

$$F_{XY}\left(x,y\right)=Pr\left[\{X\leq x\}\cap\{Y\leq y\}\right] \tag{2.13}$$

A joint probability distribution function is right continuous and possesses non-decreasing characteristics with respect to every one of the multiple dimensions. In the case of two random variables, the following relations are evident:

$$F_{XY}(-\infty,\ y)=F_{XY}(x,-\infty)=0 \tag{2.14}$$

$$F_{XY}(\infty,\infty)=1 \tag{2.15}$$

$$F_{XY}(x,\ \infty)=F_X(x) \tag{2.16}$$

$$F_{XY}(\infty,\ y)=F_Y(y) \tag{2.17}$$

The mixed derivative of the joint probability distribution function is what is meant when people talk about the joint probability density. For example, when considering a situation in two dimensions,

$$f_{XY}(x,y) = \frac{\partial^2}{\partial x \partial y} F_{XY}(x,y) \tag{2.18}$$

It should be noted that the joint probability density $f_{XY}(x,y)$ in equation (2.18) is always non-negative. Equation (2.18) can be inverted to obtain

$$F_{XY}(x,y) = \int_{-\infty}^{x}\int_{-\infty}^{y} f_{XY}(x',y')dy'dx' \tag{2.19}$$

By allowing one of the upper limits of the double integration to tend toward infinity, we get the distribution function for a single random variable, namely,

$$\int_{-\infty}^{\infty} dy' \int_{-\infty}^{x} f_{XY}(x',y')dx' = Pr(X \le x) = F_X(x) \tag{2.20}$$

In the event when both of the upper limits on the right-hand side of equation (2.19) approach infinity, the integration will then cover the entirety of the two-dimensional sample space, and the results will be as follows:

$$\int_{-\infty}^{\infty}\int_{-\infty}^{\infty} f_{XY}(x',y')dy'dx' = 1 \tag{2.21}$$

Taking the derivative of both sides of equation (2.20) with respect to x yields

$$\int_{-\infty}^{\infty} f_{XY}(x,y)dy = f_X(x) \tag{2.22}$$

In a similar manner,

$$\int_{-\infty}^{\infty} f_{XY}(x,y)dx = f_Y(y) \tag{2.23}$$

Hence, the joint probability density $f_{XY}(x,y)$ contains a greater amount of information than $f_X(x)$ and $f_Y(y)$ separately, as the latter can be derived from the former.

The two-dimensional case can be easily generalized to the multi-dimensional case. Beginning with

$$F_{X_1,X_2\cdots X_n}(x_1,x_2,\ldots,x_n) = Pr\{\{X_1 \le x_1\} \cap \ldots \cap \{X_n \le x_n\}\} \tag{2.24}$$

and

$$f_{X_1,X_2\cdots X_n}\left(x_1,x_2,\ldots,x_n\right)=\frac{\partial^n}{\partial x_1\,\partial x_2\cdots\partial x_n}F_{X_1,X_2\cdots X_n}\left(x_1,x_2,\ldots,x_n\right) \quad (2.25)$$

we are led to, for instance, the expanded form of (2.22):

$$\underbrace{\int_{-\infty}^{\infty}\cdots\cdots\int_{-\infty}^{\infty}}_{(k+1)\,\text{fold}} f_{X_1X_2\cdots X_n}\left(x_1,x_2,\ldots,x_n\right)dx_{n-k}dx_{n-k+1}\cdots dx_n = f_{X_1X_2\cdots X_{n-k-1}}\left(x_1,x_2,\ldots,x_{n-k-1}\right) \quad (2.26)$$

2.6.4 Conditional Distribution

In the event that two random variables X and Y are both discrete and jointly distributed, the probability that X equals x with the condition that Y equals y can be expressed as follows:

$$P_{X|Y}\left(x|y\right)=\frac{P_{XY}(x,y)}{P_Y(y)} \quad (2.27)$$

Equation (2.27) loses all of its significance when applied to the case of continuous random variables that are defined on an uncountable sample space because now $P_Y(y)$ is equal to zero practically everywhere. In order to get around this obstacle, we will replace (2.27) by

$$Pr\left[\{x\le X\le x+\Delta x\}\middle|\{y\le Y\le y+\Delta y\}\right]=\frac{Pr\left[\{x\le X\le x+\Delta x\}\cap\{y\le Y\le y+\Delta y\}\right]}{Pr\left[\{y\le Y\le y+\Delta y\}\right]} \quad (2.28)$$

By taking the limit as Δx and Δy approach zero on both sides of equation (2.28), the right-hand side of the equation simplifies

$$\frac{f_{XY}(x,y)dxdy}{f_Y(y)dy}=\frac{f_{XY}(x,y)dx}{f_Y(y)} \quad (2.29)$$

The limiting form corresponds to the probability of X falling within the infinitesimally small interval $[x, x + dx]$, given that Y has a certain value y. Based on this, it is reasonable to construct a conditional probability density as

$$f_{X|Y}\left(x|y\right)=\frac{f_{XY}(x,y)}{f_Y(y)}=\frac{f_{XY}(x,y)}{\int_{-\infty}^{\infty}f_{XY}(x,y)dx} \quad (2.30)$$

Once more, the definition is meaningful so long as $f_Y(y)$ is greater than zero. For the sake of completeness, we define $f_{X|Y}\left(x|y\right)$ as equal to zero whenever

$f_Y(y)$ equals zero. The third part of the expression (2.30) suggests that the conditional probability density can be derived from the joint probability density. The definition of independence for two continuous random variables can be stated as follows:

$$f_{X|Y}(x|y) = f_X(x) \tag{2.31}$$

which indicates

$$f_{XY}(x,y) = f_X(x) f_Y(y) \tag{2.32}$$

It should be noted that the expression (2.30) can be reorganized in order to compute the joint probability density, which is represented as:

$$f_{XY}(x,y) = f_Y(y)\, f_{X|Y}(x|y) \tag{2.33}$$

The solution obtained by integrating (2.33) is as follows:

$$f_X(x) = \int_{-\infty}^{\infty} f_{XY}(x,y)dy = \int_{-\infty}^{\infty} f_{X|Y}(x|y) f_Y(y)dy \tag{2.34}$$

which demonstrates the method of calculating the unconditional probability density based on the conditional probability density.

The conditional distribution function is derived by integrating equation (2.30):

$$F_{X|Y}(x|y) = \frac{\int_{-\infty}^{x'} f_{XY}(x',y)dx'}{f_Y(y)} = \frac{\int_{-\infty}^{x'} f_{XY}(x',y)dx'}{\int_{-\infty}^{\infty} f_{XY}(x',y)dx'} \tag{2.35}$$

By transposing and subsequently integrating, we obtain

$$F_X(x) = \int_{-\infty}^{\infty} F_{X|Y}(x|y) f_Y(y)\, dy \tag{2.36}$$

Moreover, the conditional PDF of a random variable X, given Y takes the value y, can be expressed as

$$f_{X|Y}(x|y) = \frac{f_{XY}(x,y)}{f_Y(y)} = \frac{f_{XY}(x,y)}{\int_{-\infty}^{\infty} f_{XY}(x,y)dx} \tag{2.37}$$

In a similar manner, the conditional PDF of a random variable Y, given X takes the value x, can be expressed as

$$f_{Y|X}(y|x) = \frac{f_{XY}(x,y)}{f_X(x)} = \frac{f_{XY}(x,y)}{\int_{-\infty}^{\infty} f_{XY}(x,y)dy} \tag{2.38}$$

2.6.5 Moments of Random Variables

One can acquire different statistical characteristics of a random variable by calculating moments and functions of moments of the probability distribution. To determine whether the PDF has a symmetric shape, one can assess the skewness parameter of the distribution. Moments not only yield information on the characteristics of a random variable but also have a crucial role in determining the probability distribution of quantities related to a random process, as will be demonstrated later in this book. The subsequent discussion will cover the moments and functions of moments of a random variable and random vector, along with the conditional moments.

Consider a discrete-type random variable X, which is associated with the probability mass function $p(x_i)$. In this case, the ***expected value*** of a function $u(x)$ is defined as follows:

$$E[u(x)] = \sum_{i \in S} u(x_i)\, p(x_i) \tag{2.39}$$

in which S stands for the sample space.

Now let us consider a continuous-type random variable X, which is associated with the PDF $f(x)$. In this case, the ***expected value*** of a function $u(x)$ is defined as follows:

$$E[u(x)] = \int_{-\infty}^{\infty} u(x) f(x) dx \tag{2.40}$$

Take into consideration a specific instance of the expected value, $u(x) = x^k$. Then, $E[x^k]$ is called the kth moment of the random variable X. The ***first moment*** of a random variable is analogous to the centroid location of a unit mass, whereas the ***second moment*** is comparable to the mass moment of inertia. Similarly, the joint second moment of two random variables can be compared to the mass product of inertia. The analogy elucidates that moments of any given order can be calculated from the probability distribution. However, in the absence of additional data, moments of all orders are typically necessary to recover the knowledge of the probability distribution. When the function $u(x)$ is equal to x^2, equation (2.40) can be expressed as:

$$E\left[x^2\right] = \int_{-\infty}^{\infty} x^2 f(x) dx \tag{2.41}$$

which defines the ***mean square value*** of x. The square root of equation (2.41) is called the ***root mean square value***, also known as the ***rms*** value.

When the function $u(x)$ is equal to x, equation (2.40) can be expressed as:

$$E[x] = \int_{-\infty}^{\infty} xf(x)dx \tag{2.42}$$

$E[x]$ is called the ***mean*** or ***average value*** of the random variable X, and is usually denoted by μ or simply $E[x]$. The mean value of a random variable is the average value of its possible outcomes. It represents the central tendency of the probability distribution and quantifies the typical or expected outcome. In the following, we present some properties of the mean value of the random variable:

$$E[a] = a, \text{ where } a \text{ is a constant.} \tag{2.43}$$

$$E[ax] = aE[X] \tag{2.44}$$

$$E[X + a] = E[X] + a \tag{2.45}$$

$$E[X + Y] = E[X] + E[Y] \tag{2.46}$$

$$E[XY] = E[X]E[Y], \ X \text{ and } Y \text{ are statistically independent random variables.} \tag{2.47}$$

In the case when μ_X represents the ***first moment*** of the random variable X, the ***n*th central moment** of X is denoted by the expression $E\left[(X-\mu_X)^n\right]$. Similarly, the $(m+n)$ th joint central moment of X and Y is denoted by $E\left[(X-\mu_X)^m (Y-\mu_Y)^n\right]$. Of the utmost significance are the ***second central moments***, which are frequently referred to by unique names and are indicated by special symbols. Specifically, the expression $E\left[(X-\mu_X)^2\right]$ is commonly referred to as the ***variance*** of X (denoted by $Var[X]$), which is calculated by using the following equation:

$$\begin{aligned} E\left[(X-\mu_X)^2\right] &= E\left[X^2 + (\mu_X)^2 - 2X\mu_X\right] = E\left[X^2\right] + E\left[(\mu_X)^2\right] - 2\mu_X E[X] \\ &= E\left[X^2\right] + (\mu_X)^2 - 2\mu_X\mu_X = E\left[X^2\right] - (\mu_X)^2 = Var[X] = (\sigma_X)^2 \end{aligned} \tag{2.48}$$

The term "**standard deviation** of X" refers to the positive square root of the variance. The standard deviation of X is denoted by the symbol σ_X. Below is a list of properties that are associated with the variance of a random variable X:

$$Var[a] = E[a^2] - (E[a])^2 = a^2 - a^2 = 0, \text{where } a \text{ is a constant.} \tag{2.49}$$

$$Var[aX] = E[a^2X^2] - (E[aX])^2 = a^2E[X^2] - a^2(E[X])^2 = a^2Var[X], \tag{2.50}$$

$$\begin{aligned} Var[X+a] &= E[(a+X)^2] - (E[a+X])^2 \\ &= E[a^2 + X^2 + 2aX] - (E[a])^2 - (E[X])^2 - 2E[aX], \\ &= E[X^2] - (E[X])^2 = Var[X] \end{aligned} \tag{2.51}$$

$$Var[aX + by] = a^2 Var[X] + b^2 Var[Y], \text{ where } X \text{ and } Y \text{ are statistically independent random variables.} \tag{2.52}$$

The mean and variance are crucial factors in probability distribution. The variance specifically quantifies the extent to which the values of a random variable deviate from its mean. A smaller variance indicates a higher level of concentration of the PDF around its mean value.

Example 2.1

▶ Calculate the mean and the variance of the random variable X, whose PDF is represented by the expression of

$$f(x) = \begin{cases} x+1 & for \;\; 0 \le x \le 1 \\ 3-x & for \;\; 1 \le x \le 3 \end{cases} \tag{2.53}$$

SOLUTION

$$E(x) = \int_0^1 x \bullet (x+1)dx + \int_1^3 x \bullet (3-x)dx \doteq \frac{5}{6} + 12 - \frac{26}{3} = 4.1667 = \mu_X \tag{2.54}$$

$$E\left(x^2\right) = \int_0^1 x^2 \bullet (x+1)dx + \int_1^3 x^2 \bullet (3-x)dx = \frac{7}{12} + 28 = 28.5833 \tag{2.55}$$

$$Var[X] = E\left[X^2\right] - (\mu_X)^2 = 28.5833 - 4.1667^2 = 11.222 \tag{2.56}$$

Thus, the mean and the variance of the random variable X are 4.1667 and 11.222, respectively. ◀

We next give the important definition of a ***standardized random variable.*** A random variable X is referred to as a ***standardized random variable*** when it has a mean of zero, denoted by $E[x] = 0$, and a variance of one, denoted by $Var[x] = 1$. Let μ and σ^2 represent the mean and variance of the random variable X, respectively. Let Z be a randomly distributed variable defined as

$$Z = \frac{X - \mu}{\sigma} \tag{2.57}$$

Then Z is called a ***standardized random variable*** of X. Z has a mean value of zero and a variance value of one, which can be demonstrated with relative ease. In other words, based on the characteristics of the mean, which are presented in equations (2.44) and (2.45), and the characteristics of the variance, which are presented in equations (2.50) and (2.51), we found that

$$E[Z]=E\left[\frac{X-\mu}{\sigma}\right]=E\left[\frac{X}{\sigma}-\frac{\mu}{\sigma}\right]=E\left[\frac{X}{\sigma}\right]-\frac{\mu}{\sigma}=\frac{\mu}{\sigma}-\frac{\mu}{\sigma}=0 \quad (2.58)$$

$$Var[Z]=Var\left[\frac{X-\mu}{\sigma}\right]=Var\left[\frac{X}{\sigma}-\frac{\mu}{\sigma}\right]=Var\left[\frac{X}{\sigma}\right]=\frac{\sigma^2}{\sigma^2}=1 \quad (2.59)$$

There are times when it is of the utmost importance to examine a number of different parameters of the probability distribution, such as those that indicate the measure of asymmetry and peakedness. In the following, we provide the definitions of each of these parameters. Consider the random variable X, which possesses a mean of μ and a variance of σ^2. The ***coefficient of variation***, which is represented by the symbol v, is expressed as the ratio of σ to μ. The ratio of the third central moment, $\mu_X{}^3$, to the third moment, σ^3, is referred to as the ***skewness*** of the distribution, which is represented by the symbol γ_1. Similarly, the ratio of the fourth central moment, $\mu_X{}^4$, to σ^4 is referred to as the ***kurtosis***, which is represented by the symbol γ_2. The mathematical equations for calculating the abovementioned parameters are summarized as follows:

$$\textit{Coefficient of variation}: v=\frac{\sigma}{\mu} \quad (2.60)$$

$$\textit{Skewness}: \gamma_1=\frac{\int_{-\infty}^{\infty}(x-\mu)^3 f(x)dx}{\sigma^3} \quad (2.61)$$

$$\textit{Kurtosis}: \gamma_2=\frac{\int_{-\infty}^{\infty}(x-\mu)^4 f(x)dx}{\sigma^4} \quad (2.62)$$

Skewness is a statistical measure that measures the asymmetry of a probability distribution. It is a measure of how concentrated the distribution is around the mean. Skewness can be positive, negative or zero. A probability density function distribution with a positive skewness has a longer tail on the right side of the distribution, and the mode of the distribution is located at a value smaller than the mean value. A probability density function distribution with a negative skewness has a longer tail on the left side, and the mode of the distribution is located at a value larger than the mean value. A probability density function distribution with zero skewness has a symmetric shape. This means that the distribution is centered around the mean without any significant tails. Zero skewness is often seen in distributions that have a normal or symmetric shape.

Kurtosis quantifies the extent to which a probability density function distribution deviates from the Gaussian distribution in terms of its peakedness. The kurtosis γ_2 for the Gaussian distribution is precisely 3. If the value of γ_2 is less than 3, the distribution is considered to be platykurtic, meaning it has a mild peak. Conversely, if the value of γ_2 is greater than 3, the distribution is considered to be leptokurtic, indicating a sharp peak.

We next present the mathematical theories of the moments of random vectors. Let's say that (X, Y) is a pair of discrete random variables, and their joint probability mass function is denoted by $p(x_i, y_j)$. In this case, the expected value of an arbitrary function $u(x, y)$ can be defined as follows:

$$E[u(x,y)] = \sum_{i,j \in S} u(x_i, y_j)\, p(x_i, y_j) \tag{2.63}$$

in which S denotes the sample space. If, for example, the function $u(x, y)$ is expressed as $x_i^k y_j^l$ (for discrete-type random variables), where k is larger than or equal to zero and l is larger than or equal to zero, then the moment of order $k + l$ of the random variables X and Y, m_{kl}, can be calculated as follows:

$$m_{kl} = \sum_{i,j \in S} x_i^k y_j^l\, p(x_i, y_j) \tag{2.64}$$

If (X, Y) is a pair of continuous random variables, and their joint PDF is denoted by $f(x,y)$. Then, in this case, the expected value of an arbitrary function $u(x, y)$ is defined as follows:

$$E[u(x,y)] = \int_{-\infty}^{\infty}\int_{-\infty}^{\infty} u(x,y) f(x,y)\, dx dy \tag{2.65}$$

If, for example, the function $u(x, y)$ is expressed as $x^k y^l$ (for continuous-type random variables), where k is larger than or equal to zero and l is larger than or equal to zero, then the moment of order $k + l$ of the random variables X and Y, m_{kl}, can be calculated as follows:

$$m_{kl} = \int_{-\infty}^{\infty}\int_{-\infty}^{\infty} x^k y^l f(x,y)\, dx dy \tag{2.66}$$

The definition of the central moment of two random variables X and Y, which is represented by the symbol μ_{kl}, is as follows:

$$\mu_{kl} = E\left[(X - \mu_X)^k (Y - \mu_Y)^l\right] \tag{2.67}$$

in which μ_X and μ_Y represent the mean values of the random variables X and Y, respectively. In particular, the central moment μ_{11} is referred to as the ***covariance*** of the two random variables X and Y, and it is represented by the symbol $\mathrm{Cov}[X, Y]$ or σ_{XY}. To clarify,

$$\begin{aligned}\mathrm{Cov}\left[X, Y\right] &= \mu_{11} = E\left[(X-\mu_X)(Y-\mu_Y)\right] \\ &= E[XY] + E[\mu_X \mu_Y] - \mu_X E[Y] - \mu_Y E[X] \\ &= E[XY] - E[X]E[Y] = \sigma_{XY}\end{aligned} \tag{2.68}$$

The distribution of two random variables is characterized by the crucial parameter known as $\mathrm{Cov}\,[X, Y]$, which was defined above. In order to make things more convenient, we may consider the continuous-type random variables in the following discussion. Let us suppose that X and Y are statistically independent. Then, we have

$$E[xy] = \int_{-\infty}^{\infty}\int_{-\infty}^{\infty} xyf(x,y)\,dx\,dy = \int_{-\infty}^{\infty} xf(x)\,dx \int_{-\infty}^{\infty} yf(y)\,dy = E[x]E[y] \tag{2.69}$$

$$\mathrm{Cov}[x,\ y] = E[xy] - E[x]E[y] = 0 \tag{2.70}$$

Thus, the value of $\mathrm{Cov}[X, Y]$ becomes to zero. If the covariance of two random variables X and Y is equal to zero (i.e., $\mathrm{Cov}[X, Y]=0$), these two random variables are then said to be ***uncorrelated***. We have therefore proved the following conclusion: If two random variables X and Y are statistically independent, then they are uncorrelated. However, the opposite is not always the case. That said, if two random variables X and Y are statistically uncorrelated, they may or may not be statistically independent.

The correlation coefficient of two random variables can be defined based on the knowledge of their variance and covariance. The ***correlation coefficient*** of two random variables, X and Y, represented as ρ_{xy}, is defined as

$$\rho_{xy} = \frac{\mathrm{Cov}[x,\ y]}{\sqrt{Var[x]\cdot Var[y]}} = \frac{\sigma_{xy}}{\sigma_x \sigma_y} \tag{2.71}$$

The correlation coefficient possesses an important characteristic in that it is limited within the following boundaries:

$$-1 \le \rho_{xy} \le 1 \tag{2.72}$$

In order to prove equation (2.72), let us examine the quadratic function, $h(s) = (Ps - Q)^2$, in which $P = X - \mu_X$ and $Q = Y - \mu_Y$. The expected value of $h(s)$ is determined by

$$E[h(s)] = E\left[(Ps-Q)^2\right] = E\left[P^2\right]s^2 - 2E[PQ]s + E\left[Q^2\right] \tag{2.73}$$

A quadratic function such as this can have two real roots, a repeated real root or two imaginary roots, depending on the value of the characteristic function of the quadratic formula (i.e., $b^2 - 4ac$). The $h(s)$ is always positive because it is a square; therefore, the roots of the equation (2.73) that was just presented above must either be imaginary or repeated. As a result, the characteristic function must be either less than zero or equal to zero. By substituting the values from equation (2.73) into the characteristic function, we obtain the inequality:

$$(2E[PQ])^2 - 4E\left[P^2\right]E\left[Q^2\right] \le 0 \tag{2.74}$$

$$(E[PQ])^2 \le E\left[P^2\right]E\left[Q^2\right] \tag{2.75}$$

Substitution for P and Q leads to

$$(E[(X-\mu_X)(Y-\mu_Y)])^2 \le E\left[(X-\mu_X)^2\right]E\left[(Y-\mu_Y)^2\right] \tag{2.76}$$

that is

$$\sigma_{xy}^2 \le \sigma_x^2\sigma_y^2 \tag{2.77}$$

This is what is referred to as Schwartz's inequity. According to this, the range of ρ_{xy} is as follows:

$$-1 \le \rho_{xy} \le 1 \tag{2.78}$$

The correlation coefficient, denoted by the symbol ρ_{xy}, is a statistical metric that serves to quantify the extent of a functional relationship between two random variables. We can write

$$\rho_{xy} = \frac{\text{cov}[X,Y]}{\sqrt{Var[X]Var[Y]}} = E\left[\left(\frac{X-\mu_X}{\sigma_X}\right)\left(\frac{Y-\mu_Y}{\sigma_Y}\right)\right] \tag{2.79}$$

Therefore, $\rho_{xy} = \pm 1$ implies that

$$(E[(X-\mu_X)(Y-\mu_Y)]) = \pm\sigma_X\sigma_Y \tag{2.80}$$

Equation (1.80) can be satisfied under the following condition:

$$\left(\frac{X-\mu_X}{\sigma_X}\right) = \pm\left(\frac{Y-\mu_Y}{\sigma_Y}\right) \tag{2.81}$$

Therefore, for $\rho_{xy} = \pm 1$, we have the following relationship:

$$Y = \mu_Y \pm \sigma_Y\left(\frac{X-\mu_X}{\sigma_X}\right) \tag{2.82}$$

Thus, when the correlation coefficient $\rho_{xy} = \pm 1$, it indicates that there is a functional relationship between the two random variables X and Y. Therefore, the statistical properties of one random variable can be fully assessed based on the statistical properties of the second random variable. Conversely, if the correlation coefficient $\rho_{xy} = 0$, it indicates that the two random variables are uncorrelated and have no functional relationship. When ρ_{xy} is a positive value between 0 and 1, there is a typical pattern where high values of X tend to be associated with high values of Y, while low values of X tend to be associated with low values of Y. When ρ_{xy} is negative, it means that there is a consistent pattern where high values of X are associated with low values of Y, and low values of X are associated with high values of Y. A value of ρ_{xy} near ± 1 signifies a strong correlation tendency, whereas a value of ρ_{xy} around 0 suggests a weak tendency.

The variance and covariance that we have just obtained for two random variables are now being extended to a random vector that is comprised of n elements. In order to accomplish this, let us express the covariance of any two elements (X_i and X_j) of a random vector $\mathbf{X}$, which is denoted by the symbol σ_{ij}, in the following manner:

$$\text{Cov}\left[X_i, Y_j\right] = E\left[\left(X_i - \mu_i\right)\left(X_j - \mu_j\right)\right] = \sigma_{ij} \tag{2.83}$$

in which μ_i and μ_j represent the mean of the random variables X_i and X_j, respectively.

The covariance matrix, which is denoted by the symbol Σ, is formed by the variance and covariance of a random vector that consists of n components. This matrix can be presented as follows:

$$\Sigma = \begin{pmatrix} \sigma_{11} & \sigma_{12} & \sigma_{13} & \sigma_{1n} \\ \sigma_{21} & \sigma_{22} & \sigma_{23} & \sigma_{2n} \\ \vdots & & & \\ \sigma_{n1} & \sigma_{n2} & \sigma_{n3} & \sigma_{nn} \end{pmatrix} \tag{2.84}$$

in which $\sigma_{ij} = \sigma_{ji}$.

2.6.6 Moment-Generating Function, Characteristic Function and Cumulants

The moments of a random variable can be computed directly using the definitions provided in the previous section. Nevertheless, the computation of moments might be challenging, depending upon the complexity of the form of the probability density function. The moment-generating function, as will be defined in this section, is highly valuable not only for the computation of moments but also for the practical implementation of probability theory. This is because a function that can generate all the moments of a random variable essentially characterizes the probability distribution law of the random variable. Nevertheless, it is worth noting that the moment-generating function may not be defined for certain random variables. Conversely, the characteristic function, which will also be introduced in this section, always exists, and there is a one-to-one relationship between the probability distribution function and the characteristic function. The characteristic function is crucial in deriving a new probability distribution function. An instance of this may be seen in the derivation of the probability density function that is appropriate for expressing the statistical characteristics of non-Gaussian random processes. This derivation is achieved by utilizing the cumulant-generating function, which is exactly the logarithm of the characteristic function. The discussion of the cumulant-generating function can be found in the later part of this section.

The function $m(\theta)=E\left[e^{\theta x}\right]$ represents the expected value of the exponential function $e^{\theta x}$. For a discrete-type random variable X, the function $m(\theta)=E\left[e^{\theta x}\right]$ where θ is a real number, is called the moment-generating function of the random variable X. That is,

$$m(\theta)=E\left[e^{\theta x}\right]=\sum_{j\in S} e^{\theta x_j} p\left(x_j\right) \tag{2.85}$$

in which S represents the sample space.

For a continuous-type random variable X, the moment-generating function of the random variable is

$$m(\theta)=E\left[e^{\theta x}\right]=\int_{-\infty}^{\infty} e^{\theta x} f(x)\,dx \tag{2.86}$$

We will differentiate the moment-generating function r times with regard to θ in order to better understand it. After that, for a discrete-type random variable, we have

$$\frac{d^r}{d\theta^r} m(\theta)=\sum_{j\in S} x_j^r e^{\theta x_j} p\left(x_j\right) \tag{2.87}$$

For a continuous-type random variable, we have

$$\frac{d^r}{d\theta^r} m(\theta)=\int_{-\infty}^{\infty} x^r e^{\theta x} f(x)\,dx \tag{2.88}$$

By letting $\theta=0$ in the above two equations, we obtain the following relationships

$$\left[\frac{d^r}{d\theta^r} m(\theta)\right]_{\theta=0}=\sum_{j\in S} x_j^{\,r} p\left(x_j\right)=E\left[x^r\right] \tag{2.89}$$

$$\left[\frac{d^r}{d\theta^r}m(\theta)\right]_{\theta=0} = \int_{-\infty}^{\infty} x^r f(x)dx = E\left[x^r\right] \tag{2.90}$$

This results in the subsequent theorem: *If a random variable X has a moment, it can be determined by performing r times of differentiation with respect to θ of the moment-generating function,* $m(\theta) = E\left[e^{\theta x}\right]$, *while letting* $\theta = 0$.

For a discrete-type random variable X, the function $\phi(\theta) = E\left[e^{i\theta x}\right]$, where θ is a real number and i is an imaginary unit, is called the characteristic function of the random variable X. That is,

$$\phi(\theta) = E\left[e^{i\theta x}\right] = \sum_{j\in S} e^{i\theta x_j} p\left(x_j\right) \tag{2.91}$$

in which S represents the sample space.

For a continuous-type random variable X, the characteristic function of the random variable is

$$\phi(\theta) = E\left[e^{i\theta x}\right] = \int_{-\infty}^{\infty} e^{i\theta x} f(x)dx \tag{2.92}$$

We will differentiate the characteristic function r times with regard to θ in order to better understand it. After that, for a discrete-type random variable, we have

$$\frac{d^r}{d\theta^r}\phi(\theta) = \sum_{j\in S}\left(ix_j\right)^r e^{i\theta x_j} p\left(x_j\right) = E\left[(ix)^r e^{i\theta x}\right] \tag{2.93}$$

For a continuous-type random variable, we have

$$\frac{d^r}{d\theta^r}\phi(\theta) = \int_{-\infty}^{\infty} (ix)^r e^{i\theta x} f(x)dx = E\left[(ix)^r e^{i\theta x}\right] \tag{2.94}$$

By letting $\theta = 0$ in the above two equations, we obtain the following relationships

$$\frac{d^r}{d\theta^r}\phi(0) = E\left[(ix)^r\right] = i^r E\left[x^r\right] \tag{2.95}$$

$$\frac{1}{i^r}\phi^r(0) = E\left[x^r\right] \tag{2.96}$$

This results in the subsequent theorem: *For a random variable, the* rth *moment, denoted as* $E[x^r]$, *if it exists, can be found by* $\frac{1}{i^r}\phi^r(0)$. *Here, i represents an imaginary unit, and* $\phi^r(0)$ *represents the* rth *derivative of the characteristic with* θ *equal to zero.*

Based on the information provided in the previously stated theorems in this section, it is possible to derive the mean and variance of a random variable by utilizing the derivatives of either the moment-generating function or the characteristic function. The cumulant-generating function is an additional function that helps evaluate the mean and variance in a straightforward manner without the need to carry out any differentiation. In spite of the fact that the cumulant-generating function does not produce the moments (with the exception of the first moment), evaluating the variance of a random variable can be a very convenient task by utilizing this function. A more important point to consider is that the cumulant-generating function plays a significant role in the process of developing the PDF that is applicable to non-Gaussian random processes.

In the vicinity of $t = 0$, let us expand the characteristic function of a random variable X in a power series. That is,

$$\phi(\theta) = E\left[e^{i\theta x}\right] = 1 + (i\theta)E[x] + \frac{(i\theta)^2}{2!}E\left[x^2\right] + \frac{(i\theta)^3}{3!}E\left[x^3\right] + \cdots = 1 + \sum_{s=1}^{\infty}(i\theta)^s\frac{m_s}{s!} \tag{2.97}$$

in which $E\left[x^s\right] = m_s$.

Next, let us generate a power series by expanding the logarithm of $\phi(\theta)$, which is represented by the symbol $\psi(\theta)$, as follows:

$$\begin{aligned}\psi(\theta) = \ln\phi(\theta) &= \ln\left\{1 + \sum_{s=1}^{\infty}(i\theta)^s\frac{m_s}{s!}\right\}\\ &= \left[\sum_{s=1}^{\infty}(i\theta)^s\frac{m_s}{s!}\right] - \frac{1}{2}\left[\sum_{s=1}^{\infty}(i\theta)^s\frac{m_s}{s!}\right]^2 + \frac{1}{3}\left[\sum_{s=1}^{\infty}(i\theta)^s\frac{m_s}{s!}\right]^3 + \cdots \\ &= m_1\frac{i\theta}{1!} + \left(m_2 - m_1^2\right)\frac{(i\theta)^2}{2!} + \left(m_3 - 3m_1m_2 + 2m_1^3\right)\frac{(i\theta)^3}{3!} + \cdots\end{aligned} \tag{2.98}$$

Let us represent the function $\psi(\theta)$ obtained from equation (2.98) as a series:

$$\psi(\theta) = \ln\phi(\theta) = \sum_{s=1}^{\infty}(i\theta)^s\frac{k_s}{s!} \tag{2.99}$$

From equation (2.98) and equation (2.99), we can obtain

$$\begin{aligned}
k_1 &= m_1 = \text{mean} \\
k_2 &= m_2 - m_1^2 = \text{variance} \\
k_3 &= m_3 - 3m_1m_2 + 2m_1^3 \\
k_4 &= m_4 - 3m_2^2 - 4m_1m_3 + 12m_1^2m_2 - 6m_1^4, \text{ and so on}
\end{aligned} \tag{2.100}$$

The function $\psi(\theta)$, defined in equation (2.99), is referred to as the ***cumulant-generating function***, while the coefficients k_s are known as ***cumulants*** or semi-invariants.

2.7 SOME IMPORTANT DISCRETE PROBABILITY DISTRIBUTIONS

This section focuses on several discrete-type probability distributions that serve as significant models for scientific events. The characteristics and practical uses of these distributions are examined. Having knowledge about the circumstances in which these random variables arise allows us to select a suitable probability distribution for a scientific topic being studied.

In statistics, a random variable is one whose value is determined by chance. There are two types of random variables: Discrete and continuous. In this section, we will explore discrete random variables and their probability distributions. A discrete random variable takes only a finite or countably infinite number of possible values. Examples of discrete random variables include the number of heads in a fair coin toss, the number of passengers on a bus, or the number of successes in a sequence of independent trials.

A probability distribution is a function that describes the likelihood of obtaining a specific value for a random variable. It provides a mathematical representation of the uncertainty associated with the value of the variable. A discrete probability distribution represents the probability of obtaining each possible value for a random variable. These distributions are characterized by a set of distinct values and their associated probabilities.

2.7.1 Binomial Distribution

One commonly used discrete probability distribution is the binomial distribution. The binomial distribution is a probability distribution that models the number of successes in a fixed number of independent trials. It is commonly used in statistics for estimating the probability of success or failure in experiments or events. The binomial distribution is binomially distributed, meaning it has two possible outcomes: Success or failure. Each trial or experiment is considered independent, meaning the outcomes of each trial are not influenced by the outcome of the previous trial.

Let us examine a random experiment where the outcome can only be one of two mutually exclusive events, for instance, success or failure, or the occurrence or non-occurrence of an event. Each trial is considered to have a probability of success p. Consequently, the probability of failure can be calculated as $q = 1 - p$. Consider a random experiment that is repeated n times independently (referred to as n Bernoulli trials with a parameter p). Let us assume that x successes are seen in these n experiments. It should be noted that there are $\binom{n}{x}$ sequences in which x instances of success and $(n - x)$ instances of failure occur. Each of these sequences has a probability of occurring equal to $p^x q^{n-x}$. Therefore, the probability of achieving x successful outcomes in n trials is

$$p(x) = \binom{n}{x} p^x q^{n-x} \tag{2.101}$$

in which $\binom{n}{x}$ represents the number of ways to choose x objects out of n objects. $\binom{n}{x}$ can be calculated as follows:

$$\binom{n}{x} = \frac{n!}{x!(n-x)!} \tag{2.102}$$

The binomial distribution can be defined as follows as a result of the above evidence: Let us say that the probability of event success is denoted by p $(0 < p < 1)$, and let us say that the probability of event failure in each trial is denoted by $q = 1 - p$. When the probability of achieving exactly x successes in n Bernoulli trials is provided by the above equation (2.101), the discrete random variable X is said to have the ***binomial probability distribution***. The binomial distribution is frequently represented by the notation $b(n, p)$. The Bernoulli probability distribution, also known as the zero-one probability distribution, is a specific type of binomial distribution that is used when the number of observations is equal to one (i.e., $n = 1$).

Example 2.2

▶ Let us consider a simple example to illustrate the use of the binomial distribution. Consider a coin flip experiment where we want to calculate the probability of observing exactly 2 heads in 5 tosses of a coin.

SOLUTION

In this example, we have: $n = 5$, $p = 0.5$ and $q = 1 - p = 0.5$. Thus, the probability of observing exactly 2 heads in 5 tosses is given by the binomial distribution as:

$$p(x) = \binom{n}{x} p^x q^{n-x} = \binom{5}{2} \times 0.5^2 \times (1 - 0.5)^{5-2} = \frac{5!}{2!(5-2)!} \times 0.25 \times 0.125 = 0.3125 \tag{2.103}$$

Therefore, the probability of observing exactly 2 heads in 5 tosses is 0.3125. ◀

2.7.2 Poisson Distribution

An assessment of the statistical characteristics of rare incidents, such as accidents, earthquakes, and floods, holds great significance in the field of applied probability within engineering and the physical sciences. Random processes, which deal with the quantification of rare events, are referred to as counting random processes and will be thoroughly examined in the next chapter. The Poisson random process is commonly observed in natural events and is the most frequently encountered among counting processes. The fundamental statistical characteristics of Poisson random processes can be explained using the Poisson probability distribution discussed in this section.

Starting with the binomial distribution, we proceed to calculate the Poisson probability distribution. Given that we are examining rare events, it is suitable to have a small probability, p, of the event occurring in each trial of the binomial distribution while having a high number of trials, n. Nevertheless, the product np, which represents the mean value of the binomial distribution, is finite. Let us denote np as μ. The binomial distribution can be represented in terms of n and μ in the following manner:

$$\begin{aligned} p(x) &= \binom{n}{x} p^x (1-p)^{n-x} = \frac{n!}{x!(n-x)!}\left(\frac{\mu}{n}\right)^x \left(1-\frac{\mu}{n}\right)^{n-x} \\ &= \frac{\mu^x}{x!}\left(1-\frac{\mu}{n}\right)^n \frac{n(n-1)(n-2)\cdots(n-x+1)}{n^x} \frac{1}{\left(1-\frac{\mu}{n}\right)^x} \\ &= \frac{\mu^x}{x!}\left(1-\frac{\mu}{n}\right)^n \frac{1\left(1-\frac{1}{n}\right)\left(1-\frac{2}{n}\right)\cdots\left(1-\frac{x-1}{n}\right)}{\left(1-\frac{\mu}{n}\right)^x} \end{aligned} \tag{2.104}$$

Because

$$\lim_{n\to\infty}\left(1-\frac{\mu}{n}\right)^n = e^{-\mu}$$

and

$$\lim_{n\to\infty} \frac{1\left(1-\frac{1}{n}\right)\left(1-\frac{2}{n}\right)\cdots\left(1-\frac{x-1}{n}\right)}{\left(1-\frac{\mu}{n}\right)^x} = 1$$

we, therefore, obtain

$$p(x) = \frac{\mu^x}{x!} e^{-\mu} \tag{2.105}$$

The above-mentioned distribution in equation (2.105) is referred to as the ***Poisson distribution***. We can see that the Poisson distribution is characterized by a single parameter, μ, which represents the average number of occurrences of events. In the Poisson distribution, x represents the count or number of occurrences of the events, $x = 0, 1, 2, \ldots, \infty$. The parameter μ determines the shape of the distribution, with a larger value of μ resulting in a higher probability of a high number of occurrences.

Example 2.3

▶ Let us suppose that the probability of a propeller manufactured by a certain factory being defective is 0.002. What is the probability that there is no defective propeller in a batch of 1000?

SOLUTION

In this example, we have: $n = 1000$, $p = 0.002$. $\mu = np = 2$. Because n is large and p is small in this example, the Poisson approximation is appropriate, and we obtain

$$p(0) = \frac{(2)^0}{0!}e^{-2} = 0.1353$$

Therefore, the probability that there is no defective propeller in a batch of 1000 is 0.1353. ◀

While the Poisson distribution offers a convenient method for analyzing count data, it has certain limitations: (1). Independence Assumption: The events under the Poisson distribution are assumed to be independent, which may not always be true in practice. (2). Non-Negativity: The Poisson distribution assumes non-negative count data, which limits its application when count data can be negative. (3). Autocorrelation: The Poisson distribution assumes no autocorrelation in the count data, which may not hold for certain real-world scenarios. In conclusion, the Poisson distribution is a widely used distribution for modeling count data. Its simplicity, flexibility, and robustness make it a valuable tool in various fields. However, it is important to be aware of its limitations and employ appropriate statistical methods when the assumptions of the Poisson distribution are not met. By understanding and applying the Poisson distribution, researchers and practitioners can gain valuable insights into their data and make informed decision-making processes.

2.8 SOME IMPORTANT CONTINUOUS PROBABILITY DISTRIBUTIONS

A continuous random variable is a variable that can take on any value within a specified range. Unlike discrete random variables, which can only take on specific values, continuous random variables can have an infinite number of possible values. The probability distribution of a continuous random variable describes the likelihood or probability that a given value will be observed. It quantifies the uncertainty and randomness associated with the variable.

In this section, let us redirect our focus towards several important continuous-type probability distributions. In situations where physical quantities such as time, length, area, temperature, pressure, load and intensity necessitate probabilistic descriptions, continuous-type random variables are employed as models. This section introduces several important continuous-type probability distributions. Similar to the previous section, our focus is on the characteristics and the scientific and engineering applications of these probability distributions.

2.8.1 Continuous Uniform Distribution

A continuous uniform distribution describes an experiment where there is an arbitrary outcome that lies between certain bounds. The bounds are defined by the parameters a and b, which represent the lowest and highest values, respectively. The interval can be classified as either closed, denoted by $[a, b]$, or open, denoted by (a, b). Thus, the distribution is commonly denoted as $U(a, b)$, where U represents the uniform distribution. The interval length is determined by the difference between the bounds. All intervals on the distribution's support that have the same length are equally probable.

A continuous random variable X is a variable that can take on any value within a certain range, rather than being limited to specific values. X follows a uniform distribution over the interval from a to b (where $b > a$) if it has an equal probability of assuming any value within this region. The PDF of X is uniformly distributed over the interval (a, b) and is characterized by a constant value and has the form as follows:

$$f_X(x) = \begin{cases} \dfrac{1}{b-a}, & for \quad a \le x \le b; \\ 0, & \text{elsewhere.} \end{cases} \tag{2.106}$$

As we can see from Figure 2.5, the PDF of the continuous uniform distribution is represented graphically as a rectangle with a base length of $(b\text{-}a)$ and a height of $1/(b\text{-}a)$. This is because every PDF integrates to 1. As the length of the base increases, the height (the probability density at a specific value inside the distribution boundaries) drops.

Example 2.4

▶ For a continuous-type random variable $X \sim U(0, 25)$, find the probability $Pr(1 < X < 15)$.

SOLUTION

$$Pr(1 < X < 15) = (15-1) \times \frac{1}{25-0} = \frac{14}{25}.$$

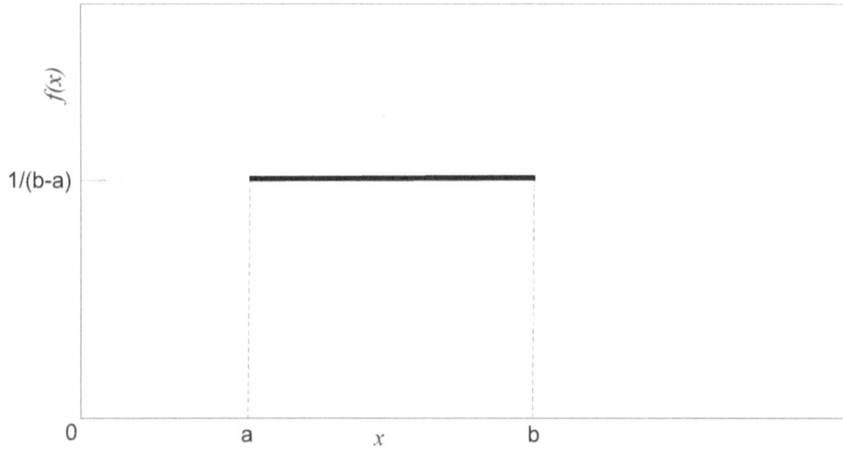

FIGURE 2.5 The probability density function of X.

The area under the curve within the given bounds, which shows the probability, is represented graphically as a rectangle in a continuous uniform distribution function. The base in the afore-mentioned case would be 14, and the height would be 1/25. ◄

Here we offer further explanations regarding the continuous uniform distribution. A continuous uniform distribution is a type of probability distribution that assigns equal probabilities to all values within a specified range. Specifically, it is a distribution where the range of possible values is continuous and the PDF is constant within the specified interval. The continuous uniform distribution is one of the simplest distributions and is frequently employed in scenarios when there is no justification for assigning unequal likelihoods to possible ranges assumed by the random variable over a specified period. For instance, the arrival time of an aircraft could be assumed to follow a uniform distribution within a specific time range. Similarly, the distribution of distances between the location of live loads on a bridge and an end support can be adequately represented by a uniform distribution over the span of the bridge. Furthermore, the continuous uniform distribution is also employed in Monte Carlo simulations to generate random variates with different probability distributions. This flexibility makes it a useful tool in simulating complex systems and analyzing uncertain outcomes. It is also worth noting that a uniform distribution is frequently assigned to a particular random variable simply due to a lack of information, beyond knowing the range of values it spans.

2.8.2 Gaussian Distribution

The ***Gaussian distribution***, also known as the ***normal distribution***, is one of the most commonly used probability distributions in statistics. It is a symmetrical distribution with a bell-shaped curve, characterized by its mean (μ) and standard deviation (σ). This section will provide an overview of the Gaussian distribution, including its properties, applications, and formulas.

A random variable X is considered to follow a ***Gaussian distribution*** with a mean μ and a variance σ^2 if its PDF is defined as

$$f_X(x) = \frac{1}{(2\pi)^{1/2}\sigma} \exp\left[-\frac{(x-\mu)^2}{2\sigma^2}\right], \quad -\infty < x < \infty \tag{2.107}$$

Its corresponding probability distribution function is:

$$F_X(x) = \frac{1}{(2\pi)^{1/2}\sigma} \int_{-\infty}^{x} \exp\left[-\frac{(x'-\mu)^2}{2\sigma^2}\right] \mathrm{d}x', \quad -\infty < x < \infty, \tag{2.108}$$

The above function in equation (2.108) is not expressible using a closed-form analytical expression, but it may be evaluated numerically for any value of x. The PDF and probability distribution function expressed by equations (2.107) and (2.108), respectively, are graphed in Figures 2.6 and 2.7, respectively, for $\mu = 0.5$ and a variance $\sigma^2 = 0.25$.

We mention again that the two parameters, μ and σ in the probability distribution equation (2.108), represent the mean and standard deviation of X, respectively. Here we also highlight a crucial characteristic of the normal distribution: Specifically, that knowing its mean and variance fully defines a normal distribution. Given the frequent reference to the normal distribution in our discussion, it is sometimes denoted by the concise notation $N(\mu, \sigma^2)$. Therefore, for instance, X: $N(1, 4)$ indicates that X follows the PDF described by equation (2.107) with a mean of 1 and a standard deviation of 2. In Figure 2.8, the higher curve follows

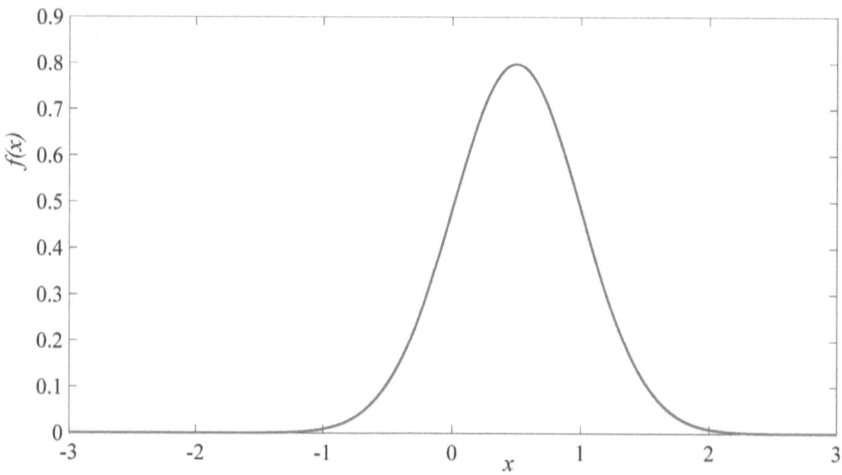

FIGURE 2.6 Probability density function of X for $\mu = 0.5$ and $\sigma^2 = 0.25$.

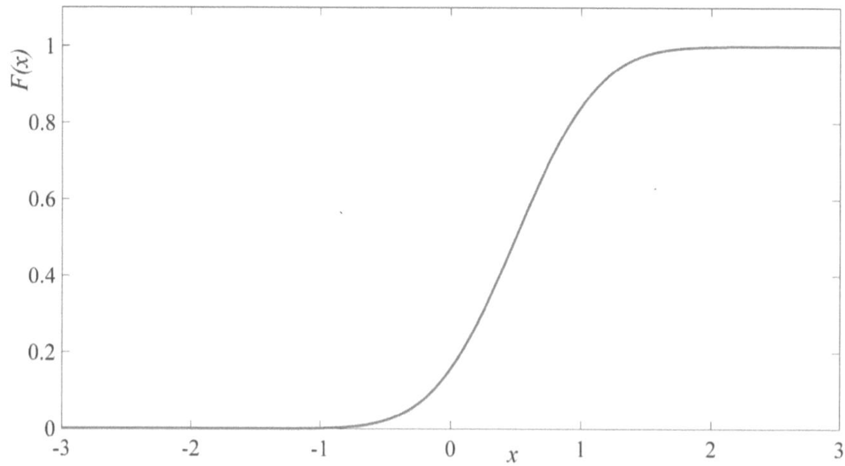

FIGURE 2.7 Probability distribution function of X for $\mu = 0.5$ and $\sigma^2 = 0.25$.

a $N(0, 0.25)$ distribution, indicating that X follows the normal PDF described by equation (2.107) with a mean of 0 and a standard deviation of 0.5. In Figure 2.8, the lower curve follows a $N(0, 1)$ distribution, indicating that X follows the normal PDF described by equation (2.107) with a mean of 0 and a standard deviation of 1. The standard deviation (SD) of a Gaussian distribution provides a measure of the central tendency and dispersion of the data points. The mean or center of a Gaussian distribution is the value of the mean, and the standard deviation indicates how spread out the data points are around the mean. A larger standard deviation

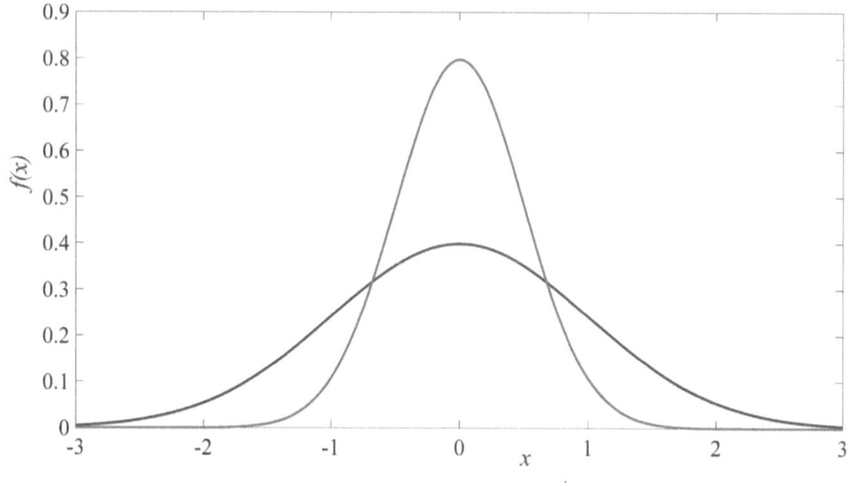

FIGURE 2.8 Probability density function of X for: (1) Higher curve: $\mu = 0$ and $\sigma^2 = 0.25$; (2) lower curve: $\mu = 0$ and $\sigma^2 = 1$.

indicates a larger spread. In Figure 2.8, the lower curve has a larger standard deviation value (i.e., $\sigma^2 = 1$), indicating the dataset X is more spread out in this case.

The standardized normal distribution, indicated by $N(0, 1)$, refers to a normal distribution having a mean (μ) of 0 and a variance (σ^2) of 1. To simplify, we will use the symbol Z to represent a random variable that follows a standardized normal distribution. The PDF of a standardized normal random variable is defined as:

$$\varphi_Z(z) = \frac{1}{(2\pi)^{1/2}} \exp\left[-\frac{z^2}{2}\right], \quad -\infty < z < \infty \tag{2.109}$$

By standardizing the random variable X, the PDF, $\varphi_Z(z)$, can be determined from the PDF, $f_X(x)$. This is accomplished by conducting the following calculation:

$$Z = \frac{X - \mu}{\sigma} \tag{2.110}$$

In Figure 2.8, the lower curve displays the PDF of the standardized normal distribution. The figure clearly illustrates that the probability of the standardized random variable falling inside the range of (–3, +3) is quite high, which implies that the probability that the value of the standardized random variable deviates from the mean value by more than 3σ is very small. This property of the Gaussian distribution is called the ***three-sigma rule***.

The standardized normal probability distribution's CDF is commonly represented as $\Phi(z)$. It is defined as

$$\Phi_Z(z) = \frac{1}{(2\pi)^{1/2}} \int_{-\infty}^{z} \exp\left[-\frac{(z')^2}{2}\right] dz' \tag{2.111}$$

By applying the definition that is presented in equation (2.111), the probability that the random variable X, which has a normal distribution $N(\mu, \sigma^2)$, falls between a and b can be evaluated using the following formula:

$$Pr\{a < X < b\} = Pr\left\{\frac{a-\mu}{\sigma} < Z < \frac{b-\mu}{\sigma}\right\} = \Phi\left(\frac{b-\mu}{\sigma}\right) - \Phi\left(\frac{a-\mu}{\sigma}\right) \tag{2.112}$$

The normal distribution holds significant practical significance due to the influential ***central limit theorem***, as stated below. Multiple formulations of the central limit theorem exist. However, this text will focus on the Lindberg–Levy theorem, as it will be used to explain the probabilistic nature of the Gaussian random process in Chapter 3.

The Central Limit Theorem: Consider a set of random variables, X_1, X_2, X_3, ..., X_n, that are statistically independent and have the same distribution. Each of these variables has a mean of μ and a variance of σ^2. Then, the sum of these random variables, denoted as $Y = X_1 + X_2 + X_3 + \cdots + X_n$, exhibits an asymptotic normal distribution with a mean of $n\mu$ and a variance of $n\sigma^2$ when n is sufficiently big. This is referred to as the **central limit theorem**. This **theorem** can also be mathematically expressed as follows:

Let

$$Y = \sum_{i=1}^{n} X_i \tag{2.113}$$

furthermore, let us define the normalized random variable Z as follows:

$$Z = \frac{Y - n\mu}{\sigma\sqrt{n}} \tag{2.114}$$

Then, as n approaches infinity, the probability distribution function of Z, denoted as $F_Z(z)$, approaches the standardized normal distribution $N(0, 1)$ for every fixed value of z.

The central limit theorem describes a broad range of random phenomena in which probability distributions can be estimated using the normal distribution. When the randomness in a physical phenomenon is the result of the accumulation of numerous small random effects that are added together, it tends to follow a normal distribution regardless of the distributions of the individual effects. However, while the central limit theorem is a powerful tool, it has its limitations as well. One key limitation is the requirement that the random variables be independent. If the variables are not independent, the theorem may not hold or may result in incorrect conclusions.

2.8.3 Multivariate Normal Distribution

Let's examine two random variables, X and Y. They are considered jointly normal if their joint PDF fits the following form

$$f_{XY}(x,y) = \frac{1}{2\pi\sigma_X\sigma_Y\left(1-\rho^2\right)^{1/2}} \exp \left\{-\frac{1}{2\left(1-\rho^2\right)}\left[\left(\frac{x-\mu_X}{\sigma_X}\right)^2 - 2\rho\frac{(x-\mu_X)(y-\mu_Y)}{\sigma_X\sigma_Y} + \left(\frac{y-\mu_Y}{\sigma_Y}\right)^2\right]\right\};$$

$$-\infty < x < \infty; \qquad -\infty < y < \infty. \tag{2.115}$$

in which μ_X and μ_Y represent the means of the random variables X and Y, respectively, whereas σ_X and σ_Y represent the standard deviations of X and Y, respectively. The correlation coefficient of X and Y is denoted by ρ.

In the context of stochastic analysis of random processes, the multivariate normal probability distribution is frequently encountered. In the studies that are described in Chapter 9, the tri-variate normal distribution of the displacement, velocity, and acceleration of a random process is taken into consideration. This is a typical example that can be met in the engineering analyses. The following is a definition of the joint PDF of the multivariate normal distribution: Let X_1, X_2, X_3, …, X_n be a series of n random variables. They are called to be jointly normal if their associated joint PDF has the following form:

$$f_{X_1X_2\ldots X_n}(x_1,x_2,\ldots,x_n) = f_{\mathbf{X}}(\mathbf{x}) = (2\pi)^{-n/2}|\Lambda|^{-1/2}\exp\left[-\frac{1}{2}(\mathbf{x}-\boldsymbol{\mu})^{\mathrm{T}}\Lambda^{-1}(\mathbf{x}-\boldsymbol{\mu})\right],$$

$$-\infty < \mathbf{x} < \infty, \tag{2.116}$$

in which $\boldsymbol{\mu}^{\mathrm{T}} = \begin{bmatrix}\mu_1 & \mu_2\ldots\mu_n\end{bmatrix} = \begin{bmatrix}E\{X_1\} & E\{X_2\}\ldots E\{X_n\}\end{bmatrix}$, and

$$\Lambda = \text{covariance matrix} = E\left[(\mathbf{x}-\boldsymbol{\mu})(\mathbf{x}-\boldsymbol{\mu})^{\mathrm{T}}\right]$$

$$= \begin{pmatrix} \sigma_1^2 & \sigma_{12} & \cdots & \sigma_{1n} \\ \sigma_{21} & \sigma_2^2 & \cdots & \sigma_{2n} \\ \vdots & & & \\ \sigma_{n1} & \sigma_{n2} & \cdots & \sigma_n^2 \end{pmatrix}$$

The superscripts T and −1 represent the matrix transpose and matrix inversion, respectively. Once again, it is evident that a joint normal distribution is fully determined by the first-order and second-order joint moments.

2.8.4 Log-Normal Distribution

A log-normal distribution is a type of continuous probability distribution in probability theory. It describes a random variable whose logarithm follows a normal distribution. If the random variable X follows a log-normal distribution, then the natural logarithm of X, denoted as $W = \ln(X)$, follows a normal distribution. Alternatively, if W follows a normal distribution, then the exponential transformation of W, denoted as $X = \exp(W)$, follows a log-normal distribution. A log-normally distributed random variable exclusively assumes positive real values.

A log-normal process is a statistical phenomenon that occurs when multiple independent random variables, all of which are positive, are multiplied together. This can be rationalized by taking into account the central limit theorem in the logarithmic domain, often known as Gibrat's law. It has been shown that normal distributions result from the accumulated sum of numerous random events. Now let us examine another common phenomenon that arises from multiple random factors that are multiplied together. An example of multiplicative phenomena can be observed in fatigue studies of materials, where the internal damage, within the material at a given stage of loading, is a random proportion of the damage observed at the preceding stage.

Consider the following

$$X = Y_1 Y_2 \cdots Y_n. \tag{2.117}$$

We are interested in studying the distribution of X as the sample size n increases, assuming that the random variables Y_j, $j = 1, 2, \ldots, n$, are restricted to positive values only.

By applying logarithms to both sides, equation (2.117) is transformed into

$$\ln X = \ln Y_1 + \ln Y_2 + \cdots + \ln Y_n \tag{2.118}$$

The random variable $\ln X$ is found to be the cumulative sum of the random variables $\ln Y_1$, $\ln Y_2$, …, and $\ln Y_n$. Consequently, according to the central limit theorem, $\ln X$ (the natural logarithm of X) approaches a normal distribution as n approaches infinity.

On occasion, variables within a system exhibit an exponential relationship, denoted by $X = \exp(W)$. $X = \exp(W)$ is a random variable with a distribution of interest if the exponent (W) is a random variable. An important special case arises when the variable W follows a normal distribution. When this occurs, the X distribution is referred to as a lognormal distribution. The nomenclature originates from the transformation $W = \ln(X)$. That is, the natural logarithm of X follows a normal distribution.

The probabilities for X are derived from the transformation of the normal distribution. If we assume that W follows a normal distribution with a mean of θ and a variance of ω^2, then the CDF for X can be expressed as:

$$\begin{aligned} F(x) &= Pr[X \le x] = Pr[\exp(W) \le x] = Pr[W \le \ln(x)] \\ &= Pr\left[Z \le \frac{\ln(x) - \theta}{\omega}\right] = \Phi\left[\frac{\ln(x) - \theta}{\omega}\right] \end{aligned} \tag{2.119}$$

For $x > 0$, where Z is a random variable following a standard normal distribution and $\Phi(\cdot)$ is the CDF of the standard normal distribution. The PDF of X can be derived from the derivative of $F(x)$. The derivative may be calculated using the fundamental theorem of calculus because $\Phi(\cdot)$ is the integral of the standard normal PDF.

$$\begin{aligned} f_X(x) &= \frac{\mathrm{d}}{\mathrm{d}x} Pr[X \le x] = \frac{\mathrm{d}}{\mathrm{d}x} Pr[W \le \ln x] = \frac{\mathrm{d}}{\mathrm{d}x} \Phi\left(\frac{\ln x - \theta}{\omega}\right) \\ &= \varphi\left(\frac{\ln x - \theta}{\omega}\right) \frac{\mathrm{d}}{\mathrm{d}x}\left(\frac{\ln x - \theta}{\omega}\right) = \varphi\left(\frac{\ln x - \theta}{\omega}\right) \frac{1}{\omega x} = \frac{1}{x\omega\sqrt{2\pi}} \exp\left(-\frac{(\ln x - \theta)^2}{2\omega^2}\right). \end{aligned} \tag{2.120}$$

in which $0 < x < \infty$.

Let us reiterate the definition of ***log-normal distribution*** as follows: Assuming that W follows a normal distribution with a mean of θ and a variance of ω^2, the random variable X, defined as $X = \exp(W)$, is a ***lognormal random variable*** with a PDF:

$$f_X(x) = \frac{1}{x\omega\sqrt{2\pi}} \exp\left(-\frac{(\ln x - \theta)^2}{2\omega^2}\right), \quad 0 < x < \infty. \tag{2.121}$$

The log-normal distribution is characterized by two parameters, namely θ and ω^2, and is commonly represented as $\Lambda(\theta, \omega^2)$. It is important to clarify that the parameters θ and ω^2 are not the mean and variance, respectively, of the random variable X. The logarithm of the random variable X follows a normal distribution with a mean of θ and a variance of ω^2. The mean and variance of the log-normal distribution are provided as follows:

$$\begin{aligned} E[x] &= \exp\left\{\theta + \frac{\omega^2}{2}\right\} \\ Var[x] &= \exp\left\{2\theta + \omega^2\right\} \cdot \left(\exp\left\{\omega^2\right\} - 1\right) \end{aligned} \tag{2.122}$$

Figure 2.9 is an illustration of lognormal distributions for a selection of the parameters' values.

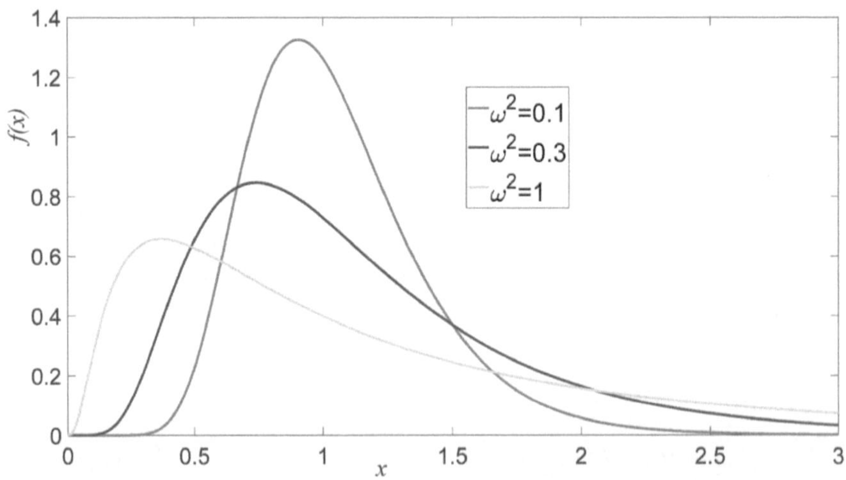

FIGURE 2.9 Lognormal probability density functions with $\theta = 0$ for selected values of ω^2: (1) The highest curve: $\theta = 0$ and $\omega^2 = 0.1$; (2) the middle curve: $\theta = 0$ and $\omega^2 = 0.3$; (3) the lowest curve: $\theta = 0$ and $\omega^2 = 1$.

2.8.5 Gamma and Related Distributions

Another category of useful one-sided distributions is referred to as the gamma distribution. The PDF that is associated with the ***gamma distribution*** is

$$f_X(x) = \frac{1}{\Gamma(m)} \lambda^m x^{m-1} e^{-\lambda x}, \quad 0 \leqslant x < \infty, \quad \lambda > 0 \tag{2.123}$$

in which $\Gamma(m)$ represents the well-known gamma function:

$$\Gamma(m) = \int_0^\infty u^{m-1} \mathrm{e}^{-u} \, du \tag{2.124}$$

The gamma distribution is characterized by two parameters, m and λ, both of which are assumed to be positive. The gamma distribution is commonly used to model physical quantities that are restricted to positive values because of its one-sided nature. Moreover, it functions as a valuable prototype due to its versatility, allowing for the generation of a diverse range of shapes for the gamma density function by manipulating the values of m and λ. The plots of equation (2.123) for various values of m and λ are depicted in Figures 2.10 and 2.11. These figures demonstrate that m determines the shape of the distribution and is therefore considered a shape parameter, whereas λ serves as a scale parameter for the distribution. Typically, the gamma density function has a single peak. For values of m less than or equal to 1, the peak is located at $x = 0$. For values of m greater than 1, the peak is located at $x = (m - 1)/\lambda$.

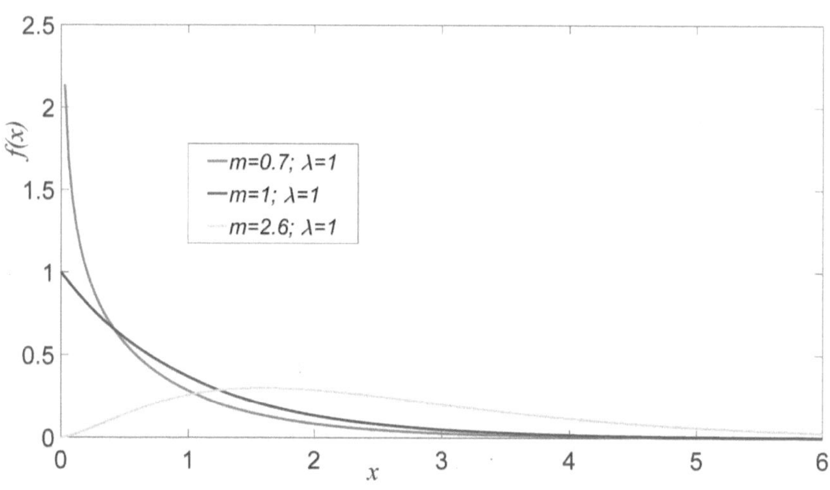

FIGURE 2.10 Gamma probability density functions with $\lambda = 1$ for selected values of m: (1) The highest curve: $\lambda = 1$ and $m = 0.7$; (2) the middle curve: $\lambda = 1$ and $m = 1$; (3) the lowest curve: $\lambda = 1$ and $m = 2.6$.

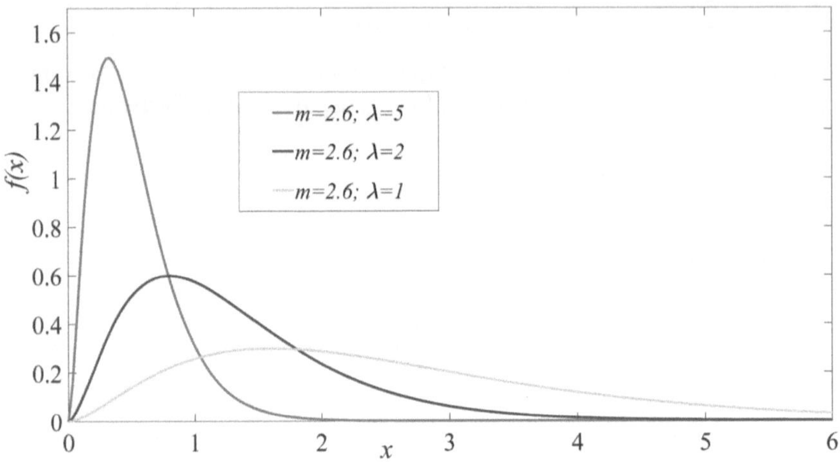

FIGURE 2.11 Gamma PDFs with $m = 2.6$ for selected values of λ: (1) The highest curve: $\lambda = 5$ and $m = 2.6$; (2) the middle curve: $\lambda = 2$ and $m = 2.6$; (3) the lowest curve: $\lambda = 1$ and $m = 2.6$.

The mean and variance of a random variable X that follows a gamma distribution have straightforward expressions. Upon completing the required integration, we acquire

$$E[x] = m_X = \frac{m}{\lambda}, \quad Var[x] = \sigma_X^2 = \frac{m}{\lambda^2} \tag{2.125}$$

There are a number of important probability distributions that are examples of the gamma distribution in their specific form. There are two of these that will be described in greater depth below. The ***exponential distribution*** is specifically referred to as the distribution for $m = 1$ in the gamma distribution, and it is defined in the following manner:

A random variable X is said to follow an ***exponential distribution*** *if its PDF is defined as:*

$$f_X(x) = \lambda e^{-\lambda x}, \quad 0 \leqslant x < \infty, \quad \lambda > 0 \tag{2.126}$$

The exponential distribution gets its name from the exponential function that is included in the PDF. Figure 2.12 displays the plots of the exponential distribution for a selection of values of λ. Regardless of the value of λ, the exponential distribution exhibits a significant degree of skewness.

The mean and variance of a random variable X that follows an ***exponential distribution*** have straightforward expressions. Upon completing the required integration, we acquire

$$E[x] = m_X = \frac{1}{\lambda}, \quad Var[x] = \sigma_X^2 = \frac{1}{\lambda^2} \tag{2.127}$$

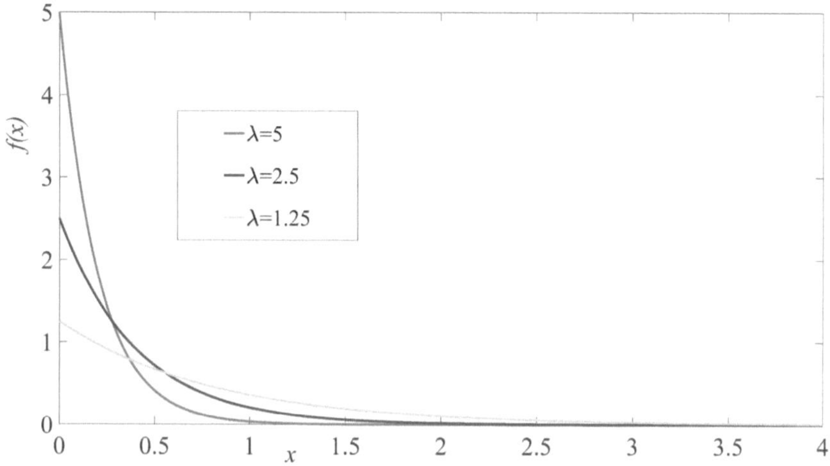

FIGURE 2.12 Exponential probability density functions for selected values of λ: (1) The highest curve: $\lambda = 5$; (2) the middle curve: $\lambda = 2.5$; (3) the lowest curve: $\lambda = 1.25$.

The ***chi-squared (χ^2) distribution***, which may be derived by putting $\lambda = 1/2$ and $m = n/2$ in equation (2.123), where n is a positive integer, is another important special case of the gamma distribution. Thus, the ***chi-squared (χ^2) distribution*** has one parameter, n, and a PDF of the following form.

$$f_X(x) = \frac{1}{\Gamma(n/2)} (1/2)^{\frac{n}{2}} x^{((n/2)-1)} e^{-x/2}, \quad 0 \leqslant x < \infty \tag{2.128}$$

The parameter n is commonly known as the degrees of freedom. The usefulness of this distribution stems from the property that the sum of the squares of n independent standardized normal random variables follows a chi-squared distribution with n degrees of freedom. In other words, if U_1, U_2, …, and U_n are independent and distributed as $N(0, 1)$, then the sum

$$X = U_1^2 + U_2^2 + \cdots + U_n^2 \tag{2.129}$$

has a ***chi-squared (χ^2) distribution*** with n degrees of freedom. Due to the fact that this relationship exists, the ***chi-squared (χ^2) distribution*** is considered to be one of the most important tools that people utilize in the field of statistical inference and hypothesis testing. The PDF $f_X(x)$ specified in equation (2.128) is plotted in Figure 2.13 for various values of n. As the value of n increases, the PDF $f_X(x)$ exhibits a greater degree of symmetry. Considering equation (2.129), given that X can be represented as a sum of identically distributed random variables, we anticipate that the ***chi-squared (χ^2) distribution*** will converge to a normal distribution as n approaches infinity, based on the central limit theorem.

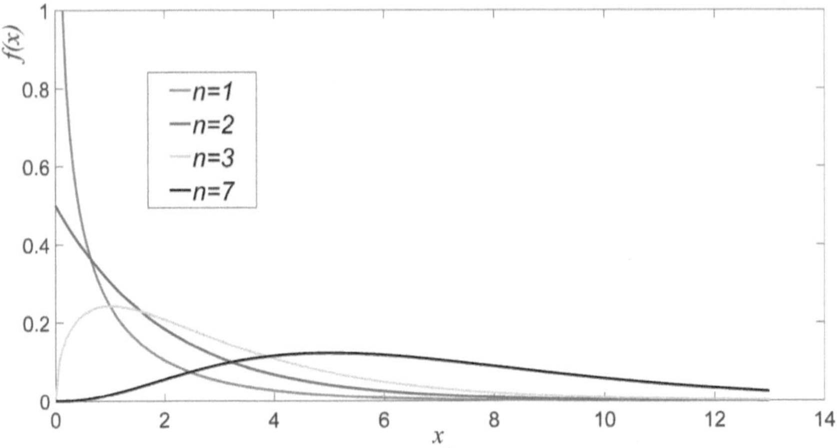

FIGURE 2.13 **Chi-squared (χ^2)** probability density functions for selected values of n: (1) The highest curve: $n = 1$; (2) the upper middle curve: $n = 2$; (3) the lower middle curve: $n = 3$; (4) the lowest curve: $n = 7$.

The mean and variance of a random variable X that follows a ***chi-squared (χ^2) distribution*** have straightforward expressions and are easily obtained from equation (2.125) as:

$$E[x] = m_X = n, \quad Var[x] = \sigma_X^2 = 2n \tag{2.130}$$

The ***chi-squared (χ^2) distribution*** is largely employed in hypothesis testing and, to a lesser degree, for constructing confidence intervals for population variance in cases where the underlying distribution is normal. Contrary to the commonly used normal and exponential distributions, the chi-squared distribution is less frequently employed in directly representing natural phenomena.

2.8.6 Weibull and Related Distributions

A random variable X is said to have a Weibull distribution if its PDF is expressed in the following manner:

$$f_X(x) = \frac{\beta}{\delta}\left(\frac{x}{\delta}\right)^{\beta-1} \exp\left[-\left(\frac{x}{\delta}\right)^{\beta}\right], \quad for \quad x > 0 \tag{2.131}$$

with a scale parameter $\delta > 0$ and a shape parameter $\beta > 0$. For the purpose of computing probabilities, the Weibull CDF is frequently utilized. If the random variable X is characterized by a Weibull distribution with parameters δ and β, then the CDF of X is

$$F_X(x) = 1 - e^{-\left(\frac{x}{\delta}\right)^{\beta}} \tag{2.132}$$

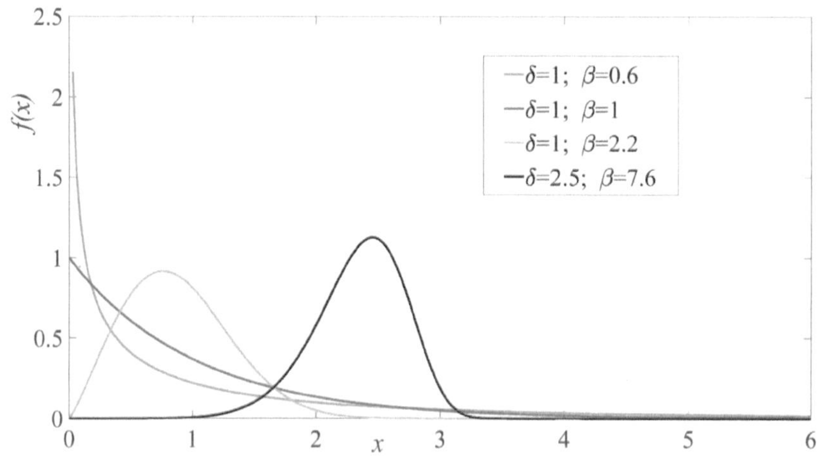

FIGURE 2.14 Weibull probability density functions for selected values of δ and β: (1) The highest curve: $\delta = 1, \beta = 0.6$; (2) the second curve: $\delta = 1, \beta = 1$; (3) the third curve: $\delta = 1, \beta = 2.2$; (4) the rightmost curve: $\delta = 2.5, \beta = 7.6$.

The Weibull distribution is frequently utilized in the modeling of the time to failure of a wide variety of various physical systems. When modeling systems in which the number of failures increases with time (bearing wear), reduces with time (certain semiconductors) or remains constant with time (failures produced by external shocks to the system), the parameters in the Weibull distribution offer a great deal of flexibility.

The flexibility of the Weibull distribution is illustrated by the graphs of selected PDFs that are presented in Figure 2.14. Upon careful examination of the PDF, it becomes evident that when the value of β is equal to one, the Weibull distribution is identical to the exponential distribution. In addition, the Rayleigh distribution is a specific example that occurs when the shape parameter is equal to 2.

A random variable X is said to follow a Rayleigh distribution if its PDF is expressed in the following manner:

$$f_X(x) = \frac{2x}{\delta^2} \exp\left[-\left(\frac{x}{\delta} \right)^2 \right], \quad for \quad x \geq 0 \tag{2.133}$$

The Rayleigh distribution can be represented as $\beta = 2$ in the Weibull distribution. Nevertheless, the Rayleigh distribution possesses a theoretical foundation, unlike the Weibull distribution. The Rayleigh distribution was created by Lord Rayleigh in 1880 to describe the combined intensity of numerous independent sounds. The subsequent theorem holds great significance in the analysis of amplitudes of a Gaussian random process.

Theorem: *Assume that both of the random variables X_1 and X_2 follow a normal distribution, with the mean being zero and the variance being σ^2, denoted as*

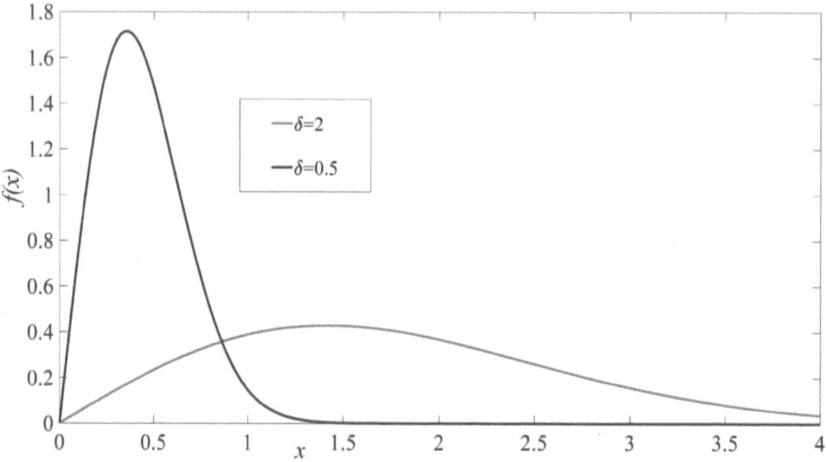

FIGURE 2.15 Rayleigh probability density functions for selected values of δ: (1) The highest curve: $\delta = 2$; (2) the lower curve: $\delta = 0.5$.

$N(0, \sigma^2)$. It can be concluded that the random variable $\sqrt{X_1^2 + X_2^2}$ follows a Rayleigh distribution as follows:

$$f_X(x) = \frac{x}{\sigma^2}\exp\left[-\frac{x^2}{2\sigma^2}\right], \quad for \quad x \ge 0 \tag{2.134}$$

The foregoing theorem can be proven by utilizing the transformation technique of random variables as outlined in Section 2.9. Therefore, the proof of the foregoing theorem is provided in Section 2.9. Figure 2.15 displays the Rayleigh probability density curves for selected values of δ in equation (2.133).

2.8.7 Probability Distribution of Extreme Values

Consider a sequence of random variables, $X_1, X_2, \ldots, X_n$, that are independent and identically distributed. Let F be the CDF of these random variables. We define M_n as the maximum value among $X_1, X_2, \ldots$, and X_n, that is, $M_n = \max(X_1, X_2, \ldots, X_n)$. The probability of M_n being less than m is equivalent to the probability of all X_i being less than m. The independence of the variables X_i implies that the product of the probabilities of each X_i being smaller than m is equal to the probability of M_n being less than m. The mathematical equations are expressed as follows:

$$\begin{aligned} F_{M_n}(m) &= Pr(M_n \le m) = Pr(X_1 \le m, X_2 \le m, \ldots, X_n \le m) \\ &= Pr(X_1 \le m)Pr(X_2 \le m)\cdots Pr(X_n \le m) \\ &= F_{X_1}(m)F_{X_2}(m)\cdots F_{X_n}(m) \\ &= [F_X(m)]^n \end{aligned} \tag{2.135}$$

The product is raised to the power of n in the final step due to the independence and identical distribution of the X_i random variables. The PDF is obtained by taking the derivative of the foregoing equation:

$$f_{M_n}(m) = n\left[F_X(m)\right]^{n-1} f_X(m) \tag{2.136}$$

The modal value of the function $f_{M_n}(m)$, represented as $\bar{M}_n$, is referred to as the probable extreme value or characteristic largest value. It is the extreme value that is most probable to occur among n observations and can be determined by solving the following equation:

$$\frac{d}{dM_n} f_{M_n}(m) = 0 \tag{2.137}$$

which yields

$$\left[F_X(m)\right] f_X'(m) + (n-1)\left[f_X(m)\right]^2 = 0 \tag{2.138}$$

We denote the m that satisfies the equation above as $\bar{M}_n$, which gives us the following:

$$1 = \frac{-(n-1)\left[f_X\left(\bar{M}_n\right)\right]^2}{\left[F_X\left(\bar{M}_n\right)\right] f_X'\left(\bar{M}_n\right)} \tag{2.139}$$

Then, after dividing the foregoing equation by $1-\left[F_X(\bar{M}_n)\right]$, we obtain

$$\frac{1}{1-\left[F_X\left(\bar{M}_n\right)\right]} = \frac{-(n-1)}{\left[F_X\left(\bar{M}_n\right)\right] f} \frac{f_X\left(\bar{M}_n\right)}{1-\left[F_X\left(\bar{M}_n\right)\right]} \frac{f_X\left(\bar{M}_n\right)}{f_X'\left(\bar{M}_n\right)} \tag{2.140}$$

Assuming the initial distribution $f_X(m)$ complies with the L'Hospital rule, which takes the form of

$$\frac{f_X\left(\bar{M}_n\right)}{1-\left[F_X(\bar{M}_n)\right]} = \frac{-f_X'\left(\bar{M}_n\right)}{f_X\left(\bar{M}_n\right)} \quad \text{for large } \bar{M}_n \tag{2.141}$$

Combining the foregoing two equations to

$$\frac{1}{1-\left[F_X\left(\bar{M}_n\right)\right]} = \frac{(n-1)}{\left[F_X\left(\bar{M}_n\right)\right]} \approx n \quad \text{for large } n \text{ and } \bar{M}_n \tag{2.142}$$

Example 2.5

▶ Determine the probable extreme value for the Rayleigh probability density distribution, which is given by

$$f_X(x) = \frac{2x}{\delta^2}\exp\left[-\left(\frac{x}{\delta}\right)^2\right], \quad for \quad x \geq 0 \tag{2.143}$$

SOLUTION

The formula for the CDF corresponding to the foregoing PDF is as follows:

$$F_X(x) = 1 - \exp\left[-\left(\frac{x}{\delta}\right)^2\right], \quad for \quad x \geq 0 \tag{2.144}$$

Therefore, from equation (2.142), we can obtain

$$\exp\left[-\left(\frac{\bar{M}_n}{\delta}\right)^2\right] = 1/n \tag{2.145}$$

and hence, the answer to this problem is given by

$$\bar{M}_n = \delta\sqrt{\ln n} \tag{2.146}$$

◀

2.9 THE TRANSFORMATION OF RANDOM VARIABLES

The topic discussed in this section is the process of determining the probability distribution of a new random variable based on the probability distribution of another random variable, given the functional relationship between them. Transforming random variables also refers to the process of deriving the joint probability distribution of two new random variables from the information on the joint probability distribution of two original random variables while being aware of the functional relationships between the new and original random variables. The methods of transformation are widely used in practical applications for various statistical predictions, including assessing failure in reliability analysis and predicting the amplitudes and periods of random processes.

2.9.1 The Transformation of a Single Random Variable

Consider X as a random variable with a probability density function $f_X(x)$. Additionally, let Y be a new random variable that is related to X through a single-valued function $Y = g(X)$. The cumulative distribution function of the random variable Y, represented as $F_Y(y)$ to differentiate it from that of the random variable X, can be expressed as

$$F_Y(y) = Pr\{Y \leqslant y\} = Pr\{g(X) \leqslant y\} = Pr\{X \leqslant h(y)\} = F_X[h(y)] \tag{2.147}$$

in which $F_X[h(y)]$ is the cumulative distribution function of the random variable X evaluated at $x = h(y)$, where $h(y)$ is the inverse function of $y = g(x)$.

Equation (2.147) can be used to derive the probability density function of the random variable Y, which can be expressed as

$$f_Y(y) = f_X[h(y)]\left|\frac{d}{dy}h(y)\right| \tag{2.148}$$

The expression $\left|\frac{d}{dy}h(y)\right|$ in equation (2.148) represents the Jacobian of the transformation. The absolute value must be taken because the probability density function should be a positive value.

Example 2.6

▸ The probability density function of a continuous random variable X is represented by the equation $f_X(x) = 5e^{-x}$, where $0 \leq x < \infty$. Let us examine the process of transforming the random variable from X to Y, where $Y = cX - d$. Here, c and d are positive constants. We have the equation $F_X(x) = 1 - 5e^{-x}$, and $X = (Y + d)/c$. Therefore, the cumulative distribution function of Y can be derived from equation (2.147) as follows:

$$F_Y(y) = 1 - 5\exp\left[-(y + d)/c\right], \quad -d \leqslant y < \infty \tag{2.149}$$

The PDF f(y) can be derived from equation (7.4) as

$$f_Y(y) = \frac{5}{c}\exp\left[-(y + d)/c\right], \quad -d \leqslant y < \infty \tag{2.150}$$

◂

2.9.2 The Transformation of Two Random Variables

It is possible to extend the idea that was discussed in the preceding section to the context of the transformation of two random variables from one joint probability density function to another. Consider X and Y as two random variables, characterized by the joint probability density function $f_{XY}(x, y)$. Now, let's take into consideration two new random variables, R and S, which have functional relationships with X and Y, as denoted by the following equations

$$R = g_1(x, y); S = g_2(x, y) \tag{2.151}$$

In this context, we assume that the functions $g_1(\bullet)$ and $g_2(\bullet)$ possess continuous partial derivatives with respect to the variables x and y. As a result, based on the definition of the cumulative distribution function, we can derive

$$\begin{aligned} F_{RS}(r,s) &= Pr\{R \leqslant r, S \leqslant s\} = Pr\{g_1(x,y) \leqslant r, g_2(x,y) \leqslant s\} \\ &= Pr\{X \leqslant h_1(r,s), Y \leqslant h_2(r,s)\} = F_{XY}\left[h_1(r,s), h_2(r,s)\right] \end{aligned} \tag{2.152}$$

in which $X = h_1(r,s)$ and $Y = h_2(r,s)$ are the inverse functions of $R = g_1(x,y)$ and $S = g_2(x,y)$, and $F_{XY}(\bullet)$ is the joint probability density function of X and Y. The joint probability density function of R and S can be derived by differentiating $F_{RS}(r,s)$ with respect to r and s.

We now provide the detailed derivation process for obtaining the joint probability density function of R and S, i.e., $f_{RS}(r,s)$.

If $g_1(x,y)$ and $g_2(x,y)$ are functions that are both continuous and differentiable, then, it is feasible to derive a method to directly extract the joint probability density function $f_{RS}(r,s)$. In light of this, consider the mathematical equations:

$$r = g_1(x,y) \quad s = g_2(x,y) \tag{2.153}$$

It is possible for the equation (2.153) to have many solutions for a given point (r,s). Let us state that these many solutions are represented by the expressions $(x_1, y_1), (x_2, y_2), \ldots, (x_n, y_n)$ in such a way that (see Figure 2.16)

$$r = g_1(x_i, y_i) \quad s = g_2(x_i, y_i) \tag{2.154}$$

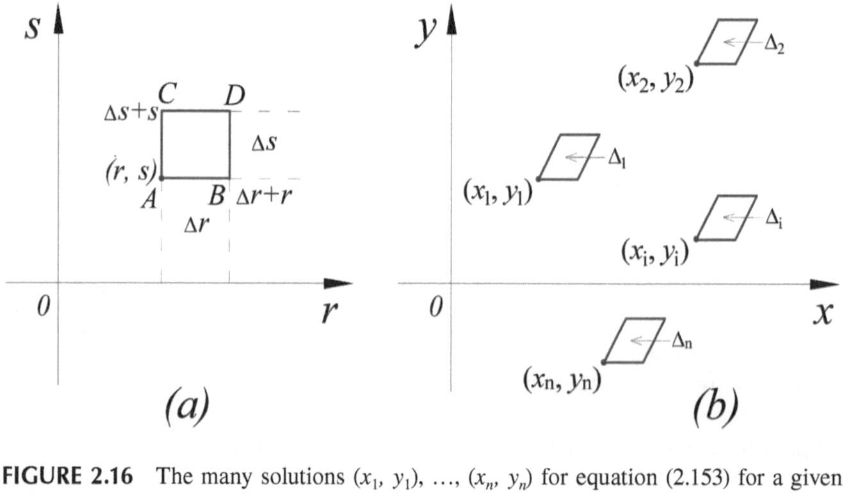

FIGURE 2.16 The many solutions $(x_1, y_1), \ldots, (x_n, y_n)$ for equation (2.153) for a given point (r,s).

Let's examine the task of assessing the probability

$$Pr\{r < R \le r+\Delta r, s < S \le s+\Delta s\} = Pr\{r < g_1(x,y) \le r+\Delta r, s < g_2(x,y) \le s+\Delta s\} \tag{2.155}$$

As it is commonly understood, the ratio

$$\frac{F(r+\Delta r, s+\Delta s) - F(r, s+\Delta s) - F(r+\Delta r, s) + F(r,s)}{\Delta r \Delta s}$$

tends to $\partial^2 F_{RS}(r,s)/\partial r \partial s$ as $\Delta r \to 0$ and $\Delta s \to 0$. Therefore, as $\Delta r \to 0$ and $\Delta s \to 0$, we have:

$$f_{RS}(r,s) = \frac{\partial^2 F_{RS}(r,s)}{\partial r \partial s} = \frac{F(r+\Delta r, s+\Delta s) - F(r, s+\Delta s) - F(r+\Delta r, s) + F(r,s)}{\Delta r \Delta s} \tag{2.156}$$

$$f_{RS}(r,s)\Delta r \Delta s = F(r+\Delta r, s+\Delta s) - F(r, s+\Delta s) - F(r+\Delta r, s) + F(r,s) \tag{2.157}$$

According to the definition of the joint probability:

$$\begin{aligned} &Pr\{r < R \le r+\Delta r,\, s < S \le s+\Delta s\} \\ &= F(r+\Delta r,\, s+\Delta s) - F(r,\, s+\Delta s) - F(r+\Delta r,\, s) + F(r,\, s) \end{aligned} \tag{2.158}$$

Combing equations (2.157) and (2.158), we obtain

$$Pr\{r < R \le r+\Delta r, s < S \le s+\Delta s\} = f_{RS}(r,s)\Delta r \Delta s \tag{2.159}$$

Therefore, we have demonstrated that the probability of (R, S) being within a differential rectangle is equivalent to the product of $f_{RS}(r,s)$ and the area $\Delta r \Delta s$ of the rectangle.

However, in order to express this probability in terms of $f_{XY}(x,y)$, it is necessary to assess the corresponding area for $\Delta r \Delta s$ in the XY plane. Regarding this matter, when examining Figure 2.17, we can note that point A, which has coordinates (r, s), is mapped onto point A', which has coordinates (x_i, y_i) (as well as to other points depicted in Figure 2.16b). As the value of r changes to $r+\Delta r$ to point B in Figure 2.17a, let B' denote its corresponding image in the XY plane. In a similar manner, when the value of s changes to $s+\Delta s$ to C, let C' be the representation of its image in the XY plane. At last, D is transferred to D', and $A'B'C'D'$ symbolizes the corresponding parallelogram in the XY plane with an area of Δ_i. With reference to Figure 2.16, due to the nonoverlapping characteristic of these regions, the probability in equation (2.159) can be represented in a different way as

$$\sum P\{(X,Y) \in \Delta_i\} = \sum f_{XY}(x_i, y_i)\Delta_i \tag{2.160}$$

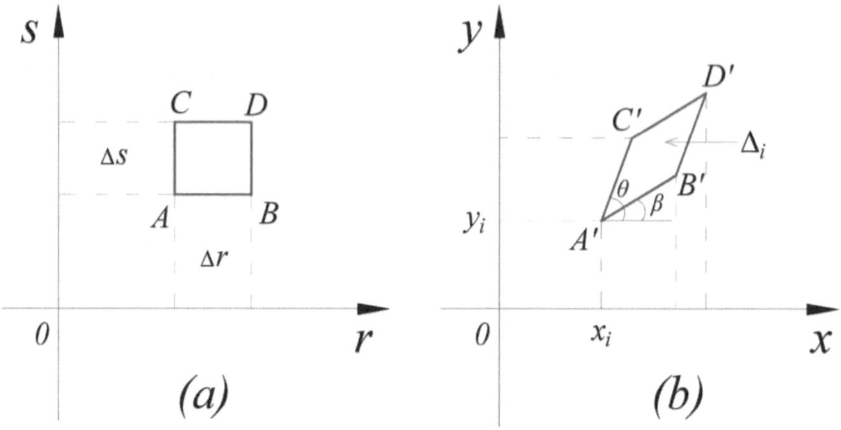

FIGURE 2.17 The corresponding area for $\Delta r \Delta s$ in the XY plane.

We get the following result by equating equation (2.159) with equation (2.160)

$$f_{RS}(r,s) = \sum_i f_{XY}(x_i, y_i) \frac{\Delta_i}{\Delta r \Delta s} \tag{2.161}$$

In order to simplify equation (2.161), it is necessary to calculate the area Δ_i of the parallelogram depicted in Figure 2.17b by evaluating it in terms of Δr and Δs. The inverse transformation in equation (2.153) is denoted by the symbols $h_1(\bullet)$ and $h_2(\bullet)$, which will be used in this context, so that

$$x_i = h_1(r,s) \; y_i = h_2(r,s) \tag{2.162}$$

While the point (r,s) is mapping onto $(x_i, y_i) \equiv A'$, the point $(r + \Delta r, s)$ is mapping onto B', the point $(r, s + \Delta s)$ is mapping onto C', and the point $(r + \Delta r, s + \Delta s)$ is mapping onto D'. The x and y coordinates of B' can therefore be obtained by implementing Taylor series expansions using the following equations

$$h_1(r + \Delta r, s) = h_1(r,s) + \frac{\partial h_1}{\partial r} \Delta r = x_i + \frac{\partial h_1}{\partial r} \Delta r \tag{2.163}$$

and

$$h_2(r + \Delta r, s) = h_2(r,s) + \frac{\partial h_2}{\partial r} \Delta r = y_i + \frac{\partial h_2}{\partial r} \Delta r \tag{2.164}$$

In a similar manner, the x and y coordinates of C' are given by the following equations

$$h_1(r, s + \Delta s) = h_1(r,s) + \frac{\partial h_1}{\partial s} \Delta s = x_i + \frac{\partial h_1}{\partial s} \Delta s \tag{2.165}$$

and

$$h_2(r,s+\Delta s)=h_2(r,s)+\frac{\partial h_2}{\partial s}\Delta s=y_i+\frac{\partial h_2}{\partial s}\Delta s \tag{2.166}$$

The area of the parallelogram $A'B'C'D'$ in Figure 2.17b can be calculated using the formula

$$\begin{aligned}\Delta_i &= (A'B')(A'C')\left|\sin(\theta-\beta)\right| \\ &= \left|(A'B'\cos\beta)(A'C'\sin\theta)-(A'B'\sin\beta)(A'C'\cos\theta)\right|\end{aligned} \tag{2.167}$$

Please note that in equation (2.167), we have taken the absolute value of $\sin(\theta-\beta)$ in order to ensure that the obtained value of the area Δ_i of the parallelogram is always positive. The value of $\sin(\theta-\beta)$ can be either positive or negative. Recall that the point $(r+\Delta r,s)$ is mapping onto B' and the point $(r,s+\Delta s)$ is mapping onto C'. After these mappings, we cannot guarantee that the ordinate of the point C' will always be larger than the ordinate of the point B'. That said, the value of the angle θ can be either larger than or smaller than the value of the angle β. However, the obtained value of the area Δ_i of the parallelogram should always be positive, so our decision to use the absolute value of $\sin(\theta-\beta)$ in equation (2.167) during the calculation is necessary.

Next, we proceed to perform the computation of the area Δ_i of the parallelogram using equation (2.167). Based on Figure 2.17b and the equations (2.163)–(2.166), we have

$$A'B'\cos\beta=\frac{\partial h_1}{\partial r}\Delta r \quad A'C'\sin\theta=\frac{\partial h_2}{\partial s}\Delta s \tag{2.168}$$

$$A'B'\sin\beta=\frac{\partial h_2}{\partial r}\Delta r \quad A'C'\cos\theta=\frac{\partial h_1}{\partial s}\Delta s \tag{2.169}$$

to ensure that

$$\begin{aligned}\Delta_i &= \left|\left(\frac{\partial h_1}{\partial r}\Delta r\right)\left(\frac{\partial h_2}{\partial s}\Delta s\right)-\left(\frac{\partial h_2}{\partial r}\Delta r\right)\left(\frac{\partial h_1}{\partial s}\Delta s\right)\right| \\ &= \left|\frac{\partial h_1}{\partial r}\frac{\partial h_2}{\partial s}-\frac{\partial h_2}{\partial r}\frac{\partial h_1}{\partial s}\right|\Delta r\Delta s\end{aligned} \tag{2.170}$$

as well as

$$\frac{\Delta_i}{\Delta r\Delta s}=\left|\frac{\partial h_1}{\partial r}\frac{\partial h_2}{\partial s}-\frac{\partial h_2}{\partial r}\frac{\partial h_1}{\partial s}\right|=\left|J(r,s)\right| \tag{2.171}$$

Please note that the $|\bullet|$ in equation (2.171) is the mathematical symbol representing the absolute value. Furthermore, $J(r, s)$ in equation (2.171) is the Jacobian determinant of the inverse transformation in equation (2.162), that is

$$J(r,s)=\begin{vmatrix} \dfrac{\partial h_1}{\partial r} & \dfrac{\partial h_1}{\partial s} \\ \dfrac{\partial h_2}{\partial r} & \dfrac{\partial h_2}{\partial s} \end{vmatrix} = \frac{\partial h_1}{\partial r}\frac{\partial h_2}{\partial s} - \frac{\partial h_2}{\partial r}\frac{\partial h_1}{\partial s} \tag{2.172}$$

Please note that the $|\bullet|$ in equation (2.172) is the mathematical symbol representing the determinant. In mathematics, the determinant is a scalar value that is a function of the entries of a square matrix. We obtain the following result by substituting the absolute value of equation (2.172) into equation (2.161):

$$f_{RS}(r,s) = \sum_i f_{XY}(x_i, y_i)\left|J(r,s)\right| = \sum_i f_{XY}(x_i, y_i)\left|\frac{\partial h_1}{\partial r}\frac{\partial h_2}{\partial s} - \frac{\partial h_2}{\partial r}\frac{\partial h_1}{\partial s}\right| \tag{2.173}$$

We have the following relationship

$$\left|J(r,s)\right| = \frac{1}{\left|J(x_i, y_i)\right|} \tag{2.174}$$

in which the determinant $J(x_i, y_i)$ represents the Jacobian of the original transformation in equation (2.151) given by

$$J(x_i, y_i) = \begin{vmatrix} \dfrac{\partial g_1}{\partial x} & \dfrac{\partial g_1}{\partial y} \\ \dfrac{\partial g_2}{\partial x} & \dfrac{\partial g_2}{\partial y} \end{vmatrix}_{x=x_i,\, y=y_i} \tag{2.175}$$

Combining equations (2.173) and equation (2.174), we obtain

$$f_{RS}(r,s) = \sum_i f_{XY}(x_i, y_i)\frac{1}{\left|J(x_i, y_i)\right|} \tag{2.176}$$

Example 2.7

▸ Suppose that X and Y are independent ***exponential*** random variables with common parameter λ. The PDFs of X and Y are, respectively, as follows:

$$f_X(x) = \lambda e^{-\lambda x}, \quad 0 \leqslant x < \infty, \qquad \lambda > 0 \tag{2.177}$$

$$f_Y(y) = \lambda e^{-\lambda y}, \quad 0 \leqslant y < \infty, \qquad \lambda > 0 \tag{2.178}$$

Consequently, the joint PDF of X and Y can be expressed as follows:

$$f_{XY}(x,y) = \lambda^2 e^{-\lambda(x+y)} \tag{2.179}$$

Define two new random variables as follows:

$$R = X + Y \tag{2.180}$$

$$S = X - Y \tag{2.181}$$

Determine the joint PDF of R and S.

Answer: Because we have $r = x + y$, $s = x - y$ and $|s| \leq r$, the transformations specified in equations (2.180)–(2.181) allow for only one solution. This is as follows:

$$x = \frac{r+s}{2},$$

$$y = \frac{r-s}{2}.$$

In addition to this, the Jacobian of the transformation can be calculated as

$$J(r,s) = \begin{vmatrix} \frac{\partial h_1}{\partial r} & \frac{\partial h_1}{\partial s} \\ \frac{\partial h_2}{\partial r} & \frac{\partial h_2}{\partial s} \end{vmatrix} = \begin{vmatrix} \frac{1}{2} & \frac{1}{2} \\ \frac{1}{2} & \frac{-1}{2} \end{vmatrix} = \frac{-1}{2} \tag{2.182}$$

We obtain the following result by substituting the absolute value of equation (2.182) into equation (2.173):

$$f_{RS}(r,s) = \sum_i f_{XY}(x_i, y_i) |J(r,s)| = \frac{\lambda^2}{2} e^{-\lambda r} \tag{2.183}$$

◀

2.10 EXERCISES

EXERCISE 2.1

a. Calculate the mode of the PDF represented by

$$f(y) = \frac{2y}{c} e^{-x^2/c}, \quad \text{in which} \quad 0 \leqslant y < \infty \quad 0 < c \tag{E.2.1}$$

b. Calculate the magnitude of the PDF corresponding to the mode of the PDF specified in equation (E.2.1).

EXERCISE 2.2

There are two random variables. These two variables, X and Y, have the joint PDF as follows

$$f(x,y) = 55/c^2, \quad 0 \leqslant x < y < c \tag{E.2.2}$$

1. Calculate the marginal PDF of X;
2. Calculate the marginal PDF of Y.

EXERCISE 2.3

X and Y, which are both random variables, have a joint PDF expressed as follows:

$$f_{XY}(x,y) = 66e^{(-2x-y)}, \quad \text{in which} \quad 0 < x < y < \infty \tag{E.2.3}$$

1. Calculate the marginal PDF of X;
2. Calculate the marginal PDF of Y;
3. Show that X and Y are not statistically independent.

EXERCISE 2.4

In this exercise, you are required to conduct an evaluation of the mean and variance of the Poisson distribution, which is the distribution that is shown below

$$p(y) = e^{-\lambda}\frac{\lambda^y}{y!}, \quad y = 0, 1, 2, \ldots \tag{E.2.4}$$

The mean value of the foregoing Poisson distribution is calculated as follows:

$$E[y] = \sum_{y=0}^{\infty} ye^{-\lambda}\frac{\lambda^y}{y!} = e^{-\lambda}\sum_{y=1}^{\infty}\frac{\lambda^y}{(y-1)!} \tag{E.2.5}$$

1. Show that: $E[y] = \lambda$.

   The mean square value of the foregoing Poisson distribution is calculated as follows:

$$E[y^2] = \sum_{y=0}^{\infty} y^2 e^{-\lambda}\frac{\lambda^y}{y!} = e^{-\lambda}\sum_{y=1}^{\infty}\frac{(y-1)\lambda^y + \lambda^y}{(y-1)!} \tag{E.2.6}$$

   The variance value of the foregoing Poisson distribution is calculated as follows:

$$Var[y] = E\left[y^2\right] - \{E[y]\}^2$$

2. Show that: $Var[y] = \lambda$.

EXERCISE 2.5

Please find the mean and variance of the following PDF which is called Rayleigh distribution:

$$f(y)=\frac{2y}{\delta}e^{-x^2/\delta},\quad 0<y<\infty \tag{E.2.7}$$

EXERCISE 2.6

In this exercise, you are required to conduct an evaluation of the mean value of the binomial distribution, which is the distribution that is being shown below

$$p(y)=\binom{n}{y}p^y(1-p)^{n-y},\quad y=0,1,2,\ldots n \tag{E.2.8}$$

The mean value of the foregoing binomial distribution is calculated as follows:

$$E[y]=\sum_{y=0}^{n}y\binom{n}{y}p^y(1-p)^{n-y} \tag{E.2.9}$$

$$E\left[y^2\right]=\sum_{y=0}^{n}y^2\binom{n}{y}p^y(1-p)^{n-y} \tag{E.2.10}$$

1. Show that: $E[y]=np$.
2. Show that: $E\left[y^2\right]=n(n-1)p^2+np$.
3. Show that: $Var[y]=np-np^2$.

EXERCISE 2.7

Please evaluate the skewness and kurtosis of the following PDF that is called Rayleigh distribution:

$$f(y)=\frac{2y}{\delta}e^{-x^2/\delta},\quad 0\le y<\infty,\quad \delta>0 \tag{E.2.11}$$

EXERCISE 2.8

Let R represent one random variable, and let S represent another random variable. The joint PDF for the variables R and S is specified as follows:

$$f_{RS}(r,s)=\frac{1}{4}\left(1-r^2+s^2\right),\quad -1\leqslant r\leqslant 1\quad -1\leqslant s\leqslant 1 \tag{E.2.12}$$

1. Calculate the marginal PDF of R;
2. Calculate the marginal PDF of S;
3. Calculate the covariance of the random variables R and S;
4. Are the random variables R and S statistically uncorrelated or not?
5. Are the random variables R and S statistically independent or not?

EXERCISE 2.9

Let R represent one random variable, and let S represent another random variable. The joint PDF for the variables R and S is specified as follows:

$$f_{RS}(r,s) = \frac{1}{2\pi} Exp\left[-\frac{\left(r^2 - 2\sqrt{3}\,rs + 4s^2\right)}{2} \right], \quad -\infty < r < \infty, \quad -\infty < s < \infty \qquad \text{(E.2.13)}$$

1. Calculate the marginal PDF of R;
2. Calculate the marginal PDF of S;
3. Calculate the covariance of the random variables R and S;
4. Are the random variables R and S statistically uncorrelated or not?
5. Are the random variables R and S statistically independent or not?
6. Calculate the correlation coefficient of the random variables R and S

EXERCISE 2.10

Suppose that X is an ***exponential*** random variable with the parameter λ. The PDF of X is as follows:

$$f_X(x) = \lambda e^{-\lambda x}, \quad 0 \leqslant x < \infty \quad \lambda > 0 \qquad \text{(E.2.14)}$$

Please find the characteristic function $\phi_X(\theta)$.

EXERCISE 2.11

Assume that the characteristic function of a random variable X is denoted by $\phi_X(\theta)$, and take into consideration that a new random variable $Y = cX + d$, where c and d are constants. Show that the characteristic function of the random variable Y is represented by the equation $\phi_Y(\theta) = \exp[id\theta]\phi_X(c\theta)$.

EXERCISE 2.12

Suppose that Z is a standardized Gaussian random variable with the PDF as follows:

$$f_Z(z) = \frac{1}{\sqrt{2\pi}} \exp\left[-\frac{z^2}{2} \right] \quad -\infty < x < \infty \qquad \text{(E.2.15)}$$

The standardized Gaussian random variable Z has a characteristic function of the form

$$\begin{aligned} \phi_Z(\theta) &= \int_{-\infty}^{\infty} \exp(i\theta z) \frac{1}{\sqrt{2\pi}} \exp\left[-z^2/2\right] dz \\ &= \exp\left[-\theta^2/2\right] \int_{-\infty}^{\infty} \frac{1}{\sqrt{2\pi}} \exp\left[-(z - i\theta)^2/2\right] dz \\ &= \exp\left[-\theta^2/2\right] \end{aligned}$$

Let us now consider a random variable X that follows the Gaussian probability distribution, with a mean of μ and a variance of σ^2. The PDF of the random variable X is as follows:

$$f_X(x) = \frac{1}{\sqrt{2\pi}\sigma}\exp\left[-\frac{(x-\mu)^2}{2\sigma^2}\right] \quad -\infty < x < \infty \tag{E.2.16}$$

Additionally, let Z be a Gaussian random variable that is standardized. The equation that describes the relationship between X and Z is $X = \mu + \sigma Z = \sigma Z + \mu$. Following this, the characteristic function of the random variable X may be obtained by applying the relationship $\phi_X(\theta) = \exp[i\mu\theta]\phi_Z(\sigma\theta)$. Show that the characteristic function of the random variable X is represented by the following equation:

$$\phi_X(\theta) = \exp\left(i\mu\theta - \frac{\sigma^2\theta^2}{2}\right) \tag{E.2.17}$$

EXERCISE 2.13

Two random variables, X and Y, are considered to have a joint Gaussian distribution if their characteristics are determined by the joint PDF, which is defined as follows:

$$f_{XY}(x,y) = \frac{1}{2\pi\sigma_X\sigma_Y\sqrt{1-\rho_{XY}^2}} \exp\left[-\frac{\sigma_Y^2(x-\mu_X)^2 - 2\sigma_X\sigma_Y\rho_{XY}(x-\mu_X)(y-\mu_Y) + \sigma_X^2(y-\mu_Y)^2}{2\sigma_X^2\sigma_Y^2(1-\rho_{XY}^2)}\right],$$
$$-\infty < x < \infty, \quad -\infty < y < \infty \tag{E.2.18}$$

By performing straightforward integrations, show that

$$\begin{aligned} &E[X] = \mu_X \\ &E[Y] = \mu_Y \\ &E\left[(X-\mu_X)^2\right] = \sigma_X^2 \\ &E\left[(Y-\mu_Y)^2\right] = \sigma_Y^2 \\ &E\left[(X-\mu_X)(Y-\mu_Y)\right] = \sigma_X\sigma_Y\rho_{XY} \end{aligned} \tag{E.2.19}$$

EXERCISE 2.14

As a result of EXERCISE 2.13, the five parameters that are involved in the joint density function of equation (E.2.18) are the expected values given by equation (E.2.19). In a more general sense, let's examine n random variables X_1, X_2,..., X_n. These random variables have expected values

$E[X_j] = \mu_{X_j}$, where j = 1, 2, ..., n. Additionally, the covariances are given by $E\left[\left(X_j - \mu_{X_j}\right)\left(X_k - \mu_{X_k}\right)\right] = \kappa_{X_j X_k}$ where j, k = 1, 2, ..., n. We state that these random variables have a jointly Gaussian distribution if the joint PDF of these variables is given by

$$f_{X_1X_2\cdots X_n}\left(x_1, x_2, \ldots, x_n\right) = \frac{1}{(2\pi)^{n/2}|S|^{0.5}} \exp\left[-\frac{1}{2|S|}\sum_{j=1}^{n}\sum_{k=1}^{n}|S|_{jk}\left(x_j - \mu_{X_j}\right)\left(x_k - \mu_{X_k}\right)\right] \tag{E.2.20}$$

in which $|S|$ represents the determinant of the matrix **S** that contains both covariances and variances, that is,

$$\mathbf{S} = \begin{bmatrix} {\sigma_{X_1}}^2 & \kappa_{X_1X_2} & & \kappa_{X_1X_n} \\ \kappa_{X_2X_1} & {\sigma_{X_2}}^2 & & \kappa_{X_2X_n} \\ \cdots & \cdots & \cdots & \cdots \\ \kappa_{X_nX_1} & \kappa_{X_nX_2} & & {\sigma_{X_n}}^2 \end{bmatrix} \tag{E.2.21}$$

in which $|S|_{jk}$ is the cofactor of the element that is located in the jth row and the kth column of **S**. At this point, we have tacitly assumed that $|S|$ is not equal to zero. Please note that the matrix **S** possesses an inverse $S^{-1} = |S|_{jk}/|S|$. As before, $|S|$ is the determinant of **S**. Show that equation (E.2.20) can be reduced to equation (E.2.18) for $n = 2$.

EXERCISE 2.15

Show that there is another way to write the equation (E.2.20) as follows:

$$f_{\mathbf{X}}(\mathbf{x}) = \frac{1}{(2\pi)^{n/2}|S|^{0.5}} \exp\left[-\frac{1}{2}\{\mathbf{x} - \boldsymbol{\mu}_{\mathbf{X}}\}' \mathbf{S}^{-1}\{\mathbf{x} - \boldsymbol{\mu}_{\mathbf{X}}\}\right] \tag{E.2.22}$$

in which $\{\mathbf{x} - \boldsymbol{\mu}_{\mathbf{X}}\}$ is the column matrix with the elements $x_j - \mu_{X_j}$, $j = 1$, 2,...,n; and the prime represents transposition. Please also note that $\mathbf{S} = E\left[\{\mathbf{x} - \boldsymbol{\mu}_{\mathbf{X}}\}\{\mathbf{x} - \boldsymbol{\mu}_{\mathbf{X}}\}'\right]$. In this notation, we consider $X_1, X_2, \ldots, X_n$ as the components of an n-dimensional random vector **X**, with a mathematical expectation represented by $\boldsymbol{\mu}_{\mathbf{X}}$. The argument of the multidimensional PDF in equation (E.2.22) is the state vector **x**.

EXERCISE 2.16

Take into consideration still another random vector with n dimensions, denoted by **Y**, which is produced from **X** by means of the linear transformation described below:

$$\mathbf{Y} = \mathbf{AX} \tag{E.2.23}$$

For the time being, we will make the assumption that the square matrix **A** is not singular; hence, the existence of $\mathbf{A}^{-1}$ is established, and

$$\mathbf{X} = \mathbf{A}^{-1}\mathbf{Y} \tag{E.2.24}$$

The relationship between the state vectors **x** and **y** is the same; that is to say,

$$\mathbf{y} = \mathbf{Ax} \quad \mathbf{x} = \mathbf{A}^{-1}\mathbf{y} \tag{E.2.25}$$

Due to the fact that $\mathbf{A}^{-1}$ is a matrix of constants, by computing the expected value of both sides of equation (E.2.24), we obtain

$$\mu_X = \mathbf{A}^{-1}\mu_Y \tag{E.2.26}$$

in which the mathematical expectation of **Y** is denoted by the symbol $\mu_\mathbf{Y}$. Utilizing equation (2.173), as well as the relations in equations (E.2.25) and (E.2.26), show that the n-dimensional PDF of **Y** can be established as follows:

$$f_\mathbf{Y}(\mathbf{y}) = \frac{1}{(2\pi)^{n/2}|S|^{0.5}} \exp\left[-\frac{1}{2}\{\mathbf{y}-\mu_Y\}'\left(\mathbf{A}^{-1}\right)'\mathbf{S}^{-1}\mathbf{A}^{-1}\{\mathbf{y}-\mu_Y\}\right]|J_n| \tag{E.2.27}$$

in which the Jacobian of the transformation, denoted by the symbol J_n, is the determinant of the matrix $\mathbf{A}^{-1}$; in other words,

$$J_n = \left|A^{-1}\right| \tag{E.2.28}$$

EXERCISE 2.17

When dealing with two or more random variables, the concept of characteristic functions can also be seen to be useful. Only continuous random variables are discussed in the following development; however, the principal findings are equally applicable to the situation of discrete random variables. A significant portion of the derivations that are involved are also eliminated due to the fact that they closely resemble those that were established for the case of a single random variable.

The joint characteristic function of two random variables, X and Y, is denoted by the expression $\phi_{XY}(t,s)$, which is defined as

$$\phi_{XY}(t,s)=E\left[e^{i(tX+sY)}\right]=\int_{-\infty}^{\infty}\int_{-\infty}^{\infty}e^{i(tx+sy)}f_{XY}(x,y)dxdy \tag{E.2.29}$$

in which t and s are two real variables that are arbitrary. It is always the case that this function exists, and it has the property of $\phi_{XY}(0,0)=1$. In a manner that is comparable to the case of one random variable, the joint characteristic function $\phi_{XY}(t,s)$ is frequently utilized to determine the joint PDF $f_{XY}(x,y)$ of X and Y as well as their joint moments. By utilizing the two-dimensional Fourier transform, show that the joint PDF $f_{XY}(x,y)$ can be uniquely calculated in terms of $\phi_{XY}(t,s)$ as follows:

$$f_{XY}(x,y)=\frac{1}{4\pi^2}\int_{-\infty}^{\infty}\int_{-\infty}^{\infty}e^{-i(tx+sy)}\phi_{XY}(t,s)dtds \tag{E.2.30}$$

Please also show that the moments $E\left\{X^nY^m\right\}=\beta_{nm}$, assuming they exist, are connected to $\phi_{XY}(t,s)$ via the following equation:

$$\left.\frac{\partial^{n+m}}{\partial t^n\,\partial s^m}\phi_{XY}(t,s)\right|_{t,s=0}=i^{n+m}\int_{-\infty}^{\infty}\int_{-\infty}^{\infty}x^ny^mf_{XY}(x,y)dxdy=i^{n+m}\beta_{nm}. \tag{E.2.31}$$

EXERCISE 2.18

In a manner that is comparable to the case of one Gaussian random variable, the joint characteristic function of n normally distributed random variables $X_1, \cdots, X_n$, for any real numbers $\theta_1,\cdots,\theta_n$, is given by

$$\phi_{X_1,\ldots,X_n}\left(\theta_1,\cdots,\theta_n\right)=\exp\left\{i\sum_{j=1}^{n}\theta_j\mu_j-\frac{1}{2}\sum_{j=1}^{n}\sum_{k=1}^{n}\theta_jK_{jk}\theta_k\right\} \tag{E.2.32}$$

in which, for $j,k=1,\ 2,\cdots,n$

$$\mu_j=E\left[X_j\right],\quad K_{jk}=\mathrm{cov}\left[X_j,X_k\right] \tag{E.2.33}$$

K_{jk} is a covariance matrix. In a manner that is comparable to the case of equation (E.2.28), we have the following relationship:

$$\left.\frac{\partial^4}{\partial\theta_1\,\partial\theta_2\,\partial\theta_3\,\partial\theta_4}\phi_{X_1X_2X_3X_4}(\theta_1,\theta_2,\theta_3,\theta_4)\right|_{\theta_1,\theta_2,\theta_3,\theta_4=0}=i^4E\left\{X_1X_2X_3X_4\right\}=E\left\{X_1X_2X_3X_4\right\}. \tag{E.2.34}$$

According to equation (E.2.32), the joint characteristic function of n normally distributed random variables X_1, ⋯, X_n, with zero means, respectively, for any real numbers $\theta_1,\ldots,\theta_n$, is given by

$$\phi_{X_1X_2X_3X_4}\left(\theta_1,\theta_2,\theta_3,\theta_4\right)=\exp\left\{-\frac{1}{2}\sum_{j=1}^{4}\sum_{k=1}^{4}\theta_jK_{jk}\theta_k\right\}=\exp\left\{-\frac{1}{2}\sum_{j=1}^{4}\sum_{k=1}^{4}\theta_j\theta_kE\left[X_jX_k\right]\right\} \tag{E.2.35}$$

For the sake of convenience in writing, we will use the symbol ϕ to represent the right-hand side of equation (E.2.35) and define

$$L_i=\sum_{j=1}^{4}\theta_jE\left[X_iX_j\right] \tag{E.2.36}$$

$$L_1=\theta_1E\left[X_1X_j\right]+\theta_2E\left[X_1X_j\right]+\theta_3E\left[X_1X_j\right]+\theta_4E\left[X_1X_j\right] \tag{E.2.37}$$

From equation (E.2.35), we have

$$\begin{aligned}\frac{\partial\phi}{\partial\theta_1}&=\phi\left\{-\frac{1}{2}\left(2\theta_1E\left[X_1X_1\right]+2\theta_2E\left[X_1X_2\right]+2\theta_3E\left[X_1X_3\right]+2\theta_4E\left[X_1X_4\right]\right)\right\}\\&=-\phi L_1\end{aligned} \tag{E.2.38}$$

In a manner as the forgoing derivation process, show that

$$\frac{\partial^2\phi}{\partial\theta_2\,\partial\theta_1}=\phi\left\{L_1L_2-E\left[X_1X_2\right]\right\}, \tag{E.2.39}$$

$$\frac{\partial^3\phi}{\partial\theta_3\,\partial\theta_2\,\partial\theta_1}=\phi\left\{-L_1L_2L_3+L_3E\left[X_1X_2\right]+L_2E\left[X_1X_3\right]+L_1E\left[X_2X_3\right]\right\} \tag{E.2.40}$$

$$\begin{aligned}\frac{\partial^4\phi}{\partial\theta_4\,\partial\theta_3\,\partial\theta_2\,\partial\theta_1}=\phi\{&L_1L_2L_3L_4-L_1L_2E\left[X_3X_4\right]-L_1L_3E\left[X_2X_4\right]\\&-L_1L_4E\left[X_2X_3\right]-L_2L_3E\left[X_1X_4\right]\\&-L_2L_4E\left[X_1X_3\right]-L_8L_4E\left[X_1X_2\right]\\&+E\left[X_1X_2\right]E\left[X_3X_4\right]+E\left[X_1X_3\right]E\left[X_2X_4\right]\\&+E\left[X_1X_4\right]E\left[X_2X_3\right]\}.\end{aligned} \tag{E.2.41}$$

Combining equations (E.2.34) and (E.2.41) and considering the fact $\phi_{X_1X_2X_3X_4}\left(0,0,0,0\right)=1$ and $L_i\big|_{\theta_1,\theta_2,\theta_3,\theta_4=0}=0$, show that

$$E\left\{X_1X_2X_3X_4\right\}=E\left[X_1X_2\right]E\left[X_3X_4\right]+E\left[X_1X_3\right]E\left[X_2X_4\right]+E\left[X_1X_4\right]E\left[X_2X_3\right] \tag{E.2.42}$$

EXERCISE 2.19

Suppose that X and Y are independent ***exponential*** random variables with common parameter λ. The PDFs of X and Y are, respectively, as follows:

$$f_X(x) = \lambda e^{-\lambda x}, \quad 0 \leqslant x < \infty, \qquad \lambda > 0 \tag{E.2.43}$$

$$f_Y(y) = \lambda e^{-\lambda y}, \quad 0 \leqslant y < \infty, \qquad \lambda > 0 \tag{E.2.44}$$

Define a new random variable as follows:

$$R = X + Y \tag{E.2.45}$$

Determine the PDF of R.

EXERCISE 2.20

Suppose that X is a random variable with a Cauchy distribution, and the PDFs of X is as follows:

$$f_X(x) = \frac{\lambda}{\pi\left(x^2 + \lambda^2\right)}, \qquad -\infty < x < \infty \tag{E.2.46}$$

Define a new random variable as follows:

$$Y = 3X + 2 \tag{E.2.47}$$

Determine the PDF of Y.

EXERCISE 2.21

Suppose that X and Y are independent and identically and normally distributed random variables. The PDFs of X and Y are, respectively, as follows:

$$f_X(x) = \frac{1}{(2\pi)^{1/2}} Exp\left(\frac{-x^2}{2}\right), \qquad -\infty < x < \infty, \tag{E.2.48}$$

$$f_Y(y) = \frac{1}{(2\pi)^{1/2}} Exp\left(\frac{-y^2}{2}\right), \qquad -\infty < y < \infty, \tag{E.2.49}$$

Define two new random variables as follows:

$$R = X + Y \tag{E.2.50}$$

$$S = X - Y \tag{E.2.51}$$

Determine the joint PDF of R and S.

EXERCISE 2.22

Suppose that X and Y are independent and identically and normally distributed random variables. The PDFs of X and Y are, respectively, as follows:

$$f_X(x) = \frac{1}{(2\pi)^{1/2}} Exp\left(\frac{-x^2}{2}\right), \qquad -\infty < x < \infty, \tag{E.2.52}$$

$$f_Y(y) = \frac{1}{(2\pi)^{1/2}} Exp\left(\frac{-y^2}{2}\right), \qquad -\infty < y < \infty, \tag{E.2.53}$$

Define two new random variables as follows:

$$R = \sqrt{X^2 + Y^2} \tag{E.2.54}$$

$$S = X/Y \tag{E.2.55}$$

Determine the joint PDF of R and S.

3 Random Processes

3.1 INTRODUCTION AND DEFINITION

An introduction to the probability theory that is applicable to issues in engineering and the physical sciences was provided in Chapter 2. It has been demonstrated that the properties of a random phenomenon can be represented by utilizing the probability distribution of a random variable that is representative of the phenomenon. Probability theory, on the other hand, does not take into account the statistical aspects of random phenomena (events) that occur in a random fashion with respect to time.

The time element is a crucial factor that cannot be disregarded while studying many random phenomena in engineering and the physical sciences. Some examples include the following: (1) The movement of particles in Brownian motion; (2) the ocean wave profile in a specific sea state; (3) the response of a wind turbine to a gust of wind; (4) the motion of a ship in a confused seaway; and (5) the vibration of a building that is caused by an earthquake. To assess the statistical properties of these random phenomena, it is essential to take into consideration the idea of an *ensemble of time functions*. The phrase "stochastic process" or "random process" refers to phenomena that are random and whose features can be determined by this idea.

To provide further elaboration on the notion of a *random process* that was presented above, let us take into consideration a collection of n recordings that indicate the time histories of a random phenomenon that is depicted in Figure 3.1.

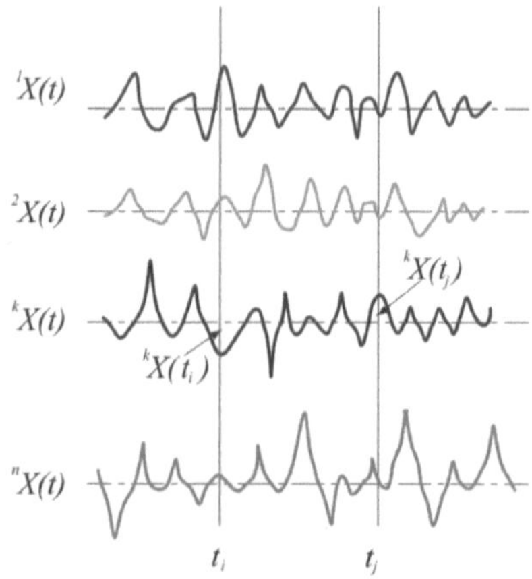

FIGURE 3.1 A wave profile random process consisting of an ensemble of time functions.

DOI: 10.1201/9781003431503-3

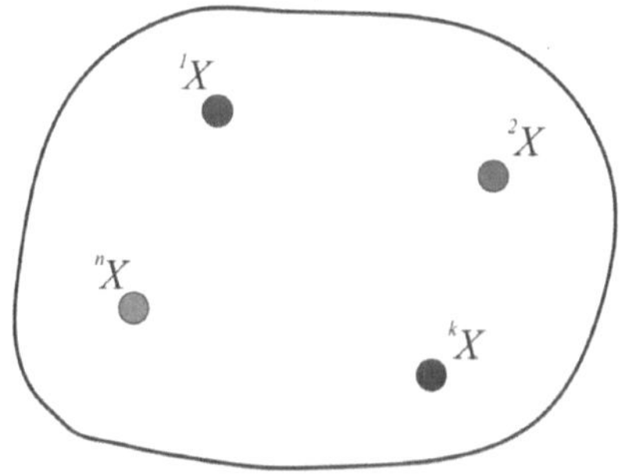

FIGURE 3.2 *n* wave recorders that are dispersed in a certain ocean area.

Specifically, in Figure 3.1, we take into consideration the wave profile in the ocean that is captured by n wave recorders (^1X, ^2X, ...,^kX, ..., ^nX) that are dispersed in a certain area as shown in Figure 3.2 as an example of a random phenomenon. Every wave recorder would capture a different time history of the wave profile due to the fact that the waves are coming from a variety of directions in a random fashion. We denote a specific time history (e.g., $^1X(t)$ as measured by the wave recorder ^1X as represented by the first curve in Figure 3.1) as a specific *time function*. In Figure 3.1, $^kX(t_i)$, in a specific time function, stands for the magnitude of deviation of the wave profile from the mean water line at a specific time t_i. The wave profile *random process* is an *ensemble of time functions* (i.e., $^1X(t)$, $^2X(t)$, ..., $^kX(t)$, ...and $^nX(t)$ altogether).

We understand that $X(t_i)$ is a *random variable* at a given time t_i and that the set consists of $\{{}^1X(t_i), {}^2X(t_i),..., {}^kX(t_i),..., {}^nX(t_i)\}$ can be viewed as a random sample with n elements. At another given time t_j, we have another *random variable* $X(t_j)$, which has a random sample with n elements $\{{}^1X(t_j), {}^2X(t_j),..., {}^kX(t_j),..., {}^nX(t_j)\}$. Viewed from this perspective, the entire wave profile *random process* can also be regarded as consisting of a *family of random variables* $X(t_1)$, $X(t_2)$, ... $X(t_i)$, ... $X(t_j)$, ... The entire family of random variables may be denoted by $\{X(t), t \in T\}$ or, more simply, by $\{X(t)\}$, which means that t is now free to assume any value belonging to a set T.

In principle, the statistical features of a random phenomenon need to be derived with regard to the *ensemble of time functions*. Nevertheless, if we assume that the random phenomenon satisfies the ergodic property, which will be addressed later, then we can extract the statistical properties from the analysis of a single *time function*. At this point, it should be noted that, in the existing literature, a *time function* is sometimes also called a sample path, a sample function, a sample record or a sample realization. The set of possible sample paths is called the *ensemble* of a random process.

3.2 SPECIFICATION OF RANDOM PROCESSES

In the following, we denote a random process by $\{X(t),\ t \in T\}$, or more simply, by $\{X(t)\}$. In this textbook, we will frequently suppress the notation "$t \in T$" in a random process and write only $\{X(t)\}$ or $\{X_t\}$. However, $X(t)$, without the curly brackets, is always the value of the process at time t, and hence a *random variable*. Because a random process $\{X(t)\}$ is a family of random variables, its specification is the same as that for jointly distributed random variables except that the number of random variables may be infinite or even uncountable.

In the order of increasing completeness, the probability structure of a random process is described by the following:

$$\begin{gathered} f_{X(t_1)}(x_1) \\ f_{X(t_1)X(t_2)}(x_1, x_2) \\ \cdots\cdots\cdots \\ f_{X(t_1)X(t_2)\cdots X(t_m)}(x_1, x_2, \ldots, x_m) \end{gathered} \tag{3.1}$$

in which $\{t_1,\ t_2, \cdots \in T\}$. These are referred to as the probability densities of the random process $\{X(t)\}$ of the first order, the second order and so on. In general, these probability density functions change when $t_1,\ t_2, \cdots$ change. Therefore, they are not just functions of the state variables $x_1,\ x_2, \ldots$ but also functions of $t_1,\ t_2, \cdots$. Please note that in equation (3.1), we understand that $X(t_1)$ is a *random variable* at a given time t_1 and that the set consisting of $\{{}^1X(t_1),\ {}^2X(t_1), \ldots,\ {}^kX(t_1), \ldots,\ {}^nX(t_1)\}$ can be viewed as a random sample with n elements. We have used a simplified notation x_1 to represent a specific element in the set $\{{}^1X(t_1),\ {}^2X(t_1), \ldots,\ {}^kX(t_1), \ldots,\ {}^nX(t_1)\}$. Please also note that in equation (3.1), we understand that $X(t_2)$ is a *random variable* at a given time t_2 and that the set consisting of $\{{}^1X(t_2),\ {}^2X(t_2), \ldots,\ {}^kX(t_2), \ldots,\ {}^nX(t_2)\}$ can be viewed as a random sample with n elements. We have used a simplified notation x_2 to represent a specific element in the set $\{{}^1X(t_2),\ {}^2X(t_2), \ldots,\ {}^kX(t_2), \ldots,\ {}^nX(t_2)\}$. In a similar manner, we have used a simplified notation x_m to represent a specific element in the set $\{{}^1X(t_m),\ {}^2X(t_m), \ldots,\ {}^kX(t_m), \ldots,\ {}^nX(t_m)\}$ corresponding to the *random variable* $X(t_m)$.

Additionally, moment functions of different orders can be used to characterize a random process. These moment functions are defined as

$$\begin{gathered} E[X(t_1)] = \int x_1 f_{X(t_1)}(x_1)\, dx_1 \\ E[X(t_1)X(t_2)] = \iint x_1 x_2 f_{X(t_1)X(t_2)}(x_1, x_2)\, dx_1\, dx_2 \end{gathered} \tag{3.2}$$

when these moments exist. These functions typically depend on variables t_1, t_2,..., etc. It is a necessity to include all the moment functions to fully characterize a random process. In the context of practical applications, the first- and second-order moment functions are especially important. Special symbols will be used to denote them as follows:

$$\mu_X(t) = E[X(t)] \quad \varphi_{XX}(t_1, t_2) = E[X(t_1)X(t_2)] \tag{3.3}$$

In equation (3.3), $\mu_X(t)$ and $\varphi_{XX}(t_1, t_2)$ are referred to as the mean value function and the autocorrelation function, respectively. The autocorrelation function is used to describe the second-order statistical property of a random process, while the mean value function is used to describe the first-order statistical property of a random process. The two random variables that were taken into consideration, $X(t_1)$ and $X(t_2)$, are both members of the same random process, as indicated by the prefix *auto* - modifying *correlation function.*

The autocovariance function can also be used to describe the second-order statistical property of a random process, and an autocovariance function is calculated as follows:

$$\begin{aligned}\kappa_2[X(t_1)X(t_2)] &= Cov[X(t_1)X(t_2)] = E\{[X(t_1)-\mu_X(t_1)][X(t_2)-\mu_X(t_2)]\} \\ &= \varphi_{XX}(t_1,t_2)-\mu_X(t_1)\mu_X(t_2)\end{aligned} \tag{3.4}$$

Previously, we have discussed the significance of the mean value functions and correlation (or covariance) functions in describing the first- and second-order statistical properties of a random process. There are several reasons for their significant importance. Firstly, for some random processes, it is possible to calculate their statistical properties beyond the second order if the properties of the first and second orders are known (it should come as no surprise that the first- and second-order statistical properties are the ones that completely characterize such random processes). The Gaussian random processes, commonly utilized in engineering analysis, are included within this category. Secondly, even in situations where the first- and second-order statistical properties do not entirely characterize a random process, they nonetheless include the most essential information regarding that process. Thirdly, when it comes to computing estimated statistical properties from the records of a random process, the higher-order results are significantly more scattered than the lower-order results. In fact, the majority of the experimental data that has been presented in the scientific literature has only been the first- and second-order estimations.

Oftentimes, the mean value function of a random process is equal to zero or another constant. In such a scenario, the correlation function or the covariance function assumes paramount importance as a means of describing the random process. This is frequently the case with problems related to structural dynamics. Special emphasis on the second-order statistical properties is, therefore, evident in Chapters 4 through 8 of this book when discussing the response of structures and systems to random excitations.

3.3 STATIONARY AND EVOLUTIONARY RANDOM PROCESSES

3.3.1 Definitions

In the theory of random processes, it has been discovered to be beneficial to divide random processes into two broad categories: Stationary and evolutionary. According to common sense, a *stationary process* is one in which the distribution of the process does not change over the course of time. This is because the random mechanism that produces the process does not undergo any changes over the course of time. A process that is not stationary is referred to as an *evolutionary process.*

3.3.2 Strictly Stationary Random Processes

When the statistical distributions of a random process $\{X(t),\ t \in T\}$ stay unchanged following a shift in the time scale, the process is referred to as being *strictly stationary.* Due to the fact that the distributions of a random process are defined by the finite-dimensional distribution functions, we are able to formulate an alternative definition of *strict stationarity,* which is as follows:

The random process $\{X(t),\ t \in T\}$ is considered to be *strictly stationary* if, for every n, every choice of times $t_1, t_2, \ldots, t_n \in T$ and every time lag τ such that $t_i + \tau \in T$, the n-dimensional random vector $\left(X(t_1+\tau),\ X(t_2+\tau), \ldots, X(t_n+\tau)\right)$ has the same multivariate probability distribution function as the vector $\left(X(t_1),\ X(t_2), \ldots, X(t_n)\right)$, then the random process is considered to be *strictly stationary.*

If a random process $\{X(t),\ t \in T\}$ has a complete probability structure that is not affected by a change in the time origin, then we say that the random process is *strictly stationary.* By using a probability-density description of a random process, *strictly stationary* implies that

$$
\begin{aligned}
f_{X(t_1)}(x_1) &= f_{X(t_1+\tau)}(x_{1\tau}) \\
f_{X(t_1)X(t_2)}(x_1, x_2) &= f_{X(t_1+\tau)X(t_2+\tau)}(x_{1\tau}, x_{2\tau}) \\
&\cdots\cdots\cdots \\
f_{X(t_1)X(t_2)\cdots X(t_n)}(x_1, x_2, \ldots, x_n) &= f_{X(t_1+\tau)X(t_2+\tau)\cdots X(t_n+\tau)}(x_{1\tau}, x_{2\tau}, \ldots, x_{n\tau})
\end{aligned}
\tag{3.5}
$$

Please note that in equation (3.5), we understand that $X(t_1)$ is a *random variable* at a given time t_1 and that the set consisting of $\{{}^1X(t_1),\ {}^2X(t_1), \ldots, {}^kX(t_1), \ldots, {}^nX(t_1)\}$ can be viewed as a random sample with n elements. We have used a simplified notation x_1 to represent a specific element in the set $\{{}^1X(t_1),\ {}^2X(t_1), \ldots, {}^kX(t_1), \ldots, {}^nX(t_1)\}$. Please also note that in equation (3.5), we understand that $X(t_1+\tau)$ is a *random variable* at a given time $t_1+\tau$ and that the set consisting of $\{{}^1X(t_1+\tau),\ {}^2X(t_1+\tau), \ldots, {}^kX(t_1+\tau), \ldots, {}^nX(t_1+\tau)\}$ can be viewed as a random sample with n elements. We have used a simplified notation $x_{1\tau}$ to represent a specific element in the set $\{{}^1X(t_1+\tau), {}^2X(t_1+\tau), \ldots, {}^kX(t_1+\tau), \ldots, {}^nX(t_1+\tau)\}$. In a similar manner, we have used a simplified notation $x_{2\tau}$ to represent a specific element in the set $\left\{ {}^1X(t_2+\tau),\ {}^2X(t_2+\tau),\ldots,\ {}^kX(t_2+\tau),\ldots,\ {}^nX(t_2+\tau) \right\}$. we have also used a simplified notation $x_{n\tau}$ to represent a specific element in the set $\left\{ {}^1X(t_n+\tau),\ {}^2X(t_n+\tau),\ldots,\ {}^kX(t_n+\tau),\ldots,\ {}^nX(t_n+\tau) \right\}$.

Take into consideration the special circumstance $\tau = -t_1$. Then, the expressions in equation (3.5) state, in effect, that the first-order probability density is independent on the time parameter t, and that the higher order probability densities are independent of the t origin but dependent on the difference in t values.

3.3.3 Weakly Stationary Random Processes

For a random process to be considered strictly stationary, it is necessary for all of its statistical properties to remain unchanged over the course of time. This is a very strict requirement, and in order to simplify life and mathematics, it is often possible to be content with a weaker condition, which is that the mean value and covariances do not change. In the case of the so-called *weakly stationary* random processes, the first- and second-order moments are time-invariant, which means that the mean value remains a constant and the covariances are only dependent on the time distance. There are a number of compelling reasons to investigate these *weakly stationary* random processes. The primary reasons are that random processes passed through linear filters are effectively handled by the first two moments of the input.

In the event where the mean value function $\mu_X(t)$ remains a constant and the covariance function is finite and solely dependent on the time difference $\tau = t - s$, the random process $\{X(t),\ t \in T\}$ is referred to be *weakly stationary*. Expressed alternatively, when the first two expressions in equation (3.5) are satisfied, the random process is said to be *weakly stationary*. When we indicate that a random process is stationary in this book, we imply that it is weakly stationary, if we do not expressly say anything else. This is something that you should keep in mind.

3.4 ERGODIC RANDOM PROCESSES

In order for the theory of random processes to be useful in practical applications, it is necessary for us to be able to estimate the characteristics of a physical random process, such as the mean and correlation functions, through measurements. Consider the sample functions of a random process $\{X(t),\ t \in T\}$, which are denoted as ${}^1X(t)$, ${}^2X(t)$, …, ${}^kX(t)$, …and ${}^nX(t)$. In this case, the mean function $E[X(t_1)]$ is the average of ${}^1X(t_1)$, ${}^2X(t_1)$, …, ${}^kX(t_1)$, …and ${}^nX(t_1)$. The correlation function $E[X(t_1)\ X(t_2)]$ is the average of the products ${}^1X(t_1)\ {}^1X(t_2)$, ${}^2X(t_1)\ {}^2X(t_2)$ and so on, and it is calculated across the whole ensemble of the sample functions. When the ensemble is countably infinite or uncountable, the limited number of sample functions that may be recorded in experiments is typically insufficient to provide trustworthy estimates. Nevertheless, if a random process is stationary, then the probabilistic structure of the process is invariant with regard to a change in the time origin. In such a scenario, we might take a sample record that is sufficiently lengthy, divide it into multiple shorter sections and then treat each section individually as if it were a new sample function. Using a lengthy record of a sample function, say ${}^1X(t)$, the aforementioned heuristic argument implies the potential to approximate the mean function, the correlation function and even the higher-order statistical properties of a stationary random process through the utilization of the following parametric averages:

$$m_1 = \frac{1}{T}\int_0^T {}^1x(t)\,dt$$

$$r_1(\tau) = \frac{1}{T-\tau}\int_0^{T-\tau} {}^1x(t+\tau)\,{}^1x(t)\,dt \tag{3.6}$$

$$\cdots\cdots\cdots\cdots\cdots\cdots$$

in which T represents the total length of the sample record. The magnitude of T must be selected in such a way that any further rise in T does not have a significant impact on the values of m_1, $r_1(\tau)$, etc.

For the m_1, $r_1(\tau)$, etc., which are acquired to be good estimates of the mean, the correlation function, etc., of a stationary random process, it is necessary that essentially the same results could be achieved if a different sample function, such as ${}^2X(t)$, were employed. Regrettably, this necessity is realized only for a subset of stationary random processes and not for all of them. The conditions under which the parametric averages m_1, $r_1(\tau)$, etc., regardless of the choice of a particular sample function, approach the corresponding ensemble averages as the sample length T increases are contained in the ergodic theorem of probability theory. The ergodic theorem was initially formulated for a stochastic process that is strictly stationary. Parzen derived the necessary and sufficient conditions for a sample mean to be ergodic and demonstrated that the condition of rigorous stationarity is not always necessary. Within the realm of mathematical statistics, the ergodic theorem is a topic that garners a lot of attention. On the other hand, it seems that the details of the ergodic theorem appear to be outside the scope of our primary interest. For this reason, we will simply state the definition of the ergodic process as follows: A random process $\{X(t),\ t \in T\}$ is referred to as an *ergodic process* if the time average of a single sample record is approximately equal to the ensemble average.

In the study of data observed in engineering and the physical sciences, it is frequently assumed that the ergodic property of a random process is valid. As a result, the statistical properties of a random process can be evaluated from the analysis of a single sample record. For the random processes that are addressed in this book, it is assumed that the ergodic property is true. As a result, the subsequent discussions will focus on a single record ${}^kX(t)$ rather than an ensemble of data.

3.5 DERIVATIVES OF RANDOM PROCESSES

For the purpose of the analytical treatment of random processes, we will be introducing the crucial notion of the convergence of a sequence of random variables in this part.

If

$$\lim_{n\to\infty} E\left[|X_n - X|^2\right] = 0 \tag{3.7}$$

Then the sequence X_n is said to converge in the mean square to X when this occurs. It is a common practice to express convergence in the mean square as

$$\underset{n\to\infty}{\text{l.i.m.}}\, X_n = X \tag{3.8}$$

in which the symbol l.i.m. reads *the limit in the mean.*

As a result of the formulation of a physical problem as a differential equation, the derivatives of the unknown quantity are included in the equation. In the event where the unknown quantity is a random process, the differential equation is only considered to be a valid description of the physical phenomena if all of the derivatives of the unknown random process that are included in the equation exist in the sense of the limitations of some sort, as mentioned in Section 3.5. The existence of a derivative of a random process can be very easily investigated if the derivative is interpreted as a limit in mean square.

Now let's focus on defining the mean-square derivative of a random process:

$$\frac{d}{dt}X(t) = X'(t) = \underset{h\to 0}{\text{l.i.m}}\frac{X(t+h)-X(t)}{h} \tag{3.9}$$

Please note that, for brevity, in equation (3.9), we have omitted the curly brackets for the random process $\{X(t)\}$. Based on the mean-square convergence criterion, it can be deduced that the equation (3.9) is valid if and only if the limit that follows is continuous and exists:

$$\lim_{h,h'\to 0} E\frac{X(t+h)-X(t)}{h}\frac{X(t+h')-X(t)}{h'} \tag{3.10}$$

It is possible to rewrite this limit as

$$\lim_{h,h'\to 0}\left\{\frac{1}{h}\left[\frac{\varphi_{XX}(t+h,t+h')-\varphi_{XX}(t+h,t)}{h'} - \frac{\varphi_{XX}(t,t+h')-\varphi_{XX}(t,t)}{h'}\right]\right\} = \frac{\partial^2}{\partial t\,\partial s}\varphi_{XX}(t,s) \quad at \quad t=s \tag{3.11}$$

As a result, in order for $X(t)$ to be differentiable in the mean square, it is essential that the correlation function $\varphi_{XX}(t,s)$ possesses a continuous mixed second derivative on the diagonal $t = s$. This is the necessary and sufficient condition.

The existence of the limit in equation (3.10) includes the special case

$$\lim_{h\to 0} E\left[\frac{X(t+h)-X(t)}{h}\right]^2 < \infty \tag{3.12}$$

which, in turn, has the implication that

$$-\infty < \lim_{h\to 0} E\frac{X(t+h)-X(t)}{h} = \frac{d}{dt}\mu_X(t) < \infty \tag{3.13}$$

Due to the fact that the operations of expectation and l.i.m. commute, the limit in equation (3.13) is just the mean function of the derivative $X'(t)$, which means that

$$\frac{d}{dt}\mu_X(t) = \mu_{X'}(t) \tag{3.14}$$

Once more, by rearranging the order of expectation and mean-square differentiation, we came up with the following:

$$\begin{aligned} E[X'(t),X'(s)] &= E\left[\frac{d}{ds}X(t),\frac{d}{dt}X(s)\right] = \frac{d}{ds}\frac{d}{dt}E[X(t),X(s)] \\ &= \frac{\partial^2}{\partial t\,\partial s}\varphi(t,s) = \varphi_{X'X'}(t,s) \\ E[X'(t),X(s)] &= E\left[\frac{d}{dt}X(t),X(s)\right] = \frac{d}{ds}E[X(t),X(s)] \\ &= \frac{\partial}{\partial t}\varphi(t,s) = \varphi_{X'X}(t,s). \end{aligned} \tag{3.15}$$

In the event where $X(t)$ is weakly stationary, then

$$\varphi_{XX}(t,s) = R_{XX}(t-s) = R_{XX}(s-t) \tag{3.16}$$

Applying the formula in equation (3.15) results in

$$\begin{aligned} \varphi_{X'X'}(t,s) &= R''_{XX}(t-s) = \frac{\partial^2}{\partial t\,\partial s}R_{XX}(t-s) = -R_{XX}(t-s) = R_{X'X'}(t-s) \\ \varphi_{X'X}(t,s) &= R'_{XX}(t-s) = \frac{\partial}{\partial t}R_{XX}(t-s) = R_{X'X}(t-s) \end{aligned} \tag{3.17}$$

in which a prime that is associated with R is a representation of one differentiation with regard to the argument. As a result of the fact that R is an even function, we have from the second of equation (3.17)

$$R_{X'X}(0) = R'_{XX}(0) = 0 \tag{3.18}$$

As a result, a random process that is weakly stationary is orthogonal to its derivative when evaluated at the same time (please note that orthogonality is defined as $E[XY]=0$).

3.6 GAUSSIAN RANDOM PROCESSES

3.6.1 Introduction

In a Gaussian process, all process values follow a normal distribution, and all linear operations, such as summation, differentiation and integration, yield random variables that are also normally distributed. Furthermore, the Gaussian process exhibits the property that its derivative is also a Gaussian process.

A Gaussian process is fully characterized by its mean value function and covariance function. As a result, it is frequently the preferred option when searching for a model for a random process, as estimating these functions is a very straightforward undertaking, including only individual data and pairs of data. Estimating higher moments, or data beyond bivariate, is considerably more complex or delicate. Another factor contributing to its widespread acceptance is the fact that numerous natural occurrences exhibit Gaussian characteristics or closely approximate a Gaussian process, or may be easily converted to a Gaussian process through a straightforward transformation.

3.6.2 Definition

A random process $\{X(t), t \in T\}$ is a Gaussian process if, for every n and all $t_1,\ldots,t_n \in T$, the vector $(X(t_1),\ldots,X(t_n))$ has an n-dimensional Gaussian distribution.

The above definition suggests that each individual value $X(t)$, for a certain t, follows a Gaussian distribution, where the mean and variance are specified by the mean value function and the covariance function:

$$X(t) \sim N(\mu_X(t), \kappa_2(t,t)) \tag{3.19}$$

Covariances and expectations are the two factors that determine the n-dimensional Gaussian distribution. As a result, the mean value function and the covariance function are the only ones that can definitively determine all of the n-dimensional distributions that pertain to a Gaussian random process:

$$\begin{aligned} &\mu_X(t_k) = \mathrm{E}\left[X(t_k)\right] \\ &\kappa(t_j,t_k) = \mathrm{Cov}\left[X(t_j),\, X(t_k)\right] = E\left[\left(X(t_j)-\mu_X(t_j)\right)\left(X(t_k)-\mu_X(t_k)\right)\right] \end{aligned} \tag{3.20}$$

We are also able to arrive at the very important conclusion that, if a mean value function and a covariance function are provided, *a priori*, then there is always a Gaussian process that has these as mean value and covariance functions.

Gaussian processes are valuable because any random process, such as $\int_0^t X(s)\,ds, X'(t)$, or $X(t+1)-X(t)$, that is derived by means of linear operations from a Gaussian process is itself a Gaussian process. In the event where the operation is feasible, the summation, differentiation and integration of Gaussian processes will invariably result in the generation of a normal random variable or a new Gaussian process. This is a straightforward result associated with the definition. Consequently, if the process $\{X(t), t \in T\}$ is a Gaussian random process, then, for instance,

$$Z_1 = X(t+h) + X(t)$$
$$Z_2 = (X(t+h) - X(t))/h$$
$$Z_3 = \left(X(t_1) + \cdots + X(t_n)\right)/n$$

all have Gaussian distributions. In addition to this, the distribution of (Z_1, Z_2, YZ_3) is a Gaussian distribution in three dimensions. Even limit of linear combinations is normally distributed, when it exists, for instance,

$$X'(t) = \lim_{h \to 0} \frac{X(t+h) - X(t)}{h},$$

is also normally distributed.

3.6.3 Stationary Gaussian Random Processes

A Gaussian process that exhibits weak stationarity is also characterized by strict stationarity. If $\{X(t), t \in T\}$ is weakly stationary, then for each given n and time points $\tau, t_1, \ldots, t_n$, weak stationarity implies that the vectors $\left(X(t_1+\tau), \ldots, X(t_n+\tau)\right)$ and $\left(X(t_1), \ldots, X(t_n)\right)$ have identical means and covariances. Given that a Gaussian distribution is defined by its means and covariances, it can be inferred that the distributions of $\left(X(t_1+\tau), \ldots, X(t_n+\tau)\right)$ and $\left(X(t_1), \ldots, X(t_n)\right)$ are identical. Therefore, the process $\{X(t)\}$ is strictly stationary. If $\{X(t)\}$ is a stationary Gaussian random process with a mean of μ_X and a covariance function of $\kappa(\tau)$, then all $X(t)$ have the same marginal distribution, which is the Gaussian distribution with a mean of μ_X and a variance of $\kappa(0)$. The probability density function can be expressed as:

$$f_{X(t)}(x) = \frac{1}{\sqrt{2\pi\kappa(0)}} \exp\left\{-\frac{(x-\mu_X)^2}{2\kappa(0)}\right\} \tag{3.21}$$

The density of $X(t_1)$ and $X(t_2)$ in two dimensions, at a distance of $\tau = t_1 - t_2$ from each other, can be described using the variance $\kappa(0)$ and the correlation coefficient $\rho = \rho(\tau) = \kappa(\tau)/\kappa(0)$,

$$f_{X(t_1),X(t_2)}(x_1, x_2)$$
$$= \frac{1}{2\pi\kappa(0)\sqrt{1-\rho^2}} \exp\left\{-\frac{1}{2\kappa(0)\left(1-\rho^2\right)}\left\{(x_1-\mu_X)^2 - 2\rho(x_1-\mu_X)(x_2-\mu_X) + (x_2-\mu_X)^2\right\}\right\} \tag{3.22}$$

3.6.4 Properties of Gaussian Random Processes

The way the normal or Gaussian process behaves with regard to linear systems is one of its most crucial characteristics. A linear system's response will typically be a significantly different random process, but it will still be normal, when the excitation is a Gaussian process. As a result, the probability distributions' forms are invariant under linear operations. The magnitudes of the parameters, such as those in equations (3.21) and (3.22), are the only variations.

A unique characteristic of Gaussian random processes relates to specific higher order expectations. Here are a few of the findings. These only apply in the scenario where the Gaussian random process has zero mean at all times; i.e., $E[X(t)] \equiv 0$,

$$\begin{aligned} &E\left[X(t_1)X(t_2)X(t_3)\right]=0 \\ &E\left[X(t_1)X(t_2)X(t_3)X(t_4)\right]=E\left[X(t_1)X(t_2)\right]E\left[X(t_3)X(t_4)\right] \\ &\quad +E\left[X(t_3)X(t_2)\right]E\left[X(t_1)X(t_4)\right]+E\left[X(t_1)X(t_3)\right]E\left[X(t_2)X(t_4)\right] \\ &E\left[X(t_1)X(t_2)X(t_3)X(t_4)X(t_5)\right]=0 \end{aligned} \tag{3.23}$$

For obtaining the relationships in equation (3.23), please refer to the derivation processes from EXERCISE 2.12 to EXERCISE 2.18.

For the purpose of facilitating computations with Gaussian random processes, here we list certain commonly occurring integrals as follows:

$$\int_0^\infty x^n \exp\left(-\frac{x^2}{2\sigma^2}\right)dx = \frac{\left(\sqrt{2}\sigma\right)^{n+1}}{2}\Gamma\left(\frac{n+1}{2}\right) \tag{3.24}$$

in which $\Gamma(x)$ is the gamma function. In particular

$$\begin{aligned} &\int_0^\infty \exp\left(-\frac{x^2}{2\sigma^2}\right)dx = \frac{\sqrt{2\pi}\sigma}{2} \\ &\int_0^\infty x\exp\left(-\frac{x^2}{2\sigma^2}\right)dx = \sigma^2 \\ &\int_0^\infty x^2\exp\left(-\frac{x^2}{2\sigma^2}\right)dx = \frac{\sqrt{2\pi}\sigma^3}{2} \\ &\int_0^\infty x^3\exp\left(-\frac{x^2}{2\sigma^2}\right)dx = 2\sigma^4 \\ &\int_0^\infty x^4\exp\left(-\frac{x^2}{2\sigma^2}\right)dx = \frac{3\sqrt{2\pi}\sigma^5}{2} \end{aligned} \tag{3.25}$$

3.7 EXERCISES

EXERCISE 3.1

There is a great variety of random processes that emerge as functions of a finite number of random variables. Imagine, for instance, that the position of a particle that is moving at a constant velocity is represented by the symbol $X(t)$. A reasonable assumption would be that $X(t)$ is of the form

$$X(t) = X_0 + Vt \tag{E.3.1}$$

in which X_0 and V are random variables, representing the initial position and velocity of the particle, respectively.

a. Calculate the mean value function of $\{X(t),\ t \geq 0\}$;
b. Show that the covariance function of $\{X(t),\ t \geq 0\}$ is given by

$$\kappa(s,t) = \text{cov}[X(s), X(t)] = Var[X_0] + (s+t)\text{cov}[X_0, V] + stVar[V].$$

EXERCISE 3.2

The equation $X(t) = P+Qt$ represents a mathematical relationship for a random process $\{X(t), t \geq 0\}$, where P and Q are random variables that are independent of each other. Both P and Q are uniformly distributed on the unit interval.

1. Calculate the mean value function $m(t) = E[X(t)]$;
2. Calculate the covariance function $\kappa(s,t) = \text{cov}[X(s), X(t)]$;
3. Is this random process weakly stationary?

EXERCISE 3.3

The equation $X(t) = P+Qt + Rt^2$ represents a mathematical relationship for a random process $\{X(t), t \geq 0\}$, where P, Q and R are separate random variables that are independent of each other. Each of these variables has a mean of 1 and a variance of 1.

1. Calculate the mean value function $m(t) = E[X(t)]$ for this random process;
2. Calculate the covariance function $\kappa(s,t) = \text{cov}[X(s), X(t)]$ for this random process;
3. Is this random process weakly stationary?

EXERCISE 3.4

The equation $X(t) = f(t+\theta)$ represents a mathematical relationship for a random process $\{X(t), t \geq 0\}$, in which $f(t)$ is a periodic function with a period L and θ is uniformly distributed on the interval of 0 to L.

1. Calculate the mean value function $m(t) = E[X(t)]$ for this random process;
2. Calculate the covariance function $\kappa(s,t) = \text{cov}[X(s), X(t)]$ for this random process;
3. Is this random process weakly stationary?

EXERCISE 3.5

Suppose that $P_1,\ldots,P_n$ are jointly normally distributed random variables. Let $Q_1,\ldots,Q_m$ be linear functions of $P_1,\ldots,P_n$:

$$\begin{aligned} Q_1 &= c_{11}P_1 + c_{12}P_2 + \cdots + c_{1n}P_n \\ &\vdots \\ Q_m &= c_{m1}P_1 + c_{m2}P_2 + \cdots + c_{mn}P_n \end{aligned} \tag{E.3.2}$$

Show that $Q_1,\ldots,Q_m$ are jointly normally distributed with joint characteristic function

$$\phi_{Q_1,\ldots,Q_m}(u_1,\ldots,u_m) = \exp\left\{ i\sum_{j=1}^{m} u_j E\left[Q_j\right] - \frac{1}{2}\sum_{j,k=1}^{m} u_j u_k \,\text{cov}\left[Q_j, Q_k\right] \right\} \tag{E.3.3}$$

in which for $j,k = 1,2,\ldots,m$

$$\begin{aligned} E\left[Q_j\right] &= c_{j1}E\left[P_1\right] + \cdots + c_{jn}E\left[P_n\right] \\ Cov\left[Q_j, Q_k\right] &= \sum_{s,t=1}^{n} c_{js}c_{kt}Cov\left[P_s, P_t\right] \end{aligned} \tag{E.3.4}$$

EXERCISE 3.6

Random processes $\{Y(t),\ t > 0\}$ that arise from Gaussian random processes through nonlinear operations are highly relevant in applications that deal with the impact of feeding a random process into a device that executes a nonlinear operation on its input. In this discussion, we will limit ourselves to the most elementary issue related to nonlinear operations on Gaussian random processes, determining the mean value function and covariance function of a random process $\{Y(t),\ t > 0\}$ that is derived from a Gaussian random process by means of quadratic operations.

Suppose that $\{X(t),\ t \geq 0\}$ is a Gaussian random process with zero means and covariance function $Cov(s,t)$. Show that, for any non-negative numbers s, t and h,

$$E\left[X^2(t)\right] = Cov(t,t), \tag{E.3.5}$$

$$E[X(t)X(t+h)] = Cov(t, t+h) \tag{E.3.6}$$

EXERCISE 3.7

Suppose that X and Y are independent Gaussian random variables, $N\left(0,\ \sigma^2\right)$. Then the random process, $\{X(t)\}$, defined by

$$X(t) = X\cos 2\pi ft + Y\sin 2\pi ft \tag{E.3.7}$$

is a Gaussian random process. Calculate the mean value function and covariance function of this Gaussian random process.

4 Spectral Analysis of Random Processes

4.1 INTRODUCTION

Before delving into the specifics of spectral analysis, it would be beneficial to provide a concise summary of the fundamentals of the analysis in order to make the theory easier to comprehend.

We take into consideration a random process with a continuous state, denoted by $\{X(t),\ t \in T\}$, that is weakly stationary and has a mean value of zero. Using the time history of a single record, we are able to define the autocorrelation function (ACF) as well as the spectral density function, with the latter being a function of frequency. It is possible to demonstrate that these two functions constitute a pair of Fourier transforms, which is referred to as the Wiener–Khintchine theorem. Therefore, in order to elucidate the theories for the spectral analysis of random processes, the theoretical backgrounds of Fourier series and Fourier transform will first be presented.

4.2 FOURIER SERIES

4.2.1 Introduction

When it comes to Fourier analysis, the Fourier series serves as the central starting point. These series are infinite in nature and were developed with the purpose of representing general periodic functions by means of basic functions, specifically sines and cosines. This trigonometric system is orthogonal, which enables the determination of the coefficients of the Fourier series by making use of the well-known Euler formulas. With the ability to solve ordinary differential equations in relation to forced vibrations and to approximate periodic functions, Fourier series is of great significance to both engineers and physicists. Fourier series is, in a certain sense, more universal than the more common Taylor series in calculus. This is due to the fact that many discontinuous periodic functions that are encountered in applications can be developed in Fourier series, but they do not have expansions in the Taylor series.

The fundamental concept of the Fourier series can be expanded in two significant manners. It is possible to substitute the trigonometric system with alternative families of orthogonal functions, such as Bessel functions, in order to get the Sturm–Liouville expansions. The second expansion involves the application of Fourier series to nonperiodic phenomena, resulting in the derivation of Fourier integrals and Fourier transforms.

DOI: 10.1201/9781003431503-4

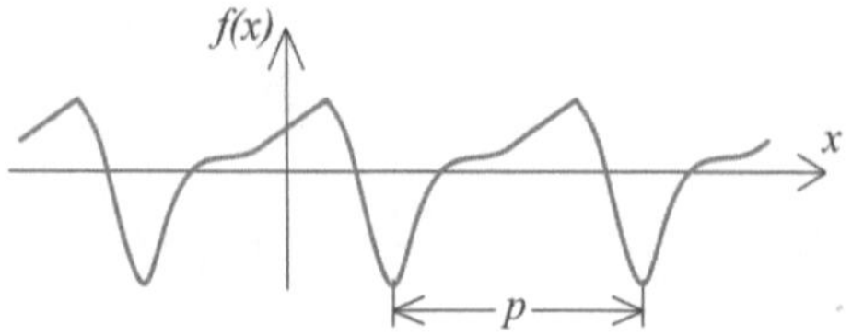

FIGURE 4.1 A periodic function of period p.

The type of series, known as a Fourier series, is an infinite series that represents periodic functions using cosines and sines. Therefore, the engineer and the applied mathematician are the ones who benefit the most from an understanding of Fourier series. To begin, we will need some prior information in order to define the Fourier series. It is said that a function $f(x)$ is a periodic function if it is defined for all real x, with the possible exception of certain points, and if there is a positive number p, which is referred to as a period of $f(x)$, such that

$$f(x+p)=f(x) \tag{4.1}$$

As shown in Figure 4.1, the graph of a periodic function is characterized by the fact that it can be generated by performing periodic repetitions of the graph of the function in any interval of length p.

4.2.2 Fourier Series Representation of a Function with a Period of 2π

In this section, we will be focusing on the representation of different functions $f(x)$ with a period of 2π using the simple functions:

$$1, \quad \cos x, \quad \sin x, \quad \cos 2x, \quad \sin 2x, \cdots, \quad \cos nx, \quad \sin nx, \cdots \tag{4.2}$$

Each one of these functions has a period of 2π. They form the so-called trigonometric system. Figure 4.2 depicts the first few of them (except for the constant 1, whose period can be any value).

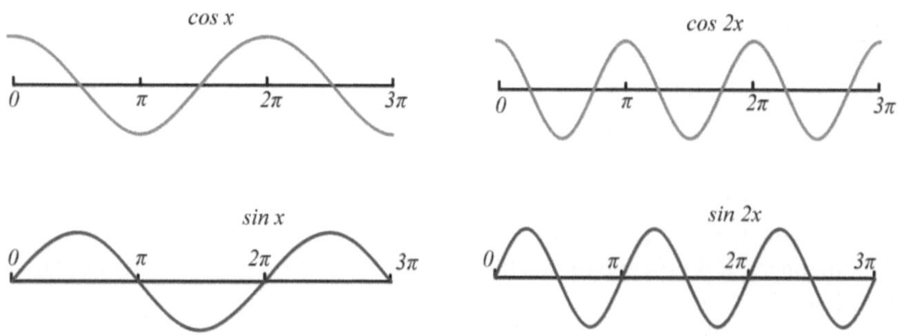

FIGURE 4.2 Sine and cosine functions having the period 2π.

The series that will be obtained is going to be a trigonometric series, which means that it will be a series of the form:

$$\begin{aligned} & a_0 + a_1 \cos x + b_1 \sin x + a_2 \cos 2x + b_2 \sin 2x + \cdots \\ & = a_0 + \sum_{n=1}^{\infty} (a_n \cos nx + b_n \sin nx) \end{aligned} \tag{4.3}$$

The coefficients of the series in equation (4.3) are constants, and they are denoted by the notations a_0, a_1, b_1, a_2, b_2,.... It can be observed that every term in equation (4.3) possesses the period 2π. Therefore, in the event when the coefficients in equation (4.3) are found to be such that the series converges, the sum of the series will be a function of the period 2π.

Now, let us assume that $f(x)$ is a function that is given and has a period of 2π. It is such that it can be represented by a series as shown in equation (4.3), which means that the series shown in equation (4.3) converges and, in addition, has the sum of $f(x)$. Following that, we can write by utilizing the equality sign as follows:

$$f(x) = a_0 + \sum_{n=1}^{\infty} (a_n \cos nx + b_n \sin nx) \tag{4.4}$$

and call equation (4.4) the Fourier series expression of $f(x)$. We will demonstrate that in this scenario the coefficients in equation (4.4) are the so-called Fourier coefficients of $f(x)$, given by the Euler formulas:

$$\begin{aligned} &\text{(i)} \quad a_0 = \frac{1}{2\pi} \int_{-\pi}^{\pi} f(x)\, dx \\ &\text{(ii)} \quad a_n = \frac{1}{\pi} \int_{-\pi}^{\pi} f(x) \cos nx\, dx, \qquad n = 1, 2, \ldots \\ &\text{(iii)} \quad b_n = \frac{1}{\pi} \int_{-\pi}^{\pi} f(x) \sin nx\, dx, \qquad n = 1, 2, \ldots. \end{aligned} \tag{4.5}$$

We next derive the Euler formulas in equation (4.5). For deriving the formula (i) in equation (4.5), we perform integration on both sides of equation (4.4) from $-\pi$ to π and obtain

$$\begin{aligned} \int_{-\pi}^{\pi} f(x)\, dx &= \int_{-\pi}^{\pi} \left[a_0 + \sum_{n=1}^{\infty} (a_n \cos nx + b_n \sin nx) \right] dx \\ &= \int_{-\pi}^{\pi} a_0\, dx + \sum_{n=1}^{\infty} \left\{ a_n \int_{-\pi}^{\pi} \cos nx\, dx + b_n \int_{-\pi}^{\pi} \sin nx\, dx \right\} \\ &= 2\pi a_0 + \sum_{n=1}^{\infty} \{ a_n \times 0 + b_n \times 0 \} = 2\pi a_0 \end{aligned} \tag{4.6}$$

Therefore, division by 2π on both sides of equation (4.5) gives the formula (**i**) in equation (4.5). For deriving the formula (**ii**) in equation (4.5), we multiply equation (4.4) on both sides by cos mx with any fixed positive integer m, and integrating from $-\pi$ to π, we obtain:

$$\int_{-\pi}^{\pi} f(x)\cos mx\,dx = \int_{-\pi}^{\pi}\left[a_0 + \sum_{n=1}^{\infty}(a_n\cos nx + b_n\sin nx)\right]\cos mx\,dx$$

$$= \int_{-\pi}^{\pi} a_0\cos mx\,dx + \sum_{n=1}^{\infty}\left\{a_n\int_{-\pi}^{\pi}\cos nx\cos mx\,dx + b_n\int_{-\pi}^{\pi}\sin nx\cos mx\,dx\right\}$$

$$= 0 + \sum_{n=1}^{\infty}\left\{a_n\int_{-\pi}^{\pi}\cos nx\cos mx\,dx + b_n\int_{-\pi}^{\pi}\sin nx\cos mx\,dx\right\}$$

$$= \sum_{n=1}^{\infty}\left\{\frac{a_n}{2}\int_{-\pi}^{\pi}\left[\cos(n+m)x + \cos(n-m)x\right]dx + \frac{b_n}{2}\int_{-\pi}^{\pi}\left[\sin(n-m)x - \sin(n+m)x\right]dx\right\}$$

$$= \sum_{n=1}^{\infty}\left\{\frac{a_n}{2}\int_{-\pi}^{\pi}\left[\cos(n+m)x + \cos(n-m)x\right]dx + 0\right\}$$

$$= \sum_{n=1}^{\infty}\left\{\frac{a_n}{2}\int_{-\pi}^{\pi}\left[\cos(n+m)x + \cos(n-m)x\right]dx\right\} = \pi a_m \tag{4.7}$$

because in the above equation the integral of $\left[\cos(n+m)x + \cos(n-m)x\right]$ is $a_m\pi$ for $n = m$ and 0 for $n \neq m$. Therefore, division by π on both sides of equation (4.7) gives the formula (**ii**) in equation (4.5) (with m instead of n).

For deriving the formula (**iii**) in equation (4.5), we multiply equation (4.4) on both sides by sin mx with any fixed positive integer m, and integrating from $-\pi$ to π, we obtain:

$$\int_{-\pi}^{\pi} f(x)\sin mx\,dx = \int_{-\pi}^{\pi}\left[a_0 + \sum_{n=1}^{\infty}(a_n\cos nx + b_n\sin nx)\right]\sin mx\,dx$$

$$= \int_{-\pi}^{\pi} a_0\sin mx\,dx + \sum_{n=1}^{\infty}\left\{a_n\int_{-\pi}^{\pi}\cos nx\sin mx\,dx + b_n\int_{-\pi}^{\pi}\sin nx\sin mx\,dx\right\}$$

$$= 0 + \sum_{n=1}^{\infty}\left\{a_n\int_{-\pi}^{\pi}\cos nx\sin mx\,dx + b_n\int_{-\pi}^{\pi}\sin nx\sin mx\,dx\right\}$$

$$= \sum_{n=1}^{\infty} \left\{ \frac{a_n}{2} \int_{-\pi}^{\pi} [\sin(n+m)x - \sin(n-m)x] dx + \frac{b_n}{2} \int_{-\pi}^{\pi} [\cos(n-m)x - \cos(n+m)x] dx \right\}$$

$$= \sum_{n=1}^{\infty} \left\{ 0 + \frac{b_n}{2} \int_{-\pi}^{\pi} [\cos(n-m)x - \cos(n+m)x] dx \right\} \tag{4.8}$$

$$= \sum_{n=1}^{\infty} \left\{ \frac{b_n}{2} \int_{-\pi}^{\pi} [\cos(n-m)x - \cos(n+m)x] dx \right\} = \pi b_m$$

because in the above equation the integral of $[\cos(n-m)x - \cos(n+m)x]$ is $b_m \pi$ for $n = m$ and 0 for $n \neq m$. Therefore, division by π on both sides of equation (4.8) gives the formula (**iii**) in equation (4.5) (with m instead of n). The proof of the Euler formulae (4.5) for the Fourier coefficients is now finished.

4.2.3 Fourier Series Representation of a Function with an Arbitrary Period

Obviously, periodic functions in real-world applications may have any period and not just 2π as in the forgoing derivations. In the following, we derive an alternate expansion of the Fourier series, which approximates a function of period $2L$ from a sum of sine and cosines:

$$f(t) = \frac{a_0}{2} + \sum_{n=1}^{\infty} a_n \cos\left(\frac{n\pi t}{L}\right) + b_n \sin\left(\frac{n\pi t}{L}\right). \tag{4.9}$$

Please note that we have used an independent variable t in equation (4.9). Convergence of the Fourier series to the function in the mean occurs throughout the entirety of the interval $(-L, L)$. An increase in the number of sines and cosines (harmonics) that are included typically results in an improvement in this convergence.

The series in equation (4.9), which is referred to as a Fourier series, converges to the value of the function $f(t)$ at every particular point in the interval $[-L, L]$. On account of the fact that every term in the series has a period of $2L$, the sum of the series also has a period of $2L$. The $n = 1$ term is the fundamental of the periodic function $f(t)$, and the harmonics are the remaining terms whose frequencies are integer multiples of the fundamental. To compute the coefficients a_n and b_n for a given function $f(t)$, we need to identify a straightforward approach that we can use. As a first attempt, we do term-by-term integration of equation (4.9) over the interval from $-L$ to L as follows:

$$\int_{-L}^{L} f(t)\,dt = \int_{-L}^{L} \frac{a_0}{2} dt + \int_{-L}^{L} \left[\sum_{n=1}^{\infty} a_n \cos\left(\frac{n\pi t}{L}\right) + b_n \sin\left(\frac{n\pi t}{L}\right) \right] dt. \tag{4.10}$$

We have

$$\int_{-L}^{L} \frac{a_0}{2} dt = a_0 L \tag{4.11}$$

$$\begin{aligned}\int_{-L}^{L}\left[\sum_{n=1}^{\infty} a_n \cos\left(\frac{n\pi t}{L}\right)\right]dt &= \sum_{n=1}^{\infty} a_n \left[(L/n\pi)\sin(n\pi t/L)\right]_{-L}^{L} \\ &= \sum_{n=1}^{\infty} a_n \left\{(L/n\pi)\left[\sin(n\pi)-\sin(-n\pi)\right]\right\} \\ &= \sum_{n=1}^{\infty} a_n \left\{(L/n\pi)[0-0]\right\} = 0\end{aligned} \tag{4.12}$$

$$\begin{aligned}\int_{-L}^{L}\left[\sum_{n=1}^{\infty} b_n \sin\left(\frac{n\pi t}{L}\right)\right]dt &= \sum_{n=1}^{\infty} b_n \left[(-L/n\pi)\cos(n\pi t/L)\right]_{-L}^{L} \\ &= \sum_{n=1}^{\infty} b_n \left\{(-L/n\pi)\left[\cos(n\pi)-\cos(-n\pi)\right]\right\} \\ &= \sum_{n=1}^{\infty} b_n \left\{(-L/n\pi)\left[\cos(n\pi)-\cos(n\pi)\right]\right\} = 0\end{aligned} \tag{4.13}$$

Combining equations (4.10), (4.11), (4.12) and (4.13), we are left with the following:

$$a_0 = \frac{1}{L}\int_{-L}^{L} f(t)dt \tag{4.14}$$

As a consequence, the value of a_0 is twice the average value of $f(t)$ over a single period.

After that, we proceed to multiply each side of equation (4.10) by $\cos(m\pi t/L)$, in which m is a fixed integer. Integrating from $-L$ to L,

$$\begin{aligned}\int_{-L}^{L} f(t)\cos\left(\frac{m\pi t}{L}\right)dt &= \frac{a_0}{2}\int_{-L}^{L}\cos\left(\frac{m\pi t}{L}\right)dt + \sum_{n=1}^{\infty} a_n \int_{-L}^{L}\cos\left(\frac{n\pi t}{L}\right)\cos\left(\frac{m\pi t}{L}\right)dt \\ &+ \sum_{n=1}^{\infty} b_n \int_{-L}^{L}\sin\left(\frac{n\pi t}{L}\right)\cos\left(\frac{m\pi t}{L}\right)dt\end{aligned}$$

$$\int_{-L}^{L} f(t)\cos\left(\frac{m\pi t}{L}\right)dt = \int_{-L}^{L} \frac{a_0}{2}\cos\left(\frac{m\pi t}{L}\right)dt$$
$$+\int_{-L}^{L}\left[\sum_{n=1}^{\infty} a_n\cos\left(\frac{m\pi t}{L}\right)\cos\left(\frac{n\pi t}{L}\right)+b_n\sin\left(\frac{n\pi t}{L}\right)\cos\left(\frac{m\pi t}{L}\right)\right]dt. \tag{4.15}$$

For arbitrary m and n values, we have

$$\begin{aligned}\int_{-L}^{L} \frac{a_0}{2}\cos\left(\frac{m\pi t}{L}\right)dt &= \left(\frac{a_0}{2}\right)\left[\left(L/m\pi\right)\sin\left(m\pi t/L\right)\right]_{-L}^{L}\\ &= \left(\frac{a_0}{2}\right)\left(L/m\pi\right)\left[\sin\left(m\pi\right)-\sin\left(-m\pi\right)\right]\\ &= \left(\frac{a_0}{2}\right)\left(L/m\pi\right)\left[0-0\right]=0\end{aligned} \tag{4.16}$$

$$\begin{aligned}&\int_{-L}^{L}\left[\sum_{n=1}^{\infty} b_n\sin\left(\frac{n\pi t}{L}\right)\cos\left(\frac{m\pi t}{L}\right)\right]dt\\ &= \int_{-L}^{L}\left\{\sum_{n=1}^{\infty}\frac{b_n}{2}\left[\sin\left(\frac{n\pi t}{L}+\frac{m\pi t}{L}\right)+\sin\left(\frac{n\pi t}{L}-\frac{m\pi t}{L}\right)\right]\right\}dt\\ &= \sum_{n=1}^{\infty}\frac{b_n}{2}\left[\left(\frac{-L}{(n+m)\pi}\right)\cos((n+m)\pi t/L)\right]_{-L}^{L}\\ &\quad+\sum_{n=1}^{\infty}\frac{b_n}{2}\left[\left(\frac{-L}{(n+m)\pi}\right)\cos((n-m)\pi t/L)\right]_{-L}^{L}\\ &= \sum_{n=1}^{\infty}\frac{b_n}{2}\left\{\left(\frac{-L}{(n+m)\pi}\right)\left[\cos(n+m)\pi-\cos(-(n+m)\pi)\right]\right\}\\ &\quad+\sum_{n=1}^{\infty}\frac{b_n}{2}\left\{\left(\frac{-L}{(n+m)\pi}\right)\left[\cos(n-m)\pi-\cos(-(n-m)\pi)\right]\right\}\\ &= \sum_{n=1}^{\infty}\frac{b_n}{2}\left\{\left(\frac{-L}{(n+m)\pi}\right)\left[\cos(n+m)\pi-\cos((n+m)\pi)\right]\right\}\\ &\quad+\sum_{n=1}^{\infty}\frac{b_n}{2}\left\{\left(\frac{-L}{(n+m)\pi}\right)\left[\cos(n-m)\pi-\cos((n-m)\pi)\right]\right\}=0+0=0\end{aligned} \tag{4.17}$$

For $n \neq m$, we have

$$\begin{aligned}
&\int_{-L}^{L}\left[\sum_{n=1}^{\infty} a_n \cos\left(\frac{m\pi t}{L}\right)\cos\left(\frac{n\pi t}{L}\right)\right]dt \\
&= \int_{-L}^{L}\left\{\sum_{n=1}^{\infty}\frac{a_n}{2}\left[\cos\left(\frac{n\pi t}{L}+\frac{m\pi t}{L}\right)+\cos\left(\frac{n\pi t}{L}-\frac{m\pi t}{L}\right)\right]\right\}dt \\
&= \sum_{n=1}^{\infty}\frac{a_n}{2}\left[\left(\frac{L}{(n+m)\pi}\right)\sin((n+m)\pi t/L)\right]_{-L}^{L} \\
&\quad + \sum_{n=1}^{\infty}\frac{a_n}{2}\left[\left(\frac{L}{(n+m)\pi}\right)\sin((n-m)\pi t/L)\right]_{-L}^{L} \\
&= \sum_{n=1}^{\infty}\frac{a_n}{2}\left\{\left(\frac{L}{(n+m)\pi}\right)\left[\sin(n+m)\pi - \sin(-(n+m)\pi)\right]\right\} \\
&\quad + \sum_{n=1}^{\infty}\frac{a_n}{2}\left\{\left(\frac{L}{(n+m)\pi}\right)\left[\sin(n-m)\pi - \sin(-(n-m)\pi)\right]\right\} = 0+0=0
\end{aligned} \tag{4.18}$$

For $n = m$, we have

$$\begin{aligned}
&\int_{-L}^{L}\left[\sum_{n=1}^{\infty} a_n \cos\left(\frac{m\pi t}{L}\right)\cos\left(\frac{n\pi t}{L}\right)\right]dt \\
&= \int_{-L}^{L}\frac{a_m}{2}\left[\cos\left(\frac{m\pi t}{L}+\frac{m\pi t}{L}\right)+\cos\left(\frac{m\pi t}{L}-\frac{m\pi t}{L}\right)\right]dt \\
&= \frac{a_m}{2}\left[\left(\frac{L}{(m+m)\pi}\right)\sin((m+m)\pi t/L)\right]_{-L}^{L} + \int_{-L}^{L}\frac{a_m}{2}dt \\
&= \frac{a_m}{2}\left\{\left(\frac{L}{(m+m)\pi}\right)\left[\sin(m+m)\pi - \sin(-(m+m)\pi)\right]\right\} + a_m L \\
&= 0 + a_m L = a_m L = a_n L
\end{aligned} \tag{4.19}$$

Combining equations (4.15), (4.16), (4.17), (4.18) and (4.19), we are left with the following:

$$\int_{-L}^{L} f(t)\cos\left(\frac{m\pi t}{L}\right)dt = a_m L, \tag{4.20}$$

or equivalently

$$\int_{-L}^{L} f(t)\cos\left(\frac{n\pi t}{L}\right)dt = a_n L \tag{4.21}$$

From equation (4.21), we obtain the following relationship

$$a_n = \frac{1}{L}\int_{-L}^{L} f(t)\cos\left(\frac{n\pi t}{L}\right)dt \tag{4.22}$$

After that, we proceed to multiply each side of equation (4.10) by $\sin(m\pi t/L)$, in which m is again a fixed integer. Integrating from $-L$ to L,

$$\int_{-L}^{L} f(t)\cos\left(\frac{m\pi t}{L}\right)dt = \frac{a_0}{2}\int_{-L}^{L}\cos\left(\frac{m\pi t}{L}\right)dt + \sum_{n=1}^{\infty} a_n \int_{-L}^{L}\cos\left(\frac{n\pi t}{L}\right)\cos\left(\frac{m\pi t}{L}\right)dt$$
$$+\sum_{n=1}^{\infty} b_n \int_{-L}^{L}\sin\left(\frac{n\pi t}{L}\right)\cos\left(\frac{m\pi t}{L}\right)dt$$

$$\begin{aligned}\int_{-L}^{L} f(t)\sin\left(\frac{m\pi t}{L}\right)dt &= \int_{-L}^{L}\frac{a_0}{2}\sin\left(\frac{m\pi t}{L}\right)dt\\ &+\int_{-L}^{L}\left[\sum_{n=1}^{\infty} a_n\sin\left(\frac{m\pi t}{L}\right)\cos\left(\frac{n\pi t}{L}\right)+b_n\sin\left(\frac{n\pi t}{L}\right)\sin\left(\frac{m\pi t}{L}\right)\right]dt.\end{aligned} \tag{4.23}$$

For arbitrary m and n values, we have

$$\begin{aligned}&\int_{-L}^{L}\frac{a_0}{2}\sin\left(\frac{m\pi t}{L}\right)dt = \left(\frac{a_0}{2}\right)\left[(-L/m\pi)\cos(m\pi t/L)\right]_{-L}^{L}\\ &=\left(\frac{a_0}{2}\right)(-L/m\pi)\left[\cos(m\pi)-\cos(-m\pi)\right]=\left(\frac{a_0}{2}\right)(-L/m\pi)\times 0 = 0\end{aligned} \tag{4.24}$$

$$\int_{-L}^{L}\left[\sum_{n=1}^{\infty} a_n\sin\left(\frac{m\pi t}{L}\right)\cos\left(\frac{n\pi t}{L}\right)\right]dt$$
$$=\int_{-L}^{L}\left\{\sum_{n=1}^{\infty}\frac{a_n}{2}\left[\sin\left(\frac{m\pi t}{L}+\frac{n\pi t}{L}\right)+\sin\left(\frac{m\pi t}{L}-\frac{n\pi t}{L}\right)\right]\right\}dt$$

$$\begin{aligned}
&=\sum_{n=1}^{\infty}\frac{a_n}{2}\left[\left(\frac{-L}{(n+m)\pi}\right)\cos((n+m)\pi t/L)\right]_{-L}^{L}\\
&\quad+\sum_{n=1}^{\infty}\frac{a_n}{2}\left[\left(\frac{-L}{(m-n)\pi}\right)\cos((m-n)\pi t/L)\right]_{-L}^{L}\\
&=\sum_{n=1}^{\infty}\frac{a_n}{2}\left\{\left(\frac{-L}{(n+m)\pi}\right)\left[\cos(n+m)\pi-\cos(-(n+m)\pi)\right]\right\}\\
&\quad+\sum_{n=1}^{\infty}\frac{a_n}{2}\left\{\left(\frac{-L}{(m-n)\pi}\right)\left[\cos(m-n)\pi-\cos(-(m-n)\pi)\right]\right\}\\
&=\sum_{n=1}^{\infty}\frac{a_n}{2}\left\{\left(\frac{-L}{(n+m)\pi}\right)\left[\cos(n+m)\pi-\cos((n+m)\pi)\right]\right\}\\
&\quad+\sum_{n=1}^{\infty}\frac{a_n}{2}\left\{\left(\frac{-L}{(m-n)\pi}\right)\left[\cos(m-n)\pi-\cos((m-n)\pi)\right]\right\}=0+0=0
\end{aligned}\tag{4.25}$$

For $n \neq m$, we have

$$\begin{aligned}
&\int_{-L}^{L}\left[\sum_{n=1}^{\infty}b_n\sin\left(\frac{n\pi t}{L}\right)\sin\left(\frac{m\pi t}{L}\right)\right]dt\\
&=\int_{-L}^{L}\left\{\sum_{n=1}^{\infty}\frac{b_n}{2}\left[\cos\left(\frac{n\pi t}{L}-\frac{m\pi t}{L}\right)-\cos\left(\frac{n\pi t}{L}+\frac{m\pi t}{L}\right)\right]\right\}dt\\
&=\sum_{n=1}^{\infty}\frac{b_n}{2}\left[\left(\frac{L}{(n-m)\pi}\right)\sin((n-m)\pi t/L)\right]_{-L}^{L}\\
&\quad-\sum_{n=1}^{\infty}\frac{b_n}{2}\left[\left(\frac{L}{(n+m)\pi}\right)\sin((n+m)\pi t/L)\right]_{-L}^{L}\\
&=\sum_{n=1}^{\infty}\frac{b_n}{2}\left\{\left(\frac{L}{(n-m)\pi}\right)\left[\sin(n-m)\pi-\sin(-(n-m))\pi\right]\right\}\\
&\quad+\sum_{n=1}^{\infty}\frac{b_n}{2}\left\{\left(\frac{L}{(n+m)\pi}\right)\left[\sin(n+m)\pi-\sin(-(n+m)\pi)\right]\right\}=0+0=0
\end{aligned}\tag{4.26}$$

For $n = m$, we have

$$\int_{-L}^{L}\left[\sum_{n=1}^{\infty} b_n \sin\left(\frac{n\pi t}{L}\right)\sin\left(\frac{m\pi t}{L}\right)\right]dt = \int_{-L}^{L}\frac{b_m}{2}\left[\cos\left(\frac{m\pi t}{L}-\frac{m\pi t}{L}\right)-\cos\left(\frac{m\pi t}{L}+\frac{m\pi t}{L}\right)\right]dt$$

$$= \int_{-L}^{L}\frac{b_m}{2}dt - \int_{-L}^{L}\frac{b_m}{2}\left[-\cos\left(\frac{m\pi t}{L}+\frac{m\pi t}{L}\right)\right]dt = b_m L - \int_{-L}^{L}\frac{b_m}{2}\left[-\cos\left(\frac{m\pi t}{L}+\frac{m\pi t}{L}\right)\right]dt$$

$$= b_m L - \frac{b_m}{2}\left[\left(\frac{-L}{(2m)\pi}\right)\sin((m+m)\pi t/L)\right]_{-L}^{L} = b_m L - \frac{b_m}{2}(0-0) = b_m L \tag{4.27}$$

Combining equations (4.23), (4.24), (4.25), (4.26), and (4.27), we are left with the following:

$$\int_{-L}^{L} f(t)\sin\left(\frac{m\pi t}{L}\right)dt = b_m L, \tag{4.28}$$

or equivalently

$$\int_{-L}^{L} f(t)\sin\left(\frac{n\pi t}{L}\right)dt = b_n L, \tag{4.29}$$

From equation (4.29), we obtain the following relationship

$$b_n = \frac{1}{L}\int_{-L}^{L} f(t)\sin\left(\frac{n\pi t}{L}\right)dt \tag{4.30}$$

Although equations (4.14), (4.22) and (4.30) provide us with the values a_0, a_n and b_n for periodic functions over the interval $[-L, L]$, there are specific circumstances in which it is more advantageous to utilize the interval $[\tau, \tau + 2L]$, where τ can be any real number. In that scenario, equation (4.9) continues to provide the Fourier series of the function $f(t)$, and

$$a_0 = \frac{1}{L}\int_{\tau}^{\tau+2L} f(t)\,dt \tag{4.31}$$

$$a_n = \frac{1}{L}\int_{\tau}^{\tau+2L} f(t)\cos\left(\frac{n\pi t}{L}\right)dt \tag{4.32}$$

$$b_n = \frac{1}{L}\int_{\tau}^{\tau+2L} f(t)\sin\left(\frac{n\pi t}{L}\right)dt \tag{4.33}$$

After recalling that the function $f(t)$ is a periodic function that extends from negative infinity to positive infinity, we are able to arrive at these results. Therefore, when we move from the interval $[-L,L]$ to the new interval $[\tau,\tau+2L]$, it is imperative that the results must remain unchanged.

4.2.4 Complex Fourier Series

A Fourier series has been expressed in terms of sines and cosines up until this point in our discussion. Now that we have reached this point, we are prepared to re-express a Fourier series as a series of complex exponentials. There are two factors that contribute to this. In the first place, the expansion of a function in terms of complex exponentials leads to coefficients that are remarkably straightforward and easy to understand in particular applications of Fourier series in the fields of engineering and physical science. In addition, these complex Fourier series serve as a precursor to the forthcoming discussion on the Fourier transform in the subsequent section.

To begin, we will introduce the variable $\omega_n = n\pi/L$, in which n can be any of the following values: 0, ±1, ±2,... Through the utilization of Euler's formula, we are able to substitute exponentials for the sine and cosine in the Fourier series. At this point, we first provide some background knowledge regarding the celebrated Euler's formula.

Euler's formula is a fundamental mathematical expression that elegantly connects various mathematical concepts. It is named after Leonhard Euler, the renowned Swiss mathematician and physicist. Euler's formula was first derived by Euler in the late 18th century and has become a cornerstone of mathematics. Specifically, Euler's formula is a mathematical equation in complex analysis that defines the essential connection between trigonometric functions and the complex exponential function. Euler's formula asserts that, for any real number x, the following relationship holds

$$e^{ix} = \cos x + i\sin x \tag{4.34}$$

in which e represents the base of the natural logarithm (e is a mathematical constant approximately equal to 2.71828), i stands for the imaginary unit and cos and sin refer to the trigonometric functions cosine and sine, respectively. Utilizing equation (4.34) and considering that $\omega_n = n\pi/L$, equation (4.9) can now be written as

$$\begin{aligned} f(t) &= \frac{a_0}{2} + \sum_{n=1}^{\infty} a_n \cos\left(\frac{n\pi t}{L}\right) + b_n \sin\left(\frac{n\pi t}{L}\right) \\ &= \frac{a_0}{2} + \sum_{n=1}^{\infty} (a_n/2)\left(e^{i\omega_n t} + e^{-i\omega_n t}\right) + (b_n/2i)\left(e^{i\omega_n t} - e^{-i\omega_n t}\right) \\ &= \frac{a_0}{2} + \sum_{n=1}^{\infty} \left(\frac{a_n}{2} - \frac{b_n i}{2}\right) e^{i\omega_n t} + \left(\frac{a_n}{2} + \frac{b_n i}{2}\right) e^{-i\omega_n t} \end{aligned} \tag{4.35}$$

In the event that we define $c_n = \frac{1}{2}(a_n - ib_n)$, then we obtain

$$c_n = \frac{1}{2}(a_n - ib_n)$$
$$= \frac{1}{2L}\int_{\tau}^{\tau+2L} f(t)\left[\cos(\omega_n t) - i\sin(\omega_n t)\right]dt = \frac{1}{2L}\int_{\tau}^{\tau+2L} f(t)e^{-i\omega_n t}\,dt \tag{4.36}$$

In a similar manner, the complex conjugate of c_n, denoted as c_n^*, is equal to

$$c_n{}^* = \frac{1}{2}(a_n + ib_n)$$
$$= \frac{1}{2L}\int_{\tau}^{\tau+2L} f(t)\left[\cos(\omega_n t) + i\sin(\omega_n t)\right]dt = \frac{1}{2L}\int_{\tau}^{\tau+2L} f(t)e^{i\omega_n t}\,dt \tag{4.37}$$

Considering that $\omega_{-n} = \frac{-n\pi}{L} = -\omega_n$, then we obtain

$$c_{-n} = \frac{1}{2L}\int_{\tau}^{\tau+2L} f(t)e^{-i\omega_{-n}t}\,dt = \frac{1}{2L}\int_{\tau}^{\tau+2L} f(t)e^{-i(-\omega_n)t}\,dt = \frac{1}{2L}\int_{\tau}^{\tau+2L} f(t)e^{i\omega_n t}\,dt \tag{4.38}$$

Through comparing equations (4.37) and (4.38), we obtain $c_n{}^* = c_{-n}$. Therefore, equation (4.35) can be written as

$$f(t) = \frac{a_0}{2} + \sum_{n=1}^{\infty}\left(\frac{a_n}{2} - \frac{b_n i}{2}\right)e^{i\omega_n t} + \left(\frac{a_n}{2} + \frac{b_n i}{2}\right)e^{-i\omega_n t}$$
$$= \frac{a_0}{2} + \sum_{n=1}^{\infty} c_n e^{i\omega_n t} + c_n{}^* e^{-i\omega_n t} = \frac{a_0}{2} + \sum_{n=1}^{\infty} c_n e^{i\omega_n t} + c_{-n}e^{i(-\omega_n)t}$$
$$= \frac{a_0}{2} + \sum_{n=1}^{\infty} c_n e^{i\omega_n t} + c_{-n}e^{i\omega_{-n}t} = \frac{a_0}{2} + \sum_{n=1}^{\infty} c_n e^{i\omega_n t} \tag{4.39}$$
$$+ \sum_{n=1}^{\infty} c_{-n}e^{i\omega_{-n}t} = \frac{a_0}{2} + \sum_{n=1}^{\infty} c_n e^{i\omega_n t} + \sum_{n=-\infty}^{-1} c_n e^{i\omega_n t}$$

On the other hand, based on equations (4.31) and (4.36) and considering the fact that $\omega_0 = 0\pi/L = 0$, we can derive

$$\frac{a_0}{2} = \frac{1}{2L}\int_{\tau}^{\tau+2L} f(t)\,dt = \frac{1}{2L}\int_{\tau}^{\tau+2L} f(t)e^{-i\omega_0 t}\,dt = c_0 = c_0 e^{i\omega_0 t} \tag{4.40}$$

Then, equation (4.39) can be written as

$$f(t) = \frac{a_0}{2} + \sum_{n=1}^{\infty} c_n e^{i\omega_n t} + \sum_{n=-\infty}^{-1} c_n e^{i\omega_n t}$$
$$= c_0 e^{i\omega_0 t} + \sum_{n=1}^{\infty} c_n e^{i\omega_n t} + \sum_{n=-\infty}^{-1} c_n e^{i\omega_n t} = \sum_{n=-\infty}^{\infty} c_n e^{i\omega_n t} \tag{4.41}$$

and $n = 0, \pm 1, \pm 2, \ldots$. Take note that the summation equation (4.41) always results in a real-valued function $f(t)$, despite the fact that c_n is known to be complex in general.

4.3 FOURIER TRANSFORM

4.3.1 Derivation of the Fourier Transform Pair

From the explanations in the previous sections, we can find that the Fourier series is a powerful tool for modeling and transforming periodic functions. It provides a concise and efficient representation of periodic signals, making analysis and synthesis relatively straightforward. However, despite its numerous advantages, there is still a limitation to using the Fourier series that needs to be considered. This limitation of the Fourier series is its inability to model non-periodic functions. Signals that exhibit non-periodic behavior, such as non-sinusoidal waves, cannot be accurately represented using the Fourier series. It is important for users to be aware of this limitation and consider alternative approaches when necessary.

The majority of the functions that are encountered in engineering are aperiodic. As will become clear in the following discussion, extending the Fourier series to these functions results in the Fourier transform. The Fourier transform is the natural extension of the Fourier series to a function $f(t)$ that has an infinite period. Equation (4.41) is referred to as the complex form of a Fourier series expansion of a periodic function $f(t)$, and this expression will be used for the Fourier transform of a non-periodic function. We note that equation (4.36) can be re-expressed as follows:

$$c_n = \frac{1}{2L} \int_{-L}^{L} f(t) e^{-i\omega_n t} \, dt \tag{4.42}$$

By combining equations (4.41) and (4.42), we can write $f(t)$ as follows:

$$f(t) = \sum_{n=-\infty}^{\infty} \left[\frac{1}{2L} \int_{-L}^{L} f(t) e^{-i\omega_n t} \, dt \right] e^{i\omega_n t} \tag{4.43}$$

Recall that we have $\omega_n = n\pi/L$ so that $\Delta\omega_n = \omega_{n+1} - \omega_n = \pi/L$. Then,

$$f(t) = \sum_{n=-\infty}^{\infty} \left[\frac{\Delta\omega_n}{2\pi} \int_{-L}^{L} f(t) e^{-i\omega_n t}\, dt \right] e^{i\omega_n t} = \frac{1}{2\pi} \sum_{n=-\infty}^{\infty} \left[\int_{-L}^{L} f(t) e^{-i\omega_n t}\, dt \right] e^{i\omega_n t} \Delta\omega_n \qquad (4.44)$$

For a non-periodic function, let $L \to \infty$, then $\Delta\omega_n \to 0$, and $\Delta\omega_n$ may be interpreted as the infinitesimal $d\omega_n$. Thus, equation (4.44) can be written in an integrable form

$$f(t) = \frac{1}{2\pi} \int_{-\infty}^{\infty} \left[\int_{-L}^{L} f(t) e^{-i\omega_n t}\, dt \right] e^{i\omega_n t}\, d\omega_n = \frac{1}{2\pi} \int_{-\infty}^{\infty} F(\omega) e^{i\omega_n t}\, d\omega_n \qquad (4.45)$$

in which

$$F(\omega) = \int_{-L}^{L} f(t) e^{-i\omega_n t}\, dt \qquad (4.46)$$

Because as $L \to \infty$, ω_n approaches a continuous variable, we can therefore write ω_n as ω. Consequently, equations (4.45) and (4.46) can, respectively, be written as

$$f(t) = \frac{1}{2\pi} \int_{-\infty}^{\infty} F(\omega) e^{i\omega t}\, d\omega \qquad (4.47)$$

$$F(\omega) = \int_{-L}^{L} f(t) e^{-i\omega t}\, dt \qquad (4.48)$$

Equations (4.47) and (4.48) are known as the Fourier transform pair, represented by $f(t) \leftrightarrow F(\omega)$. Here, $F(\omega)$ is referred to as the Fourier transform of $f(t)$, which can be represented by $F(\omega) = \mathcal{F}\{f(t)\}$. On the other hand, $f(t)$ is referred to as the inverse Fourier transform of $F(\omega)$, which can be represented by $f(t) = \mathcal{F}^{-1}\{F(\omega)\}$.

At this point, it can be noted that equation (4.45) can also be written as

$$f(t) = \frac{1}{2} \int_{-\infty}^{\infty} \frac{1}{\pi} \left[\int_{-L}^{L} f(t) e^{-i\omega_n t}\, dt \right] e^{i\omega_n t} d\omega_n = \frac{1}{2} \int_{-\infty}^{\infty} F(\omega) e^{i\omega_n t}\, d\omega_n \qquad (4.49)$$

in which

$$F(\omega) = \frac{1}{\pi} \left[\int_{-L}^{L} f(t) e^{-i\omega_n t}\, dt \right] \qquad (4.50)$$

Equations (4.49) and (4.50) are also known as a Fourier transform pair. In a similar manner, equation (4.45) can also be written as

$$f(t)=\frac{1}{\sqrt{2\pi}}\int_{-\infty}^{\infty}\frac{1}{\sqrt{2\pi}}\left[\int_{-L}^{L}f(t)e^{-i\omega_{n}t}\,dt\right]e^{i\omega_{n}t}d\omega_{n}=\frac{1}{\sqrt{2\pi}}\int_{-\infty}^{\infty}F(\omega)e^{i\omega_{n}t}\,d\omega_{n} \tag{4.51}$$

in which

$$F(\omega)=\frac{1}{\sqrt{2\pi}}\left[\int_{-L}^{L}f(t)e^{-i\omega_{n}t}\,dt\right] \tag{4.52}$$

Equations (4.51) and (4.52) are also known as a Fourier transform pair.

Example 4.1

▶ Determine the Fourier transform for

$$f(t)=\begin{cases}5, & |t|<b\\ 0, & |t|>b\end{cases} \tag{4.53}$$

SOLUTION

According to the Fourier transform's definition, we obtain

$$F(\omega)=\int_{-\infty}^{-b}0e^{-i\omega t}\,dt+\int_{-b}^{b}5e^{-i\omega t}\,dt+\int_{b}^{\infty}0e^{-i\omega t}\,dt=\frac{5\left(e^{\omega bi}-e^{-\omega bi}\right)}{\omega i}=\frac{10\sin(\omega b)}{\omega} \tag{4.54}$$

◀

4.3.2 Properties of Fourier Transform

The Fourier transform is a mathematical operation that transforms a signal or function from a time domain to a corresponding frequency domain. It is widely used in various scientific and engineering fields to analyze and process signals. It has several important properties that are essential for its applications in various fields. In this section, we will explore some of these properties.

1. Linearity

   One of the fundamental properties of the Fourier transform is its linearity. This means that if a function f is transformed by the Fourier transform F, then the result of transforming the sum of two functions f_1 and f_2 is equal to the sum of their Fourier transforms ($F_1 + F_2$). This property allows the application of the Fourier transform to a wide range of signals and systems.

2. Time shifting
   Suppose that $f(t)$ is a function that has a Fourier transform $F(\omega)$, then we have $\mathcal{F}\left[f(t-\tau)\right]=e^{-i\omega\tau}F(\omega)$.
3. Scaling factor
   Suppose that $f(t)$ is a function with a Fourier transform $F(\omega)$ and k is a real, non-zero constant. Then, we have the relationship: $\mathcal{F}\left[f(kt)\right]=F(\omega/k)/|k|$.
4. Symmetry
   Suppose that $f(t)$ has the Fourier transform $F(\omega)$, then, we have the relationship: $\mathcal{F}\left[F(t)\right]=2\pi f(-\omega)$.

4.4 SPECTRAL ANALYSIS OF A SINGLE RANDOM PROCESS

4.4.1 Introduction

A random process with a continuous state, denoted by $\{X(t), t\in T\}$, is taken into consideration here. This random process is weakly stationary and has a mean value of zero. Using the time history of a single sample record, we are able to define the autocorrelation function as well as the spectral density function, with the latter being a function of frequency. It is possible to demonstrate that these two functions constitute a pair of Fourier transforms, which is referred to as the Wiener–Khintchine theorem. The idea of spectral analysis makes it possible for us to forecast the statistical properties of the response of a system that is induced by random excitations, as will be covered in the following chapters. The spectral analysis of random processes, which is performed in the frequency domain, is therefore a necessary prerequisite for the prediction of the properties of random processes. The notion of transition from a random process in the time domain to the spectrum in the frequency domain is discussed in this section. This transformation is explained for single random processes.

4.4.2 Autocorrelation Function

Due to the fact that the autocorrelation function, which will be defined below, plays a significant role in spectral analysis, we will first provide the definition in a general form that refers to the ensemble of a random process.

The *autocorrelation function*, denoted by $R_{XX}(t_1,t_2)$, calculates the ensemble average of the product of a random process at times t_1 and t_2. That is,

$$R_{XX}(t_1,t_2)=E\left[X(t_1)X(t_2)\right]=\frac{1}{n}\sum_{k=1}^{n}{}^{k}x(t_1)^{k}\,x(t_2) \tag{4.55}$$

in which E represents the expected value. Remember that in Chapter 3, we specified that for the random process $\{X(t),\ t\in T\}$, $X(t_1)$ is a *random variable* at a given time t_1, and that the set consisting of $\left\{{}^{1}X(t_1),\ {}^{2}X(t_1),\ldots,\ {}^{k}X(t_1),\ldots,\ {}^{n}X(t_1)\right\}$

can be viewed as a random sample with n elements. In equation (4.55), we have used a lowercase notation ${}^k x(t_1)$ to represent the specific element ${}^k X(t_1)$ in the set $\{{}^1X(t_1), {}^2X(t_1),\ldots, {}^kX(t_1),\ldots, {}^nX(t_1)\}$.

Please also note that in Chapter 3, we specified that, for the random process $\{X(t), t \in T\}$, $X(t_2)$ is a *random variable* at a given time t_2, and that the set consisting of $\{{}^1X(t_2), {}^2X(t_2),\ldots, {}^kX(t_2),\ldots, {}^nX(t_2)\}$ can be viewed as a random sample with n elements. In equation (4.55), we have used a lowercase notation ${}^k x(t_2)$ to represent the specific element ${}^k X(t_2)$ in the set $\{{}^1X(t_2), {}^2X(t_2),\ldots, {}^kX(t_2),\ldots, {}^nX(t_2)\}$.

The autocovariance function $C_{XX}(t_1,t_2)$ is calculated as follows:

$$\begin{aligned} C_{XX}(t_1,t_2) &= E\Big[\big(X(t_1)-E[X(t_1)]\big)\big(X(t_2)-E[X(t_2)]\big)\Big] \\ &= \frac{1}{n}\sum_{k=1}^{n}\{{}^k x(t_1)-\bar{x}(t_1)\}\{{}^k x(t_2)-\bar{x}(t_2)\} \end{aligned} \tag{4.56}$$

It can be deduced from equations (4.55) and (4.56) that the autocorrelation function is equal to the autocovariance function in the event when the mean value of a random process is zero.

The autocorrelation function is represented by the notation $R_{XX}(t_1, t_2)$ in order to distinguish it from the cross-correlation function that is used to describe the relationship between two random processes $\{X(t), t \in T\}$ and $\{Y(t), t \in T\}$, which is represented by the notation $R_{XY}(t_1, t_2)$.

Once the weakly stationary stochastic process that was developed in the prior chapter is taken into consideration, the autocorrelation function $R_{XX}(t_1, t_2)$ transforms into a function that is solely dependent on the time difference. By expressing t_2 as $t_2 = t_1 + \tau$, we can obtain that

$$R_{XX}(\tau) = E\left[{}^k x(t)\,{}^k x(t+\tau)\right] = \frac{1}{n}\sum_{k=1}^{n}{}^k x(t)\,{}^k x(t+\tau) \tag{4.57}$$

Furthermore, by making the assumption that a random process $\{X(t), t \in T\}$ is ergodic, it is possible to evaluate the ensemble average by calculating the time average of a single sample record by shifting the record by a time τ (see Figure 4.1 for further clarification), that is

$$R_{XX}(\tau) = E\left[x(t)x(t+\tau)\right] = \lim_{T\to\infty}\frac{1}{2T}\int_{-T}^{T} x(t)x(t+\tau)\,dt \tag{4.58}$$

From equation (4.57), we can find that the autocorrelation function, also known as ACF, calculates the average correlation between the current observations and the corresponding observations lagged by a time length of $-\tau$. In the following, some properties of the autocorrelation function $R_{XX}(\tau)$ are derived and summarized.

$$R_{XX}(-\tau) = \lim_{T\to\infty}\frac{1}{2T}\int_{-T}^{T} x(t)x(t-\tau)\,dt = \lim_{T\to\infty}\frac{1}{2T}\int_{-T}^{T} x(u+\tau)x(u)\,du = R_{XX}(\tau) \quad (4.59)$$

Equation (4.59) demonstrates that one property of the autocorrelation function $R_{XX}(\tau)$ is that it is an even function.

We next prove another property of the autocorrelation function $R_{XX}(\tau)$, i.e., $R_{XX}(\tau)$ has its maximum value when $\tau = 0$. The derivation process is as follows:

$$\begin{aligned}&\lim_{T\to\infty}\frac{1}{2T}\int_{-T}^{T}\{x(t)-x(t+\tau)\}^2\,dt\\ &= \lim_{T\to\infty}\frac{1}{2T}\int_{-T}^{T}\{x(t)\}^2\,dt - 2\lim_{T\to\infty}\frac{1}{2T}\int_{-T}^{T}x(t)x(t+\tau)\,dt + \lim_{T\to\infty}\frac{1}{2T}\int_{-T}^{T}\{x(t+\tau)\}^2\,dt\\ &= R_{XX}(0) - 2R_{XX}(\tau) + R_{XX}(0) \geqslant 0 \end{aligned} \quad (4.60)$$

Equation (4.60) demonstrates that $R_{XX}(0) \ge R_{XX}(\tau)$, i.e., $R_{XX}(\tau)$ has its maximum value when $\tau = 0$.

We next prove another property of the autocorrelation function $R_{XX}(\tau)$, i.e., when the value of τ is equal to zero, the magnitude of the autocorrelation function, denoted as $R_{XX}(0)$, is equal to the variance of the random process with zero mean. This value also represents the temporal average of the power or energy of the random process. When we set τ equal to zero in equation (4.58), we get the following:

$$R_{XX}(0) = E\left[x(t)^2\right] = \lim_{T\to\infty}\frac{1}{2T}\int_{-T}^{T} x(t)^2\,dt \quad (4.61)$$

Obviously, equation (4.60) demonstrates that $R_{XX}(0)$ is equal to the variance of the random process, and this value also represents the temporal average of the energy of the random process.

Figure 4.3 depicts an example of the autocorrelation function computed from the time history of a measured wave elevation dataset (as shown in Figure 4.4) at a deep-sea site.

4.4.3 Spectral Density Function

In equation (4.61), it was demonstrated that the magnitude of the autocorrelation function for $\tau = 0$, which is denoted by $R_{XX}(0)$, is the time average of the power or energy of a random process $\{X(t)\}$. Let us take a closer look at this property in further depth.

Let us assume that the random process $\{X(t)\}$ is a wave profile in the ocean, then the right side of equation (4.61) can be considered the average wave energy,

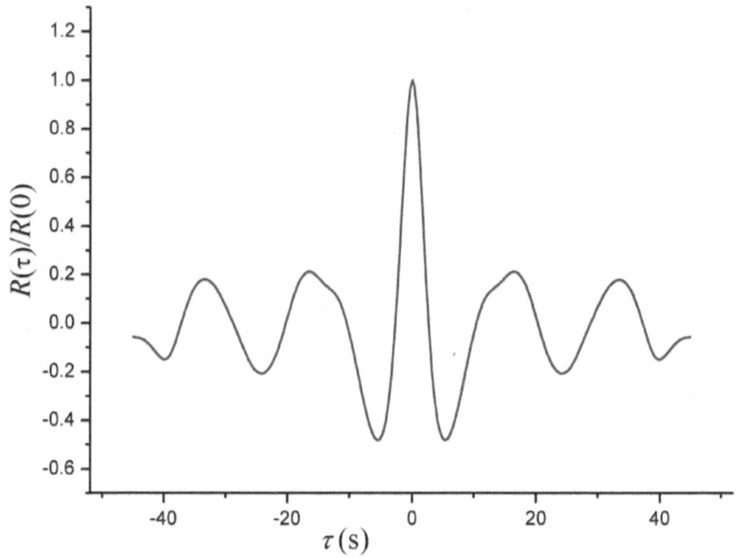

FIGURE 4.3 An example of the autocorrelation function computed from a measured dataset.

disregarding the factor ρg. Here, ρ denotes the mass density of water, and g is the gravitational constant. The temporal average of the power or energy of $\{X(t)\}$ is denoted by the symbol $\bar{P}_x$, which allows us to obtain the following:

$$\bar{P}_x = \lim_{T\to\infty} \frac{1}{2T} \int_{-T}^{T} \{x(t)\}^2\, dt \tag{4.62}$$

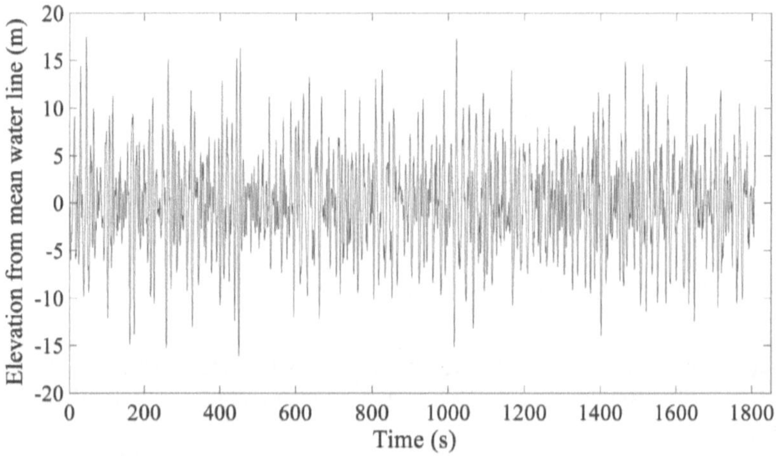

FIGURE 4.4 The time history of a measured wave elevation dataset at a deep-sea site.

We will now perform the Fourier transform in order to take into consideration the average energy of $\{X(t)\}$ in the frequency domain. In order to accomplish this, it is necessary to investigate the condition that must be met in order for the Fourier transform of $x(t)$ to exist. In general, the Fourier transform of a function $x(t)$ is considered to exist if the integration of $x(t)$ with respect to time over its whole domain, specifically $-\infty < t < \infty$, is finite. In order to strictly satisfy this criterion, it is important to take into consideration the following truncated function $x_T(t)$, which is defined as

$$x_T(t) = \begin{cases} x(t) & \text{for} \quad -T < t < T \\ 0 & \text{otherwise} \end{cases} \tag{4.63}$$

It is possible to express the average energy of the truncated function $x_T(t)$ that has been defined in equation (4.63) in the following manner:

$$\bar{P}_x = \lim_{T\to\infty} \frac{1}{2T} \int_{-T}^{T} \{x_T(t)\}^2 \, dt \tag{4.64}$$

By adopting the following Parseval theorem (whose derivation is given in Appendix A), let us proceed to express the average energy, denoted by $\bar{P}_x$, in terms of frequency

$$\int_{-\infty}^{\infty} \{x(t)\}^2 \, dt = \frac{1}{2\pi} \int_{-\infty}^{\infty} |X(\omega)|^2 \, d\omega$$

in which $X(\omega)$ is the Fourier transform of $x(t)$. As a result of the application of the Parseval theorem, the average energy that is represented by the equation (4.64) can be expressed as follows:

$$\bar{P}_x = \lim_{T\to\infty} \frac{1}{2T} \frac{1}{2\pi} \int_{-\infty}^{\infty} |X(\omega)|^2 \, d\omega = \lim_{T\to\infty} \frac{1}{4\pi T} \int_{-\infty}^{\infty} |X(\omega)|^2 \, d\omega \tag{4.65}$$

The definition of the ***spectral density function*** of a random process $\{X(t)\}$ is given by

$$S_{XX}(\omega) = \lim_{T\to\infty} \frac{1}{2\pi T} |X_T(\omega)|^2 \tag{4.66}$$

With the use of equations (4.65) and (4.66) and taking into account the fact that the spectral density function is an even function, the average energy of the random process $\{X(t)\}$ may be represented as

$$\bar{P}_x = \frac{1}{2} \int_{-\infty}^{\infty} S_{XX}(\omega) \, d\omega = \int_{0}^{\infty} S_{XX}(\omega) \, d\omega \tag{4.67}$$

Accordingly, the average energy of a random process with respect to time is equal to the area under the spectral density function that is defined in equation (4.66).

The spectral density function can also be defined as the Fourier transform of the autocorrelation function based on the Wiener–Khintchine theorem (whose derivation is given in Appendix B). For a weakly stationary ergodic random process $\{X(t)\}$, its autocorrelation function, $R_{XX}(\tau)$, and the spectral density function, $S_{XX}(\omega)$, are related by the Fourier transform. That is,

$$S_{XX}(\omega)=\frac{1}{\pi}\int_{-\infty}^{\infty} R_{XX}(\tau)e^{-i\omega\tau}\,d\tau \tag{4.68}$$

$$R_{XX}(\tau)=\frac{1}{2}\int_{-\infty}^{\infty} S_{XX}(\omega)e^{i\omega\tau}\,d\omega \tag{4.69}$$

where

$$S_{XX}(\omega)=\lim_{T\to\infty}\frac{1}{2\pi T}\left|X(\omega)\right|^2 \tag{4.70}$$

$$R_{XX}(\tau)=\lim_{T\to\infty}\frac{1}{2T}\int_{-T}^{T} x(t)x(t+\tau)\,dt \tag{4.71}$$

Equations (4.68) and (4.69) are called the Wiener–Khintchine theorem.

Starting from equations (4.68) and (4.69), we can also derive the following relationships:

$$\begin{aligned}R_{XX}(\tau)&=\frac{1}{2}\int_{-\infty}^{\infty} S_{XX}(\omega)e^{i\omega\tau}\,d\omega=\frac{1}{2}\int_{-\infty}^{\infty}\left[\frac{1}{\pi}\int_{-\infty}^{\infty} R_{XX}(\tau)e^{-i\omega\tau}\,d\tau\right]e^{i\omega\tau}d\omega\\&=\int_{-\infty}^{\infty}\left[\frac{1}{2\pi}\int_{-\infty}^{\infty} R_{XX}(\tau)e^{-i\omega\tau}\,d\tau\right]e^{i\omega\tau}\,d\omega=\int_{-\infty}^{\infty} S(\omega)e^{i\omega\tau}\,d\omega\end{aligned} \tag{4.72}$$

$$S(\omega)=\frac{1}{2\pi}\int_{-\infty}^{\infty} R_{XX}(\tau)e^{-i\omega\tau}\,d\tau \tag{4.73}$$

Equations (4.72) and (4.73) show that the power spectral density $S(\omega)$ of a stationary random process can also be defined as the Fourier transform of $R_{XX}(\tau)/2\pi$. In the field of random vibration analysis, the power spectral density is utilized substantially more frequently than the autocorrelation function.

When it comes to defining the spectral density function, the conventions that are utilized in the literature are not exactly standardized. In the field of Naval

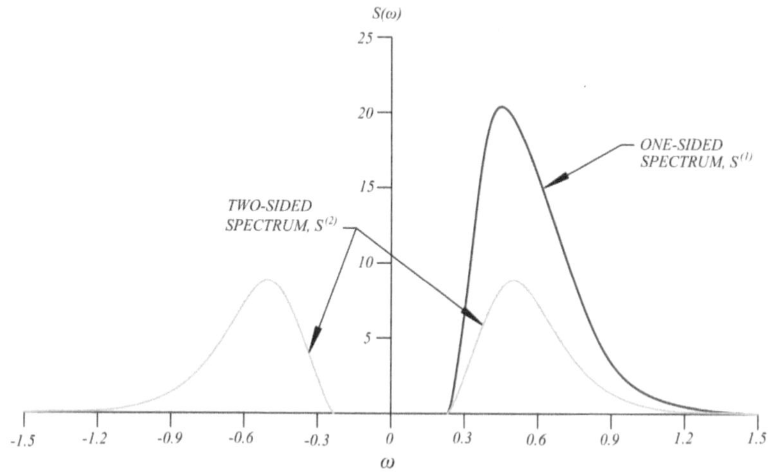

FIGURE 4.5 Definition of the one-sided and two-sided spectrum.

Architecture and Ocean Engineering, it is a common practice to depict the spectral density function as a one-sided spectrum, denoted by $S_{XX}{}^{(1)}(\omega)$ (or simply write it as $S^{(1)}(\omega)$), rather than a two-sided spectrum, as denoted by $S_{XX}(\omega)$ in equation (4.69). This is done in order to ensure that the spectral density is only present on the positive side of the frequency, ω. It is evident from Figure 4.5 that the result of this is fairly significant. The spectral density ordinates, in effect, have their value increased by a factor of two.

Figure 4.6 depicts an example of the one-sided spectral density function (spectrum) computed from the time history of the measured wave elevation dataset (as shown in Figure 4.4) at a deep-sea site.

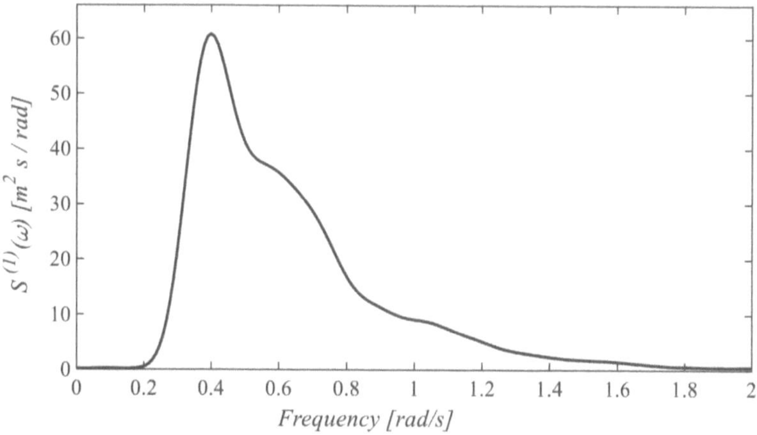

FIGURE 4.6 An example of the spectral density function computed from a measured dataset.

4.4.4 Wide-band and Narrow-band Random Processes

Although it is typically an inadequate description of the process, the spectral density function of a stationary random process does constitute a partial description of the process. Here, this description is analyzed for two extreme scenarios: a wide-band spectrum and a narrow-band spectrum. The terms "wide" and "narrow" are qualitative, and they are not clearly differentiated from one another. One type of stationary random process is known as a wide-band random process. This type of random process is characterized by having significant values for its spectral density function $S_{XX}(\omega)$ over a band or range of frequencies that is of roughly the same order of magnitude as the center frequency of the band. Within the typical sample functions of such a random process, a wide range of frequencies can be observed. Figure 4.7 shows an illustration of a wide-band spectral density function, and Figure 4.8 shows a possible sampling function corresponding to the spectrum in Figure 4.7. Random processes that are typically wide-band include the surface elevations of the irregular waves in a shallow water sea site.

Analytical investigations often simplify the spectrum of a wide-band random process by assuming a uniform spectral density S_C, as depicted in Figure 4.9. A random process that has such a spectrum is referred to as a white noise process. The ideal white noise process is expected to have a density that is uniform across all frequencies. This is a concept that cannot be realized in the physical

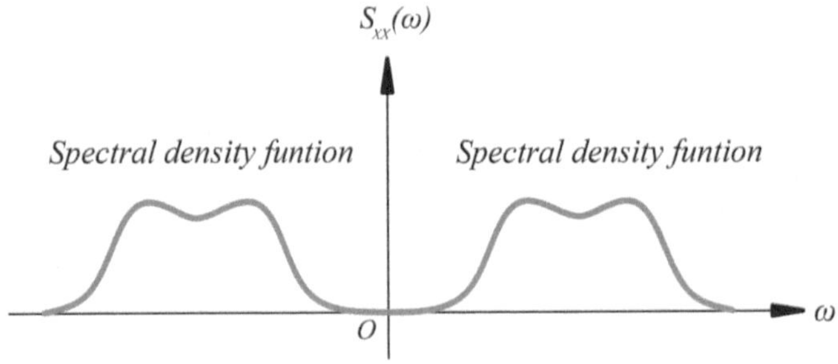

FIGURE 4.7 An example of the spectral density function of a wide band process.

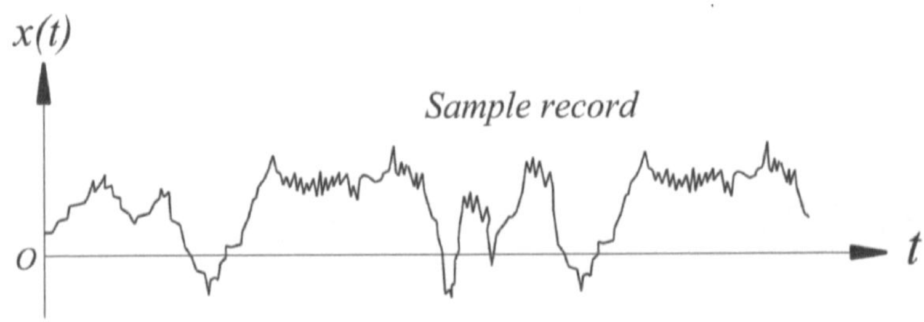

FIGURE 4.8 A possible sampling function corresponding to the spectrum in Figure 4.7.

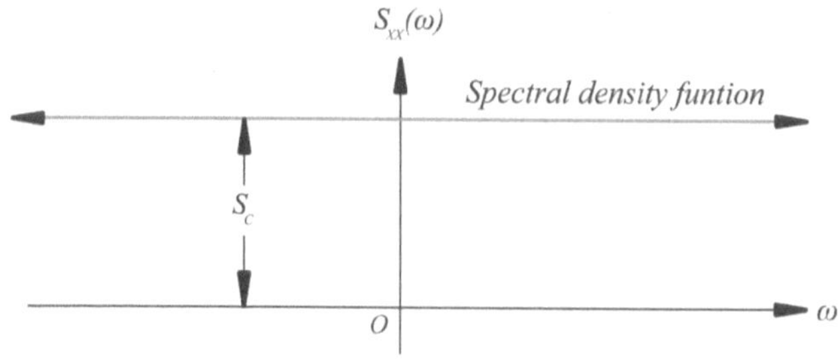

FIGURE 4.9 An example of the spectral density function of a white noise process.

world since the mean square value of such a process would be infinite because there is an infinite area under the spectrum. However, the ideal white noise model might occasionally be utilized to deliver results that are physically meaningful in a straightforward approach. For the purpose of ensuring that the outcomes of the simplified analysis do, in fact, represent a near approximation of what would be produced in the event of a physically conceivable excitation, a little bit of additional care is required.

Figure 4.10 depicts a band-limited white noise spectrum that is a close approximation to a large number of random processes that are physically realizable. In this particular scenario, the mean square value of the random process can be expressed as $2S_C(\omega_2 - \omega_1)$, which equals the whole area under the spectrum.

Another type of stationary random process is known as a narrow-band process. This type of random process is characterized by a spectral density function $S_{XX}(\omega)$ that only exhibits significant values within a frequency band whose width is small in comparison to the magnitude of the center frequency of the band. In representative samples of such a random process, only a limited frequency range is observed to be present. In Figure 4.11, we have shown an illustration of a narrow-band spectrum. In Figure 4.12, we show a possible sample record corresponding to the spectrum in Figure 4.11.

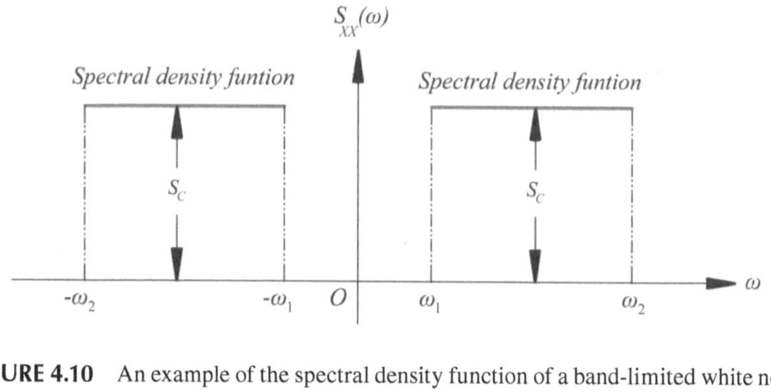

FIGURE 4.10 An example of the spectral density function of a band-limited white noise process.

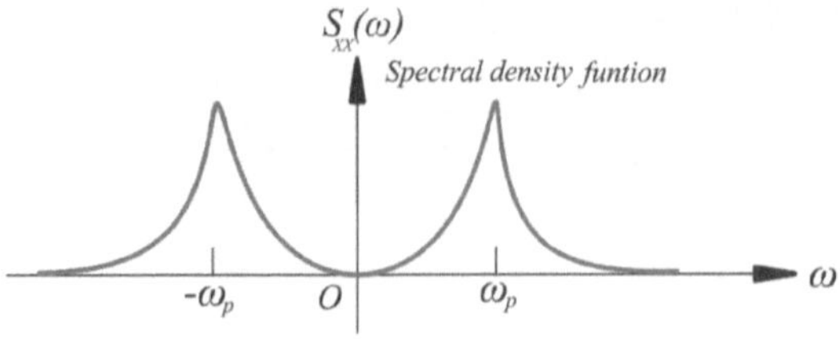

FIGURE 4.11 An example of the spectral density function of a narrow band process.

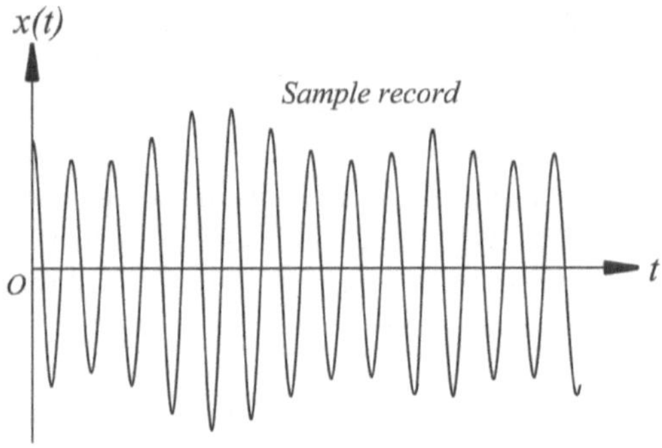

FIGURE 4.12 A possible sampling record corresponding to the spectrum in Figure 4.1.

4.5 SPECTRAL ANALYSIS OF TWO RANDOM PROCESSES

The techniques of spectrum analysis that were developed for a single random process are capable of being used for two random processes, namely $\{X(t)\}$ and $\{Y(t)\}$. As opposed to the autospectral analysis, which is used for a single random process, this type of analysis is referred to as cross-spectral analysis. By performing a cross-spectral analysis between two random processes, $\{X(t)\}$ and $\{Y(t)\}$, we are able to get a variety of extra information regarding the stochastic characteristics of the random processes. These include: (1) the correlation and phase relationship between the two random processes $\{X(t)\}$ and $\{Y(t)\}$, (2) the linearity of a system if the two random processes $\{X(t)\}$ and $\{Y(t)\}$, respectively represent the input and output of the system, (3) stochastic characteristics of the random process compound of $\{X(t)\}$ and $\{Y(t)\}$, and so on.

4.5.1 Cross-correlation Function

The *cross-correlation function*, denoted by $R_{XY}(t_1,t_2)$, calculates the ensemble average of the product of two random processes $\{X(t)\}$ and $\{Y(t)\}$ at times t_1 and t_2. That is,

$$R_{XY}(t_1,t_2) = E[X(t_1)Y(t_2)] = \frac{1}{n}\sum_{k=1}^{n}{}^{k} x(t_1)^k\, y(t_2) \tag{4.74}$$

in which E represents the expected value. Remember that in Chapter 3, we specified that for the random process $\{X(t),\ t \in T\}$, $X(t_1)$ is a *random variable* at a given time t_1, and that the set consisting of $\left\{{}^1X(t_1),\ {}^2X(t_1),\ldots,\ {}^kX(t_1),\ldots,\ {}^nX(t_1)\right\}$ can be viewed as a random sample with n elements. In equation (4.74), we have used a lowercase notation ${}^kx(t_1)$ to represent the specific element ${}^kX(t_1)$ in the set $\left\{{}^1X(t_1),\ {}^2X(t_1),\ldots,\ {}^kX(t_1),\ldots,\ {}^nX(t_1)\right\}$.

Analogously, for the random process $\{Y(t),\ t \in T\}$, $Y(t_2)$ is a *random variable* at a given time t_2, and that the set consisting of $\left\{{}^1Y(t_2),\ {}^2Y(t_2),\ldots,\ {}^kY(t_2),\ldots,\ {}^nY(t_2)\right\}$ can be viewed as a random sample with n elements. In equation (4.74), we have used a lowercase notation ${}^ky(t_2)$ to represent the specific element ${}^kY(t_2)$ in the set $\left\{{}^1Y(t_2),\ {}^2Y(t_2),\ldots,\ {}^kY(t_2),\ldots,\ {}^nY(t_2)\right\}$.

The cross-covariance function $C_{XY}(t_1,t_2)$ is calculated as follows:

$$\begin{aligned} C_{XY}(t_1,t_2) &= E\Big[\big(X(t_1)-E[X(t_1)]\big)\big(Y(t_2)-E[Y(t_2)]\big)\Big] \\ &= \frac{1}{n}\sum_{k=1}^{n}\left\{{}^kx(t_1)-\bar{x}(t_1)\right\}\left\{{}^ky(t_2)-\bar{y}(t_2)\right\} \end{aligned} \tag{4.75}$$

It can be deduced from equations (4.74) and (4.75) that the cross-correlation function is equal to the cross-covariance function in the event when the mean values of the two random processes $\{X(t)\}$ and $\{Y(t)\}$ are both equal to zero.

Once the weakly stationary random process that was developed in the prior chapter is taken into consideration, the cross-correlation function $R_{XY}(t_1,t_2)$ transforms into a function that is solely dependent on the time difference. By expressing t_2 as $t_2 = t_1 + \tau$, we can obtain that

$$R_{XY}(\tau) = E\left[{}^kx(t)\,{}^ky(t+\tau)\right] = \frac{1}{n}\sum_{k=1}^{n}{}^kx(t)\,{}^ky(t+\tau) \tag{4.76}$$

Furthermore, by making the assumption that the two random processes $\{X(t)\}$ and $\{Y(t)\}$ are both ergodic, it is possible to evaluate the ensemble average by calculating the time average based on a single sample record of $\{X(t)\}$ and a single sample record of $\{Y(t)\}$, that is

$$R_{XY}(\tau) = E\left[x(t)\,y(t+\tau)\right] = \lim_{T\to\infty}\frac{1}{2T}\int_{-T}^{T}x(t)y(t+\tau)\,dt \tag{4.77}$$

It is necessary to shift the record $y(t)$ backward by time τ in order to evaluate the cross-correlation function $R_{XY}(\tau)$, which is described in equation (4.77). This is illustrated in Figure 4.13. In the event when $y(t)$ is shifted forward by time τ, the cross-correlation function is represented by the symbol $R_{YX}(\tau)$. In the

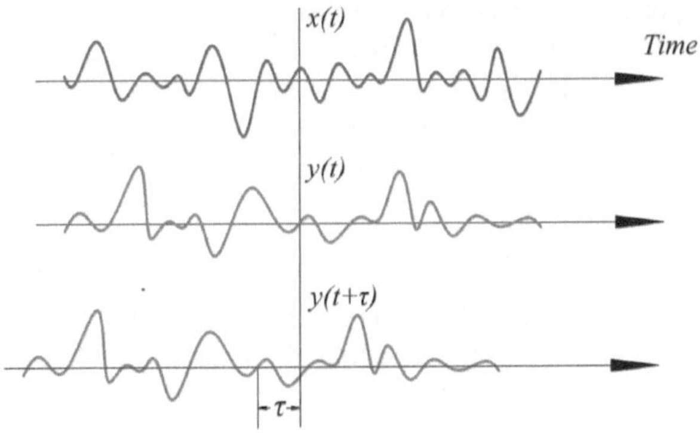

FIGURE 4.13 Definition of the cross-correlation function $R_{XY}(\tau) = E\left[x(t)\, y(t+\tau)\right]$.

following, some properties of the cross-correlation function $R_{XY}(\tau)$ are derived and summarized.

In accordance with the definition, we have

$$R_{XY}(-\tau) = \lim_{T\to\infty} \frac{1}{2T} \int_{-T}^{T} x(t)y(t-\tau)\,dt = \lim_{T\to\infty} \frac{1}{2T} \int_{-T}^{T} x(u+\tau)y(u)\,du = R_{YX}(\tau) \tag{4.78}$$

Equation (4.78) demonstrates that one property of the cross-correlation function $R_{XY}(\tau)$ is that $R_{YX}(\tau) = R_{XY}(-\tau)$.

Another property of the cross-correlation function $R_{XY}(\tau)$ is that it does not always reach its maximum value when $\tau = 0$. Furthermore, in contrast to $R_{XX}(0)$, which would be equal to the variance of the random process $\{X(t)\}$ if its mean value was zero, $R_{XY}(0)$ does not have any significant meaning.

4.5.2 Cross-Spectral Density Function

In a manner that is comparable to the definition of the autospectral density function for a single random process $\{X(t)\}$, the cross-spectral density function of two random processes $\{X(t)\}$ and $\{Y(t)\}$ is defined as follows: The frequency function

$$S_{XY}(\omega) = \lim_{T\to\infty} \frac{1}{2\pi T} X^*(\omega)Y(\omega) \tag{4.79}$$

is called the cross-spectral density function of the random processes $\{X(t)\}$ and $\{Y(t)\}$, where $X^*(\omega)$ is the conjugate function of $X(\omega)$. $X(\omega)$ is the Fourier transform of $x(t)$, and $Y(\omega)$ is the Fourier transform of $y(t)$.

It is also possible to apply the Wiener–Khintchine theorem to the cross-correlation function and the cross-spectral density function. In the case of weakly stationary ergodic random processes $\{X(t)\}$ and $\{Y(t)\}$, the cross-correlation function $R_{XY}(\tau)$ and the cross-spectral density function $S_{XY}(\omega)$ constitute a pair of Fourier transforms. That is to say,

$$S_{XY}(\omega)=\frac{1}{\pi}\int_{-\infty}^{\infty}R_{XY}(\tau)e^{-i\omega\tau}\,d\tau$$
$$R_{XY}(\tau)=\frac{1}{2}\int_{-\infty}^{\infty}S_{XY}(\omega)e^{i\omega\tau}\,d\omega \tag{4.80}$$

The relationships between the cross-correlation function $R_{XY}(\tau)$ and the cross-spectral density function $S_{XY}(\omega)$ in terms of a Fourier transform pair can also be expressed as follows:

$$S_{XY}(\omega)=\frac{1}{2\pi}\int_{-\infty}^{\infty}R_{XY}(\tau)e^{-i\omega\tau}\,d\tau$$
$$R_{XY}(\tau)=\int_{-\infty}^{\infty}S_{XY}(\omega)e^{i\omega\tau}\,d\omega \tag{4.81}$$

There is a difference between the auto-spectral density function, which is a real-valued function, and the cross-spectral density function, which is a complex function. It is important to understand the difference between the two. For the purpose of elucidating the manner in which the imaginary part of the function is introduced into the cross-spectral density function, let us conduct a detailed evaluation of $S_{XY}(\omega)$ by following the Wiener–Khintchine theorem. That is,

$$\begin{aligned}
S_{XY}(\omega)&=\frac{1}{\pi}\int_{-\infty}^{\infty}R_{XY}(\tau)e^{-i\omega\tau}\,d\tau=\frac{1}{\pi}\left\{\int_{-\infty}^{0}R_{XY}(\tau)e^{-i\omega\tau}\,d\tau+\int_{0}^{\infty}R_{XY}(\tau)e^{-i\omega\tau}\,d\tau\right\}\\
&=\frac{1}{\pi}\left\{\int_{-\infty}^{\infty}R_{YX}(-\tau)e^{i\omega(-\tau)}\,d\tau+\int_{0}^{\infty}R_{XY}(\tau)e^{-i\omega\tau}\,d\tau\right\}\\
&=\frac{1}{\pi}\left\{\int_{0}^{\infty}R_{YX}(-\tau)e^{i\omega(-\tau)}d(-\tau)+\int_{0}^{\infty}R_{XY}(\tau)e^{-i\omega\tau}\,d\tau\right\}\\
&=\frac{1}{\pi}\left\{\int_{0}^{\infty}R_{YX}(\tau)e^{i\omega(\tau)}d(\tau)+\int_{0}^{\infty}R_{XY}(\tau)e^{-i\omega\tau}\,d\tau\right\}\\
&=\frac{1}{\pi}\int_{0}^{\infty}\{R_{XY}(\tau)+R_{YX}(\tau)\}\cos\omega\tau\,d\tau+i\frac{1}{\pi}\\
&\quad+\int_{0}^{\infty}\{-R_{XY}(\tau)+R_{YX}(\tau)\}\sin\omega\tau\,d\tau=G_{XY}(\omega)+iP_{XY}(\omega)
\end{aligned} \tag{4.82}$$

in which

$$G_{XY}(\omega) = \frac{1}{\pi}\int_0^\infty \{R_{XY}(\tau) + R_{YX}(\tau)\}\cos\omega\tau\, d\tau \tag{4.83}$$

$$P_{XY}(\omega) = \frac{1}{\pi}\int_0^\infty \{-R_{XY}(\tau) + R_{YX}(\tau)\}\sin\omega\tau\, d\tau \tag{4.84}$$

When examining equation (4.82), it is evident that the cross-spectral density function contains an imaginary component. This is due to the fact that the cross-correlation function $R_{XY}(\tau)$ is not equal to the cross-correlation function $R_{YX}(\tau)$. The real part $G_{XY}(\omega)$ in equation (4.82) is known as the co-spectrum and the imaginary part $P_{XY}(\omega)$ in equation (4.82) is known as the quadrature spectrum.

If we consider the co-spectrum $G_{XY}(\omega)$, we find that it is an even function, whereas the quadrature spectrum $P_{XY}(\omega)$ is an odd function. Below is a summary of the various properties that are associated with the co-spectrum $G_{XY}(\omega)$ and the quadrature spectrum $P_{XY}(\omega)$.

$$\begin{aligned} G_{XY}(-\omega) &= \frac{1}{\pi}\int_0^\infty \{R_{XY}(\tau) + R_{YX}(\tau)\}\cos(-\omega)\tau\, d\tau \\ &= \frac{1}{\pi}\int_0^\infty \{R_{XY}(\tau) + R_{YX}(\tau)\}\cos(\omega)\tau\, d\tau = G_{XY}(\omega) \end{aligned} \tag{4.85}$$

$$\begin{aligned} G_{YX}(\omega) &= \frac{1}{\pi}\int_0^\infty \{R_{YX}(\tau) + R_{XY}(\tau)\}\cos\omega\tau\, d\tau \\ &= \frac{1}{\pi}\int_0^\infty \{R_{XY}(\tau) + R_{YX}(\tau)\}\cos\omega\tau\, d\tau = G_{XY}(\omega) \end{aligned} \tag{4.86}$$

$$G_{YX}(-\omega) = G_{XY}(-\omega) = G_{XY}(\omega) \tag{4.87}$$

$$\begin{aligned} P_{XY}(-\omega) &= \frac{1}{\pi}\int_0^\infty \{-R_{XY}(\tau) + R_{YX}(\tau)\}\sin(-\omega\tau)\, d\tau \\ &= -\frac{1}{\pi}\int_0^\infty \{-R_{XY}(\tau) + R_{YX}(\tau)\}\sin(\omega\tau)\, d\tau = -P_{XY}(\omega) \end{aligned} \tag{4.88}$$

$$P_{YX}(\omega) = \frac{1}{\pi}\int_0^\infty \{-R_{YX}(\tau) + R_{XY}(\tau)\}\sin\omega\tau\,d\tau$$
$$= -\frac{1}{\pi}\int_0^\infty \{R_{YX}(\tau) - R_{XY}(\tau)\}\sin\omega\tau\,d\tau = -P_{XY}(\omega) \tag{4.89}$$

$$P_{YX}(-\omega) = -P_{XY}(-\omega) = P_{XY}(\omega) \tag{4.90}$$

We are able to deduce the following relationships by making use of the properties that are presented in the equations that were presented above:

$$S_{XY}(-\omega) = G_{XY}(-\omega) + iP_{XY}(-\omega) = G_{XY}(\omega) - iP_{XY}(\omega) = S_{XY}*(\omega) \tag{4.91}$$

$$S_{YX}(\omega) = G_{YX}(\omega) + iP_{YX}(\omega) = G_{XY}(\omega) - iP_{XY}(\omega) = S_{XY}*(\omega) \tag{4.92}$$

in which $S_{XY}*(\omega)$ is the complex conjugate of $S_{XY}(\omega)$.

We are able to define an important parameter known as the coherency function by using the information that we have on the autospectral density function and cross-spectral density function of two random processes. The coherency function is defined as the following parameter, which conveys the degree of correlation that exists between two random processes:

$$\gamma_{XY}^2(\omega) = \frac{|S_{XY}(\omega)|^2}{S_{XX}(\omega)S_{YY}(\omega)} = \frac{S_{XY}(\omega)S_{XY}^*(\omega)}{S_{XX}(\omega)S_{YY}(\omega)} = \frac{\{G_{XY}(\omega)\}^2 + \{P_{XY}(\omega)\}^2}{S_{XX}(\omega)S_{YY}(\omega)} \tag{4.93}$$

Let's say that $x(t)$ and $y(t)$ are the input and output of a random vibration system, then the coherency function plays a vital role in clarifying the linearity of the system.

In a manner analogous to that of the correlation coefficient, the coherence function offers a non-dimensional measurement of the linear dependency that exists between $x(t)$ and $y(t)$ at each frequency. The coherence function can be shown to have values that range from 0 to 1:

$$0 \le \gamma_{XY}^2(\omega) \le 1 \tag{4.94}$$

At a frequency ω, values that are close to 1 indicate that there is a linear relationship between $x(t)$ and $y(t)$. In the event where $x(t)$ and $y(t)$ are uncorrelated, their cross-correlation, cross-spectral density and coherence functions will be equal to zero.

4.5.3 Autocorrelation and Spectral Density Functions of Derived Random Processes

The random process $\{X(t)\}$ is assumed to be weakly stationary and ergodic. Let $x(t)$ be a sample record of the random process $\{X(t)\}$. We will evaluate the autocorrelation function and the spectral density function of the velocity and acceleration by using the displacement autocorrelation function $R_{XX}(\tau)$ and the spectral density function $S_X(\omega)$. This will be done on the assumption that the time derivatives of the random process, such as the velocity $\dot{x}(t)$ and acceleration $\ddot{x}(t)$, exist.

In the beginning, we obtain the auto-correlation function. For this purpose, it could be more convenient to take into consideration the cross-correlation function of $x(t_1)$ and $x(t_2)$. That is,

$$\begin{aligned} R_{X\dot{X}}(t_1, t_2) &= E\left[x(t_1)\dot{x}(t_2)\right] \\ &= E\left[x(t_1)\frac{x(t_2+\Delta t)-x(t_2)}{\Delta t}\right] = \frac{1}{\Delta t}\left\{R_{XX}(t_1,t_2+\Delta t)-R_{XX}(t_1,t_2)\right\} \end{aligned} \tag{4.95}$$

For $\Delta t \to 0$, and for a weakly stationary and ergodic random process for which $t_1 - t_2 = \tau$, we have

$$R_{X\dot{X}}(\tau) = \frac{\partial}{\partial t_2} R_{XX}(t_1, t_2) = -\frac{d}{d\tau} R_{XX}(\tau) \tag{4.96}$$

In a similar manner, we can obtain

$$\begin{aligned} R_{\dot{X}\dot{X}}(t_1, t_2) &= E\left[\dot{x}(t_1)\dot{x}(t_2)\right] \\ &= E\left[\frac{x(t_1+\Delta t)-x(t_1)}{\Delta t}\,\frac{x(t_2+\Delta t)-x(t_2)}{\Delta t}\right] \\ &= E\left[\frac{x(t_1+\Delta t)x(t_2+\Delta t)-x(t_2+\Delta t)x(t_1)-x(t_1+\Delta t)x(t_2)+x(t_1)x(t_2)}{\Delta t\Delta t}\right] \end{aligned} \tag{4.97}$$

For $\Delta t \to 0$, and for a weakly stationary and ergodic random process for which $t_1 - t_2 = \tau$, we have

$$R_{\dot{X}\dot{X}}(\tau) = \frac{\partial^2}{\partial t_1\,\partial t_2} R_{XX}(t_1,t_2) = -\frac{d^2}{d\tau^2} R_{XX}(\tau) \tag{4.98}$$

In a similar manner, we can obtain

$$R_{\ddot{X}\ddot{X}}(\tau) = -\frac{d^4}{d\tau^4} R_{XX}(\tau) \tag{4.99}$$

For the purpose of acquiring the spectral density functions of $\dot{x}(t)$ and $\ddot{x}(t)$ from the displacement spectral density function $S_{XX}(\omega)$, we can utilize the properties that pertain to the Fourier transform of the time derivative of a function; these properties are provided below. That is, if we expressed the Fourier transform of a random process $x(t)$ as

$$\mathcal{F}\{x(t)\} = X(\omega)$$

then, we have

$$\mathcal{F}\{\dot{x}(t)\} = \dot{X}(\omega) = i\omega X(\omega)$$

as well as

$$\mathcal{F}\{\ddot{x}(t)\} = \ddot{X}(\omega) = -\omega^2 X(\omega)$$

As a result, as per the definition, the spectral density functions of velocity, denoted as $S_{\dot{X}\dot{X}}(\omega)$, and acceleration, denoted as $S_{\ddot{X}\ddot{X}}(\omega)$, can be expressed in the following manner,

$$S_{\dot{X}\dot{X}}(\omega) = \lim_{T\to\infty} \frac{1}{2\pi T}\left|\dot{X}(\omega)\right|^2 = \omega^2 \lim_{T\to\infty} \frac{1}{2\pi T}\left|X(\omega)\right|^2 = \omega^2 S_{XX}(\omega) \tag{4.100}$$

$$S_{\ddot{X}\ddot{X}}(\omega) = \lim_{T\to\infty} \frac{1}{2\pi T}\left|\ddot{X}(\omega)\right|^2 = \omega^4 \lim_{T\to\infty} \frac{1}{2\pi T}\left|X(\omega)\right|^2 = \omega^4 S_{XX}(\omega) \tag{4.101}$$

By examining equations (4.100) and (4.101), it becomes evident that the spectral density functions of velocity and acceleration may be evaluated by utilizing the displacement spectral density function $S_{XX}(\omega)$ without requiring any prior knowledge of the temporal history of velocity and acceleration. Conversely, the displacement spectral density function can be calculated based on the velocity or acceleration spectral density function. As a practical example, the displacement spectrum of ocean waves is often acquired by measuring the wave acceleration with a buoy that is floating in the seawater.

4.6 EXERCISES

EXERCISE 4.1

Suppose that $f(t), g(t)$ and $h(t)$ are functions with Fourier transforms $F(\omega)$, $G(\omega)$ and $H(\omega)$, respectively, show that

$$\mathcal{F}\left[c_1 f(t) + c_2 g(t) + c_3 h(t)\right] = c_1 F(\omega) + c_2 G(\omega) + c_3 H(\omega), \tag{E.4.1}$$

in which c_1, c_2 and c_3 are (real or complex) constants.

EXERCISE 4.2

Suppose that $f(t)$ is a function with a Fourier transform $F(\omega)$, show that

$$\mathcal{F}\left[f(t-\tau)\right]=e^{-i\omega\tau}F(\omega) \tag{E.4.2}$$

EXERCISE 4.3

Suppose that $f(t)$ is a function with a Fourier transform $F(\omega)$ and c is a real, non-zero constant, show that

$$\mathcal{F}\left[f(ct)\right]=F(\omega/c)/|c|. \tag{E.4.3}$$

EXERCISE 4.4

Suppose that the function $f(t)$ has the Fourier transform $F(\omega)$, show that

$$\mathcal{F}\left[F(t)\right]=2\pi f(-\omega). \tag{E.4.4}$$

EXERCISE 4.5

Suppose that the function $f(t)$ has the Fourier transform $F(\omega)$, and $\omega=\omega_r+i\omega_i$, then we have

$$\begin{aligned}\mathcal{F}[f'(t)]&=\int_{-\infty}^{\infty}f'(t)e^{-i\omega t}\,dt=\int_{-\infty}^{\infty}f'(t)e^{-i(\omega_r+i\omega_i)t}\,dt\\&=\int_{-\infty}^{\infty}f'(t)e^{\omega_i t}e^{-i(\omega_r)t}\,dt=\int_{-\infty}^{\infty}f'(t)e^{\omega_i t}\left[\cos(\omega_r t)-i\sin(\omega_r t)\right]dt\end{aligned} \tag{E.4.5}$$

Based on the relationship in equation (E.4.5), show that

$$\mathcal{F}[f'(t)]=i\omega F(\omega) \tag{E.4.6}$$

$$\mathcal{F}\left[f^{(n)}(t)\right]=(i\omega)^n F(\omega) \tag{E.4.7}$$

EXERCISE 4.6

Amplitude modulation, sometimes known as AM, is a common technique used in the field of communications for the purpose of transmitting information. In this process, the signal is carried according to the expression $f(t)e^{i\omega_0 t}$, in which ω_0 represents the carrier frequency and $f(t)$ is an arbitrary function of time. The amplitude spectrum of $f(t)$ reaches its peak at a frequency that is usually low in comparison to ω_0. Suppose that the function $f(t)$ has the Fourier transform $F(\omega)$, show that

$$\mathcal{F}\left[f(t)e^{i\omega_0 t}\right] = F(\omega - \omega_0) \tag{E.4.8}$$

EXERCISE 4.7

Suppose that the function $f(t)$ has the Fourier transform $F(\omega)$, show that

$$F(-\omega) = \text{complex conjugate of } F(\omega) = F^*(\omega) \tag{E.4.9}$$

EXERCISE 4.8

Utilizing a different approach, show that the Fourier transform of the derivative of $f(t)$ is equal to the product of the Fourier transform of $f(t)$ times $i\omega$. Hint: from the definition of a derivative as a limit, we have

$$\begin{aligned}\mathcal{F}\left\{\frac{d}{dt}f(t)\right\} &= \frac{1}{2\pi}\int_{-\infty}^{\infty}\frac{d}{dt}f(t)e^{-i\omega t}\,dt = \frac{1}{2\pi}\int_{-\infty}^{\infty}\lim_{h\to 0}\frac{f(t+h)-f(t)}{h}e^{-i\omega t}\,dt \\ &= \lim_{h\to 0}\frac{1}{h}\left\{\frac{1}{2\pi}\int_{-\infty}^{\infty}f(t+h)e^{-i\omega t}\,dt - \frac{1}{2\pi}\int_{-\infty}^{\infty}f(t)e^{-i\omega t}\,dt\right\} \\ &= \lim_{h\to 0}\frac{1}{h}\left\{F(\omega)e^{i\omega h} - F(\omega)\right\}\end{aligned} \tag{E.4.10}$$

show that

$$\mathcal{F}\left\{\frac{d}{dt}f(t)\right\} = i\omega F(\omega) \tag{E.4.11}$$

EXERCISE 4.9

An ergodic stationary random process $x(t)$ is given by $x(t) = A\cos(\omega t + \varepsilon)$, and A and ω are constants, and ε is a random variable. The ACF of $x(t)$ becomes

$$\begin{aligned}R_{xx}(\tau) &= \lim_{T\to\infty}\frac{1}{2T}\int_{-T}^{T}A^2\cos(\omega t+\varepsilon)\cos\left(\omega(t+\tau)+\varepsilon\right)dt \\ &= \lim_{T\to\infty}\frac{1}{2T}\int_{-T}^{T}\frac{A^2}{2}\left\{\cos\omega\tau + \cos(2\omega t + \omega\tau + 2\varepsilon)\right\}dt\end{aligned} \tag{E.4.12}$$

show that

$$R_{xx}(\tau) = \frac{A^2}{2}\cos\omega\tau \tag{E.4.13}$$

EXERCISE 4.10

The ACF of a specific random process is provided by

$$R(\tau)=\begin{cases}1-|\tau|/T & \text{for } |\tau|<T \\ 0 & \text{otherwise}\end{cases} \tag{E.4.14}$$

Please calculate the spectral density function $S(\omega)$ corresponding to the above ACF.

5 Random Vibration of Single-Degree-of-Freedom Linear Systems

After going over the principles of a theory of random variables and random processes, we will now move on to the application of this theory to problems involving random vibrations. Beginning with the current chapter, which is devoted to systems with a single degree of freedom (SDOF), the issues that are associated with linear systems will be discussed in three individual chapters that will follow one another. In order to get an understanding of the fundamental physical relationships of linear systems, it is reasonable to begin by limiting oneself to the most basic configuration and delaying the analysis of the aspects of more complex systems until the subsequent two chapters. The significance of having a complete comprehension of this simplest case cannot be overemphasized, as it is typically possible to find analogies in it to the results that are expected from cases that are more complicated.

5.1 A REVIEW OF THE DETERMINISTIC THEORY

On the assumption that the reader is already familiar with the deterministic theory of engineering vibrations, this section will proceed. Nevertheless, in order to make the subsequent discussion of the probabilistic theory more manageable, it is necessary to quickly go over a few fundamental principles that are associated with the deterministic theory of vibrations.

5.1.1 Free Vibration of Undamped Single-Degree-of-Freedom Systems

Let's have a look at the spring-mass system with a single degree of freedom represented in Figure 5.1. In the beginning, the spring is in a position where it is not extended out. We make the presumption that Hooke's law is followed by the spring. The proportionality constant, also known as the spring constant, is equal to k. This means that the force in the spring is proportional to the displacement. The spring will deflect to a point of static equilibrium, denoted by the symbol $l_0 + \delta_{st}$, when the mass, whose weight W is equal to mg, is applied. δ_{st} is the static deflection—the elongation due to the weight W of the mass m.

From Figure 5.1, we can find that, for the static equilibrium condition,

$$W = mg = k\delta_{st} \tag{5.1}$$

DOI: 10.1201/9781003431503-5

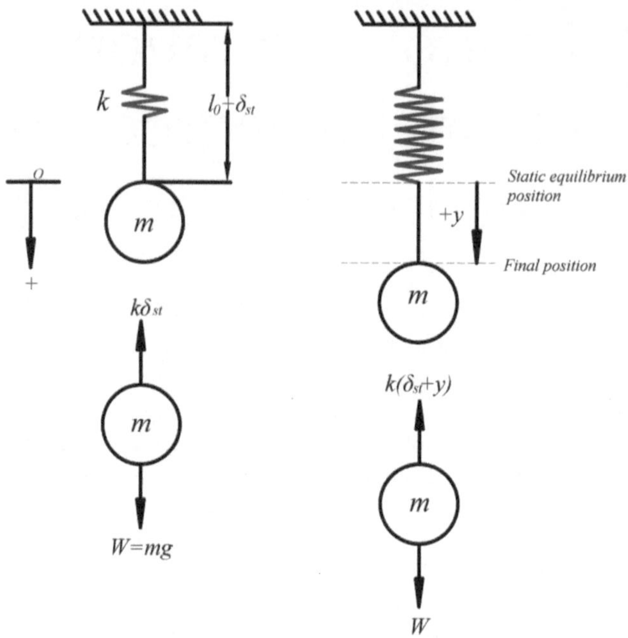

FIGURE 5.1 Undamped single-degree-of-freedom system.

in which g is the gravitational acceleration. Let the mass be deflected a distance $+y$ from its static equilibrium position; then the spring force is equal to $-k\left(y+\delta_{\text{st}}\right)$, as shown in Figure 5.1. Applying Newton's second law of motion to the mass m leads to

$$m\ddot{y}=-k\left(y+\delta_{\text{st}}\right)+W \tag{5.2}$$

and because $k\delta_{\text{st}}=W$, we have

$$m\ddot{y}+ky=0 \tag{5.3}$$

By making the assumption that $y(t)=Ce^{st}$, where C and s are constants that need to be calculated, it is possible to find the solution to equation (5.3). Substituting $y(t)=Ce^{st}$ into equation (5.3) results in:

$$C\left(ms^2+k\right)=0 \tag{5.4}$$

Because C cannot be zero, we obtain

$$ms^2+k=0 \tag{5.5}$$

and therefore,

$$s = \pm(-k/m)^{1/2} = \pm i\omega_n = \pm i(k/m)^{1/2} \tag{5.6}$$

Here in equation (5.6), $\omega_n = \sqrt{k/m}$ defines the natural frequency of the mass. Equation (5.5) is referred to as the auxiliary equation or the characteristic equation that corresponds to the differential equation (5.3). The two solution values of s provided by equation (5.6) are the roots of the characteristic equation, sometimes referred to as the eigenvalues or characteristic values of the problem. Due to the fact that both of the values of s satisfy equation (5.5), the general solution of equation (5.3) can be represented as

$$y(t) = C_1 e^{i\omega_n t} + C_2 e^{-i\omega_n t} \tag{5.7}$$

in which C_1 and C_2 are constants. We write equation (5.7) in another form as follows:

$$\begin{aligned} y(t) &= C_1 e^{i\omega_n t} + C_2 e^{-i\omega_n t} = C_1(\cos\omega_n t + i\sin\omega_n t) + C_2(\cos\omega_n t - i\sin\omega_n t) \\ &= (C_1 + C_2)\cos\omega_n t + i(C_1 - C_2)\sin\omega_n t = A\cos\omega_n t + B\sin\omega_n t \end{aligned} \tag{5.8}$$

in which A and B are new constants that have been introduced. The values of the constants C_1, C_2, A and B can be determined from the initial conditions of the vibration system. It is necessary to specify two conditions in order to evaluate these constants in a unique manner. It is important to keep in mind that the order of the differential equation that governs the system is the same as the number of conditions that need to be given. In the present case, if the values of displacement $y(t)$ and velocity $\dot{y}(t) = (dy/dt)(t)$ are specified as y_0 and $\dot{y}_0$ at $t = 0$, then we can deduce, based on equation (5.8), that

$$\begin{aligned} y(t=0) &= A = y_0 \\ \dot{y}(t=0) &= \omega_n B = \dot{y}_0 \end{aligned} \tag{5.9}$$

Therefore, subject to the initial conditions of equation (5.9), the solution to equation (5.3) is written as follows:

$$y(t) = y_0 \cos\omega_n t + \frac{\dot{y}_0}{\omega_n}\sin\omega_n t \tag{5.10}$$

5.1.2 Free Vibration of Damped Single-Degree-of-Freedom Systems

Let us take into consideration a spring-mass system that is connected to a mechanism that dissipates energy, which is characterized by the viscous damping force, as depicted in Figure 5.2.

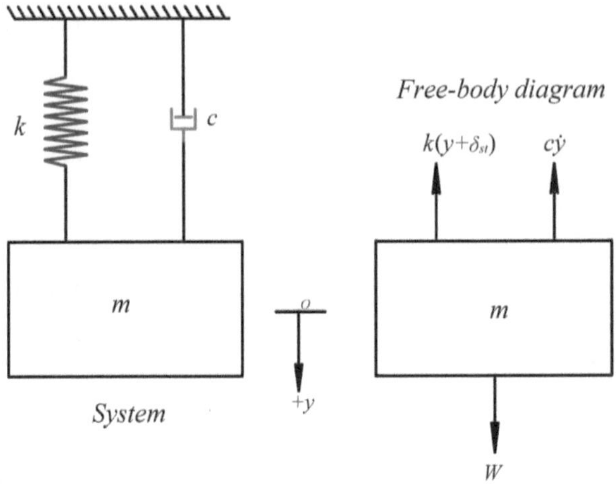

FIGURE 5.2 A damped single-degree-of-freedom system.

According to the information presented in Appendix C, the viscous damping force F is proportional to the velocity $\dot{y}$ or v and can be expressed as $F = -c\dot{y}$, in which c represents the damping constant or coefficient of viscous damping. The presence of a negative sign shows that the damping force is applied in the opposite direction as the velocity. The example of a system with a viscous damper and a single degree of freedom is depicted in Figure 5.2. Applying Newton's second law to this system in which y is measured from the position of static equilibrium of the mass m results in the equation of motion, which appears as follows:

$$m\ddot{y} = -k\left(y + \delta_{\text{st}}\right) - c\dot{y} + W \tag{5.11}$$

and because $k\delta_{\text{st}} = W$, we have

$$m\ddot{y} + c\dot{y} + ky = 0 \tag{5.12}$$

By assuming that the solution to the above homogeneous linear differential equation is $y(t) = Ce^{st}$, one can arrive to the characteristic equation

$$ms^2 + cs + k = 0 \tag{5.13}$$

the roots of which are

$$s_{1,2} = -\frac{-c \pm \sqrt{c^2 - 4mk}}{2m} = -\frac{c}{2m} \pm \sqrt{\left(\frac{c}{2m}\right)^2 - \frac{k}{m}} \tag{5.14}$$

These roots provide two solutions to equation (5.12), which are as follows:

$$y_1(t) = C_1 e^{s_1 t} \quad \text{and} \quad y_2(t) = C_2 e^{s_2 t} \tag{5.15}$$

The general solution of equation (5.12), therefore, can be obtained by combining the two solutions, $y_1(t)$ and $y(t)$, as follows:

$$y(t) = C_1 e^{s_1 t} + C_2 e^{s_2 t} = C_1 e^{\left\{-\frac{c}{2m}+\sqrt{\left(\frac{c}{2m}\right)^2-\frac{k}{m}}\right\}t} + C_2 e^{\left\{-\frac{c}{2m}-\sqrt{\left(\frac{c}{2m}\right)^2-\frac{k}{m}}\right\}t} \tag{5.16}$$

in which the values of the arbitrary constants C_1 and C_2 are to be determined based on the initial conditions of the system.

Define the *critical damping coefficient* c_c as the value of c that results in the radical in equation (5.14) being equal to zero. Therefore, we have

$$\left(\frac{c_c}{2m}\right)^2 - \frac{k}{m} = 0 \tag{5.17}$$

Solving equation (5.14) leads to

$$c_c = 2m\sqrt{\frac{k}{m}} = 2\sqrt{km} = 2m\omega_n \tag{5.18}$$

The damping ratio ζ for a damped system is defined as the ratio of the damping constant to the critical damping constant, that is, $\zeta = c/c_c$. Therefore, based on equation (5.17), we have

$$\frac{c}{2m} = \frac{c}{c_c} \cdot \frac{c_c}{2m} = \zeta\omega_n \tag{5.19}$$

Therefore, equation (5.14) can be rewritten as

$$s_{1,2} = -\frac{c}{2m} \pm \sqrt{\left(\frac{c}{2m}\right)^2 - \frac{k}{m}} = -\zeta\omega_n \pm \sqrt{(\zeta\omega_n)^2 - (\omega_n)^2} = \left(-\zeta \pm \sqrt{\zeta^2 - 1}\right)\omega_n \tag{5.20}$$

The solution, represented by equation (5.16), can thus be expressed as

$$y(t) = C_1 e^{s_1 t} + C_2 e^{s_2 t} = C_1 e^{\left(-\zeta+\sqrt{\zeta^2-1}\right)\omega_n t} + C_2 e^{\left(-\zeta-\sqrt{\zeta^2-1}\right)\omega_n t} \tag{5.21}$$

The magnitude of damping is a factor that determines the nature of the roots s_1 and s_2, and consequently, the behavior of the solution, which is represented by the equation (5.21). The undamped vibrations that were explained in Section 5.1.1

can be observed to be the result of the scenario where ζ equals zero. As a result, we make the assumption that ζ is not equal to zero and take into consideration the three potential scenarios that are presented below.

Scenario 1. Light damping: ($0<\zeta<1$, or $0<c<c_c$, or $c/2m<\sqrt{k/m}$). For this scenario, ζ^2-1 is negative. The roots s_1 and s_2 are complex and can be rewritten as

$$\begin{aligned} s_1 &= \left(-\zeta + i\sqrt{1-\zeta^2}\right)\omega_n \\ s_2 &= \left(-\zeta - i\sqrt{1-\zeta^2}\right)\omega_n \end{aligned} \tag{5.22}$$

Moreover, the solution, given by the equation (5.21), can be expressed in different forms:

$$y(t) = C_1 e^{\left(-\zeta + i\sqrt{1-\zeta^2}\right)\omega_n t} + C_2 e^{\left(-\zeta - i\sqrt{1-\zeta^2}\right)\omega_n t} \tag{5.23}$$

$$y(t) = e^{-\zeta\omega_n t}\left\{C_1 e^{i\sqrt{1-\zeta^2}\omega_n t} + C_2 e^{-i\sqrt{1-\zeta^2}\omega_n t}\right\} \tag{5.24}$$

$$y(t) = e^{-\zeta\omega_n t}\left\{(C_1+C_2)\cos\sqrt{1-\zeta^2}\,\omega_n t + i(C_1-C_2)\sin\sqrt{1-\zeta^2}\,\omega_n t\right\} \tag{5.25}$$

$$y(t) = e^{-\zeta\omega_n t}\left\{R\cos\sqrt{1-\zeta^2}\,\omega_n t + I\sin\sqrt{1-\zeta^2}\,\omega_n t\right\} \tag{5.26}$$

in which R and I are arbitrary constants that can be determined from the initial conditions. In order to determine the constants R and I, equation (5.26) is first differentiated to find the velocity, $\dot{x}(t)$, as

$$\begin{aligned} \dot{y}(t) = &-\zeta\omega_n e^{-\zeta\omega_n t}\left\{R\cos\sqrt{1-\zeta^2}\,\omega_n t + I\sin\sqrt{1-\zeta^2}\,\omega_n t\right\} \\ &+ e^{-\zeta\omega_n t}\left\{-\sqrt{1-\zeta^2}\,\omega_n R\sin\sqrt{1-\zeta^2}\,\omega_n t + \sqrt{1-\zeta^2}\,\omega_n I\cos\sqrt{1-\zeta^2}\,\omega_n t\right\} \end{aligned} \tag{5.27}$$

By using the initial conditions $y(t=0)=y_0$ and $\dot{y}(t=0)=\dot{y}_0$, respectively, in equations (5.26) and (5.27), we obtain the following results:

$$y_0 = R, \quad \dot{y}_0 = -\zeta\omega_n R + I\sqrt{1-\zeta^2}\,\omega_n = -\zeta\omega_n y_0 + I\sqrt{1-\zeta^2}\,\omega_n \tag{5.28}$$

From equation (5.28), we obtain

$$R = y_0, \quad I = \frac{\dot{y}_0 + \zeta\omega_n y_0}{\sqrt{1-\zeta^2}\,\omega_n} \tag{5.29}$$

Consequently, the solution that is provided by equation (5.26) can be represented as

$$y(t)=e^{-\zeta\omega_n t}\left\{y_0\cos\sqrt{1-\zeta^2}\,\omega_n t+\frac{\dot{y}_0+\zeta\omega_n y_0}{\sqrt{1-\zeta^2}\,\omega_n}\sin\sqrt{1-\zeta^2}\,\omega_n t\right\} \tag{5.30}$$

The motion that is given by equation (5.30) is a damped harmonic motion with an angular frequency of $\sqrt{1-\zeta^2}\,\omega_n$. However, due to the factor $e^{-\zeta\omega_n t}$, the amplitude decreases exponentially with time, as demonstrated in Figure 5.3.

The quantity, denoted as $\omega_d=\sqrt{1-\zeta^2}\,\omega_n$, is commonly referred to as the frequency of damped vibration. It can be observed that the frequency of the damped vibration, denoted as ω_d, is consistently lower than the natural frequency, denoted as ω_n, which is for the undamped vibration. When it comes to the study of engineering vibrations, the underdamped case is of utmost significance because it is the only scenario that results in an oscillatory motion.

Scenario 2. Critical damping: ($\zeta=1$, or $c=c_c$, or $\frac{c}{2m}=\sqrt{k/m}$). For this scenario, the roots s_1 and s_2 in equation (5.14) are equal, that is, $s_1=s_2=-\frac{c_c}{2m}=-\omega_n$. Let us proceed to set ζ approach unity in the limit in equation (5.26). As $\zeta\to 1$, $\sqrt{1-\zeta^2}\,\omega_n t\to 0$; thus $\cos\sqrt{1-\zeta^2}\,\omega_n t\to 1$ and $\sin\sqrt{1-\zeta^2}\,\omega_n t\to\sqrt{1-\zeta^2}\,\omega_n t$. Hence equation (5.26) yields

$$y(t)=e^{-\omega_n t}\left\{R+I\sqrt{1-\zeta^2}\,\omega_n t\right\}=(R+Qt)e^{-\omega_n t} \tag{5.31}$$

in which $Q=I\sqrt{1-\zeta^2}\,\omega_n$ is a new constant. After applying the initial conditions $y(t=0)=y_0$ and $\dot{y}(t=0)=\dot{y}_0$ for this particular scenario, the result is as follows:

$$\begin{aligned}R&=y_0\\ Q&=\dot{y}_0+\omega_n R=\dot{y}_0+\omega_n y_0\end{aligned} \tag{5.32}$$

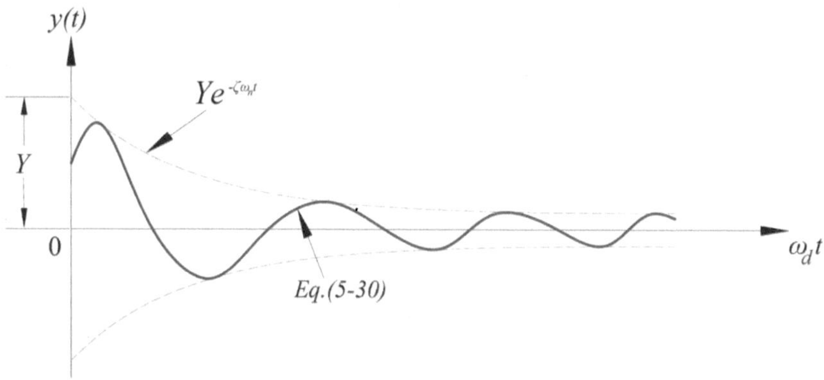

FIGURE 5.3 Light damping solution.

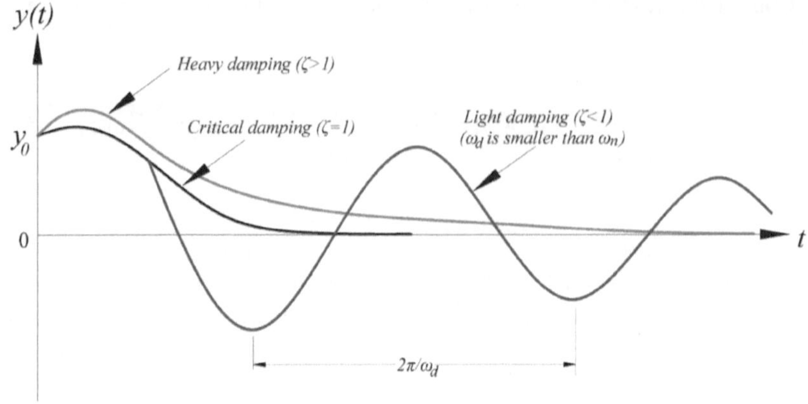

FIGURE 5.4 Comparing motions with different kinds of damping.

and the solution now becomes

$$y(t) = \left[y_0 + \left(\dot{y}_0 + \omega_n y_0 \right) t \right] e^{-\omega_n t} \tag{5.33}$$

As can be observed, the motion that is described by equation (5.33) is aperiodic (i.e., it is not periodic). As the value of $e^{-\omega_n t}$ approaches zero as t approaches infinity, it can be observed that the motion will finally decrease to zero, as depicted in Figure 5.4.

Scenario 3. Heavy damping: ($\zeta > 1$, or $c > c_c$, or $\frac{c}{2m} > \sqrt{k/m}$). In the context of this scenario, equation (5.20) demonstrates that the s_1 and s_2 are different and real, and they are provided by

$$\begin{aligned} s_1 &= \left(-\zeta + \sqrt{\zeta^2 - 1} \right) \omega_n < 0 \\ s_2 &= \left(-\zeta - \sqrt{\zeta^2 - 1} \right) \omega_n < 0 \end{aligned} \tag{5.34}$$

where $s_2 \ll s_1$. The solution, represented by equation (5.21), can be expressed as follows.

$$y(t) = C_1 e^{\left(-\zeta + \sqrt{\zeta^2 - 1}\right)\omega_n t} + C_2 e^{\left(-\zeta - \sqrt{\zeta^2 - 1}\right)\omega_n t} \tag{5.35}$$

With the initial conditions $y(t = 0) = y_0$ and $\dot{y}(t = 0) = \dot{y}_0$, it is possible to obtain the constants C_1 and C_2 as follows:

$$C_1 = \frac{y_0 \left(\zeta + \sqrt{\zeta^2 - 1} \right) \omega_n + \dot{y}_0}{2\omega_n \sqrt{\zeta^2 - 1}} \tag{5.36}$$

$$C_2 = \frac{-y_0 \left(\zeta - \sqrt{\zeta^2 - 1} \right) \omega_n - \dot{y}_0}{2\omega_n \sqrt{\zeta^2 - 1}} \tag{5.37}$$

Equation (5.35) demonstrates that the motion is not periodic regardless of the initial conditions imposed on the system. The motion decreases exponentially with time, as demonstrated in Figure 5.4, due to the fact that both of the roots, s_1 and s_2, are negative.

5.1.3 Harmonically Excited Vibration of Single-Degree-of-Freedom Systems

When an external energy source is introduced into a mechanical or structural system while it is in the process of vibrating, the system is said to be subjected to forced vibration. It is possible to obtain energy from the outside world by means of an applied force. It is possible for the applied force to be harmonic, nonharmonic but periodic, non-periodic, or random in nature. The response of a system to a harmonic excitation is referred to as the system's harmonic response. There is a possibility that the duration of the nonperiodic excitation could be either lengthy or short.

In this section, we will examine the dynamic response of a system with a single degree of freedom when subjected to harmonic excitations. These excitations can be expressed as $F(t) = F_0 e^{i(\omega t+\phi)}$ or $F(t) = F_0 \cos(\omega t + \phi)$ or $F(t) = F_0 \sin(\omega t + \phi)$, in which F_0 represents the amplitude, ω represents the frequency and ϕ represents the phase angle of the harmonic excitation. The value of ϕ is determined by the value of $F(t)$ at $t = 0$, and it is typically assumed to be 0. When the system is subjected to harmonic excitation, the response of the system will also be harmonic. In the event when the frequency of the excitation coincides with the natural frequency of the system, the response will turn out to be quite large. In order to prevent the system from failing, it is necessary to avoid this circumstance, which is referred to as resonance. Examples of harmonically excited vibration include the rolling motion of a ship model in regular waves in a model testing water tank, the oscillations of a tall chimney caused by vortex shedding in a steady wind and the vertical movement of a car over a sinusoidal road surface.

Let us take into consideration a viscously damped spring-mass system that is under an external excitation force, $F(t)$, as depicted in Figure 5.5.

Applying Newton's second law to this system in which y is measured from the position of static equilibrium of the mass m results in the equation of motion, which appears as follows:

$$m\ddot{y} = -k(y + \delta_{st}) - c\dot{y} + W + F(t) \tag{5.38}$$

and because $k\delta_{st} = W$, we have

$$m\ddot{y} + c\dot{y} + ky = F(t) \tag{5.39}$$

Due to the fact that this equation is non-homogeneous, its general solution, denoted by $y(t)$, is obtained by adding the homogeneous solution, $y_h(t)$, and the

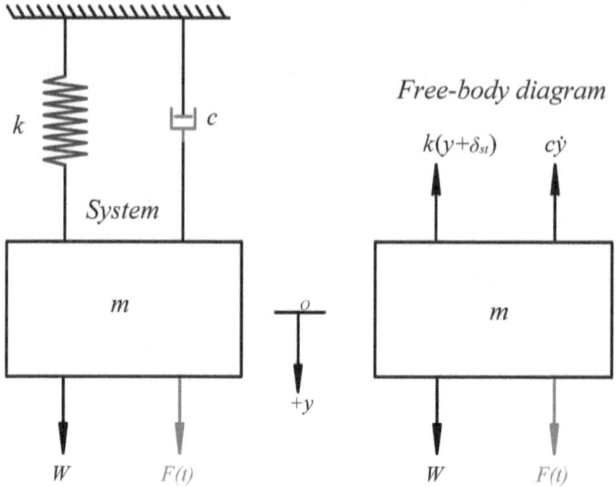

FIGURE 5.5 A harmonically excited damped single-degree-of-freedom system.

particular solution, $y_p(t)$. The homogeneous solution, which is the expression that represents the solution to the homogeneous equation

$$m\ddot{y} + c\dot{y} + ky = 0 \tag{5.40}$$

represents the free vibration of the system, which was subjected to discussion in the foregoing Section 5.1.3. As was seen in Section 5.1.3, this free vibration eventually ceases to exist over the course of time for each of the three conceivable damping situations (light damping, critical damping and heavy damping), as well as under all of the possible initial conditions. Therefore, the general solution of equation (5.39) will eventually reduce to the particular solution $y_p(t)$, which serves as a representation of the *steady-state* vibration. As long as the forcing function is present, the steady-state motion will continue to exist. Figure 5.6 illustrates the variations that occur over time in the homogeneous, particular and general solutions for a representative case. It is evident that $y_h(t)$ eventually disappears, and $y(t)$ transforms into $y_p(t)$ after a certain amount of time (T in Figure 5.6). The portion of the motion that diminishes because of damping (the free-vibration portion) is referred to as *transient*. The decay rate of the transient motion is contingent upon the system parameters k,c and m. This section focuses on deriving the particular solution of the equation (5.39) for the steady-state response under harmonic forcing functions.

If we assume that the forcing function is represented by the equation $F(t) = F_0 \cos\omega t$, then the equation of motion becomes

$$m\ddot{y} + c\dot{y} + ky = F_0 \cos\omega t \tag{5.41}$$

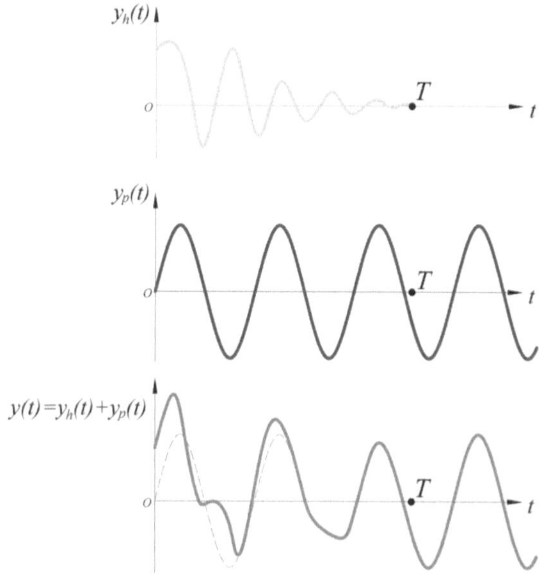

FIGURE 5.6 Homogeneous, particular and general solutions of equation (5.39) for a light damping system.

With the complex representation of the harmonic forcing function, denoted as $F(t) = F_0 e^{i\omega t}$, the equation of motion can also be expressed as follows:

$$m\ddot{y} + c\dot{y} + ky = F_0 e^{i\omega t} \tag{5.42}$$

Due to the fact that equation (5.42) is non-homogeneous, its general solution, denoted by $y(t)$, is obtained by adding the homogeneous solution, $y_h(t)$, and the particular solution, $y_p(t)$. The homogeneous solution $y_h(t)$ dies out soon; therefore, our interest will focus on $y_p(t)$, the particular solution. In the literature regarding the vibration theory, the particular solution is also referred to as the *steady-state* solution. Due to the fact that the actual excitation is given only by the real part of $F(t)$ (i.e., $F_0 e^{i\omega t}$), the *steady-state* response will also be provided only by the real part of $y(t)$, where $y(t)$ is a complex quantity satisfying the differential equation (5.42). F_0 in equation (5.42) is a real constant and i in equation (5.42) is the imaginary unit. Let's assume that the response occurs at the same frequency as the force but is typically out of phase with the force,

$$y(t) = y_p(t) = Y_0 e^{i(\omega t + \phi)} = \left(Y_0 e^{i\phi}\right) e^{i\omega t} = \tilde{Y} e^{i\omega t} \tag{5.43}$$

Substituting equation (5.43) into the differential equation (5.42) leads to

$$\left(-m\omega^2 + ic\omega + k\right)\tilde{Y} e^{i\omega t} = F_0 e^{i\omega t} \tag{5.44}$$

The transfer function, also known as the frequency response function, $H(\omega)$, is defined as the complex displacement that occurs as a result of a force of unit magnitude $(F_0 = 1)$. As a result,

$$H(\omega) = \frac{1}{(k - m\omega^2) + ic\omega} \tag{5.45}$$

The *steady-state* response can be obtained as

$$y(t) = \frac{F_0}{(k - m\omega^2) + ic\omega} e^{i\omega t} \tag{5.46}$$

5.1.4 Arbitrarily Excited Vibration of Single-Degree-of-Freedom Systems

Periodic excitation forces of any general waveform can be represented by Fourier series as a superposition of harmonic components of different frequencies. After that, the response of a linear system subject to periodic excitation forces can be determined by superposing the harmonic response to each of the exciting forces. When the exciting force $F(t)$ is not periodic, as it is in the case of the blast that is caused by an explosion, a different technique of calculating the response is required. The response of the system to an arbitrary excitation can be determined through the use of the method of convolution integral.

In most cases, the magnitude of a non-periodic exciting force varies over the course of time; it acts for a specified period of time and then stops. The simplest form is the impulsive force-a force that has a very large magnitude F and acts for a very short time duration Δt. Based on the principles of dynamics, we are aware that impulse can be measured by finding the change it causes in the momentum of the system. Under the assumption that $\dot{y}_1$ and $\dot{y}_2$ represent the velocities of the mass m prior to and after the application of the impulse, we have the following relationship:

$$\text{Impulse} = \text{F}\Delta t = m\dot{y}_2 - m\dot{y}_1 \tag{5.47}$$

The response of a system with a single degree of freedom to an impulse excitation is the first thing that we take into consideration. This particular case is of vital importance for the study of the response under more general excitations. Take into consideration a viscously damped spring-mass system that is exposed to a unit impulse at $t = 0$, as seen in Figure 5.7. For a system with light damping, the solution to the equation of motion

$$m\ddot{y} + c\dot{y} + ky = 0 \tag{5.48}$$

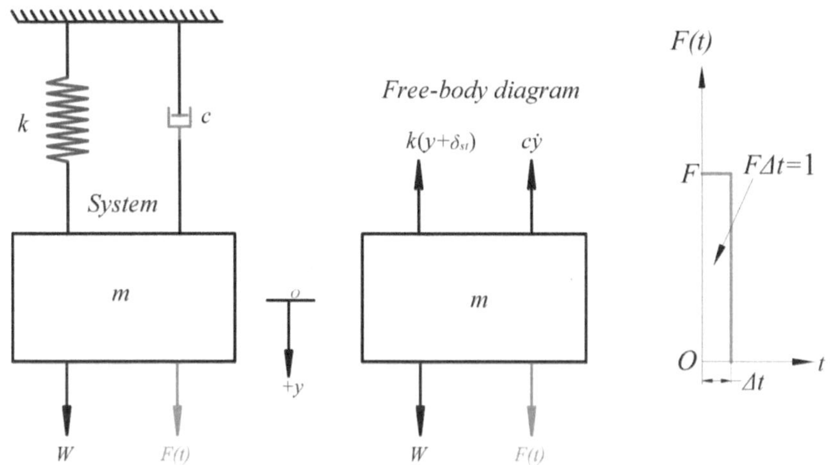

FIGURE 5.7 A single-degree-of-freedom system subjected to a unit impulse.

is given by equation (5.30) as

$$y(t) = e^{-\zeta\omega_n t}\left\{ y_0 \cos\sqrt{1-\zeta^2}\,\omega_n t + \frac{\dot{y}_0 + \zeta\omega_n y_0}{\sqrt{1-\zeta^2}\,\omega_n} \sin\sqrt{1-\zeta^2}\,\omega_n t \right\} \quad (5.49)$$

When the mass is in a state of rest prior to the application of the unit impulse $(y = \dot{y} = 0$ for $t < 0)$, we may derive the following result from the relation between the impulse and the momentum

$$\text{Impulse} = F\Delta t = \hat{f} = 1 = m\dot{y}(t=0) - m\dot{y}(t<0) = m\dot{y}_0 \quad (5.50)$$

The above result has been obtained based on the assumption that the magnitude of F is infinity and the time duration Δt tends to 0 and the unit impulse was applied at the time instant of $t = 0^-$.

Therefore, the initial conditions can be presented as follows:

$$y(t=0) = y_0 = 0 \quad (5.51)$$

$$\dot{y}(t=0) = \dot{y}_0 = \frac{1}{m} \quad (5.52)$$

The initial conditions expressed in equations (5.51) and (5.52) have been obtained based on the assumption that the imposing of the unit impulse at the time instant of $t = 0^-$ has only caused a change in velocity at the time instant of

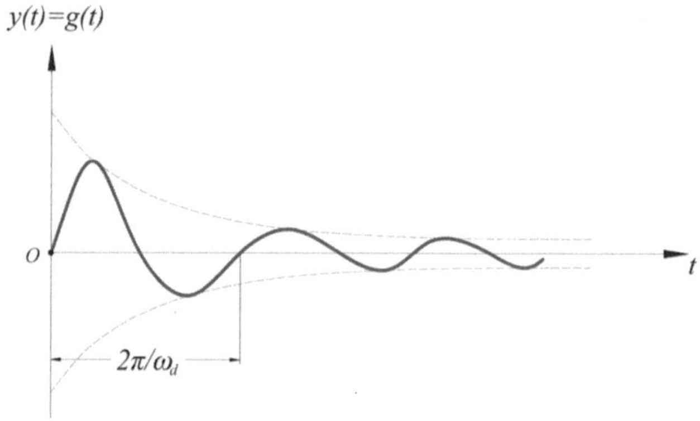

FIGURE 5.8 The impulse response function *g(t)*.

$t = 0$, not yet a change in displacement at the time instant of $t = 0$. Equation (5.49) can be simplified by considering equations (5.51) and (5.52) as follows:

$$y(t) = g(t) = e^{-\zeta\omega_n t}\left\{\frac{\sin\sqrt{1-\zeta^2}\,\omega_n t}{m\sqrt{1-\zeta^2}\,\omega_n}\right\} \tag{5.53}$$

Equation (5.53), which is also referred to as the impulse response function and is represented by the symbol *g*(*t*), allows one to determine the response of a system with a single degree of freedom to a unit impulse. Equation (5.53), which represents the function *g*(*t*), is depicted in Figure 5.8.

In the event when the magnitude of the impulse is D rather than unity, the initial velocity $\dot{y}_0$ is equal to D/*m*, and the response of the system is obtained as follows:

$$y(t) = g(t) = \mathrm{D}e^{-\zeta\omega_n t}\left\{\frac{\sin\sqrt{1-\zeta^2}\,\omega_n t}{m\sqrt{1-\zeta^2}\,\omega_n}\right\} \tag{5.54}$$

The velocity at $t = \tau$ will be altered by an amount equal to D/*m* if the impulse D is applied at an arbitrary time $t = \tau$, as depicted in Figure 5.9. If we assume that $y = 0$ until the impulse is applied, then the displacement *y* at any subsequent time *t*, which is caused by a change in the velocity at time τ, may be calculated using equation (5.54); however, the time *t* should be replaced by the amount of time that has passed after the application of the impulse, - that is, $t - \tau$. By doing so, we are able to obtain

$$y(t) = \mathrm{D}e^{-\zeta\omega_n (t-\tau)}\left\{\frac{\sin\sqrt{1-\zeta^2}\,\omega_n (t-\tau)}{m\sqrt{1-\zeta^2}\,\omega_n}\right\} \tag{5.55}$$

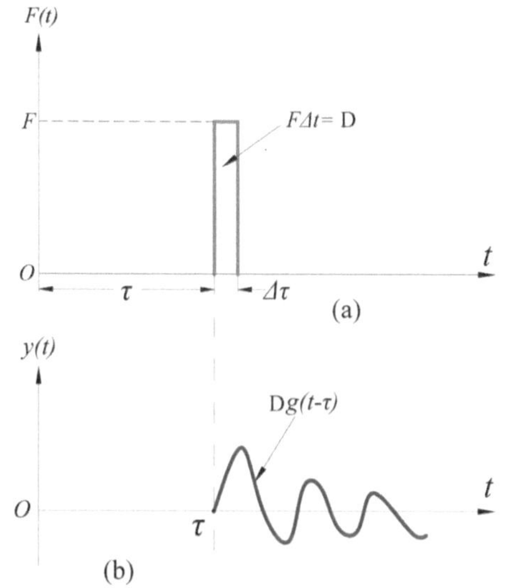

FIGURE 5.9 Impulse response.

The illustration of this can be found in Figure 5.9(b).

At this point, we will take into consideration the behavior of the system when subjected to an arbitrary external force $F(t)$, as depicted in Figure 5.10. This force can be considered to consist of a series of impulses with different magnitudes. If we assume that at the time τ the force $F(\tau)$ acts on the system for a brief time duration of $\Delta\tau$, then the impulse that acts at the time $t=\tau$ can be expressed as $F(\tau)\Delta\tau$. At every given time t, the amount of time that has elapsed since the impulse is equal to $t-\tau$. Therefore, the response of the system at t due to this impulse alone may be determined using the equation (5.55) with $\mathrm{D}=F(\tau)\Delta\tau$.

$$\Delta y(t)=F(\tau)\Delta\tau g(t-\tau) \tag{5.56}$$

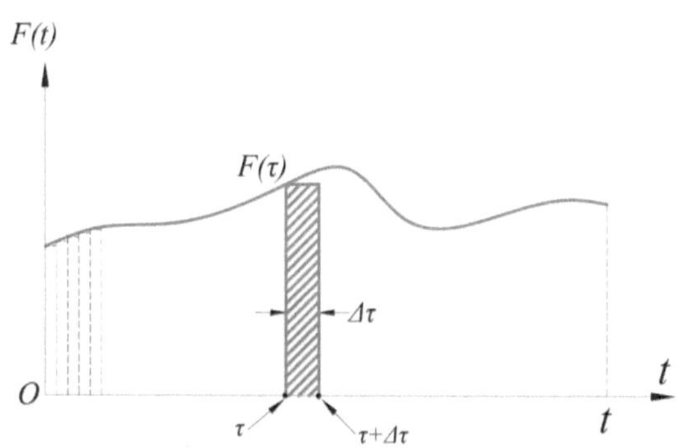

FIGURE 5.10 An aperiodic (arbitrary) forcing function.

It is possible to obtain the total response at time t by adding up all of the responses that are caused by the elementary impulses that are acting at all times τ:

$$y(t) \simeq \sum F(\tau)g(t-\tau)\Delta\tau \tag{5.57}$$

After letting $\Delta\tau \to 0$ and replacing the summation by integration, we are able to derive

$$y(t) = \int_0^t F(\tau)g(t-\tau)\,d\tau \tag{5.58}$$

Substituting equation (5.53) into equation (5.58) yields

$$y(t) = \frac{1}{m\sqrt{1-\zeta^2}\,\omega_n}\int_0^t F(\tau)e^{-\zeta\omega_n(t-\tau)}\left\{\sin\sqrt{1-\zeta^2}\,\omega_n(t-\tau)\right\}d\tau \tag{5.59}$$

which is a representation of the response of a single-degree-of-freedom system that is lightly damped to the arbitrary excitation $F(t)$. It is important to note that equation (5.59) does not take into account the effect of initial conditions of the system. This is due to the fact that equations (5.53) and (5.56) imply that the mass is considered to be at rest prior to the application of the impulse. The integral in equations (5.58) or (5.59) is referred to as the convolution integral or Duhamel integral. In many instances, the function $F(t)$ possesses a form that makes it possible to explicitly integrate equation (5.59). In the event when such integration is not feasible, we are able to perform evaluations numerically with relatively little trouble.

5.1.5 Relationship between $g(t)$ and $H(\omega)$

The input $F(t)$ in equation (5.58) can be accounted for as $t \to -\infty$, for example, $F(t)$ could be defined as $F(t) = 0$ for $t < 0$. Then equation (5.58) can be re-written as

$$y(t) = \int_{-\infty}^t F(\tau)g(t-\tau)\,d\tau \tag{5.60}$$

Please note that $g(t-\tau) = 0$ when $\tau > t$. Therefore, we can expand the limits in equation (5.60) to the interval $(-\infty, \infty)$:

$$y(t) = \int_{-\infty}^{\infty} F(\tau)g(t-\tau)\,d\tau \tag{5.61}$$

A further form that is useful can be derived by letting $\theta = t - \tau$:

$$y(t) = \int_{-\infty}^{\infty} F(t-\theta) g(\theta)\, d\theta \tag{5.62}$$

An important outcome derived from the theory of Fourier transforms is that $g(t)$ and $H(\omega)$ form a Fourier transform pair. This relationship is valuable for determining the responses of dynamic systems to random vibration excitations. Let

$$F(t) = F_0 e^{i\omega t} = e^{i\omega t} \tag{5.63}$$

Then, by considering equations (5.42), (5.43), (5.44), (5.45) and (5.61), we obtain

$$\begin{aligned} y(t) = H(\omega) e^{i\omega t} &= \int_{-\infty}^{\infty} g(t-\tau) e^{i\omega\tau}\, d\tau \\ &= \int_{-\infty}^{\infty} g(\theta) e^{i\omega(t-\theta)}\, d\theta = e^{i\omega t} \int_{-\infty}^{\infty} g(\theta) e^{-i\omega\theta}\, d\theta \end{aligned} \tag{5.64}$$

Hence, this suggests that

$$H(\omega) = \int_{-\infty}^{\infty} g(\theta) e^{-i\omega\theta}\, d\theta \tag{5.65}$$

as a result, we have

$$g(t) = \frac{1}{2\pi} \int_{-\infty}^{\infty} H(\omega) e^{i\omega t}\, d\omega \tag{5.66}$$

Consequently, $H(\omega)$ and $g(t)$ form a Fourier transform pair. From equation (5.66), we can obtain

$$g(t-\tau) = \frac{1}{2\pi} \int_{-\infty}^{\infty} H(\omega) e^{i\omega(t-\tau)}\, d\omega \tag{5.67}$$

Therefore, equation (5.61) can be re-written as

$$y(t)=\int_{-\infty}^{\infty}F(\tau)\left[\frac{1}{2\pi}\int_{-\infty}^{\infty}H(\omega)e^{i\omega(t-\tau)}\,d\omega\right]d\tau=\int_{-\infty}^{\infty}H(\omega)\left[\frac{1}{2\pi}\int_{-\infty}^{\infty}F(\tau)e^{i\omega(t-\tau)}\,d\omega\right]d\tau$$

$$=\int_{-\infty}^{\infty}H(\omega)\left[\frac{1}{2\pi}\int_{-\infty}^{\infty}F(\tau)e^{-i\omega\tau}\,d\tau\right]e^{i\omega t}d\omega=\int_{-\infty}^{\infty}H(\omega)\left[\frac{1}{2\pi}F(\omega)\right]e^{i\omega t}d\omega \tag{5.68}$$

$$=\frac{1}{2\pi}\int_{-\infty}^{\infty}F(\omega)H(\omega)e^{i\omega t}\,d\omega$$

in which $F(\omega)$ is the Fourier transform of $F(t)$. In addition to serving as a foundation for random vibration computations, this equation can be utilized for the purpose of analyzing or numerically computing the response of a system.

5.1.6 Relationship between $Y(\omega)$ and $F(\omega)$

For the purpose of deriving responses of dynamic systems to random vibration excitations, Chapters 5, 6 and 7 will make use of the relationship that exists between the Fourier transforms of $y(t)$ and $F(t)$.

Let us take the Fourier transform of both sides of equation (5.62),

$$Y(\omega)=\int_{-\infty}^{\infty}\left[\int_{-\infty}^{\infty}F(t-\theta)g(\theta)\,d\theta\right]e^{-i\omega t}\,dt \tag{5.69}$$

Now, let $\tau=t-\theta, dt=d\tau$:

$$Y(\omega)=\int_{-\infty}^{\infty}\left[\int_{-\infty}^{\infty}F(\tau)g(\theta)\,d\theta\right]e^{-i\omega(\tau+\theta)}\,d\tau=\int_{-\infty}^{\infty}g(\theta)e^{-i\omega\theta}\,d\theta\left[\int_{-\infty}^{\infty}F(\tau)e^{-i\omega\tau}\,d\tau\right] \tag{5.70}$$

$$=H(\omega)F(\omega)$$

5.2 RANDOM VIBRATION OF SINGLE-DEGREE OF FREEDOM LINEAR SYSTEMS

As was discussed in Section 5.1, the deterministic theory of a linear single-degree-of-freedom dynamic system makes the assumption that, with the exception of the motion y, every item in the governing equation of motion is known with absolute certainty. The motion can then be estimated with great accuracy for a specific set of initial conditions that have been provided. In the event when any of the given items in the governing equation is random, the motion will also be random. In practically all of the problems encountered by vibration analysts, only random external excitations are of major concern, while the other parameters of the system as well as the initial conditions, can be reasonably assumed to be deterministic. In the following, we will restrict our attention to specifically addressing such problems.

5.2.1 Response to Weakly Stationary Random Excitations

Excitations that can be fairly expected to be weakly stationary are involved in a large number of practical problems. A car, for instance, that is moving at a constant speed on an expressway that has a weakly homogeneous random roughness, is able to detect a weakly stationary excitation. In the real world, the car might be required to temporarily slow down or speed up when following or passing another vehicle. However, if only a small percentage of the total time is spent traveling at varying speeds, then it is acceptable to disregard the nonstationary portion of the excitation when evaluating, for example, the fatigue damage of a structural part of the car. There is still another illustration of a weakly stationary excitation that may be seen in the vibrations of airplane panels that are caused by jet noise. The most significant damage to the panels is known to take place during takeoff, which is the time when the jet engines are operating at their maximum capacity. It is obvious that the power of the engine must be progressively increased until it reaches its maximum level, and then gradually decreased until it reaches the level that is necessary for cruising flight. Therefore, the acoustic pressure that is generated is not stationary, according to the strictest definition. However, during the maximum engine power run, which lasts between one and two minutes, the generated acoustic pressure that is recorded at a specific place on the surface of the fuselage or wing displays a pattern that is only weakly stationary. Therefore, if the objective of the analysis is to determine the amount of fatigue damage that the panels have experienced, it is acceptable to discard the nonstationary component of the excitation that results in only a small amount of damage. Still, there is another example of a weakly stationary excitation that may be seen in the motions of a semi-submersible offshore platform moored at a deep-water sea site. When considering the long-term structural reliability of the semi-submersible platform, it is obvious that the wave loads acting on it are non-stationary because, during its life cycle, the semi-submersible platform will encounter different sea states each of which possesses a unique significant wave height. Nevertheless, when analyzing the motion responses of the semi-submersible platform in just one sea state lasting less than three hours, the significant wave height for this particular sea state will remain constant, and the sea state can be reasonably assumed to be weakly stationary. Consequently, the wave loads acting on the semi-submersible platform during this short-term sea state can be regarded as a weakly stationary random process. In the offshore industry, analyzing the motion responses of an offshore platform in a three-hour stationary sea state is called a short-term

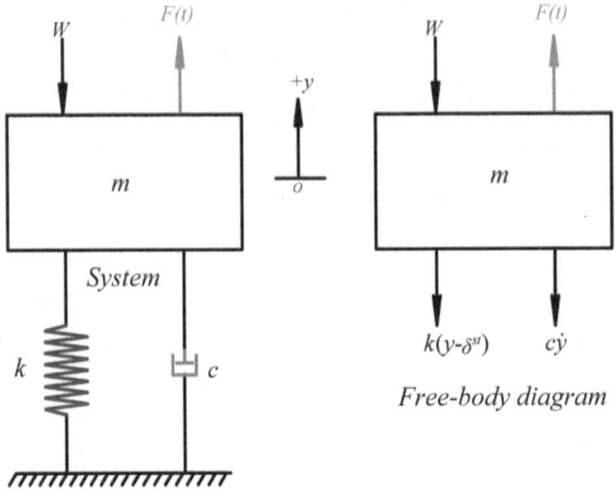

FIGURE 5.11 Single-degree-of-freedom system excited by random force *F(t)*.

analysis, and the implementation of these short-term analyses will provide a solid foundation for the long-term reliability analysis of the offshore platform.

Now, please take into consideration the system depicted in Figure 5.11, which has a single degree of freedom. The force that is applied is a random process that is ergodic and weakly stationary and has the form $F(t)$, $t > 0$. A possible sample record of the random force *F(t)* is shown in Figure 5.12. It is well known that the random process $F(t)$ possesses certain statistical properties, which include the mean μ_F, the variance σ_F^2, the autocorrelation function $R_{FF}(\tau)$, the spectral density function $S_{FF}(\omega)$, and the marginal probability density function. This section's objective is to provide a description of the statistical properties that are

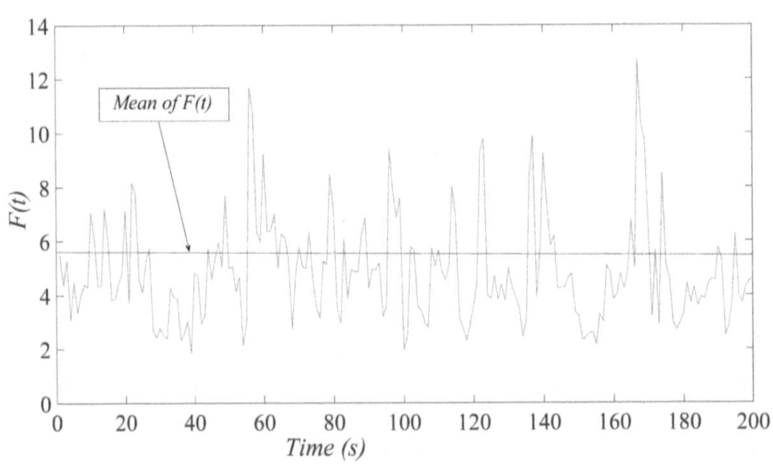

FIGURE 5.12 A possible sample record of the random force *F(t)*. in Figure 5.11.

associated with the response random process $y(t)$. Crandall and Mark (1963), Lin (1967), Newland (1984), Clough and Penzien (1993), Elishakoff (1983), and Spanos (1981,1999) are some examples of the general literature that can be referenced.

Within the entirety of this section, it is presumed, for the sake of simplicity, that the input is force, which is represented by the letter F, and the output is displacement, which is represented by the letter y. It is recommended that the reader interpret the input and output in a general sense and use mathematical intuition, in addition to the results presented in the previous sections in this chapter, in order to construct expressions for responses for other cases of input and output.

The convolution integral, which we understand as the connection between a specific sample realization of the random loading and the response to that loading, will serve as the entry point in considering each of these. This will be the point of departure. Because $F(t-\theta)=0$ for $\theta > t$, it can be deduced from equation (5.62) that

$$y(t) = \int_{-\infty}^{t} F(t-\theta)g(\theta)\,d\theta \tag{5.71}$$

Equation (5.71) will be utilized hereafter to express the moments of the response random process. The input force random process $F(t)$ is assumed to be 0 for $t < 0$, and that it has a constant mean μ_F for $t \geq 0$, and an autocorrelation function $R_{FF}(t,s) = R_{FF}(s-t)$ for $t,s \geq 0$. An additional assumption that we make is that the response has 0 initial conditions at $t = 0$, which means that $y(0) = \dot{y}(0) = 0$.

5.2.1.1 The Mean of the Response Process

A well-defined and finite mean is assumed to be present in the response random process. If it does, we will write (see equation (5.71))

$$\begin{aligned}\mu_y(t) = E[y(t)] = E\left[\int_{-\infty}^{t} F(t-\theta)g(\theta)\,d\theta\right] = \int_{-\infty}^{t} g(\theta)\mu_F\,d\theta \\ = \mu_F \int_{-\infty}^{t} g(\theta)\,d\theta \quad t \geq 0\end{aligned} \tag{5.72}$$

As long as the system remains stable, that is, if the function $g(t)$ does not diverge as $t \to \infty$, then the response mean is finite. As the value of t increases, it becomes evident that the value of $\mu_y(t)$ is getting closer to a limiting value. In fact, depending on the damping of the system, this limit is rapidly reached within a few response cycles.

Based on equation (5.65), we can derive

$$H(0) = \int_{-\infty}^{\infty} g(\theta)e^{-i0\theta}\,d\theta = \int_{-\infty}^{\infty} g(\theta)e^{0}\,d\theta = \int_{-\infty}^{\infty} g(\theta)\,d\theta \tag{5.73}$$

The integral of equation (5.72) approaches $H(0)$ as $t \to \infty$, based on equation (5.73). As a result, the mean response is

$$\mu_y = \mu_F H(0) \tag{5.74}$$

To put it another way, the static response to the mean load equals the mean response. Keep in mind that a zero mean load also results in a zero mean response. This is a highly significant result since it suggests that the static and dynamic analyses can be handled independently of one another.

As an illustrative example, let's examine the force-excited system. The transfer function represented by equation (5.45) is:

$$H(\omega) = \frac{1}{\left(k - m\omega^2\right) + ic\omega} \tag{5.75}$$

When we take into account the fact that $H(0) = 1/k$, we may deduce that the stationary state mean response is

$$\mu_y = \frac{\mu_F}{k} \tag{5.76}$$

Equation (5.76) can be re-written as

$$\mu_F = k\mu_y \tag{5.77}$$

This is a concise expression of Hooke's law. The dynamic response can be analyzed independently and then combined with the static response μ_y. If the system's initial conditions are not $y(0) = \dot{y}(0) = 0$, the homogeneous response to the initial conditions can be calculated using the results from Section 5.1.2 and then combined with the result in equation (5.72).

5.2.1.2 Autocorrelation Function of the Response Process

When it comes to characterizing the response of a single degree of freedom linear system, the autocorrelation function is a fundamental tool. In what follows, we will make the assumption that the response autocorrelation function exists in a probabilistic sense that is well defined, for instance, in a mean-square sense. When it does, we write the following, making use of the definition of R_{yy} and equation (5.71):

$$\begin{aligned}
R_{yy}(t,s) = E[y(t)y(s)] &= E\left[\left\{\int_{-\infty}^{t} g(u)F(t-u)\,du\right\}\left\{\int_{-\infty}^{s} g(v)F(s-v)\,dv\right\}\right] \\
&= E\left[\int_{-\infty}^{t} du \int_{-\infty}^{s} dv g(u)g(v)F(t-u)F(s-v)\right] \\
&= \int_{-\infty}^{t} du \int_{-\infty}^{s} dv g(u)g(v)R_{FF}(t-u,s-v) \quad t,s \geq 0
\end{aligned} \tag{5.78}$$

In situations when the excitation force is a non-stationary random process, it is necessary to utilize equation (5.78) in order to assess the response autocorrelation. It is possible to evaluate the expression in closed form for a number of simple inputs; nevertheless, in most cases, it is always necessary to evaluate it numerically. If the mean square of the response is the sole thing that is being sought, then the expression becomes

$$E\left[y^2(t)\right] = R_{yy}(t,t) = \int_{-\infty}^{t} du \int_{-\infty}^{t} dv g(u)g(v)R_{FF}(t-u,t-v) \quad t \geq 0 \tag{5.79}$$

A clear indication of the severity of the response random process as a function of time is provided by this expression. The expressions that have been presented above serve as the starting point of more advanced studies on the non-stationary response of engineering systems. The modal analysis of linear multi-degree-of-freedom systems, which will be studied in Chapter 7, is another area in which they prove to be valuable. When the excitation force $F(t)$ becomes a stationary random process for $t \geq 0$, we can write $R_{FF}(t,s) = R_{FF}(s-t)$ for $t,s \geq 0$, and equation (5.78) can be simplified to

$$R_{yy}(t,s) = \int_{-\infty}^{t} du \int_{-\infty}^{s} dv g(u)g(v)R_{FF}(s-t-(v-u)) \quad t,s \geq 0 \tag{5.80}$$

Despite the fact that the integrand is only dependent on $s-t$, $R_{yy}(t,s)$ is a function of both t and s when t and s are relatively small. The rationale for this is that dependence on t and s also enters through the limits on the integral. Nevertheless, in the case when the system is stable, that is, if $g(t)$ does not diverge as $t \to \infty$, then as $t,s \to \infty, R_{yy}(t,s)$ approaches a simple dependence on $s-t$.

The mean square of the response of the single-degree-of-freedom system is calculated, as a function of time, by evaluating equation (5.80) at $s = t$. It is

$$\sigma_y^2(t) = R_{yy}(t,t) = \int_{-\infty}^{t} du \int_{-\infty}^{t} dv g(u)g(v)R_{FF}(u-v) \quad t,s \geq 0 \tag{5.81}$$

At $t \to \infty$, this quantity is getting closer and closer to a limit, which is the quantity that is given in equation (5.83). As a matter of fact, the rate at which $\sigma_X^2(t)$ approaches its constant limit as the value of t grows is a measure of how rapidly the response approaches a stationary state, and this is a function of the damping applied to the system.

As the response is getting closer to a stationary state (when $t,s \to \infty$), we assume that $\tau = s-t$ and write

$$R_{yy}(\tau) = \int_{-\infty}^{\infty} du \int_{-\infty}^{\infty} dv g(u)g(v)R_{FF}(\tau-(v-u)) \tag{5.82}$$

In a manner analogous to that of the non-stationary case (which is represented by equation (5.78)), this expression can be evaluated in closed form over a wide range of simple inputs that are weakly stationary. In the general instance, it is much simpler to perform a numerical evaluation on it.

By evaluating equation (5.82) at a value of $\tau = 0$, one can determine the mean-square value of the weakly stationary response. It is as follows:

$$E\left[y^2(t)\right] = R_{yy}(0) = \int_{-\infty}^{\infty} du \int_{-\infty}^{\infty} dv g(u) g(v) R_{FF}(u - v) \tag{5.83}$$

In addition, the result is not a function of the time parameter. It is possible to use this formula to evaluate the response mean square; but, if this is all the analyst wants to obtain, a more effective method would be to apply the frequency-domain equations that will be established in the subsequent sections.

5.2.1.3 Spectral Density Function of the Response Process

Through the application of the Fourier transform, the autocorrelation function can be transformed into the power spectral density function.

$$\begin{aligned} S_{yy}(\omega) &= \frac{1}{2\pi} \int_{-\infty}^{\infty} R_{yy}(\tau) e^{-i\omega\tau} d\tau = \frac{1}{2\pi} \int_{-\infty}^{\infty} \left[\int_{-\infty}^{\infty} \int_{-\infty}^{\infty} g(u) g(v) R_{FF}(\tau + u - v) du\, dv \right] e^{-i\omega\tau} d\tau \\ &= \int_{-\infty}^{\infty} \int_{-\infty}^{\infty} g(u) g(v) \left[\frac{1}{2\pi} \int_{-\infty}^{\infty} R_{FF}(\tau + u - v) e^{-i\omega\tau} d\tau \right] du dv \end{aligned} \tag{5.84}$$

Let $\theta = \tau + u - v$, then we have $\tau = \theta - u + v$. Therefore, equation (5.84) can be rewritten as

$$\begin{aligned} S_{yy}(\omega) &= \int_{-\infty}^{\infty} \int_{-\infty}^{\infty} g(u) g(v) \left[\frac{1}{2\pi} \int_{-\infty}^{\infty} R_{FF}(\theta) e^{-i\omega\theta} d\tau \right] e^{+i\omega u} e^{-i\omega v} du dv \\ &= \int_{-\infty}^{\infty} \int_{-\infty}^{\infty} g(u) g(v) \left[\frac{1}{2\pi} \int_{-\infty}^{\infty} R_{FF}(\theta) e^{-i\omega\theta} d\tau \right] e^{+i\omega u} e^{-i\omega v} du dv \\ &= \left[\int_{-\infty}^{\infty} g(u) e^{+i\omega u} du \right] \left[\int_{-\infty}^{\infty} g(v) e^{-i\omega v} dv \right] \left[\frac{1}{2\pi} \int_{-\infty}^{\infty} R_{FF}(\theta) e^{-i\omega\theta} d\tau \right] \\ &= H(-\omega) H(\omega) S_{FF}(\omega) \end{aligned} \tag{5.85}$$

Because $g(\theta)$ is real and because we have the following relationships

$$\begin{aligned} H(-\omega) &= \int_{-\infty}^{\infty} g(\theta)e^{i\omega\theta}\, d\theta \\ &= \int_{-\infty}^{\infty} g(\theta)(\cos\theta + i\sin\theta)d\theta = \int_{-\infty}^{\infty} g(\theta)\cos\theta\, d\theta + i\int_{-\infty}^{\infty} g(\theta)\sin\theta\, d\theta \end{aligned} \tag{5.86}$$

$$\begin{aligned} H(\omega) &= \int_{-\infty}^{\infty} g(\theta)e^{-i\omega\theta}\, d\theta \\ &= \int_{-\infty}^{\infty} g(\theta)(\cos\theta - i\sin\theta)d\theta = \int_{-\infty}^{\infty} g(\theta)\cos\theta\, d\theta - i\int_{-\infty}^{\infty} g(\theta)\sin\theta\, d\theta \end{aligned} \tag{5.87}$$

we can derive the following relationship

$$H(-\omega) = H^*(\omega) \tag{5.88}$$

$$H(-\omega)H(\omega) = H^*(\omega)H(\omega) = |H(\omega)|^2 \tag{5.89}$$

In the above two equations, $H^*(\omega)$ is the complex conjugate of $H(\omega)$. By combining equations (5.85) and (5.89), we can obtain the following equation

$$S_{yy}(\omega) = |H(\omega)|^2\, S_{FF}(\omega) \tag{5.90}$$

To put it another way, the spectral density function of the response is equal to the spectral density function of the load multiplied by the squared modulus of the transfer function. This result is of the utmost importance.

The relationship that is established in equation (5.90) can also be deduced directly from the definition of the response spectral density function, which is something that should be taken into consideration. As a result, the response spectral density can be expressed from the definition of the spectral density function in the following manner:

$$S_{yy}(\omega) = \lim_{T\to\infty} \frac{1}{2\pi T}|Y(\omega)|^2 \tag{5.91}$$

On the other hand, from the presentations in the foregoing sections, by applying the following convolution integral $y(t) = F(t) * g(t)$ in the time domain, one can determine the output $y(t)$ of a linear system to an arbitrary random input $F(t)$. An expression that may be used to define the relationship in the frequency domain is

$$Y(\omega) = H(\omega)F(\omega) \tag{5.92}$$

in which $F(\omega)$, $Y(\omega)$ and $H(\omega)$ are the Fourier transform of $F(t)$, $y(t)$ and $g(t)$, respectively. By combining equations (5.91) and (5.92), we can obtain the following equation

$$S_{yy}(\omega) = \lim_{T\to\infty} \frac{1}{2\pi T} |F(\omega)|^2 |H(\omega)|^2 = S_{FF}(\omega)|H(\omega)|^2 \tag{5.93}$$

Based on this relation, it can be said that when both the excitation and the response are weakly stationary, the spectral density of the response is equal to the product of the spectral density of the excitation and the function $|H(\omega)|^2$. The function that dictates the fraction of energy that is to be transmitted through the system at various frequencies is referred to as the transmittancy function or the system function. It is denoted by the symbol $|H(\omega)|^2$. A linear system is sometimes referred to as a linear filter because of this reason.

5.2.1.4 Coherency Function

Up until this point, we have been talking about the stochastic relationship that exists between the input and output of a system, under the assumption that the system is linear by nature. On the other hand, it is frequently essential to investigate whether or not the system may be categorized as linear. It is possible to have a better understanding of the linearity of a system by performing an auto and cross-spectral analysis of the system through the following steps.

As a preliminary step, let us take into consideration the cross-correlation function that exists between the input and output of a linear system that possesses an impulse response function denoted by $g(t)$. More specifically,

$$\begin{aligned} R_{Fy}(\tau) &= \lim_{T\to\infty} \frac{1}{2T} \int_{-T}^{T} F(t)y(t+\tau)dt \\ &= \lim_{T\to\infty} \frac{1}{2T} \int_{-T}^{T} F(t)\left\{ \int_{-\infty}^{\infty} F(t+\tau-\theta)g(\theta)d\theta \right\} dt \\ &= \int_{-\infty}^{\infty} \left[\lim_{T} \frac{1}{2T} \int_{-T}^{T} F(t)F(t+\tau-\theta)dt \right] g(\theta)d\theta = \int_{-\infty}^{\infty} R_{FF}(\tau-\theta)g(\theta)d\theta \end{aligned} \tag{5.94}$$

When the Wiener–Khintchine theorem is utilized, the cross-spectral density function can be expressed in the following manner:

$$
\begin{aligned}
S_{Fy}(\omega) &= \frac{1}{\pi}\int_{-\infty}^{\infty}\left\{\int_{-\infty}^{\infty} R_{FF}(\tau-\theta)g(\theta)\,d\theta\right\}e^{-i\omega\tau}\,d\tau \\
&= \int_{-\infty}^{\infty}\left\{\frac{1}{\pi}\int_{-\infty}^{\infty} R_{FF}(\tau-\theta)e^{-i\omega(\tau-\theta)}\,d\tau\right\}g(\theta)e^{-i\omega\theta}\,d\theta \\
&= \left\{\frac{1}{\pi}\int_{-\infty}^{\infty} R_{FF}(\tau-\theta)e^{-i\omega(\tau-\theta)}\,d\tau\right\}\int_{-\infty}^{\infty} g(\theta)e^{-i\omega\theta}\,d\theta \\
&= S_{FF}(\omega)H(\omega)
\end{aligned}
\tag{5.95}
$$

In a similar manner, we are able to deduce

$$S_{Fy}(-\omega) = S_{FF}(\omega)H^{*}(\omega) \tag{5.96}$$

By referring to equations (5.95) and (5.96), we can deduce the following:

$$
\begin{aligned}
|H(\omega)|^2 &= \frac{S_{Fy}(\omega)S_{Fy}(-\omega)}{\{S_{FF}(\omega)\}^2} = \frac{S_{Fy}(\omega)S_{Fy}^{*}(\omega)}{\{S_{FF}(\omega)\}^2} \\
&= \frac{\{C_{Fy}(\omega)+iQ_{xy}(\omega)\}\{C_{Fy}(\omega)-iQ_{xy}(\omega)\}}{\{S_{FF}(\omega)\}^2} = \frac{\{C_{Fy}(\omega)\}^2+\{Q_{Fy}(\omega)\}^2}{\{S_{FF}(\omega)\}^2}
\end{aligned}
\tag{5.97}
$$

in which $C_{Fy}(\omega)$ is the co-spectrum of $F(t)$ and $y(t)$. $Q_{Fy}(\omega)$ is the quadrature spectrum of $F(t)$ and $y(t)$.

On the other hand, assuming the system is linear, then, the expression for $|H(\omega)|^2$ can be derived as follows:

$$|H(\omega)|^2 = \frac{S_{yy}(\omega)}{S_{FF}(\omega)} \tag{5.98}$$

Therefore, if the system is linear, it is necessary for it to satisfy both equations (5.97) and 5.98). Consequently, this leads to

$$\frac{\{C_{Fy}(\omega)\}^2+\{Q_{Fy}(\omega)\}^2}{\{S_{FF}(\omega)\}^2} = \frac{S_{yy}(\omega)}{S_{FF}(\omega)} \tag{5.99}$$

Rearranging equation (5.99) leads to

$$\{C_{Fy}(\omega)\}^2+\{Q_{Fy}(\omega)\}^2 = S_{FF}(\omega)S_{yy}(\omega) \tag{5.100}$$

In the event when the system is not linear, the relationship that is described in equation (5.100) will not be valid. As a result, it is possible for us to devise a measure for determining the linearity of a system by utilizing equation (5.100), which is referred to as the coherency function $\gamma(\omega)$ stated in equation (4.93) of Chapter 4. In this context, we may restate the definition;

$$\gamma(\omega)=\sqrt{\frac{\{C_{Fy}(\omega)\}^2+\{Q_{Fy}(\omega)\}^2}{S_{FF}(\omega)S_{yy}(\omega)}};\quad 0\leqslant\gamma(\omega)\leqslant 1 \tag{5.101}$$

Upon examining the definition of $\gamma(\omega)$, it becomes evident that the equation $\gamma(\omega)=1$ implies that the system is linear.

5.2.2 White-noise Processes as Models for Excitations

5.2.2.1 Definition of the White Noise Process

Formally speaking, the random process that is characterized by a white noise is defined by the spectral density function that is presented below

$$S_{FF}(\omega)=S_0 \qquad -\infty<\omega<\infty \tag{5.102}$$

A representation of the spectral density function may be seen in Figure 5.13. The autocorrelation function is obtained by performing a Fourier transform on $S_{FF}(\omega)$

$$R_{FF}(\tau)=2\pi S_0\delta(\tau) \tag{5.103}$$

in which $\delta(\tau)$ is the Dirac delta function. The Dirac delta function, denoted by the symbol $\delta(\tau)$, can be conceptualized in a general sense as a function on the real line. This function is zero everywhere; however, it is infinite at the origin,

$$\delta(\tau)\simeq\begin{cases}+\infty, & \tau=0\\ 0, & \tau\neq 0\end{cases} \tag{5.104}$$

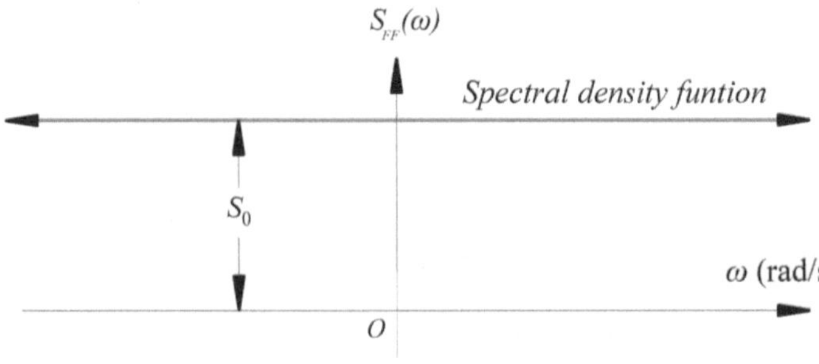

FIGURE 5.13 The spectral density function for a white noise random process.

The Dirac delta function $\delta(\tau)$ is also constrained to satisfy the identity

$$\int_{-\infty}^{\infty} \delta(\tau)\,d\tau = 1 \tag{5.105}$$

Undoubtedly, the white noise process is inherently unattainable in the physical realm. Despite the fact that the white noise process is merely an idealization, it is beneficial in dynamic analysis.

5.2.2.2 Response of Force-Excited System to White Noise Process

There are some systems that have a finite response to a white noise force. Take, for instance, the force-excited system, the transfer function of which is represented by the following equation

$$H(\omega) = \frac{1}{\left(k - m\omega^2\right) + ic\omega} \tag{5.106}$$

Assume that the mean of the process is 0. If the force is characterized by a white noise process, with a spectral density function $S_{FF}(\omega) = S_0$, then the spectral density function of the response can be determined as

$$S_{yy}(\omega) = |H(\omega)|^2 S_{FF}(\omega) = \left|\frac{1}{\left(k - m\omega^2\right) + ic\omega}\right|^2 S_0 = \frac{S_0}{\left(k - m\omega^2\right)^2 + c^2\omega^2} \tag{5.107}$$

It is possible to calculate the mean square value of the output by using the formula and by applying the method of residues (whose theory is elucidated in Appendix D):

$$\begin{aligned} E\left[y^2\right] &= \int_{-\infty}^{\infty} S_{yy}(\omega)\,d\omega = S_0 \int_{-\infty}^{\infty} \left|\frac{1}{-m\omega^2 + k + ic\omega}\right|^2 d\omega \\ &= S_0 \int_{-\infty}^{\infty} \frac{1}{\left(k - m\omega^2\right)^2 + c^2\omega^2}\,d\omega \end{aligned} \tag{5.108}$$

which can be observed to be independent of the magnitude of the mass m. We define

$$\beta = \omega/\omega_n = \omega\sqrt{m/k} \tag{5.109}$$

$$\zeta = c/c_c = \frac{c}{2m\omega_n} = \frac{c}{2m\sqrt{k/m}} = \frac{c}{2\sqrt{km}} \tag{5.110}$$

$$r_1 = i\zeta + \sqrt{1-\zeta^2} \tag{5.111}$$

$$r_2 = i\zeta - \sqrt{1-\zeta^2} \tag{5.112}$$

Then we have

$$\begin{aligned}
&(\beta - r_1)(\beta - r_2)(\beta + r_1)(\beta + r_2) = (\omega/\omega_n - r_1)(\omega/\omega_n - r_2)(\omega/\omega_n + r_1)(\omega/\omega_n + r_2)\\
&= \left(\frac{\omega}{\omega_n} - i\zeta - \sqrt{1-\zeta^2}\right)\left(\frac{\omega}{\omega_n} - i\zeta + \sqrt{1-\zeta^2}\right)\left(\frac{\omega}{\omega_n} + i\zeta + \sqrt{1-\zeta^2}\right)\left(\frac{\omega}{\omega_n} + i\zeta - \sqrt{1-\zeta^2}\right)\\
&= \left(\left(\frac{\omega}{\omega_n} - i\zeta\right)^2 - (1-\zeta^2)\right)\left(\left(\frac{\omega}{\omega_n} + i\zeta\right)^2 - (1-\zeta^2)\right)\\
&= \left(\frac{\omega^2}{\omega_n^2} - 1\right)^2 - \left(2i\zeta\frac{\omega}{\omega_n}\right)^2 = \left(\frac{\omega^2 m}{k} - 1\right)^2 - \left(2i\frac{c}{2\sqrt{km}}\omega\sqrt{\frac{m}{k}}\right)^2\\
&= \frac{1}{k^2}\left[\omega^4 m^2 - 2\omega^2 km + k^2 + c^2\omega^2\right] = \frac{1}{k^2}\left[\left(k - m\omega^2\right)^2 + c^2\omega^2\right]
\end{aligned} \tag{5.113}$$

and, we can thus obtain

$$k^2\left[(\beta - r_1)(\beta - r_2)(\beta + r_1)(\beta + r_2)\right] = \left(k - m\omega^2\right)^2 + c^2\omega^2 \tag{5.114}$$

Substituting equation (5.114) into equation (5.108) leads to

$$\begin{aligned}
E\left[y^2\right] &= S_0 \int_{-\infty}^{\infty} \frac{1}{\left(k - m\omega^2\right)^2 + c^2\omega^2} d\omega\\
&= S_0 \int_{-\infty}^{\infty} \frac{1}{k^2\left[(\beta - r_1)(\beta - r_2)(\beta + r_1)(\beta + r_2)\right]} d\omega\\
&= \frac{\omega_n S_0}{k^2} \int_{-\infty}^{\infty} \frac{1}{\left[(\beta - r_1)(\beta - r_2)(\beta + r_1)(\beta + r_2)\right]} d\beta
\end{aligned} \tag{5.115}$$

In equation (5.115), the integrand is an analytic function everywhere in the complex β plane, with the exception of the points $\beta = r_1$, $\beta = r_2$, $\beta = -r_1$ and $\beta = -r_2$, where poles of order 1 are existing. Specifically, the points $\beta = r_1$ and $\beta = r_2$ are located in the upper half plane, while the points $\beta = -r_1$ and $\beta = -r_2$ are situated on the lower half plane. The process of contour integration is carried out in the upper half plane. Within the upper half complex plane, the integrand possesses two simple poles, which are as follows:

$$\beta_1 = r_1 = i\zeta + \sqrt{1-\zeta^2} \tag{5.116}$$

$$\beta_2 = r_2 = i\zeta - \sqrt{1-\zeta^2} \tag{5.117}$$

When Cauchy's residue theorem is utilized, the integral in equation (5.115) can be easily carried out, resulting in the relation

$$
\begin{aligned}
E\left[y^2\right] &= \frac{\omega_n S_0}{k^2}\int_{-\infty}^{\infty}\frac{1}{\left[(\beta-r_1)(\beta-r_2)(\beta+r_1)(\beta+r_2)\right]}d\beta = \frac{\omega_n S_0}{k^2}\int_{-\infty}^{\infty} f(\beta)\,d\beta \\
&= \frac{\omega_n S_0}{k^2}2\pi i\left\{\sum residues\ of\ f(\beta)\ in\ \ the\ upper\ half\ complex\ plane\right\} \\
&= \frac{\omega_n S_0}{k^2}2\pi i\left\{\sum\left[(\beta-r_1)f(\beta)\right]_{\beta=r_1}+\left[(\beta-r_2)f(\beta)\right]_{\beta=r_2}\right\} \\
&= \frac{\omega_n S_0}{k^2}2\pi i\left\{\sum\frac{1}{\left[(r_1-r_2)(r_1+r_1)(r_1+r_2)\right]}+\frac{1}{\left[(r_2-r_1)(r_2+r_1)(r_2+r_2)\right]}\right\} \\
&= \frac{\omega_n S_0}{k^2}2\pi i\left\{\frac{1}{2r_2 r_1 2i\zeta}\right\} = \frac{\pi S_0}{kc}
\end{aligned}
\tag{5.118}
$$

in a manner that is not reliant on the magnitude of the mass m, as has been demonstrated. As a result of the fact that an ideal white noise corresponds to an infinite mean-square value of the random process, it has been stated that an ideal white noise is physically impossible to achieve. But equation (5.118) demonstrates that the mean-square output is finite even when the mean-square input is infinite, provided that there is some damping in the system, that is, $\xi > 0$. This holds true even if the mean-square input is infinite.

Example 5.1

▸ The equation of motion of the surge response of a semi-submersible offshore platform can be expressed as follows:

$$M\ddot{x} + D(t)\dot{x} + K_1 x = F(t) \tag{5.119}$$

We provide the following data in the analysis of the surge response of this semi-submersible offshore platform: $M = 4.15 \times 10^5$ kg. The undamped natural frequency of this offshore platform system is taken to be $\omega_n = (K_1/M)^{0.5} = 0.08$ rad/s. It is possible to simplify the time-varying wave drift force damping $D(t)$ by using a constant, $D(t) = D$, which corresponds to a damping ratio, $\xi = D/2\omega_n M = 0.06$. $S_u = 3.37 \times 10^1$ N^{2}s is the value of the spectral density function for the stochastic wave drift force ($F(t)$) per unit mass (kg) of the platform that we utilized. Determine the root mean square (RMS) value of the surge response of this semi-submersible offshore platform.

SOLUTION

The mean square value of the surge response of this semi-submersible offshore platform can be calculated as follows:

$$E\left[x^2\right] = \frac{\pi S_0}{K_1 D} \tag{5.120}$$

From the foregoing data provided, we can obtain

$$S_0 = S_u \times 4.15 \times 10^5 \text{N}^2\text{s} = 3.37 \times 10^1 \times 4.15 \times 10^5 \text{N}^2\text{s} = 1.39855 \times 10^7 \text{N}^2\text{s}$$

$$(K_1/M)^{0.5} = (K_1/4.15 \times 10^5)^{0.5} = 0.08 \,\text{rad/s}, K_1 = 2656 \,\text{N/m}$$

$$D = \xi 2\omega_n M = 2 \times 0.06 \times 0.08 \times 4.15 \times 10^5 = 3984 \,\text{N-s/m}$$

$$E\left[x^2\right] = \frac{\pi S_0}{K_1 D} = \frac{\pi \times 1.39855 \times 10^7 \times N^2 s}{2656 N/m \times 3984 N - s/m} = 4.15223 \text{m}^2$$

Therefore, the RMS value of the surge response of this semi-submersible offshore platform is

$$\sqrt{E\left[x^2\right]} = \sqrt{4.15223 m^2} = 2.0377 \text{m} \blacktriangleleft$$

5.2.2.3 Response of Base-Excited System to White Noise Process

Let us take into consideration a viscously damped spring-mass system that is under a base excitation, as depicted in Figure 5.14.

Let us first conduct the deterministic analysis. Given the base motion $b(t)$, the objective of the analysis will be to determine the absolute response $y(t)$, which normally refers to the acceleration or displacement of the mass. The mass, for instance, may be a sophisticated electronic instrument, and it would be necessary to develop a vibration isolation system that would limit the acceleration response of the mass. This problem is important in situations like these.

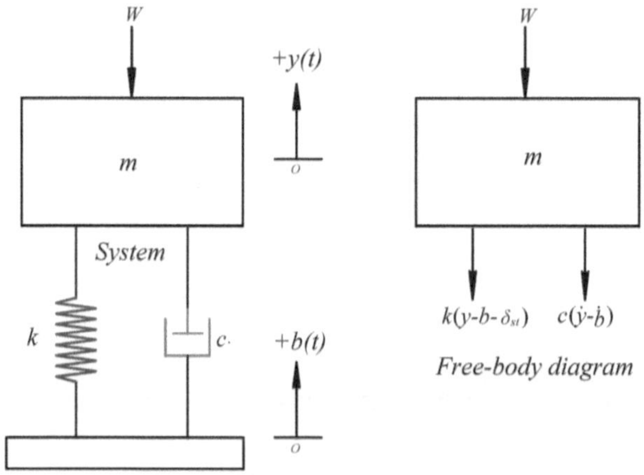

FIGURE 5.14 A base-excited damped single-degree-of-freedom system.

Applying Newton's second law to this system in which y is measured from the position of static equilibrium of the mass m results in the equation of motion, which appears as follows:

$$m\ddot{y} = -k\left(y - b - \delta_{st}\right) - c\left(\dot{y} - \dot{b}\right) - W \tag{5.121}$$

and because $k\delta_{\text{st}} = W$, we have

$$m\ddot{y} + c\dot{y} + ky = kb + c\dot{b} \tag{5.122}$$

Let us assume that the motion of the base is harmonic,

$$b(t) = B_0 e^{i\omega t} \tag{5.123}$$

and we can also make the assumption that the response has the same frequency as the base excitation, but that it is, in general, out of phase with the base excitation,

$$y(t) = Y_0 e^{i(\omega t + \phi)} = \left(Y_0 e^{i\phi}\right) e^{i\omega t} = \tilde{Y} e^{i\omega t} \tag{5.124}$$

in which the symbol Y_0 represents the amplitude of the displacement, while $\tilde{Y}$ represents the complex displacement.

By plugging in the expressions in equations (5.123) and (5.124) into the differential equation of motion (i.e., equation (5.122)), we obtain

$$\left(-m\omega^2 + ic\omega + k\right)\tilde{Y} e^{i\omega t} = \left(kB_0 + ic\omega B_0\right) e^{i\omega t} \tag{5.125}$$

Let us define the transfer function, also known as the frequency response function, $H(\omega)$, as the complex displacement that occurs as a result of a base excitation of unit magnitude $\left(B_0 = 1\right)$. As a result,

$$H(\omega) = \frac{k + ic\omega}{\left(k - m\omega^2\right) + ic\omega} \tag{5.126}$$

Let us again take into consideration the free-body diagram shown in Figure 5.14. This time, the response variable that will be taken into consideration is going to be the relative displacement, not the absolute displacement just studied. The relative displacement is calculated as follows:

$$z(t) = y(t) - b(t) \tag{5.127}$$

Within the model, the spring serves as a representation of a structural component. The stresses in that component will vary in direct proportion to z. Therefore, this problem would be pertinent to designers of structures that experience base vibrations, such as earthquakes.

Letting $z = y - b$ in the equation of motion (i.e., equation (5.122)) leads directly to

$$m\ddot{z} + c\dot{z} + kz = -m\ddot{b} \equiv p_{\text{eff}}(t) \tag{5.128}$$

In this equation, the term "$p_{\text{eff}}(t)$" represents the effective base excitation loading. To put it another way, the structural deformations that are brought about by the base acceleration $\ddot{b}(t)$ are precisely the same as those that would be brought about by an external load $p(t)$ equal to $-m\ddot{b}(t)$. Within the context of this definition of effective load, the presence of a negative sign indicates that the effective force is in opposition to the sense of the base acceleration.

Let us assume that the motion of the base is harmonic,

$$b(t) = B_0 e^{i\omega t} \tag{5.129}$$

and we can also make the assumption that the response in equation (5.128) has the same frequency as the base excitation, but that it is, in general, out of phase with the base excitation,

$$z(t) = Z_0 e^{i(\omega t + \phi)} = \left(Z_0 e^{i\phi}\right) e^{i\omega t} = \tilde{Z} e^{i\omega t} \tag{5.130}$$

in which the symbol Z_0 represents the amplitude of the displacement, while $\tilde{Z}$ represents the complex displacement.

By plugging in the expressions in equations (5.129) and (5.130) into the differential equation of motion (i.e., equation (5.128)), we obtain

$$\left(-m\omega^2 + ic\omega + k\right)\tilde{Z}e^{i\omega t} = \left(m\omega^2 B_0\right)e^{i\omega t} \tag{5.131}$$

Let us define the transfer function, also known as the frequency response function, $H(\omega)$, as the complex displacement that occurs as a result of an effective base excitation loading $\left(p_{\text{eff}}(t) = p_0 e^{i\omega t}\right)$ of unit magnitude (i.e., $p_0 = 1$). Thus, we have the relationship $m\omega^2 B_0 = 1$. As a result,

$$H(\omega) = \frac{1}{\left(k - m\omega^2\right) + ic\omega} \tag{5.132}$$

We have already mentioned that there are some systems that have a finite response to a white noise force. As another example, take, for instance, the base-excited system shown in Figure 5.14, the transfer function of which is represented by the equation (5.132). Assume that the mean of the process is zero. If the effective base excitation loading $p_{\text{eff}}(t)$ is characterized by a white noise process, with a spectral density function $S_{pp}(\omega) = S_0$, then the spectral density function of the response can be determined as

$$S_{zz}(\omega) = |H(\omega)|^2 S_{pp}(\omega) = \left|\frac{1}{\left(k - m\omega^2\right) + ic\omega}\right|^2 S_0 = \frac{S_0}{\left(k - m\omega^2\right)^2 + c^2\omega^2} \tag{5.133}$$

It is possible to calculate the response's variance by using the formula and by applying the method of residues (whose theory is elucidated in Appendix D):

$$\sigma_z^2 = \int_{-\infty}^{\infty} S_{zz}(\omega)d\omega = S_0 \int_{-\infty}^{\infty} \frac{1}{\left(k - m\omega^2\right)^2 + c^2\omega^2} d\omega = \frac{\pi S_0}{kc} \tag{5.134}$$

Making use of the notations $k^2 = \omega_n{}^4 m^2$ and $\zeta = \dfrac{c}{2m\omega_n} = \dfrac{c}{2\sqrt{km}}$, and substituting these expressions into equation (5.134) give

$$\sigma_z^2 = \frac{\pi\omega_n S_0}{2k^2\zeta} \tag{5.135}$$

in which S_0 denotes the power spectral density for a white-noise excitation $p_{\text{eff}}(t)$. The power spectral density S_0 for the white-noise excitation $p_{\text{eff}}(t)$ can be expressed as

$$S_{pp}(\omega) = S_0 = m^2 S_{\ddot{b}\ddot{b}}(\omega) \tag{5.136}$$

in which $S_{\ddot{b}\ddot{b}}(\omega)$ denotes the power spectral density for the base acceleration process. Obviously $S_{\ddot{b}\ddot{b}}(\omega)$ is also a constant and equals S_0/m^2. Substituting equation (5.136) into equation (5.135) and making use of the notation $k^2 = \omega_n{}^4 m^2$ give

$$\sigma_z^2 = \frac{\pi S_{\ddot{b}\ddot{b}}(\omega)}{2\omega_n{}^3\zeta} \tag{5.137}$$

5.2.2.4 Response of a Vehicle Model to White Noise Process

We take into consideration the system depicted in Figure 5.15, which incorporates a linear spring in tandem with a linear dashpot to suspend a single mass m from a moving support. Assume that the motion of the support is known in terms of its acceleration $\ddot{x}_0$. Furthermore, we will be focusing on the relative displacement $y = x_1 - x_0$ that exists between the mass and the support.

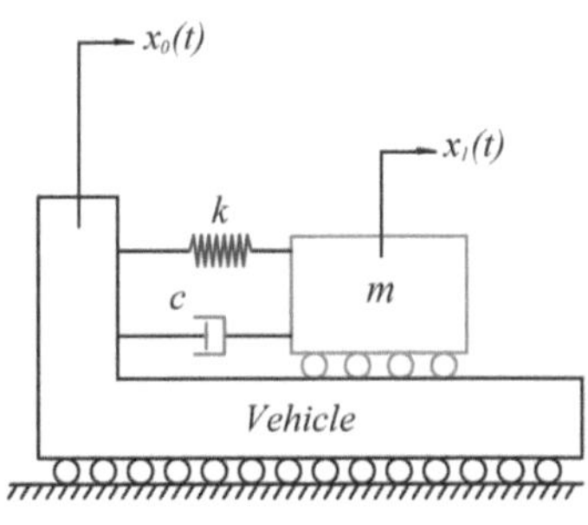

FIGURE 5.15 A vehicle model.

Applying Newton's second law to this system results in the equation of motion, which appears as follows

$$m(\ddot{x}_1 - \ddot{x}_0) = -k(x_1 - x_0) - c(\dot{x}_1 - \dot{x}_0) - m\ddot{x}_0 \tag{5.138}$$

Taking into consideration of the fact $y = x_1 - x_0$ and rearranging equation (5.138) leads to

$$m\ddot{y} + c\dot{y} + ky = -m\ddot{x}_0 \tag{5.139}$$

The above equation establishes a connection between the desired response quantity $y(t)$ and the excitation $\ddot{x}_0(t)$. In order to standardize the treatment that will follow, we introduce the following notations

$$\frac{k}{m} = \omega_n^2 \qquad \zeta = c/c_c = \frac{c}{2m\omega_n} = \frac{c}{2m\sqrt{k/m}} = \frac{c}{2\sqrt{km}} = \frac{1}{2Q} \tag{5.140}$$

regarding the natural frequency ω_n, the damping ratio ξ and the Q-factor. When these are taken into consideration, the equation of motion in equation (5.139) becomes

$$\ddot{y} + 2\omega_n\zeta\dot{y} + \omega_n{}^2 y = -\ddot{x}_0 \tag{5.141}$$

The output response quantity $y(t)$ depends indirectly on the input excitation $\ddot{x}_0(t)$ by means of a differential equation. In the following, we will demonstrate the direct representations that are generated by the utilization of the complex frequency response function and the impulse response function.

For the purpose of obtaining the complex frequency response function $H(\omega)$, the substitution $\ddot{x}_0(t) = e^{i\omega t}$ and $y = H(\omega)e^{i\omega t}$ is made in equation (5.141), and the $e^{i\omega t}$ terms are canceled out

$$H(\omega) = \frac{1}{\omega^2 - \omega_n^2 - 2i\zeta\omega_n\omega} = \frac{-1}{\omega_n^2 - \omega^2 + 2i\zeta\omega_n\omega} \tag{5.142}$$

Substituting equation (5.142) into equation (5.68) leads to

$$\begin{aligned} y(t) &= \int_{-\infty}^{\infty} H(\omega)\left[\frac{1}{2\pi}\int_{-\infty}^{\infty} F(\tau)e^{-i\omega\tau}\,d\tau\right]e^{i\omega t}\,d\omega \\ &= \int_{-\infty}^{\infty} \frac{-1}{\omega_n^2 - \omega^2 + 2i\zeta\omega_n\omega}\left[\frac{1}{2\pi}\int_{-\infty}^{\infty} \ddot{x}_0(\tau)e^{-i\omega\tau}\,d\tau\right]e^{i\omega t}\,d\omega \\ &= \frac{1}{2\pi}\int_{-\infty}^{\infty}\int_{-\infty}^{\infty} \frac{-\ddot{x}_0(\tau)e^{i\omega(t-\tau)}}{\omega_n^2 - \omega^2 + 2i\zeta\omega_n\omega}\,d\omega\,d\tau \end{aligned} \tag{5.143}$$

By solving equation (5.141), the impulse-response function $g(t)$ can be found when the value of $\ddot{x}_0(t)$ is equal to $\delta(t)$. Recall that equation (5.53) allows one to determine the response of a system with a single degree of freedom to a unit impulse when the system mass is equal to m. The impulse-response function $g(t)$ of the vehicle model corresponding to equation (5.141) can therefore be obtained based on equation (5.53) and presented as follows:

$$g(t) = -e^{-\zeta\omega_n t}\left\{\frac{\sin\sqrt{1-\zeta^2}\,\omega_n t}{\sqrt{1-\zeta^2}\,\omega_n}\right\} \quad t > 0 \tag{5.144}$$

Consequently, the general excitation-response relation in equation (5.62) for the current vehicle model becomes

$$y(t) = \int_0^{\infty} -\ddot{x}_0(t-\theta)e^{-\zeta\omega_n\theta}\frac{\sin\sqrt{1-\zeta^2}\,\omega_n\theta}{\sqrt{1-\zeta^2}\,\omega_n}d\theta \tag{5.145}$$

Equation (5.145) ought to be compared with equation (5.143). The connection between the two procedures is that $H(\omega)$ in equation (5.142) isl the Fourier transform of function $g(t)$ in equation (5.144).

The vehicle model system in Figure 5.15 is excited by the acceleration of the foundation, denoted by $\ddot{x}_0(t)$, and the response, denoted by $y(t)$, is the relative displacement between the foundation and the suspended mass. Next, we will look at the situation in which the excitation is an ideal white noise.

For all frequencies ranging from $\omega = -\infty$ to $\omega = \infty$, the excitation $\ddot{x}_0(t)$ is assumed to be a stationary random process. This process is assumed to have an ideally white spectrum. Additionally, the spectral density S_0 is uniform. The response $y(t)$ will also be a stochastic process. As a result of the fact that the excitation is stationary, the response process will also be stationary. Additionally, if it is known that the input is ergodic, then the output will also be ergodic. Likewise, if we have knowledge that the input follows a normal or Gaussian process, we can infer that the output will similarly follow a normal process. Aside from these general considerations, the foregoing results make it possible for us to make quantitative predictions about the statistical averages of the output (which are not reliant on whether the processes are ergodic or Gaussian).

Let us assume that the excitation $\ddot{x}_0(t)$ has a mean of zero. Based on the equation (5.72), it can be concluded that the response $y(t)$ also has a mean of zero. The autocorrelation function of the excitation is $2\pi S_0\delta(\tau)$, where $\delta(\tau)$ is the Dirac delta function. This is due to the fact that the excitation is an ideal white noise. By calculating equation (5.82) for $R_{FF}(\tau) = 2\pi S_0\delta(\tau)$ and for the system impulse response function $g(t)$ given in equation (5.144), the autocorrelation function of the response $R_{yy}(\tau)$ can be effectively determined.

$$R_{yy}(\tau) = \int_0^{\infty}\int_0^{\infty} 2\pi S_0\delta(\tau+\theta_1-\theta_2)e^{-\zeta\omega_n(\theta_2+\theta_1)}\frac{\sin\sqrt{1-\zeta^2}\,\omega_n\theta_1\sin\sqrt{1-\zeta^2}\,\omega_n\theta_2}{\left(1-\zeta^2\right)\omega_n^{\,2}}d\theta_1\,d\theta_2 \tag{5.146}$$

Evaluating this integral is a laborious process, but it is not impossible. It is essential to take into account the values of $\tau > 0$ and $\tau < 0$ independently. The delta function makes the first integration a relatively straightforward process. When the value of τ is greater than zero, the delta function disappears, with the exception of the case when $\theta_2 = \tau + \theta_1$ when the area of the delta function is unity. The integration over the θ_2 is instantaneous

$$R_{yy}(\tau) = \frac{2\pi S_0}{(1-\zeta^2)\omega_n^2}\int_0^\infty \left[e^{-\zeta\omega_n(\tau+2\theta_1)}\sin\sqrt{1-\zeta^2}\,\omega_n\theta_1 \sin\sqrt{1-\zeta^2}\,\omega_n(\tau+\theta_1)\right]d\theta_1 \quad (5.147)$$

Because

$$\int_0^\infty e^{-ax}\cos bx\,dx = \frac{a}{a^2+b^2} \text{ and } \int_0^\infty e^{-ax}\sin bx\,dx = \frac{b}{a^2+b^2},$$

$$\begin{aligned}
&\sin\sqrt{1-\zeta^2}\,\omega_n(\tau+\theta_1)\sin\sqrt{1-\zeta^2}\,\omega_n\theta_1 \\
&= \frac{1}{2}\Big[\cos\left(\sqrt{1-\zeta^2}\,\omega_n(\tau+\theta_1) - \sqrt{1-\zeta^2}\,\omega_n\theta_1\right) \\
&\quad - \cos\left(\sqrt{1-\zeta^2}\,\omega_n(\tau+\theta_1) + \sqrt{1-\zeta^2}\,\omega_n\theta_1\right)\Big] \\
&= \frac{1}{2}\left[\cos\left(\sqrt{1-\zeta^2}\,\omega_n\tau\right) - \cos\left(\sqrt{1-\zeta^2}\,\omega_n(\tau+2\theta_1)\right)\right],
\end{aligned}$$

$$\begin{aligned}
R_{yy}(\tau) &= \frac{2\pi S_0}{(1-\zeta^2)\omega_n^2}e^{-\zeta\omega_n\tau}\int_0^\infty \Big\{\frac{1}{2}e^{-\zeta\omega_n 2\theta_1}\Big[\cos\left(\sqrt{1-\zeta^2}\,\omega_n\tau\right) \\
&\quad - \cos\left(\sqrt{1-\zeta^2}\,\omega_n(\tau+2\theta_1)\right)\Big]\Big\}d\theta_1 \\
&= \frac{\pi S_0}{(1-\zeta^2)\omega_n^2}e^{-\zeta\omega_n\tau}\int_0^\infty \Big\{e^{-\zeta\omega_n 2\theta_1}\Big[\cos\left(\sqrt{1-\zeta^2}\,\omega_n\tau\right) \\
&\quad - \cos\left(\sqrt{1-\zeta^2}\,\omega_n(\tau+2\theta_1)\right)\Big]\Big\}d\theta_1,
\end{aligned}$$

$$\begin{aligned}
&\cos\left(\sqrt{1-\zeta^2}\,\omega_n(\tau+2\theta_1)\right) \\
&= \cos\left(\sqrt{1-\zeta^2}\,\omega_n\tau\right)\cos\left(2\sqrt{1-\zeta^2}\,\omega_n\theta_1\right) \\
&\quad - \sin\left(\sqrt{1-\zeta^2}\,\omega_n\tau\right)\sin\left(2\sqrt{1-\zeta^2}\,\omega_n\theta_1\right)
\end{aligned}$$

therefore, a reduction to standard forms can be made for the integration in equation (5.147), which results in

$$\int_0^\infty \left\{ e^{-\zeta\omega_n 2\theta_1} \left[\cos\left(\sqrt{1-\zeta^2}\,\omega_n \tau\right) \right] \right\} d\theta_1 = \cos\left(\sqrt{1-\zeta^2}\,\omega_n \tau\right) \frac{2\zeta\omega_n}{(2\zeta\omega_n)^2 + 0^2}$$

$$= \frac{1}{2\zeta\omega_n} \cos\left(\sqrt{1-\zeta^2}\,\omega_n \tau\right),$$

$$\int_0^\infty \left\{ e^{-\zeta\omega_n 2\theta_1} \left[-\cos\left(\sqrt{1-\zeta^2}\,\omega_n (\tau + 2\theta_1)\right) \right] \right\} d\theta_1$$

$$= \int_0^\infty \left\{ -e^{-\zeta\omega_n 2\theta_1} \left[\cos\left(\sqrt{1-\zeta^2}\,\omega_n \tau\right) \cos\left(2\sqrt{1-\zeta^2}\,\omega_n \theta_1\right) \right. \right.$$

$$\left. \left. - \sin\left(\sqrt{1-\zeta^2}\,\omega_n \tau\right) \sin\left(2\sqrt{1-\zeta^2}\,\omega_n \theta_1\right) \right] \right\} d\theta_1,$$

$$\int_0^\infty \left\{ -e^{-\zeta\omega_n 2\theta_1} \left[\cos\left(\sqrt{1-\zeta^2}\,\omega_n \tau\right) \cos\left(2\sqrt{1-\zeta^2}\,\omega_n \theta_1\right) \right] \right\} d\theta_1$$

$$= -\cos\left(\sqrt{1-\zeta^2}\,\omega_n \tau\right) \int_0^\infty \left\{ e^{-\zeta\omega_n 2\theta_1} \left[\cos\left(2\sqrt{1-\zeta^2}\,\omega_n \theta_1\right) \right] \right\} d\theta_1$$

$$= -\cos\left(\sqrt{1-\zeta^2}\,\omega_n \tau\right) \frac{2\zeta\omega_n}{(2\zeta\omega_n)^2 + \left(2\sqrt{1-\zeta^2}\,\omega_n\right)^2}.$$

$$= -\frac{\zeta^2}{2\zeta\omega_n} \cos\left(\sqrt{1-\zeta^2}\,\omega_n \tau\right),$$

$$\int_0^\infty \left\{ -e^{-\zeta\omega_n 2\theta_1} \left[-\sin\left(\sqrt{1-\zeta^2}\,\omega_n \tau\right) \sin\left(2\sqrt{1-\zeta^2}\,\omega_n \theta_1\right) \right] \right\} d\theta_1$$

$$= \sin\left(\sqrt{1-\zeta^2}\,\omega_n \tau\right) \int_0^\infty \left\{ e^{-\zeta\omega_n 2\theta_1} \left[\sin\left(2\sqrt{1-\zeta^2}\,\omega_n \theta_1\right) \right] \right\} d\theta_1$$

$$= \sin\left(\sqrt{1-\zeta^2}\,\omega_n \tau\right) \frac{2\sqrt{1-\zeta^2}\,\omega_n}{(2\zeta\omega_n)^2 + \left(2\sqrt{1-\zeta^2}\,\omega_n\right)^2}$$

$$= \frac{\sqrt{1-\zeta^2}}{2\omega_n} \sin\left(\sqrt{1-\zeta^2}\,\omega_n \tau\right),$$

$$
\begin{aligned}
R_{yy}(\tau) = {} & \frac{\pi S_0}{\left(1-\zeta^2\right)\omega_n^{\,2}} e^{-\zeta\omega_n \tau} \int_0^{\infty} \left\{ e^{-\zeta\omega_n 2\theta_1} \left[\cos\left(\sqrt{1-\zeta^2}\,\omega_n \tau\right) \right.\right. \\
& \left.\left. - \cos\left(\sqrt{1-\zeta^2}\,\omega_n (\tau + 2\theta_1)\right)\right]\right\} d\theta_1 \\
= {} & \frac{\pi S_0}{\left(1-\zeta^2\right)\omega_n^{\,2}} e^{-\zeta\omega_n \tau} \left[\frac{1}{2\zeta\omega_n} \cos\left(\sqrt{1-\zeta^2}\,\omega_n \tau\right) - \frac{\zeta^2}{2\zeta\omega_n} \cos\left(\sqrt{1-\zeta^2}\,\omega_n \tau\right) \right. \\
& \left. + \frac{\sqrt{1-\zeta^2}}{2\omega_n} \sin\left(\sqrt{1-\zeta^2}\,\omega_n \tau\right) \right] \\
= {} & \frac{\pi S_0}{\left(1-\zeta^2\right)\omega_n^{\,2}} e^{-\zeta\omega_n \tau} \left[\frac{1-\zeta^2}{2\zeta\omega_n} \cos\left(\sqrt{1-\zeta^2}\,\omega_n \tau\right) \right. \\
& \left. + \frac{\zeta\sqrt{1-\zeta^2}}{2\zeta\omega_n} \sin\left(\sqrt{1-\zeta^2}\,\omega_n \tau\right) \right] \\
= {} & \frac{\pi S_0}{2\zeta\omega_n^{\,3}} e^{-\zeta\omega_n \tau} \left[\cos\left(\sqrt{1-\zeta^2}\,\omega_n \tau\right) \right. \\
& \left. + \frac{\zeta}{\sqrt{1-\zeta^2}} \sin\left(\sqrt{1-\zeta^2}\,\omega_n \tau\right) \right] \quad \tau > 0.
\end{aligned}
\tag{5.148}
$$

Equation (5.148) can be utilized to calculate the autocorrelation function of the response when $\tau < 0$. In the case of $\tau < 0$, we can simply use the fact that the autocorrelation function $R_{yy}(\tau)$ is an even function of τ to write

$$
\begin{aligned}
R_{yy}(\tau) = {} & \frac{\pi S_0}{2\zeta\omega_n^{\,3}} e^{\zeta\omega_n \tau} \left[\cos\left(\sqrt{1-\zeta^2}\,\omega_n \tau\right) - \frac{\zeta}{\sqrt{1-\zeta^2}} \right. \\
& \left. \sin\left(\sqrt{1-\zeta^2}\,\omega_n \tau\right) \right] \quad \tau < 0
\end{aligned}
\tag{5.149}
$$

These results are illustrated in a sketch form in Figure 5.16 (a) for a case of light damping. When the value of τ is equal to zero, the curve exhibits continuous first and second derivatives, but the third derivative is discontinuous at that point. The correlation has a similarity to that of a sinusoid at frequency ω_n when the τ value is small, but when the τ value is big, the correlation falls to zero. It is typical for a narrow-band process to behave in this manner.

The corresponding analysis in the frequency domain includes less complex computations. The spectrum of the input is nothing more than the constant S_0. In accordance with equation (5.90), the spectrum of the output is as follows:

$$
S_{yy}(\omega) = |H(\omega)|^2 S_0 = \left| \frac{-1}{\omega_n^2 - \omega^2 + 2i\zeta\omega_n\omega} \right|^2 S_0 = \frac{S_0}{\left(\omega_n^2 - \omega^2\right)^2 + 4\zeta^2\omega_n^2\omega^2} \tag{5.150}
$$

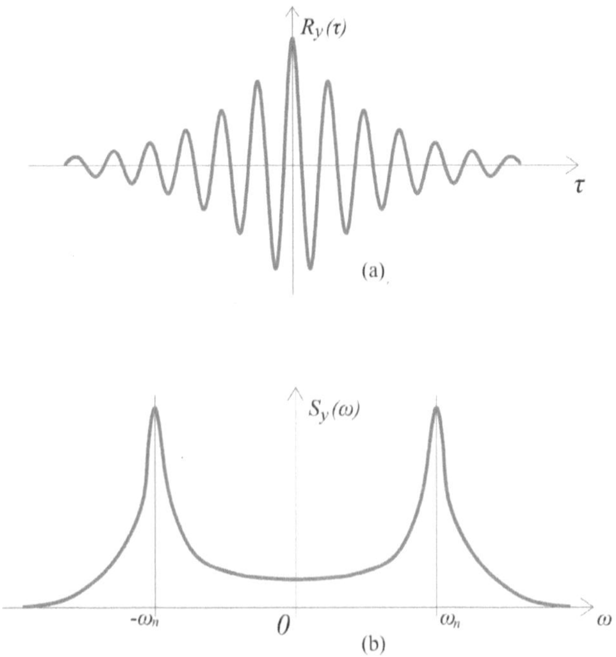

FIGURE 5.16 (a) Autocorrelation function and (b) mean square spectral density used to show the responses of the single-degree-of-freedom system in Figure 5.15.

in which $H(\omega)$ has been taken from equation (5.142). The spectrum is depicted in Figure 5.16 (b), once again representing a scenario of light damping. This is a typical example of a narrow-band spectrum. Given that both $R_{yy}(\tau)$ and $S_{yy}(\omega)$ are in accordance with the Wiener–Khintchine relations, it can be concluded that one implies the other, and both of them convey information that is essentially identical regarding the process.

It is possible to derive the mean square $E\left[y^2\right]$ of the response by utilizing either the autocorrelation function in equation (5.149) or the spectral density in equation (5.150). By utilizing equation (5.149), we may obtain the result directly from the autocorrelation function

$$E\left[y^2\right] = R_{yy}(0) = \frac{\pi S_0}{2\zeta\omega_n^{\ 3}} e^0\left[\cos(0)\right] = \frac{\pi S_0}{2\zeta\omega_n^{\ 3}}$$

In the event when the process is Gaussian or normal, its autocorrelation function (or power spectral density) provides a complete statistical description of the process. In the scenario where the mean is zero, the first-order probability distribution is described solely by the mean square value $E\left[y^2\right]$. In addition, the probability distribution of the envelope, the distribution of the peaks, and the average frequency of the narrowband response are all reliant on $E\left[y^2\right]$ alone in a Gaussian process with zero mean.

5.2.3 Application Examples

Example 5.2

▸ In a linear stochastic dynamical system, the input is a white noise spectrum denoted by S_0, and the impulse response function is represented by the equation

$$g(t) = e^{-\beta t} \qquad t \geq 0 \tag{5.151}$$

Find the system frequency response function and the system output spectral density function.

SOLUTION

The system frequency response function can be obtained as follows:

$$H(\omega) = \int_0^\infty e^{-\beta t} \cdot e^{-i\omega t}\, dt = \int_0^\infty e^{-\beta t}\left(\cos\omega t - i\sin\omega t\right) dt$$

$$= \int_0^\infty e^{-\beta t} \cos\omega t\, dt - i\int_0^\infty e^{-\beta t} \sin\omega t\, dt$$

Because

$$\int_0^\infty e^{-ax}\cos bx\, dx = \frac{a}{a^2+b^2} \text{ and } \int_0^\infty e^{-ax}\sin bx\, dx = \frac{b}{a^2+b^2},$$

we have

$$H(\omega) = \int_0^\infty e^{-\beta t}\cos\omega t\, dt - i\int_0^\infty e^{-\beta t}\sin\omega t\, dt = \frac{\beta}{\beta^2+\omega^2} - i\frac{\omega}{\beta^2+\omega^2} = \frac{1}{\beta+i\omega}$$

The output spectral density function can be calculated by utilizing equation (5.90) as follows:

$$S_{yy}(\omega) = \left|H(\omega)\right|^2 S_0 = \left|\frac{1}{\beta+i\omega}\right|^2 S_0 = \left|\frac{1}{\beta+i\omega}\right|^2 S_0 = \frac{S_0}{\beta^2+\omega^2}$$ ◂

Example 5.3

▸ Consider the SDOF system

$$m\ddot{y} + c\dot{y} + ky = F(t) \tag{5.152}$$

Over the range of $-\infty < \omega < \infty$, this system is excited by a zero mean ergodic random process denoted as $F(t)$, which possesses a constant power spectral density defined as S_0. Please determine the average rate of energy dissipation in this SDOF system with time.

SOLUTION

There is only one non-conservative force in the system, which is the damping force $c\dot{y}$. Therefore, the instantaneous rate of energy dissipation may be calculated using the formula $c\dot{y}^2$. As a result, the energy dissipation average rate, denoted by the symbol "P_{avg}", can be represented as

$$P_{avg} = c\left\langle \dot{y}^2 \right\rangle = cR_{\dot{y}\dot{y}}(0) \tag{5.153}$$

If the power spectral density and autocorrelation functions of the random variable $y(t)$ are given, it is straightforward to determine these functions for the time derivatives of $y(t)$, such as $\dot{y}(t)$. To exemplify the approach, let us examine the fundamental form of the autocorrelation function for $y(t)$

$$R_{yy}(\tau) \equiv E\left[y(t)y(t+\tau)\right] \tag{5.154}$$

Taking the derivative with respect to τ yields

$$R'_{yy}(\tau) = \frac{dR_{yy}(\tau)}{d\tau} = E\left[y(t)\dot{y}(t+\tau)\right] \tag{5.155}$$

Given that the process $y(t)$ is stationary, it follows that equation (5.155) can also be represented in the following manner:

$$R'_{yy}(\tau) = \frac{dR_{yy}(\tau)}{d\tau} = E\left[y(t-\tau)\dot{y}(t)\right] \tag{5.156}$$

Taking the derivative with respect to τ once more yields

$$R''_{yy}(\tau) = -E\left[\dot{y}(t-\tau)\dot{y}(t)\right] \tag{5.157}$$

Due to the fact that the ensemble average in equation (5.156) is, by definition, the autocorrelation function for $\dot{y}(t)$, it becomes evident that

$$R_{\dot{y}\dot{y}}(\tau) = -R''_{yy}(\tau) \tag{5.158}$$

Substituting equation (5.148) into equation (5.157) gives

$$R_{yy}(\tau) = \frac{\pi S_0}{2\zeta\omega_n^{\ 3}} e^{-\zeta\omega_n\tau}\left[\cos\left(\sqrt{1-\zeta^2}\,\omega_n\tau\right) + \frac{\zeta}{\sqrt{1-\zeta^2}} \sin\left(\sqrt{1-\zeta^2}\,\omega_n\tau\right)\right] \quad \tau > 0. \tag{5.159}$$

$$
\begin{aligned}
-R_{\dot{y}\dot{y}}(\tau) = R''_{yy}(\tau) = (\zeta\omega_n)^2 \frac{\pi S_0}{2\zeta\omega_n^{\ 3}} e^{-\zeta\omega_n\tau} \Bigg[&\cos\left(\sqrt{1-\zeta^2}\,\omega_n\tau\right) \\
&+ \frac{\zeta}{\sqrt{1-\zeta^2}} \sin\left(\sqrt{1-\zeta^2}\,\omega_n\tau\right) \Bigg] \\
&- \zeta\omega_n \frac{\pi S_0}{2\zeta\omega_n^{\ 3}} e^{-\zeta\omega_n\tau} \Bigg[-\sqrt{1-\zeta^2}\,\omega_n \sin\left(\sqrt{1-\zeta^2}\,\omega_n\tau\right) \\
&+ \sqrt{1-\zeta^2}\,\omega_n \frac{\zeta}{\sqrt{1-\zeta^2}} \cos\left(\sqrt{1-\zeta^2}\,\omega_n\tau\right) \Bigg] \\
&- \zeta\omega_n \frac{\pi S_0}{2\zeta\omega_n^{\ 3}} e^{-\zeta\omega_n\tau} \Bigg[-\sqrt{1-\zeta^2}\,\omega_n \sin\left(\sqrt{1-\zeta^2}\,\omega_n\tau\right) \\
&+ \sqrt{1-\zeta^2}\,\omega_n \frac{\zeta}{\sqrt{1-\zeta^2}} \cos\left(\sqrt{1-\zeta^2}\,\omega_n\tau\right) \Bigg] \\
&+ \frac{\pi S_0}{2\zeta\omega_n^{\ 3}} e^{-\zeta\omega_n\tau} \Bigg[-\left(\sqrt{1-\zeta^2}\,\omega_n\right)^2 \cos\left(\sqrt{1-\zeta^2}\,\omega_n\tau\right) \\
&- \left(\sqrt{1-\zeta^2}\,\omega_n\right)^2 \frac{\zeta}{\sqrt{1-\zeta^2}} \sin\left(\sqrt{1-\zeta^2}\,\omega_n\tau\right) \Bigg]
\end{aligned}
\tag{5.160}
$$

from which

$$
\begin{aligned}
-R_{\dot{y}\dot{y}}(0) = R''_{yy}(0) &= -\zeta\omega_n \frac{\pi S_0}{2\zeta\omega_n^{\ 3}} \left[\sqrt{1-\zeta^2}\,\omega_n \frac{\zeta}{\sqrt{1-\zeta^2}} \right] \\
&\quad + \frac{\pi S_0}{2\zeta\omega_n^{\ 3}} \left[-\left(\sqrt{1-\zeta^2}\,\omega_n\right)^2 \right] \\
&= -(\zeta\omega_n)^2 \frac{\pi S_0}{2\zeta\omega_n^{\ 3}} + \frac{\pi S_0}{2\zeta\omega_n^{\ 3}} \left[(\zeta^2-1)\,\omega_n^{\ 2} \right] \\
&= \frac{\pi S_0}{2\zeta\omega_n^{\ 3}} \left[(-1)\,\omega_n^{\ 2} \right] = \frac{-\pi S_0}{2\zeta\omega_n}
\end{aligned}
\tag{5.161}
$$

$$
R_{\dot{y}\dot{y}}(0) = \frac{\pi S_0}{2\zeta\omega_n} \tag{5.162}
$$

Combining equations (5.140), (5.153) and (5.162), we obtain

$$
P_{\text{avg}} = c \frac{\pi S_0}{2\zeta\omega_n} = \pi m S_0 \tag{5.163}
$$

◀

5.3 EXERCISES

EXERCISE 5.1

Calculate the complex frequency response function $H(\omega)$ for the velocity response $\dot{x}(t)$ of the SDOF system (shown in Figure 5.17), which has a mass m, a damping coefficient c, and no spring element. The excitation force is given by $mx(t)$. Please also calculate the value of $|H(\omega)|^2$ and plot the figure of $|H(\omega)|^2$.

EXERCISE 5.2

Calculate the steady-state velocity response of the SDOF system in Exercise 5.1 when the excitation is also steady-state, as defined by the equation

$$x(t) = f\cos\omega t, \quad -\infty < t < \infty$$

in which f and ω represent the amplitude and frequency, respectively.

EXERCISE 5.3

Given the vehicle model depicted in Figure 5.18, which has a mass denoted as m, a spring constant denoted as k, and a damping coefficient denoted as c, the task in this exercise is to calculate the frequency response function $Q(\omega)$ and the impulse response function $q(t)$ of the relative velocity $(\dot{x}_1 - \dot{x}_0)$ that corresponds to the acceleration excitation $\ddot{x}_0(t)$. Taking into consideration the fact $y = x_1 - x_0$, the equation of motion for this model is

$$m\ddot{y} + c\dot{y} + ky = -m\ddot{x}_0 \tag{5.164}$$

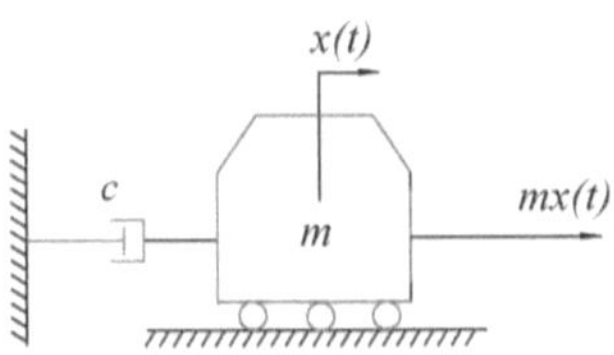

FIGURE 5.17 A single degree-of-freedom system with no spring element.

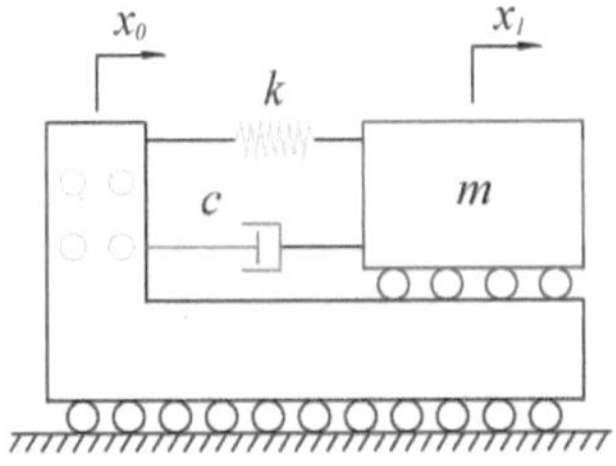

FIGURE 5.18 A single degree-of-freedom vehicle model.

EXERCISE 5.4

For the problem of Exercise 5.1, please determine the power spectral density $S_y(\omega)$ and the mean square $E\left[y^2(t)\right]$ of the stationary velocity response $y(t) = \dot{x}(t)$, with the assumption that the excitation $x(t)$ is a stationary white noise with a uniform spectral density value of S_0.

EXERCISE 5.5

The impulse response function, denoted by the symbol $q(t)$, of a linear system is expressed as

$$q(t) = \begin{cases} \mu e^{-\beta t} & \text{for } t > 0 \\ 0 & \text{for } t < 0 \end{cases} \tag{5.165}$$

in which μ is a constant, and β is also a constant.

Let's say that the input $x(t)$ is a stationary random process with zero mean, and the autocorrelation function of this process is given by

$$R_{xx}(\tau) = \sin\theta\,\delta(\tau) \tag{5.166}$$

in which θ is a constant, and $\delta(\tau)$ is a Dirac delta function. Please calculate the spectral density function and autocorrelation function of the output $y(t)$.

EXERCISE 5.6

Recall that in equation (5.90), we have the relationship

$$S_{yy}(\omega) = |H(\omega)|^2 S_{FF}(\omega) \tag{5.167}$$

This relationship shows that the spectral density function of the response displacement $y(t)$ is equal to the spectral density function of the load $F(t)$ multiplied by the squared modulus of the transfer function. If the response velocity is denoted by $\dot{y}(t)$ and the response acceleration is denoted by $\ddot{y}(t)$, show that

$$S_{y\dot{y}}(\omega) = i\omega|H(\omega)|^2 S_{FF}(\omega)$$

$$S_{\dot{y}\dot{y}}(\omega) = \omega^2|H(\omega)|^2 S_{FF}(\omega)$$

$$S_{\ddot{y}\ddot{y}}(\omega) = \omega^4|H(\omega)|^2 S_{FF}(\omega)$$

EXERCISE 5.7

Let's examine a spring-dashpot system with a negligible mass, which is governed by a differential equation

$$c\dot{y} + ky = f(t) \tag{5.168}$$

Under the assumption that $f(t)$ is a band-limited white noise, as follows:

$$S_{ff}(\omega)=\begin{cases} S_0, & |\omega|\le\omega_c \\ 0, & \text{otherwise} \end{cases} \tag{5.169}$$

Show that the mean-square value of the response displacement $y(t)$ is equal to

$$E\left(y^2\right)=\frac{2S_0}{kc}\tan^{-1}\left(\frac{c\omega_c}{k}\right) \tag{5.170}$$

Please also show that the mean-square value of the response velocity $\dot{y}(t)$ is equal to

$$E\left(\dot{y}^2\right)=\frac{2S_0}{c^2}\left[\omega_c-\frac{k}{c}\tan^{-1}\left(\frac{c\omega_c}{k}\right)\right] \tag{5.171}$$

EXERCISE 5.8

Let's examine a mass-spring-dashpot system with a light damping $0<\zeta\le 1$, which is governed by a differential equation

$$\ddot{y}+2\zeta\omega_0\dot{y}+\omega_0{}^2y=\frac{1}{m}f(t) \tag{5.172}$$

Under the assumption that $f(t)$ is a band-limited white noise, as follows:

$$S_{ff}(\omega)=\begin{cases} S_0, & |\omega|\leqslant\omega_c \\ 0, & \text{otherwise} \end{cases} \tag{5.173}$$

Show that the mean-square value of the response displacement $y(t)$ is equal to

$$E\left(y^2\right)=\frac{S_0}{2m^2\omega_0^2}\left\{\frac{1}{2\omega_0\sqrt{\left(1-\zeta^2\right)}}\ln\frac{\left(\omega_c+\omega_0\sqrt{\left(1-\zeta^2\right)}\right)^2+\omega_0^2\zeta^2}{\left(\omega_c-\omega_0\sqrt{\left(1-\zeta^2\right)}\right)^2+\omega_0^2\zeta^2}\right.$$
$$\left.+\frac{1}{\omega_0\zeta}\left[\tan^{-1}\left(\frac{\omega_c-\omega_0\sqrt{\left(1-\zeta^2\right)}}{\omega_0\zeta}\right)+\tan^{-1}\left(\frac{\omega_c+\omega_0\sqrt{\left(1-\zeta^2\right)}}{\omega_0\zeta}\right)\right]\right\} \tag{5.174}$$

6 Random Vibration of Two-Degree-of-Freedom Linear Systems

Systems are referred to as having two degrees of freedom when they require two independent coordinates in order to characterize their motion. For the purpose of providing a straightforward introduction to the behavior of systems with an arbitrarily high number of degrees of freedom, which is the topic of Chapter 7, we will only be discussing systems with two degrees of freedom in this chapter.

It is possible to adequately bring out all of the fundamental concepts and analysis methodology for the response of linear multi-degree-of-freedom systems to random excitation by conducting a comprehensive treatment of the problem for a linear two-degree-of-freedom system. Quite simply, this is due to the fact that in the case of deterministic or non-random systems, the problem of the two-degree-of-freedom systems carries all of the important features of the general multi-degree-of-freedom system, and the inclusion of random considerations does not change this fact. When it comes to comprehending the random response problem, the two-degree-of-freedom system's simplicity is quite beneficial.

6.1 A REVIEW OF THE DETERMINISTIC THEORY

For a free vibration system with two degrees of freedom, there exist two equations of motion: One for each mass (or, to be more accurate, one for each degree of freedom). In most cases, they take the form of coupled differential equations, which means that each equation involves all the coordinates. Assuming a harmonic solution for each coordinate, the free vibration equations of motion result in a frequency equation that provides two natural frequencies for the system. When the system is given appropriate initial excitation, it will vibrate at one of these natural frequencies. During free vibration at one of the natural frequencies, the amplitudes of the two degrees of freedom (coordinates) are interrelated in a certain manner. This configuration is referred to as a *normal mode*, *principal mode* or *natural mode* of vibration. Therefore, a system with two degrees of freedom possesses two normal modes of vibration, which correspond to the two natural frequencies.

In the event that we provide the system with an arbitrary initial excitation, the free vibration that is produced will be a superposition of the two normal modes of vibration. However, if the system vibrates as a result of the action of an external harmonic force, the forced harmonic vibration that is produced will occur at the

DOI: 10.1201/9781003431503-6

frequency of the force that is being applied. In the context of harmonic excitation, resonance takes place (that is, the amplitudes of the two coordinates will be at their highest) when the frequency of the forcing is equal to one of the natural frequencies of the system.

It is possible to specify the configuration of a system by using a set of independent coordinates, such as length, angle, or some other physical parameters. Generalized coordinates are a term that refers to any such set of coordinates. Even though the equations of motion of a system with two degrees of freedom are often coupled in such a manner that each equation involves all of the coordinates, it is always feasible to discover a specific set of coordinates in such a way that each equation of motion contains just one coordinate. After that, the equations of motion are decoupled, and they are able to be solved independently of one another. The term "principal coordinates" refers to such a set of coordinates that result in an equations system that is not coupled.

6.1.1 Motion Equations for Forced Vibration

Take into consideration a viscously damped spring-mass-damper system depicted in Figure 6.1(a), which has two degrees of freedom. The motion of the system is completely described by the coordinates $y_1(t)$ and $y_2(t)$, which determine the positions of the masses m_1 and m_2 at any time t from their respective equilibrium positions. The masses m_1 and m_2 are subject to the external forces $F_1(t)$ and $F_2(t)$,

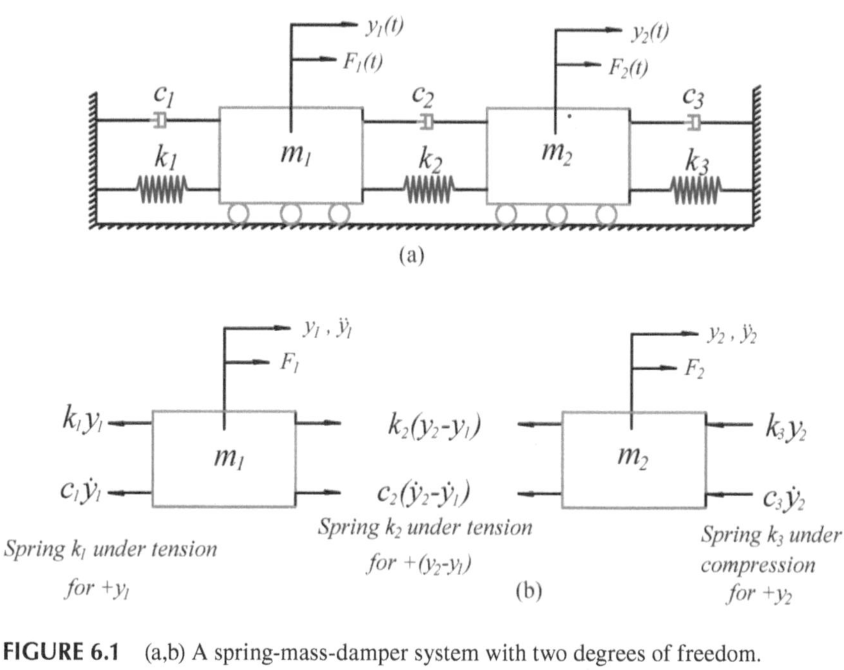

FIGURE 6.1 (a,b) A spring-mass-damper system with two degrees of freedom.

respectively. The figures in Figure 6.1(b) depict the free-body diagrams of masses m_1 and m_2. The equations of motion can be derived by applying Newton's second law of motion to each of the masses as follows:

$$m_1\ddot{y}_1 + (c_1 + c_2)\dot{y}_1 - c_2\dot{y}_2 + (k_1 + k_2)y_1 - k_2y_2 = F_1(t) \tag{6.1}$$

$$m_2\ddot{y}_2 - c_2\dot{y}_1 + (c_3 + c_2)\dot{y}_2 - k_2y_1 + (k_3 + k_2)y_2 = F_2(t) \tag{6.2}$$

It may be observed that equation (6.1) has terms that involve y_2 (namely, $-c_2\dot{y}_2$ and $-k_2y_2$), whereas equation (6.2) contains terms that involve y_1 (namely, $-c_2\dot{y}_1$ and $-k_2y_1$). Because of this, they constitute a system consisting of two coupled differential equations of the second order. Therefore, it is reasonable to anticipate that the motion of the mass m_1 will have an effect on the motion of the mass m_2, and vice versa.

One possible way to express equations (6.1) and (6.2) in matrix form is as follows:

$$\mathbf{M}\ddot{\mathbf{y}} + \mathbf{C}\dot{\mathbf{y}} + \mathbf{K}\mathbf{y} = \mathbf{F}(t) \tag{6.3}$$

in which **M**, **C** and **K** are referred to as the mass, damping and stiffness matrices, respectively, and are given by

$$\mathbf{M} = \begin{bmatrix} m_1 & 0 \\ 0 & m_2 \end{bmatrix}$$

$$\mathbf{C} = \begin{bmatrix} c_1 + c_2 & -c_2 \\ -c_2 & c_2 + c_3 \end{bmatrix}$$

$$\mathbf{K} = \begin{bmatrix} k_1 + k_2 & -k_2 \\ -k_2 & k_2 + k_3 \end{bmatrix}$$

and **y** and **F**(t) are referred to as the displacement and force vectors, respectively, and are given by

$$\mathbf{y} = \begin{Bmatrix} y_1(t) \\ y_2(t) \end{Bmatrix}$$

$$\mathbf{F}(t) = \begin{Bmatrix} F_1(t) \\ F_2(t) \end{Bmatrix}$$

It is evident that **M**, **C** and **K** are all 2×2 matrices whose elements are the values of the known masses, damping coefficients and stiffnesses of the system, respectively. Moreover, it can be observed that these matrices are symmetric, indicating that

$$\mathbf{M}^T = \mathbf{M}, \quad \mathbf{C}^T = \mathbf{C}, \quad \mathbf{K}^T = \mathbf{K}$$

in which the superscript T represents the transpose of the matrix.

It should be noted that the equations of motion in equations (6.1) and (6.2) only become uncoupled (independent of one another) when $c_2 = k_2 = 0$. This indicates that the two masses m_1 and m_2 are not physically connected to one another. When this occurs, the matrices **M**, **C** and **K** transform into diagonal matrices. For any arbitrary forces $F_1(t)$ and $F_2(t)$, it is difficult to obtain the solution of the equations of motion in equations (6.1) and (6.2). This is mainly due to the fact that the variables $y_1(t)$ and $y_2(t)$ are coupled to one another. The solution to equations (6.1) and (6.2) requires the use of four constants of integration, with two constants being required for each equation. It is common practice to specify the initial displacements and velocities of the two masses as follows: $y(t=0)=y_1(0)$, $y_1(t=0)=\dot{y}_1(0)$, $y_2(t=0)=y_2(0)$ and $\dot{y}_2(t=0)=\dot{y}_2(0)$. In the first place, we will take into consideration the free-vibration solution of equations (6.1) and (6.2).

6.1.2 Free Response of an Undamped System

We will begin by conducting an analysis of the system depicted in Figure 6.2. For the sake of clarity, external forcings and damping have been left out of the discussion; they will be discussed later in this chapter.

We will concentrate on two-degree-of-freedom problems when we are introducing and discussing multi-degree-of-freedom characteristics. This is because these two-degree-of-freedom problems are simple to comprehend, can be solved analytically and can be proved empirically by utilizing a basic spring with some masses. However, the conclusions will be equally applicable to problems with any number of degrees of freedom that the researcher desires. From the free body diagram depicted in Figure 6.2(b), we can observe that the equations of motion for this system may be expressed as

$$m_1\ddot{y}_1 + (k_1 + k_2)y_1 - k_2y_2 = 0 \tag{6.4}$$

$$m_2\ddot{y}_2 - k_2y_1 + (k_3 + k_2)y_2 = 0 \tag{6.5}$$

Equations (6.4) and (6.5) can just as easily be re-expressed as

$$\begin{bmatrix} m_1 & 0 \\ 0 & m_2 \end{bmatrix}\begin{Bmatrix} \ddot{y}_1(t) \\ \ddot{y}_2(t) \end{Bmatrix} + \begin{bmatrix} k_1 + k_2 & -k_2 \\ -k_2 & k_2 + k_3 \end{bmatrix}\begin{Bmatrix} y_1(t) \\ y_2(t) \end{Bmatrix} = \begin{Bmatrix} 0 \\ 0 \end{Bmatrix} \tag{6.6}$$

It can easily be seen that equation (6.6) can just as easily be re-expressed as

$$\mathbf{M}\ddot{\mathbf{y}} + \mathbf{K}\mathbf{y} = \mathbf{0} \tag{6.7}$$

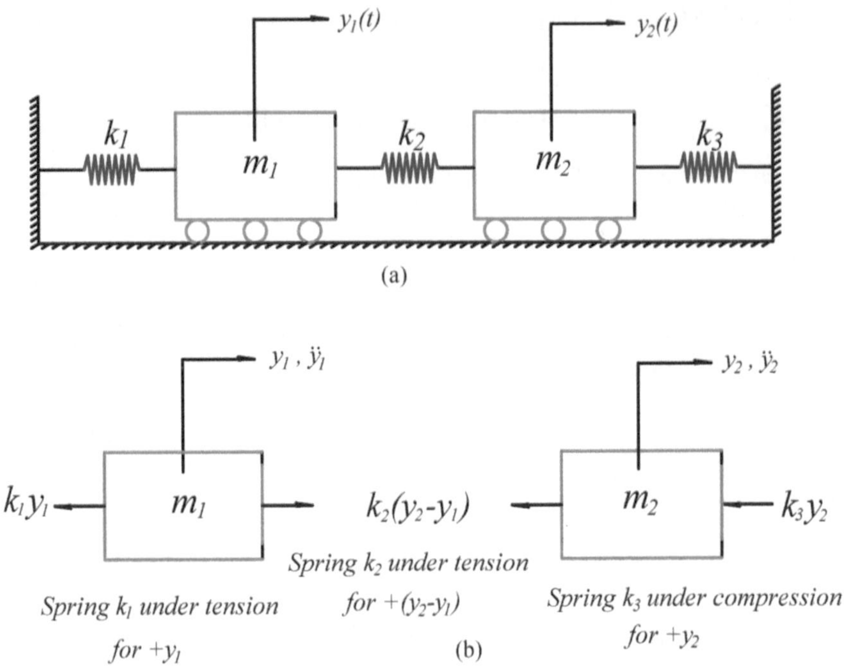

FIGURE 6.2 (a,b) A spring-mass system with two degrees of freedom.

in which **M** and **K** are referred to as the mass and stiffness matrices, respectively, and are given by

$$\mathbf{M} = \begin{bmatrix} m_1 & 0 \\ 0 & m_2 \end{bmatrix} \qquad \mathbf{K} = \begin{bmatrix} k_1 + k_2 & -k_2 \\ -k_2 & k_2 + k_3 \end{bmatrix} \tag{6.8}$$

and **y** and **0** are referred to as the displacement and force vectors, respectively, and are given by

$$\mathbf{y} = \begin{Bmatrix} y_1(t) \\ y_2(t) \end{Bmatrix} \mathbf{0} = \begin{Bmatrix} 0 \\ 0 \end{Bmatrix} \tag{6.9}$$

Therefore, we have a form in equation (6.7) that appears to be exactly the same as our single degree of freedom (SDOF) oscillator. Currently, our system consists of matrices representing the "masses" and "springs", and our dependent variable is a vector rather than a scalar. To clarify, our "masses" and "springs" are now matrices, and our dependent variable is a vector rather than a scalar. This is the difference.

Our curiosity lies in determining whether m_1 and m_2 are capable of oscillating harmonically with the same frequency and phase angle, but with amplitudes that are distinct from one another. Given the assumption that it is feasible for both m_1 and m_2 to exhibit harmonic motion at the identical frequency ω and phase angle ϕ, we take the solutions of equations (6.4) and (6.5) as:

$$\begin{aligned} y_1(t) &= Y_1\cos(\omega t+\phi) \\ y_2(t) &= Y_2\cos(\omega t+\phi) \end{aligned} \tag{6.10}$$

in which the values Y_1 and Y_2 represent the maximum amplitudes of $y_1(t)$ and $y_2(t)$, while the phase angle is denoted by the character ϕ. When we plug equation (6.10) into equations (6.4) and (6.5), we get the following:

$$\begin{aligned} \left[\left\{-m_1\omega^2+(k_1+k_2)\right\}Y_1-k_2Y_2\right]\cos(\omega t+\phi) &= 0 \\ \left[-k_2Y_1+\left\{-m_2\omega^2+(k_2+k_3)\right\}Y_2\right]\cos(\omega t+\phi) &= 0 \end{aligned} \tag{6.11}$$

In order to ensure that equation (6.11) is satisfied for all possible values of time t, the terms that are included in brackets must be equal to zero. This results in

$$\begin{aligned} \left\{-m_1\omega^2+(k_1+k_2)\right\}Y_1-k_2Y_2 &= 0 \\ -k_2Y_1+\left\{-m_2\omega^2+(k_2+k_3)\right\}Y_2 &= 0 \end{aligned} \tag{6.12}$$

which depict two simultaneous homogeneous algebraic equations in the variables Y_1 and Y_2 that are unknown. Equation (6.12) is shown to be satisfied by the trivial solution $Y_1 = Y_2 = 0$, which indicates that there is no vibration. In order to have a solution that is not trivial for Y_1 and Y_2, the determinant of the coefficients of Y_1 and Y_2 must be equal to zero:

$$\det\left(\begin{bmatrix} k_1+k_2-\omega^2m_1 & -k_2 \\ -k_2 & k_2+k_3-\omega^2m_2 \end{bmatrix}\right)=0 \tag{6.13}$$

Upon evaluating this determinant, we will obtain:

$$\left(k_1+k_2-\omega^2m_1\right)\left(k_2+k_3-\omega^2m_2\right)-k_2^2=0 \tag{6.14}$$

By expanding this expression and combining the terms with powers of ω together, we obtain

$$m_1m_2\omega^4-\omega^2\left((k_2+k_3)m_1+(k_1+k_2)m_2\right)+k_1k_2+k_1k_3+k_2k_3=0 \tag{6.15}$$

The equation (6.15) is referred to as the frequency or characteristic equation due to the fact that its solution provides the frequencies or the characteristic values of the system under consideration. All of the roots of equation (6.15) can be found using

$$\omega_1^2 = \frac{1}{2}\left\{\frac{(k_1+k_2)m_2+(k_2+k_3)m_1}{m_1 m_2}\right\} - \frac{1}{2}\left[\left\{\frac{(k_1+k_2)m_2+(k_2+k_3)m_1}{m_1 m_2}\right\}^2 - 4\left\{\frac{(k_1+k_2)(k_2+k_3)-k_2^2}{m_1 m_2}\right\}\right]^{1/2} \quad (6.16)$$

$$\omega_2^2 = \frac{1}{2}\left\{\frac{(k_1+k_2)m_2+(k_2+k_3)m_1}{m_1 m_2}\right\} + \frac{1}{2}\left[\left\{\frac{(k_1+k_2)m_2+(k_2+k_3)m_1}{m_1 m_2}\right\}^2 - 4\left\{\frac{(k_1+k_2)(k_2+k_3)-k_2^2}{m_1 m_2}\right\}\right]^{1/2} \quad (6.17)$$

This demonstrates that it is feasible for the system to possess a non-trivial harmonic solution in the form of equation (6.10), provided that ω is equal to ω_1 (or ω_2) as indicated by equation (6.16) (or equation (6.17)). We refer to ω_1 and ω_2 as the system's natural frequencies.

The values of Y_1 and Y_2 have not yet been determined. These values depend upon the natural frequencies ω_1 and ω_2. The values of Y_1 and Y_2 that correspond to ω_1 will be denoted as $Y_1^{(1)}$ and $Y_2^{(1)}$, whereas the values that correspond to ω_2 will be denoted as $Y_1^{(2)}$ and $Y_2^{(2)}$. Moreover, due to the homogeneity of equation (6.12), it is only possible to determine the ratios $r_1 = \left\{Y_2^{(1)}/Y_1^{(1)}\right\}$ and $r_2 = \left\{Y_2^{(2)}/Y_1^{(2)}\right\}$.

At this point, it should be noted that one of the most significant distinctions between a system with a two-degree-of-freedom and one with a single degree of freedom is that in the first case, we have to be concerned with an additional mass. The important consequence of this fact is that a sinusoidal response of the two-degree-of-freedom system will automatically imply a relationship between the motions of the various masses in the structure. For example, in a system with two degrees of freedom, the motion of one mass could be represented by a cosine and the motion of the other by a cosine with twice the amplitude of the first. Since the fundamental property (cosine oscillations) is the same for both masses, we should focus on the relative motion between them in order to define the motion. In the case that we have just mentioned, this would mean noting that one mass moves twice as much as the other mass. This idea of relative motion is a crucial one, and because it underlies most of the remaining parts of this chapter and the following chapter, we will study it in some detail.

For $\omega^2 = \omega_1^2$ and $\omega^2 = \omega_2^2$, equation (6.12) gives

$$r_1 = \frac{Y_2^{(1)}}{Y_1^{(1)}} = \frac{k_2}{-m_2\omega_1^2 + (k_2+k_3)}$$
$$r_2 = \frac{Y_2^{(2)}}{Y_1^{(2)}} = \frac{k_2}{-m_2\omega_2^2 + (k_2+k_3)} \quad (6.18)$$

The normal modes of vibration associated with ω_1^2 and ω_2^2 can be represented respectively, as follows:

$$\mathbf{Y}^{(1)} = \begin{Bmatrix} Y_1^{(1)} \\ Y_2^{(1)} \end{Bmatrix} = \begin{Bmatrix} Y_1^{(1)} \\ r_1 Y_1^{(1)} \end{Bmatrix} \tag{6.19}$$

$$\mathbf{Y}^{(2)} = \begin{Bmatrix} Y_1^{(2)} \\ Y_2^{(2)} \end{Bmatrix} = \begin{Bmatrix} Y_1^{(2)} \\ r_2 Y_1^{(2)} \end{Bmatrix} \tag{6.20}$$

The vectors $\mathbf{Y}^{(1)}$ and $\mathbf{Y}^{(2)}$, which represent the normal modes of vibration, are referred to as the modal vectors of the system. The solution for free vibration or the motion over time can be written using equation (6.10) as:

$$\mathbf{y}^{(1)}(t) = \begin{Bmatrix} y_1^{(1)}(t) \\ y_2^{(1)}(t) \end{Bmatrix} = \begin{Bmatrix} Y_1^{(1)} \cos(\omega_1 t + \phi_1) \\ r_1 Y_1^{(1)} \cos(\omega_1 t + \phi_1) \end{Bmatrix} = \text{first mode} \tag{6.21}$$

$$\mathbf{y}^{(2)}(t) = \begin{Bmatrix} y_1^{(2)}(t) \\ y_2^{(2)}(t) \end{Bmatrix} = \begin{Bmatrix} Y_1^{(2)} \cos(\omega_1 t + \phi_1) \\ r_1 Y_1^{(2)} \cos(\omega_1 t + \phi_1) \end{Bmatrix} = \text{first mode} \tag{6.22}$$

in which the constants $Y_1^{(1)}$, $Y_1^{(2)}$, ϕ_1 and ϕ_2 are determined by the initial conditions.

Example 6.1

▶ Please determine the natural frequencies and modal vectors for the vibration system shown in Figure 6.2(a): $m_1 = 3$ kg, $m_2 = 5$ kg, $k_1 = 50$ N/m, $k_2 = 150$ N/m and $k_3 = 250$ N/m.

SOLUTION

Performing a force balance allows one to determine the equations of motion for the system, which are as follows:

$$m_1\ddot{y}_1 + (k_1 + k_2)y_1 - k_2 y_2 = 0$$

$$m_2\ddot{y}_2 - k_2 y_1 + (k_3 + k_2)y_2 = 0$$

The result of putting this into matrix form and utilizing the parameter values that have been provided is

$$\begin{bmatrix} 3 & 0 \\ 0 & 5 \end{bmatrix}\begin{Bmatrix} \ddot{y}_1(t) \\ \ddot{y}_2(t) \end{Bmatrix} + \begin{bmatrix} 200 & -150 \\ -150 & 400 \end{bmatrix}\begin{Bmatrix} y_1(t) \\ y_2(t) \end{Bmatrix} = \begin{Bmatrix} 0 \\ 0 \end{Bmatrix}$$

Utilizing equations (6.16) and (6.17), the natural frequencies of this system can be calculated as follows:

$$\omega_1^2 = \frac{1}{2}\left\{\frac{(200)\times 5 + 400\times 3}{3\times 5}\right\} - \frac{1}{2}\left[\left\{\frac{(200)\times 5 + 400\times 3}{3\times 5}\right\}^2 - 4\left\{\frac{200\times 400 - 150^2}{3\times 5}\right\}\right]^{1/2} = 34.034$$

$$\omega_2^2 = \frac{1}{2}\left\{\frac{(200)\times 5 + 400\times 3}{3\times 5}\right\} + \frac{1}{2}\left[\left\{\frac{(200)\times 5 + 400\times 3}{3\times 5}\right\}^2 - 4\left\{\frac{200\times 400 - 150^2}{3\times 5}\right\}\right]^{1/2} = 112.633$$

Utilizing equations (6.19) and (6.20), the modal vectors for this vibration system can be calculated as follows:

$$r_1 = \frac{150}{-5\times 34.034 + (150+250)} = \frac{150}{229.83} = 0.653$$

$$r_2 = \frac{150}{-5\times 112.633 + (150+250)} = \frac{150}{-163.165} = -0.919$$

$$\mathbf{Y}^{(1)} = \begin{Bmatrix} Y_1^{(1)} \\ Y_2^{(1)} \end{Bmatrix} = \begin{Bmatrix} Y_1^{(1)} \\ 0.653Y_1^{(1)} \end{Bmatrix}$$

$$\mathbf{Y}^{(2)} = \begin{Bmatrix} Y_1^{(2)} \\ Y_2^{(2)} \end{Bmatrix} = \begin{Bmatrix} Y_1^{(2)} \\ -0.919Y_1^{(2)} \end{Bmatrix}$$

6.1.3 Forced-vibration Analysis of an Undamped System

We take into consideration the simple two-degree-of-freedom structural model that is depicted in Figure 6.3. In terms of the displacement y_j, excitations $F_j(t); j = 1, 2$, the spring constant k and mass m, the equations of motion are found to be as follows:

$$m\ddot{y}_1 + 2ky_1 - ky_2 = F_1(t) \tag{6.23}$$

$$m\ddot{y}_2 - ky_1 + 2ky_2 = F_2(t) \tag{6.24}$$

First, let us consider the scenario of a normal mode vibration, where $F(t) = 0$ and $F_2(t) = 0$. Furthermore, let us assume that the motion of the two masses is

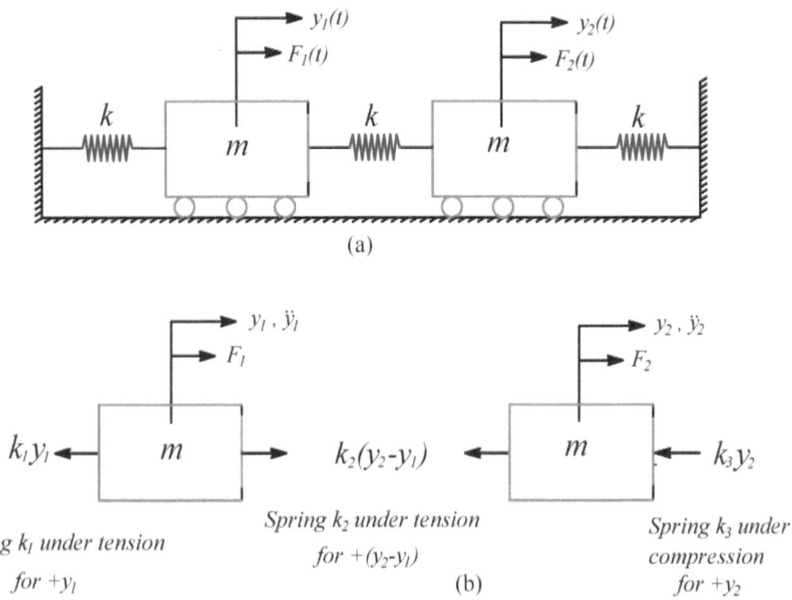

FIGURE 6.3 (a,b) A forced spring-mass system with two degrees of freedom.

harmonic, with unknown amplitudes Y_1 and Y_2, respectively, at the same unknown frequency ω.

$$y_1(t) = Y_1 \sin \omega t; \quad y_2(t) = Y_2 \sin \omega t \tag{6.25}$$

With the assumption that there is no load, the following equations are obtained by substituting equation (6.25) into equations (6.23) and (6.24):

$$-mY_1Y_2\omega^2 + 2kY_1Y_2 - kY_2{}^2 = 0 \tag{6.26}$$

$$-mY_1Y_2\omega^2 + 2kY_1Y_2 - kY_1{}^2 = 0 \tag{6.27}$$

Solving equations (6.23) and (6.24) leads to the following results:

$$Y_1 = Y_2 \quad \text{and} \quad Y_1 = -Y_2 \tag{6.28}$$

$$\omega_1 = \sqrt{k/m} \quad \text{and} \quad \omega_2 = \sqrt{3k/m} \tag{6.29}$$

Utilizing equation (6.18), we can obtain:

$$\begin{aligned} r_1 &= \frac{Y_2^{(1)}}{Y_1^{(1)}} = \frac{k_2}{-m_2\omega_1^2 + (k_2 + k_3)} = \frac{k}{-m\dfrac{k}{m} + (2k)} = 1 \\ r_2 &= \frac{Y_2^{(2)}}{Y_1^{(2)}} = \frac{k_2}{-m_2\omega_2^2 + (k_2 + k_3)} = \frac{k}{-m\dfrac{3k}{m} + (2k)} = -1 \end{aligned} \tag{6.30}$$

The normal modes of vibration associated with ω_1^2 and ω_2^2 can be represented, respectively, as follows:

$$\mathbf{Y}^{(1)} = \begin{Bmatrix} Y_1^{(1)} \\ Y_2^{(1)} \end{Bmatrix} = \begin{Bmatrix} Y_1^{(1)} \\ 1Y_1^{(1)} \end{Bmatrix} \tag{6.31}$$

$$\mathbf{Y}^{(2)} = \begin{Bmatrix} Y_1^{(2)} \\ Y_2^{(2)} \end{Bmatrix} = \begin{Bmatrix} Y_1^{(2)} \\ -1Y_1^{(2)} \end{Bmatrix} \tag{6.32}$$

The natural mode shapes can also be written as follows:

$$\frac{Y_{11}}{Y_{21}} = 1 \qquad \text{and} \qquad \frac{Y_{12}}{Y_{22}} = -1 \tag{6.33}$$

in which the mode number is indicated by the second index. We will now proceed to discuss the normal coordinates, which are defined by

$$q_1 = Y_{11}y_1 + Y_{12}y_2 = y_1 + y_2 \tag{6.34}$$

$$q_2 = Y_{21}y_1 + Y_{22}y_2 = y_1 - y_2 \tag{6.35}$$

Through the process of substituting y_1 and y_2 from equations (6.34) and (6.35) into equations (6.23) and (6.24), with the condition that $Y_{11} = Y_{12} = 1$, we are able to obtain the uncoupled and forced equations of motion

$$\ddot{q}_1 + \omega_1^2 q_1 = \frac{1}{m}\left[F_1(t) + F_2(t)\right] = b_1(t) \tag{6.36}$$

$$\ddot{q}_2 + \omega_2^2 q_2 = \frac{1}{m}\left[F_1(t) - F_2(t)\right] = b_2(t) \tag{6.37}$$

As two separate single-degree-of-freedom systems, the solutions for the responses $q_1(t)$ and $q_2(t)$ to transient excitations $b_1(t)$ and $b_2(t)$ can be determined from superposition in the time domain. These solutions are provided by Duhamel's integrals

$$q_1(t) = \int_{-\infty}^{\infty} b_1(t-\theta)g_1(\theta)\,d\theta \tag{6.38}$$

$$q_2(t) = \int_{-\infty}^{\infty} b_2(t-\theta)g_2(\theta)\,d\theta \tag{6.39}$$

in which $g_1(\theta)$ and $g_2(\theta)$ represent the impulse response functions for the uncoupled systems equations (6.36) and (6.37). These functions are given by

$$g_j(t)=\begin{cases}\dfrac{1}{\omega_j}\sin\omega_j t & \text{for} \quad t>0,\ \ j=1,\,2\\ 0 & \text{for} \quad t<0\end{cases} \tag{6.40}$$

As an alternative, the Fourier integrals can be used to derive the solutions for the responses $q_1(t)$ and $q_2(t)$ through superposition in the frequency domain

$$q_1(t)=\frac{1}{2\pi}\int_{-\infty}^{\infty}H_1(\omega)e^{i\omega t}\left[\int_{-\infty}^{\infty}b_1(\theta)e^{-i\omega\theta}\,d\theta\right]d\omega \tag{6.41}$$

$$q_2(t)=\frac{1}{2\pi}\int_{-\infty}^{\infty}H_2(\omega)e^{i\omega t}\left[\int_{-\infty}^{\infty}b_2(\theta)e^{-i\omega\theta}\,d\theta\right]d\omega \tag{6.42}$$

in which the functions $H_1(\omega)$ and $H_2(\omega)$ represent the complex frequency response functions for the uncoupled systems, as described by equations (6.36) and (6.37). These functions are given by

$$H_j(\omega)=\frac{1}{\omega_j^2-\omega^2};\quad j=1,\,2 \tag{6.43}$$

Given that the original two-degree-of-freedom system is converted into two uncoupled single-degree-of-freedom systems by the normal mode approach in equations (6.36) and (6.37), it is essential to take note of the fact that the solution strategy that is described in equations (6.38)–(6.43) is identical to the one that is used for the single-degree-of-freedom system. Similar to the solutions for the single-degree-of-freedom system, the unit solutions, denoted as $g_j(t)$ and $H_j(\omega)$, are identical. Therefore, similar to the system with a single-degree-of-freedom, $g_j(t)$ and $H_j(\omega)$ are Fourier transforms of each other with

$$g_j(t)=\frac{1}{2\pi}\int_{-\infty}^{\infty}H_j(\omega)e^{i\omega t}\,d\omega \tag{6.44}$$

$$H_j(\omega)=\int_{-\infty}^{\infty}g_j(t)e^{-i\omega t}\,dt \tag{6.45}$$

Once the normal coordinates $q_1(t)$ and $q_2(t)$ have been obtained, the response displacements $y_1(t)$ and $y_2(t)$ can be easily computed by using equations (6.34) and (6.35) as follows:

$$y_1(t) = \frac{1}{2}\left[q_1(t) + q_2(t)\right] \tag{6.46a}$$

$$y_2(t) = \frac{1}{2}\left[q_1(t) - q_2(t)\right] \tag{6.46b}$$

Therefore, the solution to the problem of the linear two-degree-of-freedom system's transient response to non-random excitations $F_1(t)$ and $F_2(t)$ has now been finished.

The decoupling of the two-degree-of-freedom system into two independent single-degree-of-freedom systems through the use of the modal superposition method is the most important part of the solution approach. Each and every analysis that was performed with the single-degree-of-freedom system is capable of being transferred over to the two-degree-of-freedom system. For the two-degree-of-freedom system, the analysis that is novel involves the determination of the natural frequencies ω_i and the natural mode shapes Y_{11}/Y_{12}, Y_{12}/Y_{22}. Additionally, the normal coordinates $q_1(t)$ and $q_2(t)$ are introduced by the analysis. Once the system is decoupled, the extension from the SDOF systems to the two-degree-of-freedom (TDOF) systems involves just simple algebraic operations. This will be demonstrated in Section 6.1.2. In addition, the random analysis will demonstrate that this extension is possible. As an additional point of interest, in the random analysis, the extension from the SDOF systems to the two-degree-of-freedom systems involves only simple algebraic operations after the system is decoupled, as we will demonstrate in Section 6.2.

Example 6.2

► In Figure 6.3, the two-degree-of-freedom system is subjected to a steady-state excitation force denoted by the symbol $F_1(t)$. Assuming that

$$F_1(t) = 58F_0 \sin \omega t$$

and $F_2(t) = 0$, please determine the response displacements $y_1(t)$ and $y_2(t)$.

SOLUTION

We are able to obtain the uncoupled and forced equations of motion

$$\ddot{q}_1 + \omega_1^2 q_1 = \frac{1}{m}\left[F_1(t) + F_2(t)\right] = b_1(t) = \frac{58F_0 \sin \omega t}{m}$$

$$\ddot{q}_2 + \omega_2^2 q_2 = \frac{1}{m}\left[F_1(t) - F_2(t)\right] = b_2(t) = \frac{58F_0 \sin \omega t}{m}$$

Therefore, it may be deduced that:

$$q_1 = \frac{58F_0 \sin \omega t}{m} H_1(\omega)$$

By substituting the expression for $H_1(\omega)$ from equation (6.43) into the equation mentioned before, we obtain

$$q_1 = \frac{58F_0 \sin \omega t}{m} \frac{1}{\omega_1^2 - \omega^2}$$

In a similar manner,

$$q_2(t) = \frac{58F_0 \sin \omega t}{m} \frac{1}{\omega_2^2 - \omega^2}$$

The steady-state response displacements from equations (6.19) and (6.20) are

$$y_1(t) = \frac{29F_0 \sin \omega t}{m} \left[\frac{1}{\omega_1^2 - \omega^2} + \frac{1}{\omega_2^2 - \omega^2} \right] \tag{6.47a}$$

$$y_2(t) = \frac{29F_0 \sin \omega t}{m} \left[\frac{1}{\omega_1^2 - \omega^2} - \frac{1}{\omega_2^2 - \omega^2} \right] \tag{6.47b}$$

It is important to observe that when the forcing frequency $\omega \to \omega_1$, the first term on the right side of the equation (6.47a) (or the equation (6.47b)) becomes the most significant, resulting in $y_1(t)$ being approximately equal to $y_2(t)$. As a result, the system with two degrees of freedom vibrates primarily in its first mode, with the ratio Y_{11}/Y_{21} being approximately 1. When the frequency ω approaches ω_2, the second term on the right side of the equation (6.47a) (or the equation (6.47b)) becomes dominant, resulting in $y_2(t)$ being almost equal to $-y_1(t)$. As a result, the system vibrates in its second mode. In a state of steady-state vibration, the frequency of the response is consistently equal to the frequency of the excitation.

6.1.4 Forced-vibration Analysis of a Damped System

In the foregoing subsection, damping has been excluded in order to maintain simplicity and clarity while addressing the distinct challenges that arise in the two-degree-of-freedom system. When damping is included, the only new problem lies in the deterministic aspect of decoupling, as we will demonstrate in the subsequent explanation. Naturally, the inclusion of the additional damping term results in lengthier expressions.

For the purpose of simplifying the derivations, we will only take into consideration a single excitation, denoted by $F_1(t)$. This is due to the fact that the problem with the multiple excitations in the damped case is basically similar to the

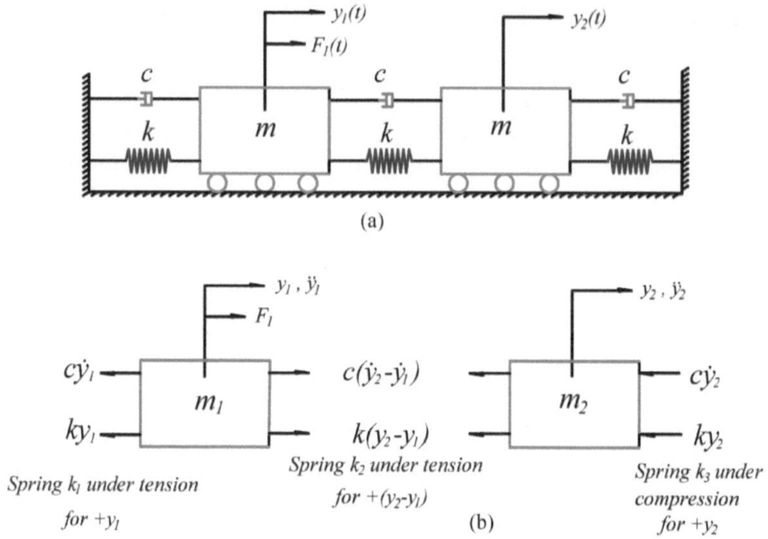

FIGURE 6.4 (a,b) A forced spring-mass-damper system with two degrees of freedom.

problem in the undamped case. In addition, we will organize all of the terms in a matrix format in anticipation of the upcoming matrix formulation and operations that will be performed on the general multi-degree-of-freedom systems.

The equations of motion for the system depicted in Figure 6.4, which has two degrees of freedom, are as follows:

$$\begin{bmatrix} m & 0 \\ 0 & m \end{bmatrix}\begin{Bmatrix} \ddot{y}_1(t) \\ \ddot{y}_2(t) \end{Bmatrix}+\begin{bmatrix} 2c & -c \\ -c & 2c \end{bmatrix}\begin{Bmatrix} \dot{y}_1(t) \\ \dot{y}_2(t) \end{Bmatrix}+\begin{bmatrix} 2k & -k \\ -k & 2k \end{bmatrix}\begin{Bmatrix} y_1(t) \\ y_2(t) \end{Bmatrix}=\begin{Bmatrix} F_1(t) \\ 0 \end{Bmatrix} \tag{6.48}$$

Another way to express equation (6.48) in matrix form is as follows:

$$\mathbf{M}\ddot{\mathbf{y}}+\mathbf{C}\dot{\mathbf{y}}+\mathbf{K}\mathbf{y}=\mathbf{F}(t) \tag{6.49}$$

in which **M**, **C** and **K** are referred to as the mass, damping and stiffness matrices, respectively, and are given by

$$\mathbf{M}=\begin{bmatrix} m & 0 \\ 0 & m \end{bmatrix}$$

$$\mathbf{C}=\begin{bmatrix} 2c & -c \\ -c & 2c \end{bmatrix}$$

$$\mathbf{K}=\begin{bmatrix} 2k & -k \\ -k & 2k \end{bmatrix}$$

and $\mathbf{y}$ and $\mathbf{F}(t)$ are referred to as the displacement and force vectors, respectively, and are given by

$$\mathbf{y} = \begin{Bmatrix} y_1(t) \\ y_2(t) \end{Bmatrix}$$

$$\mathbf{F}(t) = \begin{Bmatrix} F_1(t) \\ 0 \end{Bmatrix}$$

Once more, the undamped natural frequencies corresponding to the vibration system in Figure 6.4 are deduced from the study of the problem of the undamped free vibration

$$\omega_1 = \sqrt{k/m} \quad \text{and} \quad \omega_2 = \sqrt{3k/m} \tag{6.50}$$

The corresponding natural mode shapes are

$$\boldsymbol{\psi}_1 = \begin{Bmatrix} \psi_{11} \\ \psi_{21} \end{Bmatrix} = \begin{Bmatrix} 1 \\ 1 \end{Bmatrix}, \quad \boldsymbol{\psi}_2 = \begin{Bmatrix} \psi_{12} \\ \psi_{22} \end{Bmatrix} = \begin{Bmatrix} 1 \\ -1 \end{Bmatrix} \tag{6.51}$$

Introduce the normal coordinates $\mathbf{q}$ in such a way that the response displacements are calculated as shown below

$$\mathbf{y} = \boldsymbol{\psi}\mathbf{q} = \{\boldsymbol{\psi}_1 \ \boldsymbol{\psi}_2\}\begin{Bmatrix} q_1 \\ q_2 \end{Bmatrix} = \boldsymbol{\psi}_1 q_1 + \boldsymbol{\psi}_2 q_2 = \begin{Bmatrix} 1 \\ 1 \end{Bmatrix} q_1 + \begin{Bmatrix} 1 \\ -1 \end{Bmatrix} q_2 \tag{6.52}$$

By substituting equation (6.52) into equation (6.49), we obtain the following:

$$\mathbf{M}\boldsymbol{\psi}\ddot{\mathbf{q}} + \mathbf{C}\boldsymbol{\psi}\ddot{\mathbf{q}} + \mathbf{K}\boldsymbol{\psi}\mathbf{q} = \mathbf{F}(t) \tag{6.53}$$

By multiplying equation (6.53) with the transpose of the j th modal vector $\boldsymbol{\psi}_j^T$, we obtain

$$\boldsymbol{\psi}_j^T\mathbf{M}\boldsymbol{\psi}\ddot{\mathbf{q}} + \boldsymbol{\psi}_j^T\mathbf{C}\boldsymbol{\psi}\dot{\mathbf{q}} + \boldsymbol{\psi}_j^T\mathbf{K}\boldsymbol{\psi}\mathbf{q} = \boldsymbol{\psi}_j^T\mathbf{F}(t) \tag{6.54}$$

Please note that we have the following relationships:

$$\boldsymbol{\psi}_1^T\mathbf{M}\boldsymbol{\psi}_2 = \{1 \quad 1\}\begin{bmatrix} m & 0 \\ 0 & m \end{bmatrix}\begin{Bmatrix} 1 \\ -1 \end{Bmatrix} = 0 \tag{6.55}$$

$$\boldsymbol{\psi}_j^T\mathbf{M}\boldsymbol{\psi}_k = 0 \qquad \text{for} \qquad j \neq k \tag{6.56}$$

$$\boldsymbol{\psi}_1^T\mathbf{M}\boldsymbol{\psi}\ddot{\mathbf{q}} = \boldsymbol{\psi}_1^T\mathbf{M}\left\{\boldsymbol{\psi}_1\ \boldsymbol{\psi}_2\right\}\ddot{\mathbf{q}} = \left\{1 \quad 1\right\}\begin{bmatrix} m & 0 \\ 0 & m \end{bmatrix}\begin{bmatrix} 1 & 1 \\ 1 & -1 \end{bmatrix}\begin{Bmatrix} \ddot{q}_1 \\ \ddot{q}_2 \end{Bmatrix} = 2m\ddot{q}_1 = M_1\ddot{q}_1 \tag{6.57}$$

$$\boldsymbol{\psi}_2^T\mathbf{M}\boldsymbol{\psi}\ddot{\mathbf{q}} = \boldsymbol{\psi}_1^T\mathbf{M}\left\{\boldsymbol{\psi}_1\ \boldsymbol{\psi}_2\right\}\ddot{\mathbf{q}} = \left\{1 \quad -1\right\}\begin{bmatrix} m & 0 \\ 0 & m \end{bmatrix}\begin{bmatrix} 1 & 1 \\ 1 & -1 \end{bmatrix}\begin{Bmatrix} \ddot{q}_1 \\ \ddot{q}_2 \end{Bmatrix} = 2m\ddot{q}_2 = M_2\ddot{q}_2 \tag{6.58}$$

$$\boldsymbol{\psi}_1^T\mathbf{C}\boldsymbol{\psi}_2 = \left\{1 \quad 1\right\}\begin{bmatrix} 2c & -c \\ -c & 2c \end{bmatrix}\begin{Bmatrix} 1 \\ -1 \end{Bmatrix} = 0 \tag{6.59}$$

$$\boldsymbol{\psi}_j^T\mathbf{C}\boldsymbol{\psi}_k = 0 \qquad \text{for} \qquad j \neq k \tag{6.60}$$

$$\boldsymbol{\psi}_1^T\mathbf{C}\boldsymbol{\psi}\dot{\mathbf{q}} = \boldsymbol{\psi}_1^T\mathbf{C}\left\{\boldsymbol{\psi}_1\ \boldsymbol{\psi}_2\right\}\dot{\mathbf{q}} = \left\{1 \quad 1\right\}\begin{bmatrix} 2c & -c \\ -c & 2c \end{bmatrix}\begin{bmatrix} 1 & 1 \\ 1 & -1 \end{bmatrix}\begin{Bmatrix} \dot{q}_1 \\ \dot{q}_2 \end{Bmatrix} = 2c\dot{q}_1 = C_1\dot{q}_1 \tag{6.61}$$

$$\boldsymbol{\psi}_2^T\mathbf{C}\boldsymbol{\psi}\dot{\mathbf{q}} = \boldsymbol{\psi}_2^T\mathbf{C}\left\{\boldsymbol{\psi}_1\ \boldsymbol{\psi}_2\right\}\dot{\mathbf{q}} = \left\{1 \quad -1\right\}\begin{bmatrix} 2c & -c \\ -c & 2c \end{bmatrix}\begin{bmatrix} 11 & \\ 1 & -1 \end{bmatrix}\begin{Bmatrix} \dot{q}_1 \\ \dot{q}_2 \end{Bmatrix} = 6c\dot{q}_2 = C_2\dot{q}_2 \tag{6.62}$$

$$\boldsymbol{\psi}_1^T\mathbf{K}\boldsymbol{\psi}_2 = \left\{1 \quad 1\right\}\begin{bmatrix} 2k & -k \\ -k & 2k \end{bmatrix}\begin{Bmatrix} 1 \\ -1 \end{Bmatrix} = 0 \tag{6.63}$$

$$\boldsymbol{\psi}_j^T\mathbf{K}\boldsymbol{\psi}_k = 0 \qquad \text{for} \qquad j \neq k \tag{6.64}$$

$$\boldsymbol{\psi}_1^T\mathbf{K}\boldsymbol{\psi}\mathbf{q} = \boldsymbol{\psi}_1^T\mathbf{K}\left\{\boldsymbol{\psi}_1\ \boldsymbol{\psi}_2\right\}\mathbf{q} = \left\{1 \quad 1\right\}\begin{bmatrix} 2k & -k \\ -k & 2k \end{bmatrix}\begin{bmatrix} 1 & 1 \\ 1 & -1 \end{bmatrix}\begin{Bmatrix} q_1 \\ q_2 \end{Bmatrix} = 2kq_1 = K_1q_1 \tag{6.65}$$

$$\boldsymbol{\psi}_2^T\mathbf{K}\boldsymbol{\psi}\mathbf{q} = \boldsymbol{\psi}_2^T\mathbf{K}\left\{\boldsymbol{\psi}_1\ \boldsymbol{\psi}_2\right\}\mathbf{q} = \left\{1 \quad -1\right\}\begin{bmatrix} 2k & -k \\ -k & 2k \end{bmatrix}\begin{bmatrix} 1 & 1 \\ 1 & -1 \end{bmatrix}\begin{Bmatrix} q_1 \\ q_2 \end{Bmatrix} = 6kq_2 = K_2q_2 \tag{6.66}$$

$$\boldsymbol{\psi}_1^T\mathbf{F}(t) = \left\{1 \quad 1\right\}\begin{Bmatrix} F_1(t) \\ 0 \end{Bmatrix} = F_1(t) = F_1 \tag{6.67}$$

$$\boldsymbol{\psi}_2^T\mathbf{F}(t) = \left\{1 \quad -1\right\}\begin{Bmatrix} F_1(t) \\ 0 \end{Bmatrix} = F_1(t) = F_2 \tag{6.68}$$

Based on the relationships in equations (6.54) to (6.68), we find that equation (6.54) can be simplified to the uncoupled systems

$$M_j\ddot{q}_j + C_j\dot{q}_j + K_jq_j = F_j; \quad j = 1, 2 \tag{6.69}$$

in which the scalar quantities

$$M_j = \psi_j^T\mathbf{M}\psi; \quad M_1 = 2m, \quad M_2 = 2m \tag{6.70}$$

$$C_j = \psi_1^T\mathbf{C}\psi; \quad C_1 = 2c, \quad C_2 = 6c \tag{6.71}$$

$$K_j = \psi_j^T\mathbf{K}\psi; \quad K_1 = 2k, \quad K_2 = 6k \tag{6.72}$$

represent the generalized mass, the damping coefficient, the stiffness coefficient, respectively.

When equation (6.69) is divided by M_j, the result is:

$$\ddot{q}_j + 2\zeta_j\omega_j\dot{q}_j + \omega_j^2q_j = G_j; \quad j = 1, 2 \tag{6.73}$$

in which $2\zeta_j\,\omega_j = C_j/M_j$, $\omega_j^2 = K_j/M_j$ and $G_j = F_j/M_j$. It should be noted that ω_j represents the jth undamped natural frequency, while ζ_j represents the jth modal damping ratio.

The second term on the left-hand side of equation (6.73) represents a definition of the modal viscous damping ratio

$$\zeta_j = \frac{C_j}{2\omega_jM_j} \tag{6.74}$$

It is typically more convenient and physically justifiable to define the damping of a two-degree-of-freedom system by using the damping ratio for each mode, rather than calculating the coefficients of the damping matrix **C**. This is because the modal damping ratios ζ_j can be determined through experimentation or estimated with sufficient accuracy in many scenarios.

Example 6.3

▶ Using the frequency domain superposition method, please compute the vector $\mathbf{y}$ that represents the transient displacement response to a single excitation $F_1(t)$ for the damped two-degree-of-freedom system shown in Figure 6.4.

SOLUTION

The transient response normal coordinates $q_j(t)$ in equation (6.73) can be calculated in terms of Fourier integrals as follows

$$q_j(t) = \frac{1}{2\pi}\int_{-\infty}^{\infty} H_j(\omega)e^{i\omega t}\left[\int_{-\infty}^{\infty} G_j(\theta)e^{-i\omega\theta}\,d\theta\right]d\omega; \quad j = 1, 2$$

in which $G_1 = G_2 = F_1(t)/2m$ and the complex frequency response is

$$H_j(\omega) = \frac{1}{\omega_j^2 - \omega^2 + 2i\zeta_j\omega_j\omega}; \quad j = 1, 2$$

in which

$$2\zeta_1\omega_1 = c/m, \quad 2\zeta_2\omega_2 = 3c/m$$
$$\omega_1^2 = k/m, \quad \omega_2^2 = 3k/m$$

The displacement response vector that is required can be found using equation (6.52).

$$\mathbf{y} = \boldsymbol{\psi}\mathbf{q} = \{\boldsymbol{\psi}_1 \ \boldsymbol{\psi}_2\}\begin{Bmatrix} q_1 \\ q_2 \end{Bmatrix} = \boldsymbol{\psi}_1 q_1 + \boldsymbol{\psi}_2 q_2 = \begin{Bmatrix} 1 \\ 1 \end{Bmatrix} q_1 + \begin{Bmatrix} 1 \\ -1 \end{Bmatrix} q_2 \blacktriangleleft$$

6.2 RANDOM VIBRATIONS OF TWO-DEGREE-OF-FREEDOM SYSTEMS

6.2.1 Random Excitations of an Undamped System

We take into consideration the random vibration analysis of a two-degree-of-freedom structural model that is depicted in Figure 6.3. In terms of the displacement y_j, random excitations $F_j(t); j = 1, 2$, the spring constant k and mass m, the equations of motion can be expressed as follows:

$$m\ddot{y}_1 + 2ky_1 - ky_2 = F_1(t) \tag{6.75}$$

$$m\ddot{y}_2 - ky_1 + 2ky_2 = F_2(t) \tag{6.76}$$

Now, let us take into consideration the two excitations $F_1(t)$ and $F_2(t)$ as stationary random processes with a mean of zero and with autocorrelation functions

$$R_1(\tau) = E\left[F_1(t)F_1(t+\tau)\right] \tag{6.77}$$

$$R_2(\tau) = E\left[F_2(t)F_2(t+\tau)\right] \tag{6.78}$$

and spectral density functions

$$S_1(\omega) = \frac{1}{2\pi}\int_{-\infty}^{\infty} R_1(\tau)e^{-i\omega\tau}\, d\tau \tag{6.79}$$

$$S_2(\omega) = \frac{1}{2\pi}\int_{-\infty}^{\infty} R_2(\tau)e^{-i\omega\tau}\, d\tau \tag{6.80}$$

Furthermore, when it comes to random excitations, the cross-correlation functions are defined by

$$R_{12}(\tau)=E\left[F_1(t)F_2(t+\tau)\right] \tag{6.81}$$

$$R_{21}(\tau)=E\left[F_2(t)F_1(t+\tau)\right] \tag{6.82}$$

and the cross-spectral density functions by

$$S_{12}(\omega)=\frac{1}{2\pi}\int_{-\infty}^{\infty}R_{12}(\tau)e^{-i\omega\tau}\,d\tau \tag{6.83}$$

$$S_{21}(\omega)=\frac{1}{2\pi}\int_{-\infty}^{\infty}R_{21}(\tau)e^{-i\omega\tau}\,d\tau \tag{6.84}$$

It should be noted that, according to the stationary property,

$$R_{12}(\tau)=E\left[F_1(t-\tau)F_2(t)\right]=R_{21}(-\tau) \tag{6.85}$$

and

$$\begin{aligned} S_{12}(\omega)&=\frac{1}{2\pi}\int_{-\infty}^{\infty}R_{12}(\tau)e^{-i\omega\tau}\,d\tau=\frac{1}{2\pi}\int_{-\infty}^{\infty}R_{21}(-\tau)e^{-i\omega\tau}\,d\tau\\ &=\frac{1}{2\pi}\int_{-\infty}^{\infty}R_{21}(\tau)e^{i\omega\tau}\,d\tau=S_{21}(-\omega) \end{aligned} \tag{6.86}$$

6.2.1.1 Response Autocorrelation Function

According to equations (6.46) and (6.47), the autocorrelation function $R_{y_1}(\tau)$ for the displacement response $y_1(t)$ is as follows:

$$\begin{aligned} R_{y_1}(\tau)&=E\left[y_1(t)y_1(t+\tau)\right]=\frac{1}{4}E\left\{\left[q_1(t)+q_2(t)\right]\left[q_1(t+\tau)+q_2(t+\tau)\right]\right\}\\ &=\frac{1}{4}\left\{E\left[q_1(t)q_1(t+\tau)\right]+E\left[q_1(t)q_2(t+\tau)\right]+E\left[q_2(t)q_1(t+\tau)\right]\right.\\ &\quad\left.+E\left[q_2(t)q_2(t+\tau)\right]\right\} \end{aligned} \tag{6.87}$$

There are two autocorrelation functions for the normal coordinates $q_1(t)$ and $q_2(t)$, respectively, and two cross-correlation functions that are included on the right-hand side of equation (6.87). By utilizing equations (6.36)–(6.40) to establish a connection between the normal coordinate $q_1(t)$ and the random excitations $F_1(t)$ and $F_2(t)$, we can get the autocorrelation.

$$E\left[q_1(t)q_1(t+\tau)\right]=E\left\{\int_{-\infty}^{\infty}\frac{1}{m}\left[F_1(t-\theta_1)+F_2(t-\theta_1)\right]g_1(\theta_1)d\theta_1 \right. \\ \left. \times\int_{-\infty}^{\infty}\frac{1}{m}\left[F_1(t+\tau-\theta_2)+F_2(t+\tau-\theta_2)\right]g_1(\theta_2)d\theta_2\right\} \tag{6.88}$$

By interchanging the order of expectation and integration and utilizing equations (6.77)–(6.82), the above equation is further simplified to yield

$$E\left[q_1(t)q_1(t+\tau)\right]=\frac{1}{m^2}\int_{-\infty}^{\infty}\int_{-\infty}^{\infty}\left[R_1(\tau-\theta_2+\theta_1)+R_{12}(\tau-\theta_2+\theta_1)\right. \\ \left.+R_{21}(\tau-\theta_2+\theta_1)+R_2(\tau-\theta_2+\theta_1)\right]g_1(\theta_1)g_1(\theta_2)d\theta_1d\theta_2 \tag{6.89}$$

In a similar manner, the remaining three terms that are located on the right-hand side of equation (6.87), are

$$E\left[q_1(t)q_2(t+\tau)\right]=\frac{1}{m^2}\int_{-\infty}^{\infty}\int_{-\infty}^{\infty}\left[R_1(\tau-\theta_2+\theta_1)-R_{12}(\tau-\theta_2+\theta_1)\right. \\ \left.+R_{21}(\tau-\theta_2+\theta_1)-R_2(\tau-\theta_2+\theta_1)\right]g_1(\theta_1)g_2(\theta_2)d\theta_1d\theta_2 \tag{6.90}$$

$$E\left[q_2(t)q_1(t+\tau)\right]=\frac{1}{m^2}\int_{-\infty}^{\infty}\int_{-\infty}^{\infty}\left[R_1(\tau-\theta_2+\theta_1)+R_{12}(\tau-\theta_2+\theta_1)\right. \\ \left.-R_{21}(\tau-\theta_2+\theta_1)-R_2(\tau-\theta_2+\theta_1)\right]g_1(\theta_2)g_2(\theta_1)d\theta_1d\theta_2 \tag{6.91}$$

$$E\left[q_2(t)q_2(t+\tau)\right]=\frac{1}{m^2}\int_{-\infty}^{\infty}\int_{-\infty}^{\infty}\left[R_1(\tau-\theta_2+\theta_1)-R_{12}(\tau-\theta_2+\theta_1)\right. \\ \left.-R_{21}(\tau-\theta_2+\theta_1)+R_2(\tau-\theta_2+\theta_1)\right]g_2(\theta_1)g_2(\theta_2)d\theta_1d\theta_2 \tag{6.92}$$

Equations (6.87)–(6.92) represent the input excitation-output response relations in terms of autocorrelation and cross-correlation functions. While the determination of the autocorrelation function $R_{y_1}(\tau)$ for the stationary displacement response $y_1(t)$ involves more terms (i.e., 16 terms) compared to a single term in the case of a single-degree-of-freedom system, these terms are essentially constructed in a similar manner and therefore do not introduce any basic difficulty. A direct consequence of the multiple input excitations is that one-half of the sixteen terms involve cross-correlation functions $R_{12}(\tau)$ and $R_{21}(\tau)$ of the random excitations $F_1(t)$ and $F_2(t)$.

In the specific scenario when there is just one random excitation, denoted as $F_1(t)$, only the first term on the right side of each of the four equations ((6.89)–(6.92) stays non-zero. The 16 terms are then reduced to only four, and the cross-correlation functions are eliminated. This is expressed as:

$$R_{y_1}(\tau)=\frac{1}{4m^2}\int_{-\infty}^{\infty}\int_{-\infty}^{\infty}R_1(\tau+\theta_1-\theta_2)\big[g_1(\theta_1)g_1(\theta_2)+g_1(\theta_1)g_2(\theta_2)+g_2(\theta_1)g_1(\theta_2)+g_2(\theta_1)g_2(\theta_2)\big]d\theta_1 d\theta_2 \tag{6.93}$$

6.2.1.2 Response Spectral Density Function

After doing an analysis of the input-output autocorrelation relation in the preceding derivation process, it was discovered that there are significant parallels between it and that of the SDOF system. As a matter of fact, the input-output autocorrelation relation of the SDOF system is identical to the first term that is located on the right-hand side of equation (6.89). Due to the fact that the input-output spectral density relation for the SDOF system has previously been obtained through the connection between autocorrelation and spectral density functions, such a developed package is naturally applicable in the present situation as follows:

By definition and based on equation (6.87), the spectral density $S_{y_1}(\omega)$ of the response displacement $y_1(t)$ is described as

$$S_{y_1}(\omega)=\frac{1}{2\pi}\int_{-\infty}^{\infty}R_{y_1}(\tau)e^{-i\omega t}d\tau=\frac{1}{2\pi}\int_{-\infty}^{\infty}\frac{1}{4}\big\{E\left[q_1(t)q_1(t+\tau)\right]+E\left[q_1(t)q_2(t+\tau)\right]+E\left[q_2(t)q_1(t+\tau)\right]+E\left[q_2(t)q_2(t+\tau)\right]\big\}e^{-i\omega\tau}d\tau \tag{6.94}$$

When equation (6.89) is utilized, the first term on the right-hand side of equation (6.94), which is represented here by I_1 for the sake of simplicity, is transformed into

$$I_1=\frac{1}{8\pi m^2}\int_{-\infty}^{\infty}e^{-i\omega\tau}d\tau\int_{-\infty}^{\infty}\int_{-\infty}^{\infty}\big[R_1(\tau-\theta_2+\theta_1)+R_{12}(\tau-\theta_2+\theta_1)+R_{21}(\tau-\theta_2+\theta_1)+R_2(\tau-\theta_2+\theta_1)\big]g_1(\theta_1)g_1(\theta_2)d\theta_1 d\theta_2 \tag{6.95a}$$

Equation (6.95a) can be re-written as

$$I_1=\frac{1}{8\pi m^2}\int_{-\infty}^{\infty}e^{-i\omega\tau}d\tau\int_{-\infty}^{\infty}\int_{-\infty}^{\infty}\big[R_1(\tau-\theta_2+\theta_1)+R_{12}(\tau-\theta_2+\theta_1)+R_{21}(\tau-\theta_2+\theta_1)+R_2(\tau-\theta_2+\theta_1)\big]g_1(\theta_1)g_1(\theta_2)d\theta_1 d\theta_2$$

$$
\begin{aligned}
&= \frac{1}{8\pi m^2}\int_{-\infty}^{\infty} e^{-i\omega\tau}\,d\tau \int_{-\infty}^{\infty}\int_{-\infty}^{\infty}\left[R_1(\tau-\theta_2+\theta_1)\right]g_1(\theta_1)g_1(\theta_2)d\theta_1\,d\theta_2 \\
&+\frac{1}{8\pi m^2}\int_{-\infty}^{\infty} e^{-i\omega\tau}\,d\tau \int_{-\infty}^{\infty}\int_{-\infty}^{\infty}\left[R_{12}(\tau-\theta_2+\theta_1)\right]g_1(\theta_1)g_1(\theta_2)d\theta_1\,d\theta_2 \\
&+\frac{1}{8\pi m^2}\int_{-\infty}^{\infty} e^{-i\omega\tau}\,d\tau \int_{-\infty}^{\infty}\int_{-\infty}^{\infty}R_{21}(\tau-\theta_2+\theta_1)g_1(\theta_1)g_1(\theta_2)d\theta_1\,d\theta_2 \\
&+\frac{1}{8\pi m^2}\int_{-\infty}^{\infty} e^{-i\omega\tau}\,d\tau \int_{-\infty}^{\infty}\int_{-\infty}^{\infty}\left[R_2(\tau-\theta_2+\theta_1)\right]g_1(\theta_1)g_1(\theta_2)d\theta_1\,d\theta_2 \\
&= I_{1a}+I_{1b}+I_{1c}+I_{1d}
\end{aligned}
\tag{6.95b}
$$

I_{1a}, I_{1b}, I_{1c} and I_{1d} in the above equation are, respectively, calculated as

$$I_{1a} = \frac{1}{8\pi m^2}\int_{-\infty}^{\infty} e^{-i\omega\tau}\,d\tau \int_{-\infty}^{\infty}\int_{-\infty}^{\infty}\left[R_1(\tau-\theta_2+\theta_1)\right]g_1(\theta_1)g_1(\theta_2)d\theta_1\,d\theta_2 \tag{6.96}$$

$$I_{1b} = \frac{1}{8\pi m^2}\int_{-\infty}^{\infty} e^{-i\omega\tau}\,d\tau \int_{-\infty}^{\infty}\int_{-\infty}^{\infty}\left[R_{12}(\tau-\theta_2+\theta_1)\right]g_1(\theta_1)g_1(\theta_2)d\theta_1\,d\theta_2 \tag{6.97}$$

$$I_{1c} = \frac{1}{8\pi m^2}\int_{-\infty}^{\infty} e^{-i\omega\tau}\,d\tau \int_{-\infty}^{\infty}\int_{-\infty}^{\infty}R_{21}(\tau-\theta_2+\theta_1)g_1(\theta_1)g_1(\theta_2)d\theta_1\,d\theta_2 \tag{6.98}$$

$$I_{1d} = \frac{1}{8\pi m^2}\int_{-\infty}^{\infty} e^{-i\omega\tau}\,d\tau \int_{-\infty}^{\infty}\int_{-\infty}^{\infty}\left[R_2(\tau-\theta_2+\theta_1)\right]g_1(\theta_1)g_1(\theta_2)d\theta_1\,d\theta_2 \tag{6.99}$$

In the following step, the order of integration in equation (6.96) is inverted, and the exponential factors $e^{i\omega\theta_1}$, $e^{-i\omega\theta_2}$ and $e^{-i\omega(\theta_1-\theta_2)}$ are inserted. The product of these factors is identical to unity.

$$
\begin{aligned}
I_{1a} &= \frac{1}{8\pi m^2}\int_{-\infty}^{\infty} g_1(\theta_1)e^{i\omega\theta_1}\,d\theta_1 \int_{-\infty}^{\infty} g_1(\theta_2)e^{-i\omega\theta_2}\,d\theta_2 \int_{-\infty}^{\infty}\left[R_1(\tau-\theta_2+\theta_1)\right]e^{-i\omega(\tau+\theta_1-\theta_2)}\,d\tau \\
&= \frac{1}{4m^2}\int_{-\infty}^{\infty} g_1(\theta_1)e^{i\omega\theta_1}\,d\theta_1 \int_{-\infty}^{\infty} g_1(\theta_2)e^{-i\omega\theta_2}\,d\theta_2 \frac{1}{2\pi}\int_{-\infty}^{\infty}\left[R_1(\tau+\theta_1-\theta_2)\right]e^{-i\omega(\tau+\theta_1-\theta_2)}\,d\tau
\end{aligned}
\tag{6.100}
$$

There is only one thing left to do: Interpret this result. It is possible to assume that the variable of integration is $\xi = \tau + \theta_1 - \theta_2$ in the third integral on the right.

This is because θ_1 and θ_2 are held constant in this integral. After that, we are able to recognize the Wiener–Khintchine theorem

$$S_1(\omega) = \frac{1}{2\pi}\int_{-\infty}^{\infty} R_1(\xi) e^{(-i\omega\xi)}\, d\xi \tag{6.101}$$

According to equation (5.65), the second integral is known as the Fourier transform $H_1(\omega)$ of the impulse response $g_1(\theta_2)$. With the exception of the sign in the exponent, the first integral displays the same form. In addition to being able to be represented as $H_1(-\omega)$, it is also possible to observe that the first integral is the complex conjugate of $H_1(\omega)$. Given these interpretations, equation (6.100) can be expressed as

$$I_{1a} = \frac{1}{4m^2}\left[H_1(-\omega) H_1(\omega) S_1(\omega)\right] \tag{6.102}$$

A slightly more compact version can be created by observing that the product of $H_1(\omega)$ and its complex conjugate can be represented as the square of the magnitude of $H_1(\omega)$; that is

$$I_{1a} = \frac{|H_1(\omega)|^2}{4m^2}\left[S_1(\omega)\right] \tag{6.103}$$

Similarly, I_{1b} in equation (6.97) can be calculated as follows:

$$\begin{aligned}
I_{1b} &= \frac{1}{8\pi m^2}\int_{-\infty}^{\infty} e^{-i\omega\tau}\, d\tau \int_{-\infty}^{\infty}\int_{-\infty}^{\infty} \left[R_{12}(\tau - \theta_2 + \theta_1)\right] g_1(\theta_1) g_1(\theta_2)\, d\theta_1\, d\theta_2 \\
&= \frac{1}{8\pi m^2}\int_{-\infty}^{\infty} g_1(\theta_1) e^{i\omega\theta_1}\, d\theta_1 \int_{-\infty}^{\infty} g_1(\theta_2) e^{-i\omega\theta_2}\, d\theta_2 \int_{-\infty}^{\infty} \left[R_{12}(\tau + \theta_1 - \theta_2)\right] e^{-i\omega(\tau+\theta_1-\theta_2)}\, d\tau \\
&= \frac{1}{4m^2}\int_{-\infty}^{\infty} g_1(\theta_1) e^{i\omega\theta_1}\, d\theta_1 \int_{-\infty}^{\infty} g_1(\theta_2) e^{-i\omega\theta_2}\, d\theta_2\, \frac{1}{2\pi}\int_{-\infty}^{\infty} \left[R_{12}(\tau + \theta_1 - \theta_2)\right] e^{-i\omega(\tau+\theta_1-\theta_2)}\, d\tau \\
&= \frac{1}{4m^2}\left[H_1(-\omega) H_1(\omega) S_{12}(\omega)\right] = \frac{|H_1(\omega)|^2}{4m^2}\left[S_{12}(\omega)\right]
\end{aligned} \tag{6.104}$$

I_{1c} in equation (6.98) can be calculated as follows:

$$\begin{aligned}
I_{1c} &= \frac{1}{8\pi m^2}\int_{-\infty}^{\infty} e^{-i\omega\tau}\, d\tau \int_{-\infty}^{\infty}\int_{-\infty}^{\infty} R_{21}(\tau - \theta_2 + \theta_1) g_1(\theta_1) g_1(\theta_2)\, d\theta_1\, d\theta_2 \\
&= \frac{1}{8\pi m^2}\int_{-\infty}^{\infty} g_1(\theta_1) e^{i\omega\theta_1}\, d\theta_1 \int_{-\infty}^{\infty} g_1(\theta_2) e^{-i\omega\theta_2}\, d\theta_2 \int_{-\infty}^{\infty} \left[R_{21}(\tau + \theta_1 - \theta_2)\right] e^{-i\omega(\tau+\theta_1-\theta_2)}\, d\tau
\end{aligned}$$

$$= \frac{1}{4m^2} \int_{-\infty}^{\infty} g_1(\theta_1) e^{i\omega\theta_1} d\theta_1 \int_{-\infty}^{\infty} g_1(\theta_2) e^{-i\omega\theta_2} d\theta_2 \frac{1}{2\pi} \int_{-\infty}^{\infty} [R_{21}(\tau + \theta_1 - \theta_2)] e^{-i\omega(\tau+\theta_1-\theta_2)} d\tau$$

$$= \frac{1}{4m^2} [H_1(-\omega) H_1(\omega) S_{12}(\omega)] = \frac{|H_1(\omega)|^2}{4m^2} [S_{21}(\omega)] \tag{6.105}$$

I_{1d} in equation (6.99) can be calculated as follows:

$$I_{1d} = \frac{1}{8\pi m^2} \int_{-\infty}^{\infty} e^{-i\omega\tau} d\tau \int_{-\infty}^{\infty} \int_{-\infty}^{\infty} [R_2(\tau - \theta_2 + \theta_1)] g_1(\theta_1) g_1(\theta_2) d\theta_1 d\theta_2$$

$$= \frac{1}{8\pi m^2} \int_{-\infty}^{\infty} g_1(\theta_1) e^{i\omega\theta_1} d\theta_1 \int_{-\infty}^{\infty} g_1(\theta_2) e^{-i\omega\theta_2} d\theta_2 \int_{-\infty}^{\infty} [R_2(\tau + \theta_1 - \theta_2)] e^{-i\omega(\tau+\theta_1-\theta_2)} d\tau$$

$$= \frac{1}{4m^2} \int_{-\infty}^{\infty} g_1(\theta_1) e^{i\omega\theta_1} d\theta_1 \int_{-\infty}^{\infty} g_1(\theta_2) e^{-i\omega\theta_2} d\theta_2 \frac{1}{2\pi} \int_{-\infty}^{\infty} [R_2(\tau + \theta_1 - \theta_2)] e^{-i\omega(\tau+\theta_1-\theta_2)} d\tau$$

$$= \frac{1}{4m^2} [H_1(-\omega) H_1(\omega) S_2(\omega)] = \frac{|H_1(\omega)|^2}{4m^2} [S_2(\omega)] \tag{6.106}$$

Consequently, I_1 in equation (6.95) can be calculated as follows:

$$I_1 = I_{1a} + I_{1b} + I_{1c} + I_{1d} = \frac{|H_1(\omega)|^2}{4m^2} [S_1(\omega) + S_{12}(\omega) + S_{21}(\omega) + S_2(\omega)] \tag{6.107}$$

When comparing to the single-degree-of-freedom system, the addition of three spectral density terms is observed in the above equation due to the second excitation $F_2(t)$. A comparison with that of the single-degree-of-freedom system also shows the addition of the constant multiplier $1/4m^2$ through modal superposition.

When equation (6.90) is utilized, the second term on the right-hand side of equation (6.94), which is represented here by I_2 for the sake of simplicity, is transformed into

$$I_2 = \frac{1}{8\pi m^2} \int_{-\infty}^{\infty} e^{-i\omega\tau} d\tau \int_{-\infty}^{\infty} \int_{-\infty}^{\infty} [R_1(\tau - \theta_2 + \theta_1) - R_{12}(\tau - \theta_2 + \theta_1) \\ + R_{21}(\tau - \theta_2 + \theta_1) - R_2(\tau - \theta_2 + \theta_1)] g_1(\theta_1) g_2(\theta_2) d\theta_1 d\theta_2 \tag{6.108}$$

Equation (6.108) can be re-written as

$$I_2 = \frac{1}{8\pi m^2} \int_{-\infty}^{\infty} e^{-i\omega\tau} d\tau \int_{-\infty}^{\infty} \int_{-\infty}^{\infty} [R_1(\tau - \theta_2 + \theta_1) - R_{12}(\tau - \theta_2 + \theta_1) \\ + R_{21}(\tau - \theta_2 + \theta_1) - R_2(\tau - \theta_2 + \theta_1)] g_1(\theta_1) g_2(\theta_2) d\theta_1 d\theta_2$$

$$
\begin{aligned}
&= \frac{1}{8\pi m^2}\int_{-\infty}^{\infty} e^{-i\omega\tau}\,d\tau \int_{-\infty}^{\infty}\int_{-\infty}^{\infty}\left[R_1(\tau-\theta_2+\theta_1)\right]g_1(\theta_1)g_2(\theta_2)d\theta_1\,d\theta_2 \\
&\quad - \frac{1}{8\pi m^2}\int_{-\infty}^{\infty} e^{-i\omega\tau}\,d\tau \int_{-\infty}^{\infty}\int_{-\infty}^{\infty}\left[R_{12}(\tau-\theta_2+\theta_1)\right]g_1(\theta_1)g_2(\theta_2)d\theta_1\,d\theta_2 \\
&\quad + \frac{1}{8\pi m^2}\int_{-\infty}^{\infty} e^{-i\omega\tau}\,d\tau \int_{-\infty}^{\infty}\int_{-\infty}^{\infty} R_{21}(\tau-\theta_2+\theta_1)g_1(\theta_1)g_2(\theta_2)d\theta_1\,d\theta_2 \\
&\quad - \frac{1}{8\pi m^2}\int_{-\infty}^{\infty} e^{-i\omega\tau}\,d\tau \int_{-\infty}^{\infty}\int_{-\infty}^{\infty}\left[R_2(\tau-\theta_2+\theta_1)\right]g_1(\theta_1)g_2(\theta_2)d\theta_1\,d\theta_2 \\
&= I_{2a} - I_{2b} + I_{2c} - I_{2d}
\end{aligned}
\tag{6.109}
$$

I_{2a}, I_{2b}, I_{2c} and I_{2d} in the above equation are, respectively, calculated as

$$
I_{2a} = \frac{1}{8\pi m^2}\int_{-\infty}^{\infty} e^{-i\omega\tau}\,d\tau \int_{-\infty}^{\infty}\int_{-\infty}^{\infty}\left[R_1(\tau-\theta_2+\theta_1)\right]g_1(\theta_1)g_2(\theta_2)d\theta_1\,d\theta_2 \tag{6.110}
$$

$$
I_{2b} = \frac{1}{8\pi m^2}\int_{-\infty}^{\infty} e^{-i\omega\tau}\,d\tau \int_{-\infty}^{\infty}\int_{-\infty}^{\infty}\left[R_{12}(\tau-\theta_2+\theta_1)\right]g_1(\theta_1)g_2(\theta_2)d\theta_1\,d\theta_2 \tag{6.111}
$$

$$
I_{2c} = \frac{1}{8\pi m^2}\int_{-\infty}^{\infty} e^{-i\omega\tau}\,d\tau \int_{-\infty}^{\infty}\int_{-\infty}^{\infty} R_{21}(\tau-\theta_2+\theta_1)g_1(\theta_1)g_2(\theta_2)d\theta_1\,d\theta_2 \tag{6.112}
$$

$$
I_{2d} = \frac{1}{8\pi m^2}\int_{-\infty}^{\infty} e^{-i\omega\tau}\,d\tau \int_{-\infty}^{\infty}\int_{-\infty}^{\infty}\left[R_2(\tau-\theta_2+\theta_1)\right]g_1(\theta_1)g_2(\theta_2)d\theta_1\,d\theta_2 \tag{6.113}
$$

When the order of integration in equation (6.110) is inverted and the exponential factors $e^{i\omega\theta_1}$, $e^{-i\omega\theta_2}$ and $e^{-i\omega(\theta_1-\theta_2)}$ are inserted, I_{2a} can be calculated as

$$
\begin{aligned}
I_{2a} &= \frac{1}{8\pi m^2}\int_{-\infty}^{\infty} g_1(\theta_1)e^{i\omega\theta_1}\,d\theta_1 \int_{-\infty}^{\infty} g_2(\theta_2)e^{-i\omega\theta_2}\,d\theta_2 \int_{-\infty}^{\infty}\left[R_1(\tau-\theta_2+\theta_1)\right]e^{-i\omega(\tau+\theta_1-\theta_2)}\,d\tau \\
&= \frac{1}{4m^2}\int_{-\infty}^{\infty} g_1(\theta_1)e^{i\omega\theta_1}\,d\theta_1 \int_{-\infty}^{\infty} g_2(\theta_2)e^{-i\omega\theta_2}\,d\theta_2 \frac{1}{2\pi}\int_{-\infty}^{\infty}\left[R_1(\tau+\theta_1-\theta_2)\right]e^{-i\omega(\tau+\theta_1-\theta_2)}\,d\tau \\
&= \frac{H_1(-\omega)H_2(\omega)}{4m^2}\left[S_1(\omega)\right]
\end{aligned}
\tag{6.114}
$$

In a similar manner, I_{2b} can be calculated as

$$I_{2b}=\frac{1}{8\pi m^2}\int_{-\infty}^{\infty}g_1(\theta_1)e^{i\omega\theta_1}\,d\theta_1\int_{-\infty}^{\infty}g_2(\theta_2)e^{-i\omega\theta_2}\,d\theta_2\int_{-\infty}^{\infty}\left[R_{12}(\tau-\theta_2+\theta_1)\right]e^{-i\omega(\tau+\theta_1-\theta_2)}\,d\tau$$

$$=\frac{1}{4m^2}\int_{-\infty}^{\infty}g_1(\theta_1)e^{i\omega\theta_1}\,d\theta_1\int_{-\infty}^{\infty}g_2(\theta_2)e^{-i\omega\theta_2}\,d\theta_2\frac{1}{2\pi}\int_{-\infty}^{\infty}\left[R_{12}(\tau+\theta_1-\theta_2)\right]e^{-i\omega(\tau+\theta_1-\theta_2)}\,d\tau$$

$$=\frac{H_1(-\omega)H_2(\omega)}{4m^2}\left[S_{12}(\omega)\right] \tag{6.115}$$

I_{2c} can be calculated as

$$I_{2c}=\frac{1}{8\pi m^2}\int_{-\infty}^{\infty}g_1(\theta_1)e^{i\omega\theta_1}\,d\theta_1\int_{-\infty}^{\infty}g_2(\theta_2)e^{-i\omega\theta_2}\,d\theta_2\int_{-\infty}^{\infty}\left[R_{21}(\tau-\theta_2+\theta_1)\right]e^{-i\omega(\tau+\theta_1-\theta_2)}\,d\tau$$

$$=\frac{1}{4m^2}\int_{-\infty}^{\infty}g_1(\theta_1)e^{i\omega\theta_1}\,d\theta_1\int_{-\infty}^{\infty}g_2(\theta_2)e^{-i\omega\theta_2}\,d\theta_2\frac{1}{2\pi}\int_{-\infty}^{\infty}\left[R_{21}(\tau+\theta_1-\theta_2)\right]e^{-i\omega(\tau+\theta_1-\theta_2)}\,d\tau$$

$$=\frac{H_1(-\omega)H_2(\omega)}{4m^2}\left[S_{21}(\omega)\right] \tag{6.116}$$

I_{2d} can be calculated as

$$I_{2d}=\frac{1}{8\pi m^2}\int_{-\infty}^{\infty}g_1(\theta_1)e^{i\omega\theta_1}\,d\theta_1\int_{-\infty}^{\infty}g_2(\theta_2)e^{-i\omega\theta_2}\,d\theta_2\int_{-\infty}^{\infty}\left[R_2(\tau-\theta_2+\theta_1)\right]e^{-i\omega(\tau+\theta_1-\theta_2)}\,d\tau$$

$$=\frac{1}{4m^2}\int_{-\infty}^{\infty}g_1(\theta_1)e^{i\omega\theta_1}\,d\theta_1\int_{-\infty}^{\infty}g_2(\theta_2)e^{-i\omega\theta_2}\,d\theta_2\frac{1}{2\pi}\int_{-\infty}^{\infty}\left[R_2(\tau+\theta_1-\theta_2)\right]e^{-i\omega(\tau+\theta_1-\theta_2)}\,d\tau$$

$$=\frac{H_1(-\omega)H_2(\omega)}{4m^2}\left[S_2(\omega)\right] \tag{6.117}$$

Consequently, I_2 in equation (6.109) can be calculated as

$$I_2=I_{2a}-I_{2b}+I_{2c}-I_{2d}=\frac{H_1(-\omega)H_2(\omega)}{4m^2}\left[S_1(\omega)-S_{12}(\omega)+S_{21}(\omega)-S_2(\omega)\right] \tag{6.118}$$

The third term on the right-hand side of equation (6.94), which is represented here by I_3, is calculated as

$$I_3=\frac{1}{8\pi m^2}\int_{-\infty}^{\infty}e^{-i\omega\tau}\,d\tau\int_{-\infty}^{\infty}\int_{-\infty}^{\infty}\left[R_1(\tau-\theta_2+\theta_1)+R_{12}(\tau-\theta_2+\theta_1)\right.$$
$$\left.-R_{21}(\tau-\theta_2+\theta_1)-R_2(\tau-\theta_2+\theta_1)\right]g_1(\theta_2)g_2(\theta_1)d\theta_1 d\theta_2$$

$$
\begin{aligned}
&= \frac{1}{8\pi m^2}\int_{-\infty}^{\infty} e^{-i\omega\tau}\,d\tau \int_{-\infty}^{\infty}\int_{-\infty}^{\infty}\left[R_1(\tau-\theta_2+\theta_1)\right]g_1(\theta_2)g_2(\theta_1)d\theta_1\,d\theta_2 \\
&+ \frac{1}{8\pi m^2}\int_{-\infty}^{\infty} e^{-i\omega\tau}\,d\tau \int_{-\infty}^{\infty}\int_{-\infty}^{\infty}\left[R_{12}(\tau-\theta_2+\theta_1)\right]g_1(\theta_2)g_2(\theta_1)d\theta_1\,d\theta_2 \\
&- \frac{1}{8\pi m^2}\int_{-\infty}^{\infty} e^{-i\omega\tau}\,d\tau \int_{-\infty}^{\infty}\int_{-\infty}^{\infty} R_{21}(\tau-\theta_2+\theta_1)g_1(\theta_2)g_2(\theta_1)d\theta_1\,d\theta_2 \\
&- \frac{1}{8\pi m^2}\int_{-\infty}^{\infty} e^{-i\omega\tau}\,d\tau \int_{-\infty}^{\infty}\int_{-\infty}^{\infty}\left[R_2(\tau-\theta_2+\theta_1)\right]g_1(\theta_2)g_2(\theta_1)d\theta_1\,d\theta_2 \\
&= I_{3a} + I_{3b} - I_{3c} - I_{3d} \\
&= \frac{H_1(\omega)H_2(-\omega)}{4m^2}\left[S_1(\omega)+S_{12}(\omega)-S_{21}(\omega)-S_2(\omega)\right]
\end{aligned}
\tag{6.119}
$$

The fourth term on the right-hand side of equation (6.94), which is represented here by I_4, is calculated as

$$
\begin{aligned}
I_4 &= \frac{1}{8\pi m^2}\int_{-\infty}^{\infty} e^{-i\omega\tau}\,d\tau \int_{-\infty}^{\infty}\int_{-\infty}^{\infty}\left[R_1(\tau-\theta_2+\theta_1) - R_{12}(\tau-\theta_2+\theta_1)\right. \\
&\left. -R_{21}(\tau-\theta_2+\theta_1) + R_2(\tau-\theta_2+\theta_1)\right]g_2(\theta_1)g_2(\theta_2)d\theta_1 d\theta_2 \\
&= \frac{1}{8\pi m^2}\int_{-\infty}^{\infty} e^{-i\omega\tau}\,d\tau \int_{-\infty}^{\infty}\int_{-\infty}^{\infty}\left[R_1(\tau-\theta_2+\theta_1)\right]g_2(\theta_1)g_2(\theta_2)d\theta_1\,d\theta_2 \\
&+ \frac{1}{8\pi m^2}\int_{-\infty}^{\infty} e^{-i\omega\tau}\,d\tau \int_{-\infty}^{\infty}\int_{-\infty}^{\infty}\left[R_{12}(\tau-\theta_2+\theta_1)\right]g_2(\theta_1)g_2(\theta_2)d\theta_1\,d\theta_2 \\
&- \frac{1}{8\pi m^2}\int_{-\infty}^{\infty} e^{-i\omega\tau}\,d\tau \int_{-\infty}^{\infty}\int_{-\infty}^{\infty} R_{21}(\tau-\theta_2+\theta_1)g_2(\theta_1)g_2(\theta_2)d\theta_1\,d\theta_2 \\
&- \frac{1}{8\pi m^2}\int_{-\infty}^{\infty} e^{-i\omega\tau}\,d\tau \int_{-\infty}^{\infty}\int_{-\infty}^{\infty}\left[R_2(\tau-\theta_2+\theta_1)\right]g_2(\theta_1)g_2(\theta_2)d\theta_1\,d\theta_2 \\
&= I_{4a} + I_{4b} - I_{4c} - I_{4d} \\
&= \frac{H_2(-\omega)H_2(\omega)}{4m^2}\left[S_1(\omega)-S_{12}(\omega)-S_{21}(\omega)+S_2(\omega)\right] \\
&= \frac{|H_2(\omega)|^2}{4m^2}\left[S_1(\omega)-S_{12}(\omega)-S_{21}(\omega)+S_2(\omega)\right]
\end{aligned}
\tag{6.120}
$$

Finally, equation (6.94) gives the spectral density function $S_{y_1}(\omega)$ of the response displacement $y_1(t)$ as

$$\begin{aligned} S_{y_1}(\omega) = I_1 + I_2 + I_3 + I_4 = {} & \frac{|H_1(\omega)|^2}{4m^2}\left[S_1(\omega) + S_{12}(\omega) + S_{21}(\omega) + S_2(\omega)\right] \\ & + \frac{H_1(-\omega)H_2(\omega)}{4m^2}\left[S_1(\omega) - S_{12}(\omega) + S_{21}(\omega) - S_2(\omega)\right] \\ & + \frac{H_1(\omega)H_2(-\omega)}{4m^2}\left[S_1(\omega) + S_{12}(\omega) - S_{21}(\omega) - S_2(\omega)\right] \\ & + \frac{|H_2(\omega)|^2}{4m^2}\left[S_1(\omega) - S_{12}(\omega) - S_{21}(\omega) + S_2(\omega)\right] \end{aligned} \tag{6.121}$$

In the particular scenario of a single random stimulation $F_1(t)$, the spectral density of the response displacement $y_1(t)$ is calculated as follows:

$$S_{y_1}(\omega) = \frac{S_1(\omega)}{4m^2}\left[|H_1(\omega)|^2 + H_1(-\omega)H_2(\omega) + H_1(\omega)H_2(-\omega) + |H_2(\omega)|^2\right] \tag{6.122}$$

Example 6.4

▶ Assuming a single stationary random process $F_1(t)$ with a white noise spectrum $2\pi S_0$ as the excitation, please find the power spectral density function $S_{y_1}(\omega)$ of the stationary response displacement $y_1(t)$ for the two-degree-of-freedom structural model depicted in Figure 6.5.

SOLUTION

According to equation (6.43), the complex frequency responses are as follows:

$$H_1(\omega) = \frac{1}{\omega_1^2 - \omega^2}; \qquad H_2(\omega) = \frac{1}{\omega_2^2 - \omega^2}$$

$$|H_1(\omega)|^2 = \frac{1}{\left(\omega_1^2 - \omega^2\right)^2} \quad |H_2(\omega)|^2 = \frac{1}{\left(\omega_2^2 - \omega^2\right)^2}$$

$$H_1(\omega)H_2(-\omega) = \frac{1}{\left(\omega_1^2 - \omega^2\right)\left(\omega_2^2 - \omega^2\right)}$$

$$H_2(\omega)H_1(-\omega) = \frac{1}{\left(\omega_2^2 - \omega^2\right)\left(\omega_1^2 - \omega^2\right)}$$

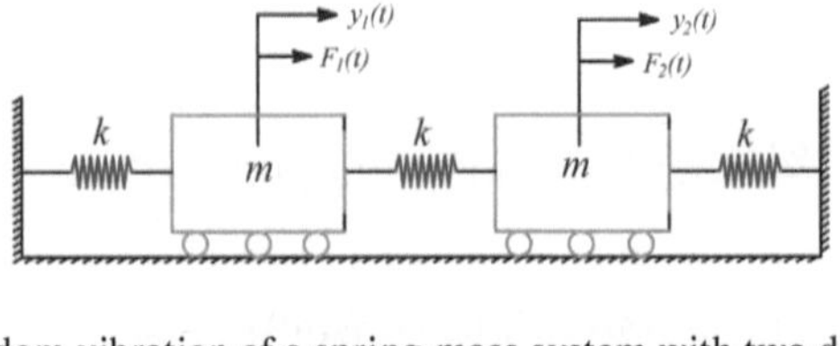

FIGURE 6.5 Random vibration of a spring-mass system with two degrees of freedom.

When the preceding equations are substituted into equation (6.122), the result is

$$S_{y_1}(\omega) = \frac{2\pi S_0}{4m^2}\left[\frac{1}{\left(\omega_1^2-\omega^2\right)^2}+\frac{2}{\left(\omega_1^2-\omega^2\right)\left(\omega_2^2-\omega^2\right)}+\frac{1}{\left(\omega_2^2-\omega^2\right)^2}\right]$$

$$= \frac{\pi S_0}{2m^2}\left[\frac{1}{\left(\omega_1^2-\omega^2\right)^2}+\frac{2}{\left(\omega_1^2-\omega^2\right)\left(\omega_2^2-\omega^2\right)}+\frac{1}{\left(\omega_2^2-\omega^2\right)^2}\right] \blacktriangleleft$$

6.2.2 Random Excitations of a Damped System

6.2.2.1 Damped Response Autocorrelation Function

The autocorrelation function for the stationary response displacement $y_1(t)$ of a damped two-degree-of-freedom system, as depicted in Figure 6.6, to a single random excitation $F_1(t)$ is, according to the definition of and derived from equation (6.52),

$$\begin{aligned} R_{y_1}(\tau) &= E\left[y_1(t)y_1(t+\tau)\right] \\ &= E\left\{\left[\psi_{11}q_1(t)+\psi_{12}q_2(t)\right]\left[\psi_{11}q_1(t+\tau)+\psi_{12}q_2(t+\tau)\right]\right\} \\ &= E\left[\psi_{11}^2q_1(t)q_1(t+\tau)+\psi_{11}\psi_{12}q_1(t)q_2(t+\tau)\right. \\ &\quad \left.+\psi_{12}\psi_{11}q_2(t)q_1(t+\tau)+\psi_{12}^2q_2(t)q_2(t+\tau)\right] \\ &= \sum_{j=1}^{2}\sum_{k=1}^{2}\psi_{1j}\psi_{1k}E\left[q_j(t)q_k(t+\tau)\right] \end{aligned} \tag{6.123}$$

in which ψ_{1j} and ψ_{1k} represent the first elements of the first and second modal vectors, respectively.

The normal coordinates can be obtained by expressing the transient response for equation (6.73) using Duhamel's integral.

$$q_j(t) = \int_{-\infty}^{\infty} G_j(t-\theta)g_j(\theta)d\theta; \quad j=1,2 \tag{6.124}$$

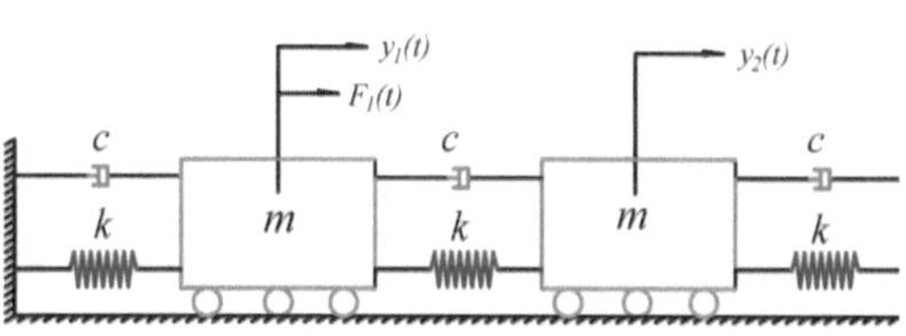

FIGURE 6.6 Random vibration of a spring-mass-damper system with two degrees of freedom.

in which

$$G_1 = G_2 = F_1(t)/2m \tag{6.125}$$

$$g_j(t) = \frac{\exp(-\zeta_j\omega_j t)}{\omega_j\sqrt{1-\zeta_j^2}}\sin\sqrt{1-\zeta_j^2}\,\omega_j t \tag{6.126}$$

$$2\zeta_1\omega_1 = c/m, \quad 2\zeta_2\omega_2 = 3c/m, \quad \omega_1^2 = k/m, \quad \omega_2^2 = 3k/m \tag{6.127}$$

Based on equation (6.124), we have

$$E\left[q_j(t)q_k(t+\tau)\right] = E\left[\int_{-\infty}^{\infty} G_j(t-\theta_1)g_j(\theta_1)d\theta_1 \int_{-\infty}^{\infty} G_k(t+\tau-\theta_2)g_k(\theta_2)d\theta_2\right] \tag{6.128}$$

Because we have

$$M_j = \boldsymbol{\psi}_j^T\mathbf{M}\boldsymbol{\psi}_j \tag{6.129}$$

$$\boldsymbol{\psi}_1^T\mathbf{F}(t) = \{1 \quad 1\}\begin{Bmatrix}F_1(t)\\0\end{Bmatrix} = F_1(t) = F_1 = \psi_{11}F_1(t) \tag{6.130}$$

$$\boldsymbol{\psi}_2^T\mathbf{F}(t) = \{1 \quad -1\}\begin{Bmatrix}F_1(t)\\0\end{Bmatrix} = F_1(t) = F_2 = \psi_{12}F_1(t) \tag{6.131}$$

we can obtain

$$G_j(t) = \frac{F_j(t)}{M_j} = \frac{\psi_{1j}F_1(t)}{\boldsymbol{\psi}_j^T\mathbf{M}\boldsymbol{\psi}_j} \tag{6.132}$$

When equation (6.132) is substituted into equation (6.128) and the order of expectation and integration is interchanged, the result is

$$E\left[q_j(t)q_k(t+\tau)\right] = \int_{-\infty}^{\infty}\int_{-\infty}^{\infty}\frac{\psi_{1j}\psi_{1k}}{M_jM_k}R_1(\tau-\theta_2+\theta_1)g_j(\theta_1)g_k(\theta_2)d\theta_1\,d\theta_2 \tag{6.133}$$

in which $R_1(\tau-\theta_2+\theta_1)$ is the autocorrelation function of the input excitation $F_1(t)$. $R_1(\tau-\theta_2+\theta_1)$ is defined by equation (6.77). Equations (6.123) and (6.133) together constitute the stationary input-output relation for the autocorrelation functions.

Example 6.5

▶ Please find the stationary input-output relationship for the autocorrelation functions of the two-degree-of-freedom system depicted in Figure 6.6.

SOLUTION

For the two-degree-of-freedom system depicted in Figure 6.6, we have

$$\psi_{11} = 1, \qquad \psi_{12} = 1$$

$$M_1 = 2m, \quad M_2 = 2n$$

The required input-output relation can be obtained by substituting these constants into equations (6.133) and (6.123) as follows:

$$\begin{aligned}
R_{y_1}(\tau) &= E\left[y_1(t)y_1(t+\tau)\right] \\
&= E\left\{\left[\psi_{11}q_1(t)+\psi_{12}q_2(t)\right]\left[\psi_{11}q_1(t+\tau)+\psi_{12}q_2(t+\tau)\right]\right\} \\
&= E\left[q_1(t)q_1(t+\tau)+q_1(t)q_2(t+\tau)\right. \\
&\quad \left. +q_2(t)q_1(t+\tau)+q_2(t)q_2(t+\tau)\right] \\
&= \frac{1}{4m^2}\int_{-\infty}^{\infty}\int_{-\infty}^{\infty} R_1\left(\tau-\theta_2+\theta_1\right)g_1(\theta_1)g_1\left(\theta_2\right)d\theta_1\,d\theta_2 \\
&\quad +\frac{1}{4m^2}\int_{-\infty}^{\infty}\int_{-\infty}^{\infty} R_1\left(\tau-\theta_2+\theta_1\right)g_1(\theta_1)g_2\left(\theta_2\right)d\theta_1\,d\theta_2 \\
&\quad +\frac{1}{4m^2}\int_{-\infty}^{\infty}\int_{-\infty}^{\infty} R_1\left(\tau-\theta_2+\theta_1\right)g_2(\theta_1)g_1\left(\theta_2\right)d\theta_1\,d\theta_2 \\
&\quad +\frac{1}{4m^2}\int_{-\infty}^{\infty}\int_{-\infty}^{\infty} R_1\left(\tau-\theta_2+\theta_1\right)g_2(\theta_1)g_2\left(\theta_2\right)d\theta_1\,d\theta_2 \\
&= \frac{1}{4m^2}\int_{-\infty}^{\infty}\int_{-\infty}^{\infty} R_1\left(\tau-\theta_2+\theta_1\right)\left[g_1(\theta_1)g_1\left(\theta_2\right)+g_1(\theta_1)g_2\left(\theta_2\right)\right. \\
&\quad \left. +g_2(\theta_1)g_1\left(\theta_2\right)+g_2(\theta_1)g_2\left(\theta_2\right)\right]d\theta_1 d\theta_2
\end{aligned} \tag{6.134}$$

◀

It is important to take note that the final result in equation (6.134) is precisely the same form as the one that is produced by equation (6.93) for the undamped two-degree-of-freedom system. The sole distinction, of course, resides in the impulse response functions $g_j(\theta)$, which are given by equation (6.126) in the current damped problem.

6.2.2.2 Damped Response Spectral Density Function

The stationary spectral density function $S_{y1}(\omega)$ for the stationary response displacement $y_1(t)$ of the damped two-degree-of-freedom system, as depicted in Figure 6.6, to a single random excitation $F_1(t)$ is related to the autocorrelation $R_{y1}(\tau)$ by the following equation

$$S_{y1}(\omega)=\frac{1}{2\pi}\int_{-\infty}^{\infty}R_{y1}(\tau)e^{-i\omega\tau}\,d\tau \tag{6.135}$$

From equations (6.123) and (6.133), we have the general expression of $R_{y1}(\tau)$ for the two-degree-of-freedom system. After substituting $R_{y1}(\tau)$ into equation (6.135), we obtain

$$S_{y1}(\omega)=\frac{1}{2\pi}\sum_{j=1}^{2}\sum_{k=1}^{2}\frac{\psi_{1j}^{2}\psi_{1k}^{2}}{M_jM_k}\int_{-\infty}^{\infty}\int_{-\infty}^{\infty}\int_{-\infty}^{\infty}R_1(\tau-\theta_2+\theta_1)g_j(\theta_1)g_k(\theta_2)e^{-i\omega\tau}\,d\theta_1\,d\theta_2\,d\tau \tag{6.136}$$

In the following step, the order of integration in equation (6.136) is inverted, and the exponential factors $e^{i\omega\theta_1}$, $e^{-i\omega\theta_2}$ and $e^{-i\omega(\theta_1-\theta_2)}$ are inserted. The product of these factors is identical to unity.

$$\begin{aligned}S_{y1}(\omega)=\frac{1}{2\pi}\sum_{j=1}^{2}\sum_{k=1}^{2}\frac{\psi_{1j}^{2}\psi_{1k}^{2}}{M_jM_k}&\int_{-\infty}^{\infty}g_j(\theta_1)e^{i\omega\theta_1}\,d\theta_1\int_{-\infty}^{\infty}g_k(\theta_2)e^{-i\omega\theta_2}\,d\theta_2\\&\int_{-\infty}^{\infty}\left[R_1(\tau-\theta_2+\theta_1)\right]e^{-i\omega(\tau+\theta_1-\theta_2)}\,d\tau\\&\sum_{j=1}^{2}\sum_{k=1}^{2}\frac{\psi_{1j}^{2}\psi_{1k}^{2}}{M_jM_k}\int_{-\infty}^{\infty}g_j(\theta_1)e^{i\omega\theta_1}\,d\theta_1\int_{-\infty}^{\infty}g_k(\theta_2)e^{-i\omega\theta_2}\,d\theta_2\frac{1}{2\pi}\\&\int_{-\infty}^{\infty}\left[R_1(\tau-\theta_2+\theta_1)\right]e^{-i\omega(\tau+\theta_1-\theta_2)}\,d\tau\end{aligned} \tag{6.137}$$

There is only one thing left to do: Interpret this result. It is possible to assume that the variable of integration is $\xi=\tau+\theta_1-\theta_2$ in the third integral on the right. This is because θ_1 and θ_2 are held constant in this integral. After that, we are able to recognize the Wiener–Khintchine theorem

$$S_1(\omega)=\frac{1}{2\pi}\int_{-\infty}^{\infty}R_1(\xi)e^{(-i\omega\xi)}\,d\xi \tag{6.138}$$

According to equation (5.65), the second integral is known as the Fourier transform $H_k(\omega)$ of the impulse response $g_k(\theta_2)$. With the exception of the sign in the exponent, the first integral displays the same form and is able to be represented as $H_j(-\omega)$. Given these interpretations, equation (6.137) can be expressed as

$$\begin{aligned} S_{y_1}(\omega) &= \sum_{j=1}^{2}\sum_{k=1}^{2}\frac{\psi_{1j}^2\psi_{1k}^2}{M_jM_k}\int_{-\infty}^{\infty} g_j(\theta_1)e^{i\omega\theta_1}\,d\theta_1\int_{-\infty}^{\infty} g_k(\theta_2)e^{-i\omega\theta_2}\,d\theta_2\frac{1}{2\pi} \\ &\quad \int_{-\infty}^{\infty}\left[R_1(\tau-\theta_2+\theta_1)\right]e^{-i\omega(\tau+\theta_1-\theta_2)}\,d\tau \\ &= S_1(\omega)\sum_{j=1}^{2}\sum_{k=1}^{2}\frac{\psi_{1j}^2\psi_{1k}^2}{M_jM_k}H_j(-\omega)H_k(\omega) \end{aligned} \tag{6.139}$$

Example 5

▶ Assuming a single stationary random process $F_1(t)$ with a white noise spectrum $2\pi S_0$ as the excitation, please find the stationary input-output relationship for the spectral density functions $S_{y_1}(\omega)$ of the stationary response displacement $y_1(t)$ for the two-degree-of-freedom structural model depicted in Figure 6.5.

SOLUTION

For the two-degree-of-freedom system depicted in Figure 6.6, we have

$$\psi_{11}=1, \qquad \psi_{12}=1$$

$$M_1=2m, \quad M_2=2m$$

The required input-output relation for the spectral density functions can be obtained by substituting these constants and the complex frequency response functions from equation (6.43) into equation (6.139) as follows:

$$\begin{aligned} &S_{y_1}(\omega) \\ &= S_1(\omega)\sum_{j=1}^{2}\sum_{k=1}^{2}\frac{\psi_{1j}^2\psi_{1k}^2}{M_jM_k}H_j(-\omega)H_k(\omega)=\frac{1}{4m^2}S_1(\omega)\sum_{j=1}^{2}\sum_{k=1}^{2}H_j(-\omega)H_k(\omega) \\ &= \frac{S_1(\omega)}{4m^2}\left[H_1(-\omega)H_1(\omega)+H_1(-\omega)H_2(\omega)+H_2(-\omega)H_1(\omega)+H_2(-\omega)H_2(\omega)\right] \\ &= \frac{S_1(\omega)}{4m^2}\left[\frac{1}{\left(\omega_1^2-\omega^2\right)^2}+\frac{2}{\left(\omega_1^2-\omega^2\right)\left(\omega_2^2-\omega^2\right)}+\frac{1}{\left(\omega_2^2-\omega^2\right)^2}\right] \\ &= \frac{\pi S_0}{2m^2}\left[\frac{1}{\left(\omega_1^2-\omega^2\right)^2}+\frac{2}{\left(\omega_1^2-\omega^2\right)\left(\omega_2^2-\omega^2\right)}+\frac{1}{\left(\omega_2^2-\omega^2\right)^2}\right] \end{aligned}$$ ◀

6.2.3 Random Excitations of a Vehicle Model

The system depicted in Figure 6.7 offers an illustration of a vehicle model as an example. Linear springs are used to provide connections between two masses and a moving foundation. In this context, the excitation refers to the motion of the foundation, which we consider to be characterized by its acceleration, denoted by $\ddot{x}_0(t)$. We will investigate four different responses, namely the relative displacements $y_1 = x_1 - x_0$ and $y_2 = x_2 - x_1$, as well as the absolute accelerations of the two masses, $\ddot{x}_1$ and $\ddot{x}_2$. In terms of the relative displacements, the equations of motion are as follows:

$$m\ddot{y}_1 + ky_1 - ky_2 = -m\ddot{x}_0 \tag{6.140}$$

$$m\left(\ddot{y}_1 + \ddot{y}_2\right) + ky_2 = -m\ddot{x}_0 \tag{6.141}$$

that establish a connection between the response qualities y_1 and y_2 and the excitation $\ddot{x}_0$ in an indirect way. After determining the values of y_1 and y_2, the remaining response quantities $\ddot{x}_1$ and $\ddot{x}_2$ can be obtained by inserting the equations $\ddot{x}_1 = \ddot{x}_0 + \ddot{y}_1$ and $\ddot{x}_2 = \ddot{x}_0 + \ddot{y}_1 + \ddot{y}_2$ in equations (6.140) and (6.141).

$$m\ddot{x}_1 = -ky_1 + ky_2 \tag{6.142}$$

$$m\ddot{x}_2 = -ky_2 \tag{6.143}$$

In order to standardize the treatment that follows, we introduce the following notation

$$\frac{k}{m} = \omega_1^2, \quad \frac{k}{m} = \omega_2^2 \tag{6.144}$$

for the uncoupled natural frequencies ω_1 and ω_2.

In order to differentiate among the four different responses, we will be using subscripts. Therefore, it can be seen that there are four complex frequency responses represented by the symbols $H_{y_1}(\omega), H_{y_2}(\omega), H_{\ddot{x}_1}(\omega)$ and $H_{\ddot{x}_2}(\omega)$. The first pair can be obtained by setting

$$x_0 = e^{i\omega t}, \quad y_1 = H_{y_1}(\omega)e^{i\omega t}, \quad y_2 = H_{y_2}(\omega)e^{i\omega t} \tag{6.145}$$

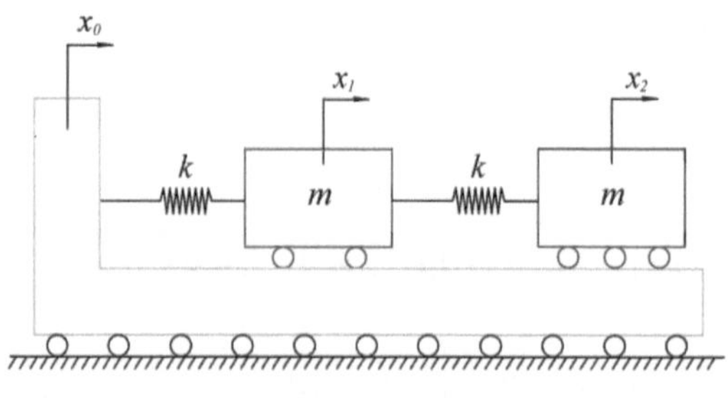

FIGURE 6.7 Vibration of a vehicle model with two degrees of freedom.

in equations (6.140) and (6.141), canceling the $e^{i\omega l}$ term and solving algebraically for

$$H_{y_1}(\omega) = \frac{2\dfrac{k}{m} - \omega^2}{\omega^4 - \dfrac{3k}{m}\omega^2 + \dfrac{k^2}{m^2}} \tag{6.146}$$

$$H_{y_2}(\omega) = \frac{\dfrac{k}{m}}{\omega^4 - \dfrac{3k}{m}\omega^2 + \dfrac{k^2}{m^2}} \tag{6.147}$$

By inserting equations (6.145), (6.146) and (6.147) into equations (6.142) and (6.142), we obtain the second set of frequency responses.

$$H_{\ddot{x}_1}(\omega) = \frac{\dfrac{k}{m}\left(\omega^2 - \dfrac{k}{m}\right)}{\omega^4 - \dfrac{3k}{m}\omega^2 + \dfrac{k^2}{m^2}} \tag{6.148}$$

$$H_{\ddot{x}_2}(\omega) = \frac{\dfrac{-k^2}{m^2}}{\omega^4 - \dfrac{3k}{m}\omega^2 + \dfrac{k^2}{m^2}} \tag{6.149}$$

This last pair of response functions provides the amplitudes of acceleration at the two masses respectively caused by a unit acceleration $\ddot{x}_0$ (all at frequency ω). Consequently, they are sometimes referred to as transmissibility functions. Because they represent ratios of like quantities, they are dimensionless. In a broader sense, the frequency response functions that we have referred to are sometimes known as transfer functions. They provide the response amplitude at a certain point resulting from a unit amplitude of excitation at a different point.

As an additional application of the findings presented in this sub-section, we will take into consideration the ideal white noise excitation of the two-mass system that was presented in this sub-section and visually represented in Figure 6.7. In a general sense, the nature of the problem is comparable to that of the system with a single degree of freedom that had been addressed previously.

Consider the excitation $\ddot{x}_0(t)$ to be a stationary random process with a mean of zero and a spectrum of ideally white noise of constant density S_0, spanning all frequencies from $\omega = -\infty$ to $\omega = \infty$. Then, the four response quantities $y_1(t)$, $y_2(t)$, $\ddot{x}_1(t)$ and $\ddot{x}_2(t)$ will also follow the characteristics of zero-mean stationary random processes. The general input-output relationship $S_y(\omega) = |H(\omega)|^2 S_x(\omega)$, as well as equations (6.146)–(6.149), which describe the frequency responses, can be utilized to yield the response spectra. For instance, we have the relative motion spectral densities

$$S_{y1}(\omega) = |H_{y1}(\omega)|^2 S_0 = \left| \frac{2\frac{k}{m} - \omega^2}{\omega^4 - \frac{3k}{m}\omega^2 + \frac{k^2}{m^2}} \right|^2 S_0 = \left(\frac{2km - m\omega^2}{m^2\omega^4 - 3km\omega^2 + k^2} \right)^2 S_0 \tag{6.150}$$

$$S_{y2}(\omega) = |H_{y2}(\omega)|^2 S_0 = \left| \frac{\frac{k}{m}}{\omega^4 - \frac{3k}{m}\omega^2 + \frac{k^2}{m^2}} \right|^2 S_0 = \left(\frac{km}{m^2\omega^4 - 3km\omega^2 + k^2} \right)^2 S_0 \tag{6.151}$$

6.3 EXERCISES

EXERCISE 6.1

Determine the natural frequencies and mode shapes of a spring-mass system, which is depicted in Figure 6.8, and which is only allowed to move in the vertical direction.

EXERCISE 6.2

Determine the necessary initial conditions for the system depicted in Figure 6.8 in order to induce vibration in (*a*) the first mode and (*b*) the second mode.

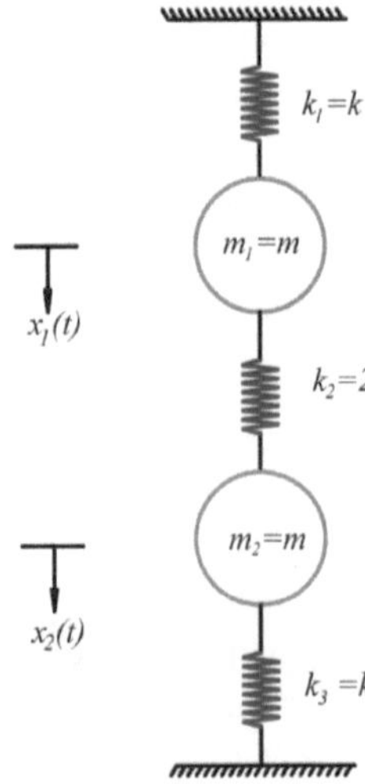

FIGURE 6.8 The vibration of a two degrees of freedom system.

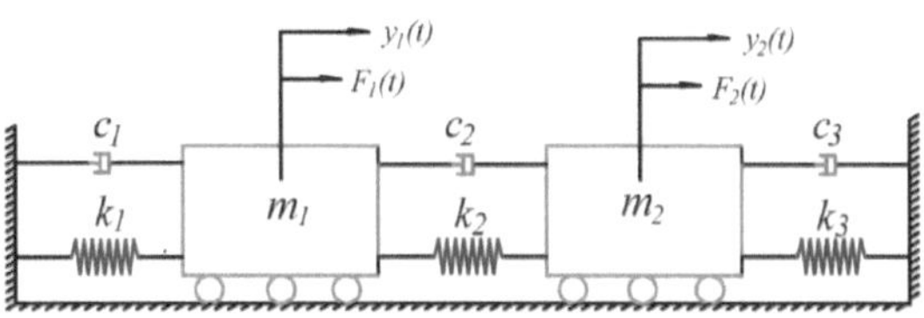

FIGURE 6.9 A spring-mass-damper system with two degrees of freedom.

EXERCISE 6.3

Take into consideration a viscously damped spring-mass-damper system depicted in Figure 6.9, which has two degrees of freedom. The motion of the system is completely described by the coordinates $y_1(t)$ and $y_2(t)$, which determine the positions of the masses m_1 and m_2 at any time t from their respective equilibrium positions. The masses m_1 and m_2 are not subject to any external forces, i.e., $F_1(t) = 0$ and $F_2(t) = 0$. Please determine the free-vibration response of the system shown in Figure 6.9 with $k_1 = 300$, $k_2 = 50$, $k_3 = 0$, $m_1 = 100$, $m_2 = 10$ and $c_1 = c_2 = c_3 = 0$ for the initial conditions $y_1(0) = 10$, $\dot{y}_1(0) = y_2(0) = \dot{y}_2(0) = 0$.

EXERCISE 6.4

Determine the steady-state response of the system depicted in Figure 6.10 when the mass m_1 is excited because the force $F_1(t) = F_{10}\cos\omega t$ is applied to it.

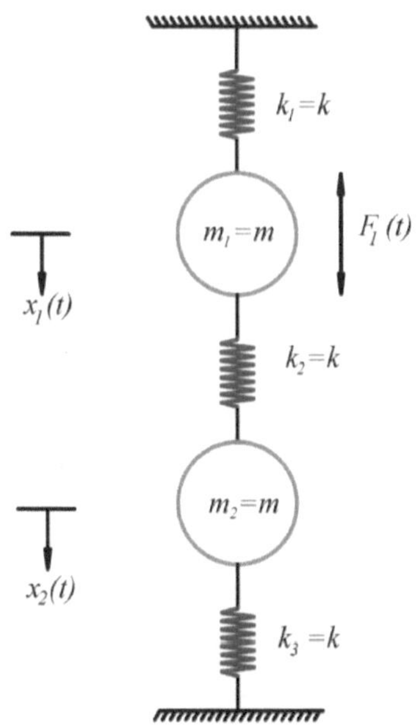

FIGURE 6.10 Vibration of a two-degree-of-freedom system.

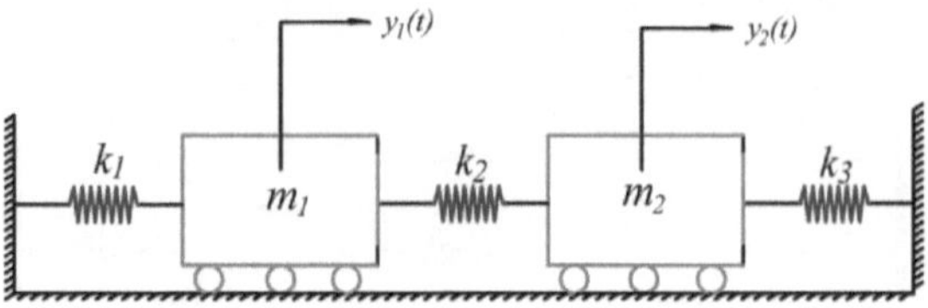

FIGURE 6.11 A spring-mass system with two degrees of freedom.

EXERCISE 6.5

In this exercise you are required to conduct an analysis of the system depicted in Figure 6.11. Find the natural frequencies and eigenvectors for the system depicted in Figure 6.11, which has the following values: $m_1 = 3$ kg, $m_2 = 6$ kg, $k_1 = 50$ N/m, $k_2 = 150$ N/m and $k_3 = 350$ N/m.

EXERCISE 6.6

For a spring-mass system with two degrees of freedom with the following equations of motion

$$\begin{bmatrix} m_1 & 0 \\ 0 & m_2 \end{bmatrix} \begin{Bmatrix} \ddot{y}_1(t) \\ \ddot{y}_2(t) \end{Bmatrix} + \begin{bmatrix} k_1 + k_2 & -k_2 \\ -k_2 & k_2 + k_3 \end{bmatrix} \begin{Bmatrix} y_1(t) \\ y_2(t) \end{Bmatrix} = \begin{Bmatrix} F_1(t) \\ F_2(t) \end{Bmatrix},$$

$$\begin{bmatrix} m_1 & 0 \\ 0 & m_2 \end{bmatrix} = \begin{bmatrix} 2 & 0 \\ 0 & 4 \end{bmatrix} kg,$$

$$k_1 = 200N/m, \quad k_2 = 800N/m, \quad k_3 = 1200N/m,$$

$$\begin{bmatrix} k_1 + k_2 & -k_2 \\ -k_2 & k_2 + k_3 \end{bmatrix} = \begin{bmatrix} 1000 & -800 \\ -800 & 2000 \end{bmatrix} N/m,$$

a. Find natural frequencies and eigenvectors for this system;
b. Please determine the steady state displacement response $\begin{Bmatrix} y_1(t) \\ y_2(t) \end{Bmatrix}$ when a steady state harmonic load vector

$$\begin{Bmatrix} F_1(t) \\ F_2(t) \end{Bmatrix} = \begin{Bmatrix} 1 \\ 1 \end{Bmatrix} F_0 \sin\omega t$$

is applied. Use $F_0 = 20N$ and $\omega = 20rad/\sec$.

EXERCISE 6.7

For a spring-mass system with two degrees of freedom with the following equations of motion

$$\begin{bmatrix} m_1 & 0 \\ 0 & m_2 \end{bmatrix}\begin{Bmatrix} \ddot{y}_1(t) \\ \ddot{y}_2(t) \end{Bmatrix}+\begin{bmatrix} k_1+k_2 & -k_2 \\ -k_2 & k_2+k_3 \end{bmatrix}\begin{Bmatrix} y_1(t) \\ y_2(t) \end{Bmatrix}=\begin{Bmatrix} F_1(t) \\ F_2(t) \end{Bmatrix},$$

$$\begin{bmatrix} m_1 & 0 \\ 0 & m_2 \end{bmatrix}=\begin{bmatrix} 4 & 0 \\ 0 & 4 \end{bmatrix}kg,$$

$$k_1=1000N/m, \quad k_2=1000N/m, \quad k_3=1000N/m,$$

$$\begin{bmatrix} k_1+k_2 & -k_2 \\ -k_2 & k_2+k_3 \end{bmatrix}=\begin{bmatrix} 2000 & -1000 \\ -1000 & 2000 \end{bmatrix}N/m,$$

Consider the excitation load vector

$$\begin{Bmatrix} F_1(t) \\ F_2(t) \end{Bmatrix}=\begin{Bmatrix} 1 \\ 0 \end{Bmatrix}F_0(t)$$

$F_0(t)$ is a stationary random process with a mean value of zero mean and a white noise spectral density S_0.

a. Please determine the autocorrelation function $R_{y_1}(\tau)$ for the response displacement $y_1(t)$;
b. Please determine the spectral density function $S_{y_1}(\omega)$ for the response displacement $y_1(t)$.

EXERCISE 6.8

The system depicted in Figure 6.12 offers an illustration of a vehicle model. Linear springs and linear dashpots are used to provide connections between two masses and a moving foundation. In this context, the excitation refers to

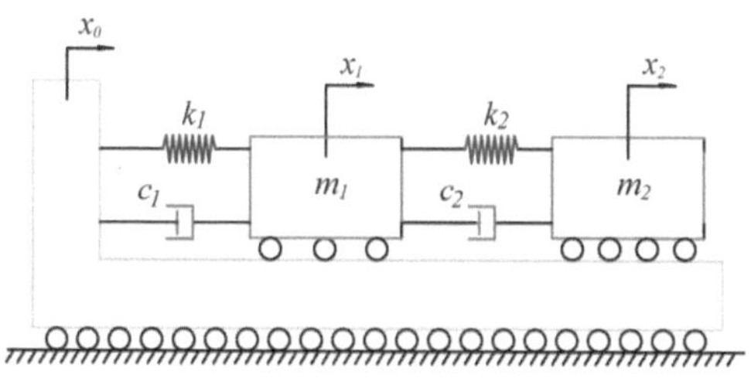

FIGURE 6.12 Vibration of a vehicle model with two degrees of freedom.

the motion of the foundation, which is considered to be characterized by its acceleration, denoted by $\ddot{x}_0(t)$. You are required to investigate four different responses, namely the relative displacements $y_1 = x_1 - x_0$ and $y_2 = x_2 - x_1$, as well as the absolute accelerations of the two masses, $\ddot{x}_1$ and $\ddot{x}_2$. In terms of the relative displacements, the equations of motion are as follows:

$$m_1\ddot{y}_1 + c_1\dot{y}_1 - c_2\dot{y}_2 + k_1y_1 - k_2y_2 = -m_1\ddot{x}_0$$

$$m_2\left(\ddot{y}_1 + \ddot{y}_2\right) + c_2\dot{y}_2 + k_2y_2 = -m_2\ddot{x}_0$$

that establish a connection between the response qualities y_1 and y_2 and the excitation $\ddot{x}_0$ in an indirect way. After determining the values of y_1 and y_2, the remaining response quantities $\ddot{x}_1$ and $\ddot{x}_2$ can be obtained by inserting the equations $\ddot{x}_1 = \ddot{x}_0 + \ddot{y}_1$ and $\ddot{x}_2 = \ddot{x}_0 + \ddot{y}_1 + \ddot{y}_2$ in the above two equations:

$$m\ddot{x}_1 = -c_1\dot{y}_1 + c_2\dot{y}_2 - k_1y_1 + k_2y_2$$

$$m\ddot{x}_2 = -c_2\dot{y}_2 - k_2y_2$$

a. Please determine the four complex frequency response functions represented by the symbols $H_{y_1}(\omega), H_{y_2}(\omega), H_{\ddot{x}_1}(\omega)$ and $H_{\ddot{x}_2}(\omega)$;
b. You will next be required to take into consideration the ideal white noise excitation of the two-mass system that was presented in this exercise and visually represented in Figure 6.12. Consider the excitation $\ddot{x}_0(t)$ to be a stationary random process with a mean of zero and a spectrum of ideally white noise of constant density S_0, spanning all frequencies from $\omega = -\infty$ to $\omega = \infty$. Then the four response quantities $y_1(t)$, $y_2(t)$, $\ddot{x}_1(t)$ and $\ddot{x}_2(t)$ will also follow the characteristics of zero-mean stationary random processes. Using the general input-output relationship $S_y(\omega) = |H(\omega)|^2 S_x(\omega)$, as well as the two complex frequency response functions $H_{y_1}(\omega)$ and $H_{y_2}(\omega)$ you just obtained in this exercise to derive the closed-form expressions for the relative motion spectral densities $S_{y_1}(\omega)$ and $S_{y_2}(\omega)$.

7 Random Vibration of Multi-Degree-of-Freedom Linear Systems

Almost all engineering systems are continuous and have an infinite number of degrees of freedom. The analysis of vibrations in continuous systems necessitates solving partial differential equations, a task that is notably challenging. Analytical solutions are non-existent for many partial differential equations. In contrast, the study of a system with multiple degrees of freedom involves solving a set of ordinary differential equations, which is rather straightforward. Therefore, in order to make the analysis simpler, continuous systems are frequently modeled as systems with multiple degrees of freedom. Systems are referred to as having multiple degrees of freedom when they require multiple independent coordinates in order to characterize their motion. For example, a system is called having three degree of freedom when it requires three independent coordinates in order to characterize its motion.

The concepts presented in the previous chapter can be directly applied to multi-degree-of-freedom (MDOF) systems. For instance, there exists one equation of motion corresponding to each degree of freedom. Either by utilizing the influence coefficients or by utilizing Newton's second law of motion, one can obtain the equations of motion. Because the problem of the deterministic two-degree-of-freedom system carries all of the important features of the general deterministic MDOF system, and because the inclusion of random considerations does not change this fact, we can then bring out the basic concepts and analysis methodology for the response of linear MDOF systems to random excitation based on the knowledge already developed in the previous chapter for the treatment of the problem for a linear two-degree-of-freedom system.

A vibration system with n degrees of freedom has n natural frequencies, each corresponding to its own mode shape. The technique of calculating the natural frequencies by setting the determinant to zero and solving the resulting characteristic equation can also be used for these systems. It is important to note that the solution of the characteristic equation becomes more complicated as the number of degrees of freedom grows. The mode shapes have a property that is known as orthogonality, and this property can be employed for the purpose of solving problems of undamped forced vibration by employing a method that is known as modal analysis. By utilizing a concept known as proportional damping, it is also possible to conveniently find a solution to a forced-vibration problem that is associated with a viscously damped system.

DOI: 10.1201/9781003431503-7

7.1 A REVIEW OF THE DETERMINISTIC THEORY

7.1.1 Representing Continuous Systems as Multi-Degree-of-Freedom Systems

There are a variety of approaches that can be utilized in order to approximate a continuous system as a MDOF system. Replacement of the system's distributed mass or inertia with a finite number of lumped masses or rigid bodies is an example of a straightforward method. In this scenario, it is assumed that the lumped masses are connected by massless elastic and damping members. In order to describe the motion of lumped masses (or rigid bodies), linear coordinates (or angular coordinates), are utilized. These models are referred to as lumped-parameter systems, lumped-mass systems or discrete-mass systems. The minimal number of coordinates that are required to describe the motion of the lumped masses and rigid bodies defines the number of degrees of freedom of the vibration system. It stands to reason that the accuracy of the analysis that is produced will be higher when the number of lumped masses that are incorporated into the model is increased.

Certain problems automatically suggest the appropriate lumped-parameter model to be employed. In the case of an industrial building that has five horizontal floors, for instance, the utilization of a five-lumped-mass model is automatically suggested. This model assumes that the system's inertia is concentrated as five-point masses positioned at the floor levels, and the elasticities of the vertical columns supporting the horizontal floors are substituted with springs.

Another common approach to approximate a continuous system as a MDOF system is to replace the geometry of the system with a high number of small elements. An approximation of the solution to the original system is obtained by applying the principles of compatibility and equilibrium under the assumption of a simple solution within each element. This technique, which is referred to as the finite element method, is a powerful numerical technique used to solve complex problems in the field of engineering. One of the applications of the finite element method is vibration analysis, which involves studying the behavior of vibrating systems. The finite element method is based on the principle of dividing the vibration system being analyzed into a large number of small elements, called finite elements. These elements have assigned material properties, such as mass and stiffness that are used to model the behavior of the system. To analyze the vibration of a structure, the finite element method solves the governing equations, which are based on Newton's laws of motion. These equations describe the forces and displacements acting on the system. By solving these equations, the finite element method calculates the response of the system in terms of vibration modes, frequencies and amplitudes. However, the detailed theories of the finite element method will not be included in this book. Interested readers can refer to the many excellent textbooks published recently in this field.

7.1.2 Motion Equations for Forced Vibration by Newton's Second Law

Newton's second law of motion, also known as the law of acceleration, states that the net force acting on an object is equal to the product of the mass of the object and the magnitude of the acceleration it produces. Newton's second law of motion is based on the principle that a force is required to accelerate an object. It quantifies the relationship between force and acceleration, stating that the greater the force applied, the greater the acceleration produced. In order to obtain the equations of motion for a MDOF system utilizing Newton's second law of motion, one may employ the subsequent procedure:

1. In order to describe the positions of the various point masses and rigid bodies that are present in the system, appropriate coordinates need to be established. Assign appropriate positive directions for the displacements, velocities and accelerations of the masses and rigid bodies.
2. The configuration of the system that is in a state of static equilibrium should be determined, and the displacements of the masses and rigid bodies should be measured from their respective positions in the state of static equilibrium.
3. Create a visual representation, known as a free-body diagram, for each individual mass or rigid body inside the system. Specify the spring, damping and external forces exerted on each mass or rigid body when positive displacement and velocity are applied to that mass or rigid body.
4. For every mass or rigid body depicted by the free-body diagram, use Newton's second law of motion as

$$m_i \ddot{y}_i = \sum_j F_{ij} \left(\text{for mass } m_i\right) \tag{7.1}$$

in which $\Sigma_j F_{ij}$ represents the sum of all forces acting on the mass m_i.

Let us consider a viscously damped spring-mass-damper system depicted in Figure 7.1(a), which has many degrees of freedom. The motion of the system is completely described by the coordinates $y_1(t)$, $y_2(t)$, ..., $y_i(t)$, ... $y_j(t)$, ..., $y_n(t)$, which determine the positions of the masses m_1, m_2, ..., m_i,..., m_j,..., m_n at any time t from their respective equilibrium positions. The masses m_1, m_2,..., m_i,..., m_j,..., m_n are subject to the external forces $F_1(t)$, $F_2(t)$,..., $F_i(t)$,..., $F_j(t)$,..., $F_n(t)$, respectively. Figure 7.1(b) depicts the free-body diagram of a typical interior mass m_i, together with the assigned positive directions for its displacement, velocity and acceleration. The equations of motion can be derived by applying Newton's second law of motion to this typical interior mass m_i as follows:

$$\begin{aligned} m_i \ddot{y}_i &= -k_i\left(y_i - y_{i-1}\right) + k_{i+1}\left(y_{i+1} - y_i\right) \\ &\quad - c_i\left(\dot{y}_i - \dot{y}_{i-1}\right) + c_{i+1}\left(\dot{y}_{i+1} - \dot{y}_i\right) + F_i(t), \qquad i = 2, 3, \ldots, n-1 \end{aligned}$$

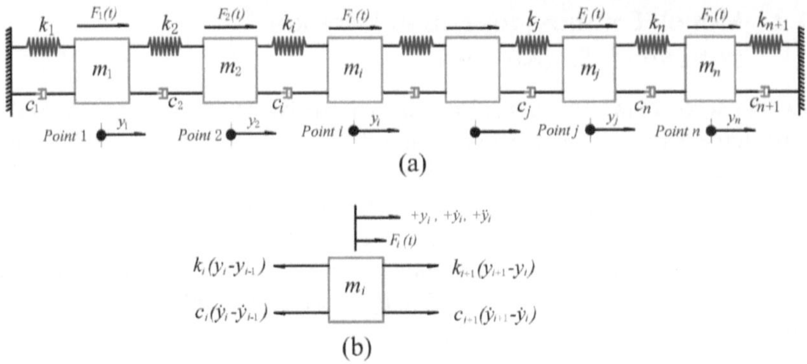

FIGURE 7.1 (a,b) A spring-mass-damper system with many degrees of freedom.

or

$$\begin{aligned} &m_i\ddot{y}_i - c_i\dot{y}_{i-1} + (c_i + c_{i+1})\dot{y}_i - c_{i+1}\dot{y}_{i+1} \\ &-k_i y_{i-1} + (k_i + k_{i+1}) y_i - k_{i+1} y_{i+1} = F_i(t), \qquad i = 2, 3, \ldots, n-1 \end{aligned} \tag{7.2}$$

Using equation (7.2) as a starting point, one can obtain the equations of motion for the masses m_1 and m_n by setting $i = 1$ along with $y_0 = 0$ and $i = n$ along with $y_{n+1} = 0$, respectively:

$$m_1\ddot{y}_1 + (c_1 + c_2)\dot{y}_1 - c_2\dot{y}_2 + (k_1 + k_2) y_1 - k_2 y_2 = F_1(t) \tag{7.3}$$

$$m_n\ddot{y}_n - c_n\dot{y}_{n-1} + (c_n + c_{n+1})\dot{y}_n - k_n y_{n-1} + (k_n + k_{n+1}) y_n = F_n(t) \tag{7.4}$$

It may be observed that equation (7.3) has terms that involve y_2 (namely, $-c_2\dot{y}_2$ and $-k_2 y_2$), whereas equation (7.4) contains terms that involve y_{n-1} (namely, $-c_n\dot{y}_{n-1}$ and $-k_n y_{n-1}$). It may also be observed that equation (7.2) has terms that involve y_{i-1} (namely, $-c_i\dot{y}_{i-1}$ and $-k_i y_{i-1}$) and y_{i+1} (namely, $-c_{i+1}\dot{y}_{i+1}$ and $-k_{i+1} y_{i+1}$). Because of this, they constitute a system consisting of n coupled differential equations of the second order.

Equations (7.2)–(7.4) can be expressed in matrix form is as follows:

$$\mathbf{M}\ddot{\mathbf{y}} + \mathbf{C}\dot{\mathbf{y}} + \mathbf{K}\mathbf{y} = \mathbf{F}(t) \tag{7.5}$$

where **M**, **C** and **K** are referred to as the mass, damping and stiffness matrices, respectively, and are given by

$$\mathbf{M} = \begin{bmatrix} m_1 & 0 & 0 & \cdots & 0 & 0 \\ 0 & m_2 & 0 & \cdots & 0 & 0 \\ 0 & 0 & m_3 & \cdots & 0 & 0 \\ \vdots & & & & & \\ 0 & 0 & 0 & \cdots & 0 & m_n \end{bmatrix} \tag{7.6}$$

$$\mathbf{C} = \begin{bmatrix} (c_1+c_2) & -c_2 & 0 & \cdots & 0 & 0 \\ -c_2 & (c_2+c_3) & -c_3 & \cdots & 0 & 0 \\ 0 & -c_3 & (c_3+c_4) & \cdots & 0 & 0 \\ \cdot & \cdot & \cdot & \cdots & \cdot & \cdot \\ \cdot & \cdot & \cdot & \cdots & \cdot & \cdot \\ \cdot & \cdot & \cdot & \cdots & \cdot & \cdot \\ 0 & 0 & 0 & \cdots & -c_n & (c_n+c_{n+1}) \end{bmatrix} \tag{7.7}$$

$$\mathbf{K} = \begin{bmatrix} (k_1+k_2) & -k_2 & 0 & \cdots & 0 & 0 \\ -k_2 & (k_2+k_3) & -k_3 & \cdots & 0 & 0 \\ 0 & -k_3 & (k_3+k_4) & \cdots & 0 & 0 \\ \cdot & & & & & \\ \cdot & & & & & \\ \cdot & & & \cdots & -k_n & (k_n+k_{n+1}) \end{bmatrix} \tag{7.8}$$

and $\mathbf{y}$, $\dot{\mathbf{y}}$, $\ddot{\mathbf{y}}$, and $\mathbf{F}(t)$ are referred to as the displacement, velocity, acceleration and force vectors, respectively, and are given by

$$\mathbf{y} = \begin{Bmatrix} y_1(t) \\ y_2(t) \\ \cdot \\ \cdot \\ \cdot \\ y_n(t) \end{Bmatrix}, \qquad \dot{\mathbf{y}} = \begin{Bmatrix} \dot{y}_1(t) \\ \dot{y}_2(t) \\ \cdot \\ \cdot \\ \cdot \\ \dot{y}_n(t) \end{Bmatrix}, \qquad \ddot{\mathbf{y}} = \begin{Bmatrix} \ddot{y}_1(t) \\ \ddot{y}_2(t) \\ \cdot \\ \cdot \\ \cdot \\ \ddot{y}_n(t) \end{Bmatrix} \tag{7.9}$$

$$\mathbf{F}(t) = \begin{Bmatrix} F_1(t) \\ F_2(t) \\ \cdot \\ \cdot \\ \cdot \\ F_n(t) \end{Bmatrix} \tag{7.10}$$

The spring-mass-damper system just mentioned is a specific example of a general spring-mass-damper system with n degrees of freedom. In their most general form, the mass, damping and stiffness matrices are expressed as where **M**, **C**, and **K**

$$\mathbf{M} = \begin{bmatrix} m_{11} & m_{12} & m_{13} & \ldots & m_{1n} \\ m_{21} & m_{22} & m_{23} & \ldots & m_{2n} \\ \cdot & & & & \\ \cdot & & & & \\ \cdot & & & & \\ m_{n1} & m_{n2} & m_{n3} & \ldots & m_{nn} \end{bmatrix} \tag{7.11}$$

$$\mathbf{C} = \begin{bmatrix} c_{11} & c_{12} & c_{13} & \ldots & c_{1n} \\ c_{21} & c_{22} & c_{23} & \ldots & c_{2n} \\ \cdot & \cdot & \cdot & \ldots & \cdot \\ \cdot & \cdot & \cdot & \ldots & \cdot \\ \cdot & \cdot & \cdot & \ldots & \cdot \\ c_{n1} & c_{n2} & c_{n3} & \cdots & c_{nn} \end{bmatrix} \tag{7.12}$$

$$\mathbf{K} = \begin{bmatrix} k_{11} & k_{12} & k_{13} & \cdots & k_{1n} \\ k_{21} & k_{22} & k_{23} & \cdots & k_{2n} \\ \cdot & & & & \\ \cdot & & & & \\ \cdot & & & & \\ k_{n2} & k_{n2} & k_{n3} & \cdots & k_{nn} \end{bmatrix} \tag{7.13}$$

When the mass matrix is not diagonal, the system is referred to as having mass or inertia coupling. If the damping matrix is not diagonal, it indicates that the system exhibits damping or velocity coupling. If the stiffness matrix is not diagonal, the system is considered to have elastic or static coupling. Both mass and damping coupling are also known as dynamic coupling.

In the spring-mass system that has been investigated in Figure 7.1(a), it is possible to observe that the differential equations are coupled; each equation involves more than one coordinate. Consequently, it is not possible to solve the equations individually one at a time; rather, it is only possible to solve them simultaneously. Additionally, it is possible to observe that the system is statically coupled due to the fact that the stiffnesses are coupled, which means that the stiffness matrix has at least one non-zero off-diagonal term. When the mass matrix, on the other hand, contains at least one off-diagonal term that is not zero, the system is said to be dynamically coupled. In addition, if the stiffness and mass matrices both have non-zero off-diagonal terms, then the system is said to be coupled both statically and dynamically.

7.1.3 Motion Equations Written in Terms of Influence Coefficients

It is also possible to write the equations of motion of a system with many degrees of freedom in terms of influence coefficients, which are widely utilized in the field of structural engineering design. One set of influence coefficients can be associated with each of the matrices that are involved in the equations of motion. This is the fundamental idea. Inertia influence coefficients and stiffness influence coefficients are the names given to the influence coefficients that are associated with the mass matrices and the stiffness matrices, respectively. The influence coefficients associated with the damping matrix are known as the damping influence coefficients.

For the purpose of this development of the equations of motion of a general multi-degree-of-freedom system in terms of influence coefficients, it will be appropriate to refer to the general simple beam depicted in Figure 7.2 as an example of a typical situation. The discussion applies equally to any type of structure and system.

It is going to be assumed that the motion of this structure is characterized by the displacements of a set of discrete points on the beam: $y_1(t)$, $y_2(t)$,..., $y_i(t)$,..., $y_N(t)$. In theory, these points can be placed anywhere on the structure; but, in practice, they should be associated with specific features of the physical properties that may be significant, and they should be distributed in such a way as to provide a good definition of the shape that has been deflected. Although increasing numbers of degrees of freedom (displacement components) provide better approximations of the underlying dynamic behavior, in many circumstances, excellent results can be produced with only two or three degrees of freedom. The number of degrees of freedom that are to be considered is left up to the discretion of the analyst. One displacement component has been associated with each nodal point on the beam depicted in Figure 7.2.

It is possible to formulate the equations of motion for the system depicted in Figure 7.2 by expressing the equilibrium of the effective forces that are associated with each of the system's degrees of freedom. At any given point i, there are, in

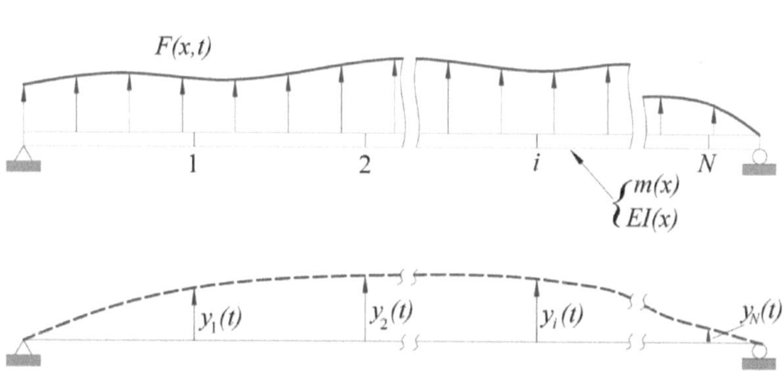

FIGURE 7.2 A general beam-type structure that has been discretized.

general, four types of forces involved: The externally applied load $F_i(t)$, as well as the forces caused by motion, namely inertia f_{Ii}, damping f_{Di} and elastic f_{Si}. Therefore, for each of the multiple degrees of freedom, the state of dynamic equilibrium can be mathematically represented as

$$f_{I1} + f_{D1} + f_{S1} = F_1(t)$$

$$f_{I2} + f_{D2} + f_{S2} = F_2(t)$$

$$\cdots\cdots\cdots\cdots\cdots\cdots$$

$$f_{Ii} + f_{Di} + f_{Si} = F_i(t) \tag{7.14}$$

$$\cdots\cdots\cdots\cdots\cdots\cdots$$

$$f_{In} + f_{Dn} + f_{Sn} = F_n(t)$$

Alternatively, when the force vectors are expressed using matrix notation,

$$\mathbf{f}_I + \mathbf{f}_D + \mathbf{f}_S = \mathbf{F}(t) \tag{7.15}$$

For the most convenient representation of each resisting force, a suitable set of influence coefficients is utilized. Let us examine the elastic-force component that is developed at point 1. In general, this is dependent on the displacement components that are developed at all points of the structure:

$$f_{S1} = k_{11}y_1 + k_{12}y_2 + k_{13}y_3 + \cdots + k_{1N}y_N \tag{7.16}$$

In a similar manner, the elastic force that corresponds to the degree of freedom y_2 is

$$f_{S2} = k_{21}y_1 + k_{22}y_2 + k_{23}y_3 + \cdots + k_{2N}y_N \tag{7.17}$$

and, in general,

$$f_{Si} = k_{i1}y_1 + k_{i2}y_2 + k_{i3}y_3 + \cdots + k_{iN}y_N \tag{7.18}$$

In these equations, it has been tacitly assumed that the structural behavior is linear, which means that the principle of superposition is applicable. For a simple linear spring, the force that is required to cause a unit elongation is referred to as the stiffness of the spring. When dealing with more complicated systems, we can express the relationship between the displacement at a particular point and the forces that are acting at various other points of the system through the utilization of stiffness influence coefficients. The stiffness influence coefficient, represented by k_{ij}, is defined as the force at the point i due to a unit displacement at the point j when all the points other than the point j are fixed.

When represented in matrix form, the complete set of elastic-force relationships can be expressed as

$$\begin{Bmatrix} f_{S1} \\ f_{S2} \\ \cdot \\ f_{Si} \\ \cdot \\ f_{SN} \end{Bmatrix} = \begin{bmatrix} k_{11} & k_{12} & k_{13} & \cdots & k_{1N} \\ k_{21} & k_{22} & k_{23} & \cdots & k_{2N} \\ \cdots & \cdots & \cdots & \cdots & \cdots \\ k_{i1} & k_{i2} & k_{i3} & \cdots & k_{iN} \\ \cdots & \cdots & \cdots & \cdots & \cdots \\ k_{N1} & k_{N2} & k_{N3} & \cdots & k_{NN} \end{bmatrix} \begin{Bmatrix} y_1 \\ y_2 \\ \cdot \\ y_i \\ \cdot \\ y_N \end{Bmatrix} \tag{7.19}$$

Alternatively, in a symbolic manner,

$$\mathbf{f}_S = \mathbf{K}\mathbf{y} \tag{7.20}$$

where the matrix of stiffness coefficients **K** is referred to as the stiffness matrix of the structure (for the particular set of displacement coordinates) and v is the displacement vector that represents the displaced shape of the structure.

It is important to take note of the following aspects of the stiffness influence coefficients: (1) Because the force that must be applied at point i in order to cause a unit deflection at point j and zero deflection at all other points is the same as the force that must be applied at point j in order to cause a unit deflection at point i and zero deflection at all other points (according to Maxwell's reciprocity theorem), we may conclude that $k_{ij} = k_{ji}$. (2) Through the application of the principles of statics and solid mechanics, it is possible to compute the stiffness influence coefficients. (3) When referring to torsional systems, the stiffness influence coefficients can be defined in terms of the unit angular displacement as well as the torque that is responsible for causing the angular displacement. For instance, the torque at point i (rotor i) resulting from a unit angular displacement at point j and zero angular displacement at all other points can be expressed as k_{ij} in a multi-rotor torsional system.

If we assume that the damping is velocity dependent, namely of the viscous type, we can represent the damping forces corresponding to the chosen degrees of freedom by means of damping influence coefficients in a similar manner. The complete set of damping forces can be determined by drawing an analogy with equation (7.19)

$$\begin{Bmatrix} f_{D1} \\ f_{D2} \\ \cdot \\ f_{Di} \\ \cdot \\ f_{DN} \end{Bmatrix} = \begin{bmatrix} c_{11} & c_{12} & c_{13} & \cdots & c_{1N} \\ c_{21} & c_{22} & c_{23} & \cdots & c_{2N} \\ \cdots & \cdots & \cdots & \cdots & \cdots \\ c_{i1} & c_{i2} & c_{i3} & \cdots & c_{iN} \\ \cdots & \cdots & \cdots & \cdots & \cdots \\ c_{N1} & c_{N2} & c_{N3} & \cdots & c_{NN} \end{bmatrix} \begin{Bmatrix} \dot{y}_1 \\ \dot{y}_2 \\ \cdot \\ \dot{y}_i \\ \cdot \\ \dot{y}_N \end{Bmatrix} \tag{7.21}$$

where the coefficients c_{ij} are known as damping influence coefficients and $\dot{y}_i$ stands for the time rate of change (velocity) of the i displacement coordinate. The damping influence coefficients, represented by c_{ij}, is defined as the force at the point i due to a unit velocity at the point j when all the points other than the point j have 0 velocities.

Alternatively, equation (7.21) can be written in a symbolic manner,

$$\mathbf{f}_D = \mathbf{C}\dot{\mathbf{y}} \tag{7.22}$$

where the matrix of damping coefficients $\mathbf{C}$ is referred to as the damping matrix of the structure (for the specified degrees of freedom) and $\dot{\mathbf{y}}$ is the velocity vector.

The inertial forces can be described in a similar manner by a set of influence coefficients that are referred to as the mass coefficients. These represent the relationship between the accelerations of the degrees of freedom and the resulting inertial forces. The complete set of inertial forces can be determined by drawing an analogy with equation (7.19)

$$\begin{Bmatrix} f_{I1} \\ f_{I2} \\ \cdot \\ f_{Ii} \\ \cdot \\ f_{IN} \end{Bmatrix} = \begin{bmatrix} m_{11} & m_{12} & m_{13} & \cdots & m_{1N} \\ m_{21} & m_{22} & m_{23} & \cdots & m_{2N} \\ \cdots & \cdots & \cdots & \cdots & \cdots \\ m_{i1} & m_{i2} & m_{i3} & \cdots & m_{iN} \\ \cdots & \cdots & \cdots & \cdots & \cdots \\ m_{N1} & m_{N2} & m_{N3} & \cdots & m_{NN} \end{bmatrix} \begin{Bmatrix} \ddot{y}_1 \\ \ddot{y}_2 \\ \cdot \\ \ddot{y}_i \\ \cdot \\ \ddot{y}_N \end{Bmatrix} \tag{7.23}$$

in which the coefficients m_{ij} are known as mass influence coefficients and $\ddot{y}_i$ stands for the acceleration of the i displacement coordinate. The mass influence coefficients, represented by m_{ij}, is defined as the force at the point i due to a unit acceleration at the point j when all the points other than the point j have 0 acceleration. For a linear system, it can be readily proven that the inertia influence coefficients are symmetric, that is, $m_{ij} = m_{ji}$.

Alternatively, equation (7.23) can be written in a symbolic manner,

$$\mathbf{f}_I = \mathbf{M}\ddot{\mathbf{y}} \tag{7.24}$$

where the matrix of mass coefficients $\mathbf{M}$ is referred to as the mass matrix of the structure (for the specified set of degrees of freedom) and $\ddot{\mathbf{y}}$ is its acceleration vector defined for the specified set of displacement coordinates.

By substituting equations (7.20), (7.22) and (7.24) into equation (7.15), the complete dynamic equilibrium of the structure may be determined, taking into account all degrees of freedom:

$$\mathbf{M}\ddot{\mathbf{y}} + \mathbf{C}\dot{\mathbf{y}} + \mathbf{K}\mathbf{y} = \mathbf{F}(t) \tag{7.25}$$

The order of the matrix in equation (7.25) corresponds to the number of degrees of freedom that are employed in characterizing the displacements of the structure. Therefore, equation (7.25) is the expression of the N equations of motion that are used to define the response of the MDOF system.

Example 7.1

▶ Please determine the stiffness influence coefficients of the system shown in Figure 7.3.

SOLUTION

Let y_1, y_2 and y_3 represent the displacements that masses m_1, m_2 and m_3 have moved, respectively. The stiffness influence coefficients k_{ij} of the system can be calculated based on the spring stiffnesses k_1, k_2 and k_3 using the following method. First, we assign a displacement value of one to m_1 ($y_1 = 1$), whereas m_2 and m_3 have displacements of zero ($y_2 = y_3 = 0$), as depicted in Figure 7.4(a). The set of forces $k_{i1}\,(i = 1, 2, 3)$ is assumed to maintain the system in this current configuration. The free-body diagrams of the masses corresponding to the configuration shown in Figure 7.4(a) are depicted in Figure 7.4(b).

The equilibrium of the forces for the mass m_1 in the horizontal direction gives $k_1 = -k_2 + k_{11}$. The equilibrium of the forces for the mass m_2 in the horizontal direction gives $k_{21} = -k_2$. The equilibrium of the forces for the mass m_3 in the horizontal direction gives $k_{31} = 0$. Consequently, we can obtain the following relations

$$k_{11} = k_1 + k_2, \quad k_{21} = -k_2, \quad k_{31} = 0 \tag{7.26}$$

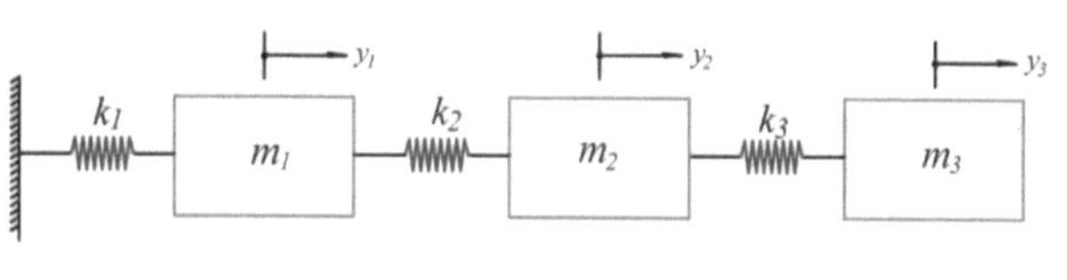

FIGURE 7.3 A three-degree-of-freedom system.

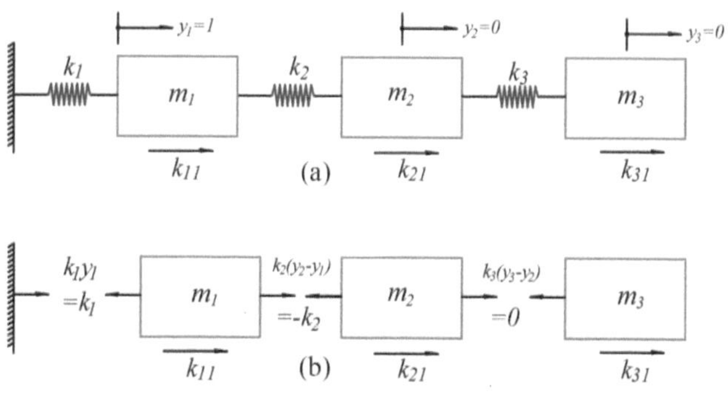

FIGURE 7.4 (a,b) The first configuration of a three-degree-of-freedom system.

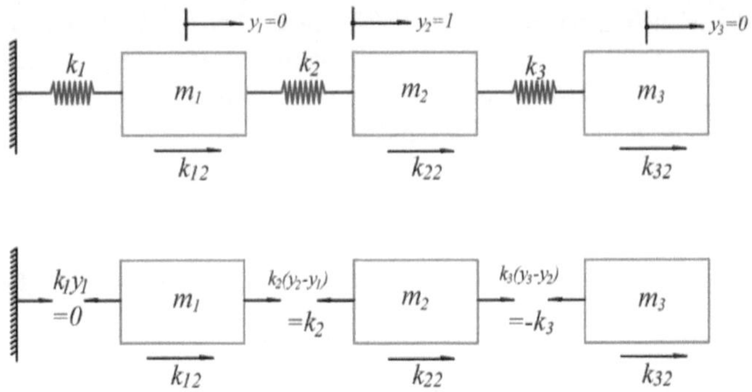

FIGURE 7.5 The second configuration of a three-degree-of-freedom system.

Subsequently, the displacements of the masses are assumed as $y_1 = 0$, $y_2 = 1$ and $y_3 = 0$, as depicted in Figure 7.5(a). Because the forces $k_{i2}\,(i = 1, 2, 3)$ are assumed to maintain the vibration system in this configuration, we may create free-body diagrams for the masses as shown in Figure 7.5(b). The equilibrium of the forces for the mass m_1 in the horizontal direction provides $0 = k_2 + k_{12}$. The equilibrium of the forces for the masse m_2 in the horizontal direction gives $k_{22} - k_3 = k_2$. The equilibrium of the forces for the masse m_3 in the horizontal direction yields $k_{32} = -k_3$. Consequently, we can obtain the following relations

$$k_{12} = -k_2, \quad k_{22} = k_2 + k_3, \quad k_{32} = -k_3 \tag{7.27}$$

Finally, the displacements of the masses are assumed to be $y_1 = 0$, $y_2 = 0$ and $y_3 = 1$, as shown in Figure 7.6(a). Since the forces $k_{i3}\,(i = 1, 2, 3)$ are assumed to maintain the vibration system in this configuration, we can draw free-body diagrams for the masses as shown in Figure 7.6(b). The equilibrium of the forces for the mass m_1 in the horizontal direction provides $k_{13} = 0$. The equilibrium of the forces for the masse m_2 in the horizontal direction gives $k_{23} + k_3 = 0$.

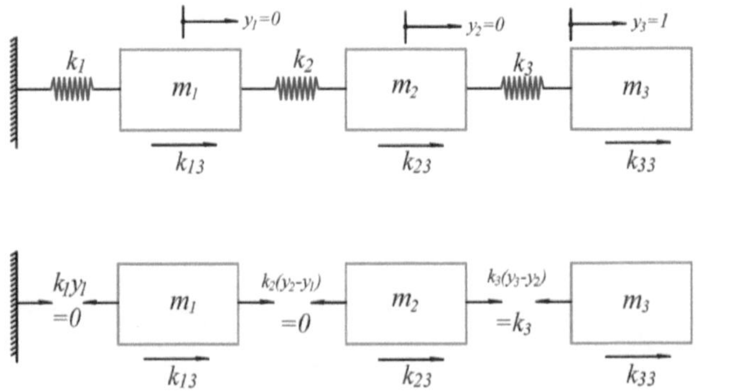

FIGURE 7.6 The third configuration of a three-degree-of-freedom system.

The equilibrium of the forces for the masse m_3 in the horizontal direction yields $k_{33} = k_3$. Consequently, we can derive the following relations

$$k_{13} = 0, \quad k_{23} = -k_3, \quad k_{33} = k_3 \tag{7.28}$$

Therefore, the stiffness matrix of the system is provided as follows:

$$\mathbf{K} = \begin{bmatrix} (k_1 + k_2) & -k_2 & 0 \\ -k_2 & (k_2 + k_3) & -k_3 \\ 0 & -k_3 & k_3 \end{bmatrix} \tag{7.29}$$

Example 7.2

▶ Please determine the mass influence coefficients of the system shown in Figure 7.7.

SOLUTION

Let $\ddot{y}_1$, $\ddot{y}_2$ and $\ddot{y}_3$ represent the accelerations of the masses m_1, m_2 and m_3, respectively. The mass influence coefficients m_{ij} of the system can be calculated based on the spring stiffnesses k_1, k_2 and k_3 using the following method. First, we assign an acceleration value of one to m_1 ($\ddot{y}_1 = 1$), whereas m_2 and m_3 have accelerations of zero, respectively ($\ddot{y}_2 = \ddot{y}_3 = 0$), as depicted in Figure 7.8(a). The set of forces m_{i1} $(i = 1, 2, 3)$ is assumed to maintain the system in this current configuration. The free-body diagrams of the masses corresponding to the configuration shown in Figure 7.8(a) are depicted in Figure 7.8(b).

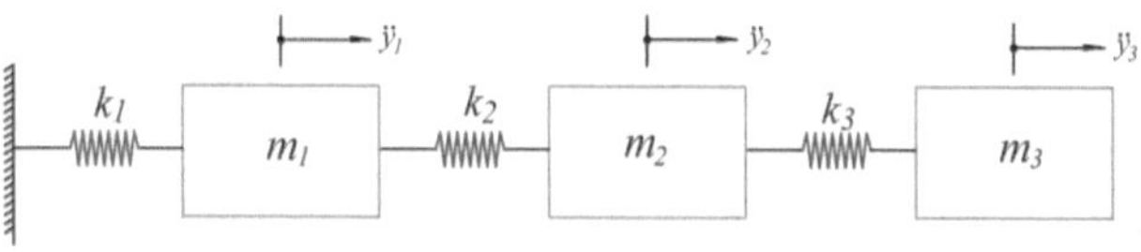

FIGURE 7.7 A three-degree-of-freedom system.

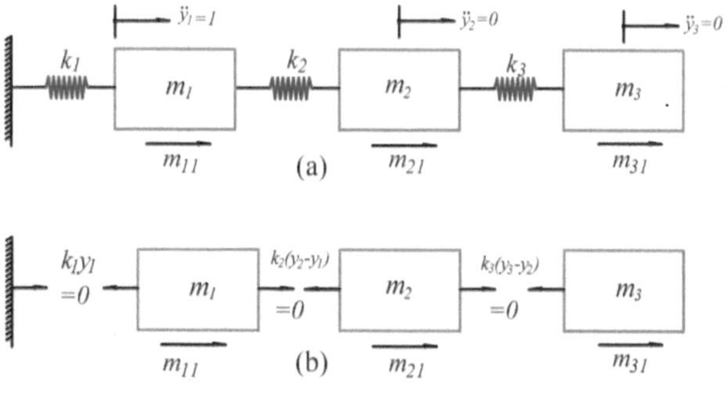

FIGURE 7.8 The fourth configuration of a three-degree-of-freedom system.

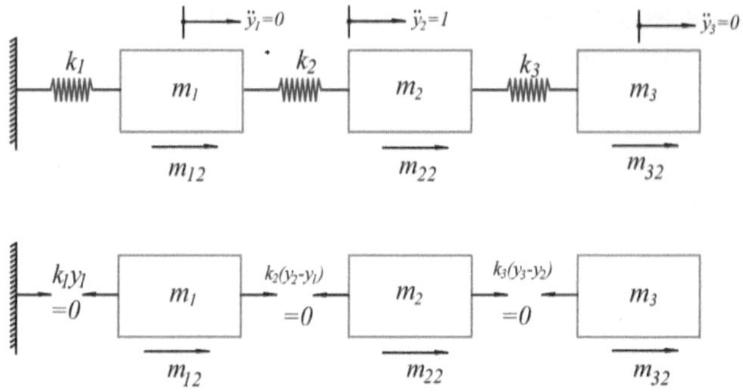

FIGURE 7.9 The fifth configuration of a three-degree-of-freedom system.

The equilibrium of the forces for the mass m_1 in the horizontal direction gives $m_{11} = m_1\ddot{y}_1 = m_1$. The equilibrium of the forces for the mass m_2 in the horizontal direction gives $m_{21} = m_2\ddot{y}_2 = 0$. The equilibrium of the forces for the mass m_3 in the horizontal direction gives $m_{31} = m_3\ddot{y}_3 = 0$. Consequently, we can obtain the following relations

$$m_{11} = m_1, \quad m_{21} = 0, \quad m_{31} = 0 \tag{7.30}$$

Next, the accelerations of the masses are assumed as $\ddot{y}_1 = \ddot{y}_3 = 0$ and $\ddot{y}_2 = 1$, as depicted in Figure 7.9(a). Because the forces $m_{i2}\left(i = 1, 2, 3\right)$ are assumed to maintain the vibration system in this configuration, we may create free-body diagrams for the masses as shown in Figure 7.9(b). The equilibrium of the forces for the mass m_1 in the horizontal direction provides $m_{12} = m_1\ddot{y}_1 = 0$. The equilibrium of the forces for the masse m_2 in the horizontal direction gives $m_{22} = m_2\ddot{y}_2 = m_2$. The equilibrium of the forces for the masse m_3 in the horizontal direction yields $m_{32} = m_3\ddot{y}_3 = 0$. Consequently, we can obtain the following relations

$$m_{12} = 0, \quad m_{22} = m_2, \quad m_{32} = 0 \tag{7.31}$$

Finally, the accelerations of the masses are assumed to be $\ddot{y}_1 = \ddot{y}_2 = 0$ and $\ddot{y}_3 = 1$, as shown in Figure 7.10(a). Since the forces $m_{i3}\left(i = 1, 2, 3\right)$ are assumed to maintain the vibration system in this configuration, we can draw free-body diagrams for the masses as shown in Figure 7.10(b). The equilibrium of the forces for the mass m_1 in the horizontal direction provides $m_{13} = m_1\ddot{y}_1 = 0$. The equilibrium of the forces for the masse m_2 in the horizontal direction gives $m_{23} = m_2\ddot{y}_2 = 0$. The equilibrium of the forces for the masse m_3 in the horizontal direction yields $m_{33} = m_3\ddot{y}_3 = m_3$. Consequently, we can derive the following relations

$$m_{13} = 0, \quad m_{23} = 0, \quad m_{33} = m_3 \tag{7.32}$$

Consequently, the mass matrix of the system is provided as follows:

$$\mathbf{M} = \begin{bmatrix} m_1 & 0 & 0 \\ 0 & m_2 & 0 \\ 0 & 0 & m_3 \end{bmatrix} \tag{7.33}$$

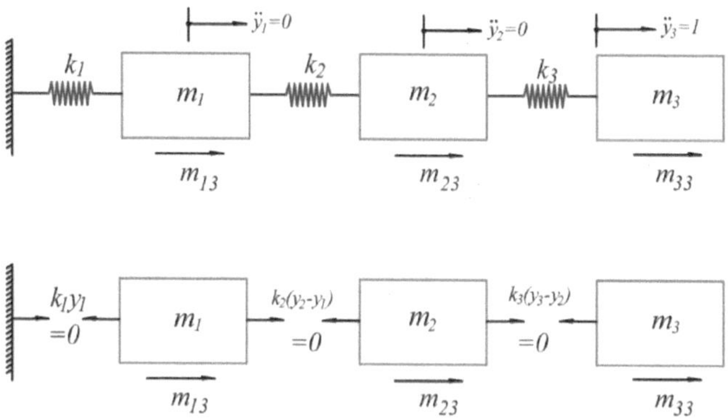

FIGURE 7.10 The sixth configuration of a three-degree-of-freedom system.

7.1.4 Free Response of an Undamped System

By eliminating the damping matrix and the applied-loads vector from equation (7.25), it is possible to construct the equations of motion for a system that is freely vibrating:

$$\mathbf{M}\ddot{\mathbf{y}} + \mathbf{K}\mathbf{y} = 0 \tag{7.34}$$

in which **M** and **K** are called the mass and stiffness matrices, respectively. **y** and **0** are called the displacement and force vectors, respectively, and are given by

$$\mathbf{y} = \begin{Bmatrix} y_1(t) \\ y_2(t) \\ \cdot \\ y_i(t) \\ \cdot \\ y_N(t) \end{Bmatrix} \quad \mathbf{0} = \begin{Bmatrix} 0 \\ 0 \\ \cdot \\ 0 \\ \cdot \\ 0 \end{Bmatrix} \tag{7.35}$$

Figuring out the conditions in which the equilibrium condition that is described by equation (7.34) will be satisfied is the task that is involved in vibration analysis. It shall be assumed, by analogy with the behavior of the single-degree-of-freedom (SDOF) system, that the motion of the free vibration of the multiple-degree-of-freedom system is simple harmonic, which can be expressed as

$$\mathbf{y}(t) = \mathbf{Y}\cos(\omega t + \phi) \tag{7.36}$$

In equation (7.36), the variable **Y** denotes the shape of the system (which does not change with time; only the amplitude varies) and ϕ denotes a phase angle.

The accelerations in free vibration can be obtained by taking the second time derivative of equation (7.36):

$$\ddot{\mathbf{y}} = -\omega^2 \mathbf{Y} \cos(\omega t + \phi) = -\omega^2 \mathbf{y} \tag{7.37}$$

When equations (7.36) and (7.37) is substituted into equation (7.34), the result is

$$-\omega^2 \mathbf{MY} \cos(\omega t + \phi) + \mathbf{KY} \cos(\omega t + \phi) = \mathbf{0} \tag{7.38}$$

whose expression can be represented as (given that the cosine term is arbitrary and can be left out)

$$\left[\mathbf{K} - \omega^2 \mathbf{M}\right] \mathbf{Y} = \mathbf{0} \tag{7.39}$$

Equation (7.39) represents a specific form of an eigenvalue or characteristic value problem. The quantities ω^2 represent the eigenvalues or characteristic values, which indicate the square of the free-vibration frequencies. The corresponding displacement vectors **Y** represent the corresponding shapes of the vibrating system, known as the eigenvectors or mode shapes.

Given that equation (7.39) describes a system of n linear homogeneous equations with unknowns Y_i $(i = 1, 2,\ldots,n)$, the trivial solution is when all Y_i values are equal to zero ($Y_1 = Y_2 = \cdots = Y_n = 0$). In order to find a solution that is not trivial for equation (7.39), the determinant Δ of the coefficient matrix must be equal to zero. That is,

$$\Delta = \left|\mathbf{K} - \omega^2 \mathbf{M}\right| = 0 \tag{7.40}$$

Equation (7.40) is referred to as the characteristic equation. and ω is referred to as the natural frequency of the system.

By expanding equation (7.40), one can obtain a polynomial equation of the Nth order in ω^2. This polynomial or characteristic equation has solutions, which are the roots, which give n values of ω^2. The fact that all of the n roots are real and positive can be demonstrated when the matrices **K** and **M** are symmetric and positive definite. If we consider the n roots in increasing order of magnitude, denoted by $\omega_1^2, \omega_2^2,\ldots, \omega_N^2$, then the n natural frequencies of the system $(\omega_1 \le \omega_2 \le \cdots \le \omega_N)$ are given by the positive square roots of these roots. The value that is the lowest, denoted by ω_1, is referred to as the fundamental or first natural frequency. Although it is possible for two natural frequencies to have the same value in some circumstances, in general, all of the natural frequencies ω_i are distinct from one another.

Equation (7.40) is also referred to as the frequency equation of the system. In the case of a system with n degrees of freedom, expanding the determinant will result in the generation of an algebraic equation that represents the n th degree in the frequency parameter ω^2. The frequencies of the n modes of vibration that are possible in the system are represented by the n roots of this frequency equation $\left(\omega_1^2, \omega_2^2, \omega_3^2, \ldots, \omega_N^2\right)$. In this context, the mode with the lowest frequency ω_1 is referred to as the first mode (this mode shape is denoted as $\mathbf{Y}^{(1)}$). The mode with the next higher frequency ω_2 is referred to as the second mode (this mode shape is denoted as $\mathbf{Y}^{(2)}$). The mode with the ith higher frequency ω_i is referred to as the ith mode (this mode shape is denoted as $\mathbf{Y}^{(i)}$), and so on. We shall refer to the vector that is composed of the entire set of modal frequencies, which are grouped in a sequential manner, as the frequency vector $\boldsymbol{\omega}$:

$$\boldsymbol{\omega} = \begin{Bmatrix} \omega_1 \\ \omega_2 \\ \omega_3 \\ \vdots \\ \omega_N \end{Bmatrix} \tag{7.41}$$

The obtained mode shapes $\mathbf{Y}^{(1)}$, $\mathbf{Y}^{(2)}$ and $\mathbf{Y}^{(i)}$ through the solution of equation (7.39) are respectively expressed as follows:

$$\mathbf{Y}^{(1)} = \begin{Bmatrix} Y_1^{(1)} \\ Y_2^{(1)} \\ \cdot \\ Y_i^{(1)} \\ \cdot \\ Y_N^{(1)} \end{Bmatrix} \mathbf{Y}^{(2)} = \begin{Bmatrix} Y_1^{(2)} \\ Y_2^{(2)} \\ \cdot \\ Y_i^{(2)} \\ \cdot \\ Y_N^{(2)} \end{Bmatrix} \mathbf{Y}^{(i)} = \begin{Bmatrix} Y_1^{(i)} \\ Y_2^{(i)} \\ \cdot \\ Y_i^{(i)} \\ \cdot \\ Y_N^{(i)} \end{Bmatrix} \tag{7.42}$$

For the sake of convenience, each vector in equation (7.42) is typically written in a standard form by dividing all the components in it by one reference component, which is typically the first of the components. The vectors obtained in this way are written as

$$\boldsymbol{\psi}_1 = \begin{Bmatrix} \psi_{11} \\ \psi_{21} \\ \cdot \\ \psi_{i1} \\ \cdot \\ \psi_{N1} \end{Bmatrix} = \begin{Bmatrix} 1 \\ Y_2^{(1)}/Y_1^{(1)} \\ \cdot \\ Y_i^{(1)}/Y_1^{(1)} \\ \cdot \\ Y_N^{(1)}/Y_1^{(1)} \end{Bmatrix} \boldsymbol{\psi}_2 = \begin{Bmatrix} \psi_{12} \\ \psi_{22} \\ \cdot \\ \psi_{i2} \\ \cdot \\ \psi_{N2} \end{Bmatrix} = \begin{Bmatrix} 1 \\ Y_2^{(2)}/Y_1^{(2)} \\ \cdot \\ Y_i^{(2)}/Y_1^{(2)} \\ \cdot \\ Y_N^{(2)}/Y_1^{(2)} \end{Bmatrix} \boldsymbol{\psi}_i = \begin{Bmatrix} \psi_{1i} \\ \psi_{2i} \\ \cdot \\ \psi_{ii} \\ \cdot \\ \psi_{Ni} \end{Bmatrix} = \begin{Bmatrix} 1 \\ Y_2^{(i)}/Y_1^{(i)} \\ \cdot \\ Y_i^{(i)}/Y_1^{(i)} \\ \cdot \\ Y_N^{(i)}/Y_1^{(i)} \end{Bmatrix} \tag{7.43}$$

In the foregoing discussions, we developed a technique that involves determining the n natural frequencies, denoted as ω_i, and the normal modes or modal vectors that correspond to them, denoted as $\mathbf{Y}^{(i)}$. Now, we will examine a significant characteristic of the normal modes, which is orthogonality. The natural frequency ω_i and the corresponding modal vector $\mathbf{Y}^{(i)}$ satisfy equation (7.39) so that

$$\left[\mathbf{K}-\omega_i{}^2\mathbf{M}\right]\mathbf{Y}^{(i)}=\mathbf{0} \tag{7.44}$$

$$\omega_i{}^2\mathbf{M}\mathbf{Y}^{(i)}=\mathbf{K}\mathbf{Y}^{(i)} \tag{7.45}$$

When we take into account another natural frequency ω_j and its corresponding modal vector $\mathbf{Y}^{(j)}$, they also satisfy equation (7.39) so that

$$\omega_j{}^2\mathbf{M}\mathbf{Y}^{(j)}=\mathbf{K}\mathbf{Y}^{(j)} \tag{7.46}$$

Through the process of pre-multiplication of equations (7.45) and (7.46) by $\mathbf{Y}^{(j)T}$ and $\mathbf{Y}^{(i)T}$, respectively, we are able to derive, taking into consideration the symmetry of the matrices $\mathbf{M}$ and $\mathbf{K}$

$$\omega_i{}^2\mathbf{Y}^{(j)T}\mathbf{M}\mathbf{Y}^{(i)}=\mathbf{Y}^{(j)T}\mathbf{K}\mathbf{Y}^{(i)}\equiv\mathbf{Y}^{(i)T}\mathbf{K}\mathbf{Y}^{(j)} \tag{7.47}$$

$$\omega_j{}^2\mathbf{Y}^{(i)T}\mathbf{M}\mathbf{Y}^{(j)}\equiv\omega_j{}^2\mathbf{Y}^{(j)T}\mathbf{M}\mathbf{Y}^{(i)}=\mathbf{Y}^{(i)T}\mathbf{K}\mathbf{Y}^{(j)} \tag{7.48}$$

By subtracting equation (7.48) from equation (7.47), we acquire

$$\left(\omega_i^2-\omega_j^2\right)\mathbf{Y}^{(j)T}\mathbf{M}\mathbf{Y}^{(i)}=\mathbf{0} \tag{7.49a}$$

Typically, $\omega_i^2\neq\omega_j^2$, hence equation (7.49a) results in

$$\mathbf{Y}^{(j)T}\mathbf{M}\mathbf{Y}^{(i)}=\mathbf{0},\qquad i\neq j \tag{7.49b}$$

Consequently, from equation (7.47), we can derive

$$\omega_i{}^2\mathbf{Y}^{(j)T}\mathbf{M}\mathbf{Y}^{(i)}=\omega_i{}^2\mathbf{0}=\mathbf{Y}^{(j)T}\mathbf{K}\mathbf{Y}^{(i)}=\mathbf{0},\qquad i\neq j \tag{7.50}$$

The modal vectors $\mathbf{Y}^{(i)}$ and $\mathbf{Y}^{(j)}$ are orthogonal with respect to the mass and stiffness matrices, as indicated by equations (7.49b) and (7.50). Typically, it is more convenient to express the conditions of orthogonality using the standard mode-shape vectors $\boldsymbol{\psi}_i$ instead of the arbitrary amplitudes $\mathbf{Y}^{(i)}$. It is evident that equations (7.49b) and (7.50) are equally valid when divided by any reference displacement value; hence, the orthogonality conditions become

$$\boldsymbol{\psi}_j{}^T\mathbf{M}\boldsymbol{\psi}_i=\mathbf{0},\qquad i\neq j \tag{7.51}$$

$$\boldsymbol{\psi}_j{}^T\mathbf{K}\boldsymbol{\psi}_i=\mathbf{0},\qquad i\neq j \tag{7.52}$$

7.1.5 Vibration-response Analysis Using Superposition

Previously, when we were talking about an arbitrary linear system with n degrees of freedom, the displaced position was defined by the n components that were contained in the vector $\mathbf{y}$. However, while conducting vibration-response analysis, it is frequently beneficial to express this position in terms of the free-vibration mode shapes. These mode shapes constitute n independent displacement patterns, the amplitudes of which can serve as generalized coordinates to express any set of displacements. Therefore, the mode shapes serve the same purpose as the trigonometric functions in a Fourier series, and they are utilized for the same reasons. This is due to the fact that (1) they possess orthogonality properties, and (2) they are efficient in the sense that they typically have the ability to describe all n displacements with sufficient accuracy using only a few shapes.

Please take into consideration, for instance, the cantilever column depicted in Figure 7.11, for which the deflected shape is expressed in terms of translational displacements at three different levels. It is possible to develop any displacement vector $\mathbf{y}$, whether it be static or dynamic, for this structure by superposing suitable amplitudes of the normal modes, as shown. For each given modal component $\mathbf{y}_{in}$, the displacements can be determined by multiplying the mode-shape vector ψ_n with the modal amplitude q_n. Therefore

$$\mathbf{y}_{in} = \psi_n q_n \tag{7.53}$$

One can then derive the total displacement vector $\mathbf{y}$ by adding up the modal vectors, which are written as

$$\mathbf{y} = \psi \mathbf{q} = \left\{ \psi_1 \ \psi_2 \cdots \psi_n \right\} \begin{Bmatrix} q_1 \\ q_2 \\ \cdots \\ q_n \end{Bmatrix} = \psi_1 q_1 + \psi_2 q_2 + \cdots + \psi_n q_n \tag{7.54}$$

In the equation shown above, it is evident that the $n \times n$ mode-shape matrix ψ is responsible for transforming the generalized coordinate vector $\mathbf{q}$ into the

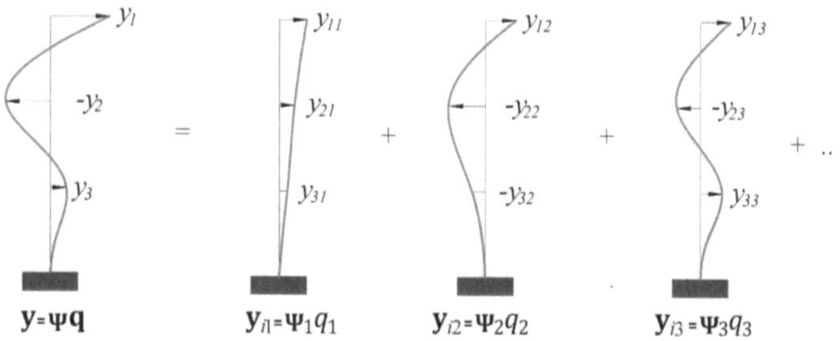

FIGURE 7.11 Representing deflections as the sum of modal components.

geometric coordinate vector $\mathbf{y}$. The generalized components in vector $\mathbf{q}$ are referred to as the *normal coordinates* of the structure.

Due to the fact that the mode-shape matrix is composed of n independent modal vectors, denoted as $\psi = \{\psi_1\ \psi_2 \cdots \psi_n\}$, it is non-singular and can be inverted. In light of this, it is always possible to solve equation (7.54) directly for the normal-coordinate amplitudes in $\mathbf{q}$ that are associated with any given displacement vector $\mathbf{y}$. In doing so, however, it is not necessary to solve a set of simultaneous equations, thanks to the orthogonality property of the mode shapes.

For the purpose of evaluating any arbitrary normal coordinate, such as q_n, one can pre-multiply equation (7.54) by $\psi_n{}^T\mathbf{M}$ to obtain

$$\psi_n{}^T\mathbf{M}\mathbf{y} = \psi_n{}^T\mathbf{M}\psi_1 q_1 + \psi_n{}^T\mathbf{M}\psi_2 q_2 + \cdots + \psi_n{}^T\mathbf{M}\psi_n q_n \tag{7.55}$$

Due to the orthogonality property with respect to mass, specifically $\psi_n{}^T\mathbf{M}\psi_m = 0$ for $m \neq n$, all terms on the right-hand side of this equation disappear, except for the term that includes $\psi_n{}^T\mathbf{M}\psi_n$, resulting in

$$\psi_n{}^T\mathbf{M}\mathbf{y} = \psi_n{}^T\mathbf{M}\psi_n q_n \tag{7.56}$$

from which

$$q_n = \frac{\psi_n{}^T\mathbf{M}\mathbf{y}}{\psi_n{}^T\mathbf{M}\psi_n} \qquad n = 1, 2, \ldots, N \tag{7.57}$$

At this point, it is of interest to investigate the conditions under which this normal coordinate transformation will be able to serve the purpose of uncoupling the damped equations of motion. These equations (equation (7.25)) are

$$\mathbf{M}\ddot{\mathbf{y}} + \mathbf{C}\dot{\mathbf{y}} + \mathbf{K}\mathbf{y} = \mathbf{F}(t)$$

Given that the mode shapes do not change with time, introducing equation (7.54) and its second time derivative, $\ddot{\mathbf{y}} = \psi\ddot{\mathbf{q}}$, leads to the following:

$$\mathbf{M}\psi\ddot{\mathbf{q}} + \mathbf{C}\psi\dot{\mathbf{q}} + \mathbf{K}\psi\mathbf{q} = \mathbf{F}(t) \tag{7.58a}$$

When equation (7.58a) is pre-multiplied by the transpose of the nth mode-shape vector $\psi_n{}^T$, the result is that the equation becomes

$$\psi_n{}^T\mathbf{M}\psi\ddot{\mathbf{q}} + \psi_n{}^T\mathbf{C}\psi\dot{\mathbf{q}} + \psi_n{}^T\mathbf{K}\psi\mathbf{q} = \psi_n{}^T\mathbf{F}(t) \tag{7.58b}$$

However, if the three terms on the left side are expanded as indicated in equation (7.55), then all of the terms, with the exception of the nth, will disappear due to the orthogonality properties of the mode-shape; hence, the result is

$$\psi_n{}^T\mathbf{M}\psi_n\ddot{q}_n + \psi_n{}^T\mathbf{C}\psi_n\dot{q}_n + \psi_n{}^T\mathbf{K}\psi_n q_n = \psi_n{}^T\mathbf{F}(t) \tag{7.59}$$

At this time, new symbols will be defined in the following manner:

$$M_n \equiv \psi_n^T \mathbf{M} \psi_n \tag{7.60}$$

$$C_n = \psi_n^T \mathbf{C} \psi_n \tag{7.61}$$

$$K_n \equiv \psi_n^T \mathbf{K} \psi_n \tag{7.62}$$

$$F_n(t) \equiv \psi_n^T \mathbf{F}(t) \tag{7.63}$$

that are referred to as the normal-coordinate generalized mass, generalized damping, generalized stiffness and generalized load for mode n, respectively. With them, equation (7.59) can be written as

$$M_n \ddot{q}_n + C_n \dot{q}_n + K_n q_n = F_n, \qquad n = 1, 2, \cdots n \tag{7.64}$$

which is a single-degree-of-freedom (SDOF) equation of motion for mode n.

The aforementioned method can be employed to obtain an independent SDOF equation for every mode of oscillation of the structure with damping. Therefore, the utilization of normal coordinates facilitates the transformation of the equations of motion from a set of n simultaneous differential equations, which are coupled by the off-diagonal terms in the mass, damping and stiffness matrices, into a set of n independent equations in normal coordinates. To obtain the dynamic response, one can solve for the response of each normal (modal) coordinate separately and then superpose these using equation (7.54) to yield the response in the original geometric coordinates. This technique is referred to as the mode-superposition method, or more precisely, the mode displacement superposition method.

Multiplying both sides of the equation $\mathbf{K}\psi_n = \omega^2 \mathbf{M} \psi_n$ by ψ_n^T, we may establish a relationship between the generalized stiffness for mode n and the generalized mass, based on the vibration frequency:

$$\psi_n^T \mathbf{K} \psi_n = \omega^2 \psi_n^T \mathbf{M} \psi_n = K_n = \omega^2 M_n \tag{7.65}$$

If equation (7.64) is divided by the generalized mass M_n, the modal equation of motion can be expressed in an alternative form:

$$\ddot{q}_n + \frac{C_n}{M_n} \dot{q}_n + \omega_n^2 q_n = \frac{F_n}{M_n} = \ddot{q}_n + 2\zeta_n \omega_n \dot{q}_n + \omega_n^2 q_n \tag{7.66}$$

in which equation (7.65) has been used to rewrite the stiffness term and where ξ_n is the modal viscous damping ratio defined as follows:

$$\zeta_n = \frac{C_n}{2\omega_n M_n} \tag{7.67}$$

In order to proceed with the solution of the uncoupled equations of motion (equation (7.64)), one must first solve the eigenvalue problem

$$\left[\mathbf{K} - \omega^2 \mathbf{M}\right] \mathbf{Y} = \mathbf{0} \tag{7.68}$$

to obtain the required mode shapes ψ_n ($n = 1, 2, \cdots$) and corresponding natural frequencies ω_n. Typically, the modal damping ratios, denoted as ζ_n, are presumed to be based on the evidence obtained from experiments.

The total response of the MDOF system can be determined by solving the n uncoupled modal equations and superposing their effects, as shown in equation (7.54). Each of the equations (7.64) is a standard SDOF equation of motion. These equations can be solved using either time domain or frequency domain methods. The solution in the time domain is expressed by the Duhamel integral

$$q_n(t) = \frac{1}{M_n \omega_{Dn}} \int_0^t F_n(\tau) \exp\left[-\zeta_n \omega_n (t-\tau)\right] \sin \omega_{Dn}(t-\tau) \, d\tau \tag{7.69}$$

in which $\omega_{Dn} = \omega_n \sqrt{1-\zeta_n^2}$. In the frequency domain, the dynamic response is obtained from

$$q_n(t) = \frac{1}{2\pi} \int_{-\infty}^{\infty} H_n(i\omega) L_n(i\omega) \exp[i\omega t] \, d\omega \tag{7.70}$$

in which the complex load function $L_n(i\omega)$ is the Fourier transform of the modal loading $F_n(t)$, and it is given by

$$L_n(i\omega) = \int_{-\infty}^{\infty} F_n(t) \exp[-i\omega t] \, dt \tag{7.71}$$

Also, in equation (7.70), the complex frequency response function, $H_n(i\omega)$, can be expressed as shown below:

$$H_n(i\omega) = \frac{1}{M_n \omega_n^2} \frac{1}{\left(1-\beta_n^2\right) + 2i\zeta_n \beta_n}, \qquad \zeta_n \geq 0 \tag{7.72}$$

in which $\beta_n \equiv \omega/\omega_n$. For any general modal loading, the modal response $q_n(t)$ for $t \geq 0$ can be obtained by solving equations (7.69) or (7.70), assuming zero initial conditions, i.e., $q_n(0) = \dot{q}_n(0) = 0$. In the event that the initial conditions do not equal zero, the damped free-vibration response

$$q_n(t) = \left[q_n(0) \cos \omega_{Dn} t + \left(\frac{\dot{q}_n(0) + q_n(0) \xi_n \omega_n}{\omega_{Dn}} \right) \sin \omega_{Dn} t \right] \exp(-\zeta_n \omega_n t) \tag{7.73}$$

needs to be added to the forced-vibration response provided by equations (7.69) or (7.70).

The initial conditions $q_n(0)$ and $\dot{q}_n(0)$ in this equation are determined from $\mathbf{y}(0)$ and $\dot{\mathbf{y}}(0)$ using equation (7.57) in the forms

$$q_n(0)=\frac{\psi_n{}^T\mathbf{M}\mathbf{y}(0)}{\psi_n{}^T\mathbf{M}\psi_n} \tag{7.74}$$

$$\dot{q}_n(0)=\frac{\psi_n{}^T\mathbf{M}\dot{\mathbf{y}}(0)}{\psi_n{}^T\mathbf{M}\psi_n} \tag{7.75}$$

Once the total response for each mode $q_n(t)$ has been generated by employing either equation (7.69) or equation (7.70), as well as equation (7.73), the displacements that are expressed in the geometric coordinates can be derived by employing equation (7.54), that is,

$$\mathbf{y}(t)=\psi_1 q_1(t)+\psi_2 q_2(t)+\cdots+\psi_n q_n(t) \tag{7.76}$$

which superposes the separate modal displacement contributions, which is why this technique is frequently referred to as the mode superposition method. In general, the displacement contributions are greatest for the lower modes, and they tend to decrease for the higher modes. This is something that should be taken into consideration for the majority of the different types of loadings. As a consequence of this, it is typically not required to incorporate all of the higher modes of vibration into the superposition process. The series can be truncated once the response has been acquired to whatever degree of accuracy is desired. Furthermore, it is important to keep in mind that the mathematical idealization of any complicated structural system has a tendency to be less trustworthy in predicting the higher modes of vibration. For this reason, it is also believed to be beneficial to limit the number of modes that are taken into consideration in a dynamic-response analysis.

Example 7.3

▸ Please calculate the steady-state response of the system depicted in Figure 7.12 when the masses are subjected to simple harmonic forces $F_1 = F_2 = F_3 = F_0\cos\omega t$, where $\omega = 0.4\sqrt{k/m}$. Let's assume that the masses are equal, $m_1 = m_2 = m_3 = m$, the damp coefficients are equal, $c_1 = c_2 = c_3 = c$, the stiffnesses are equal, $k_1 = k_2 = k_3 = k$, and the damping ratio in each normal mode is given by $\zeta_i = 0.02$, where $i = 1, 2, 3$.

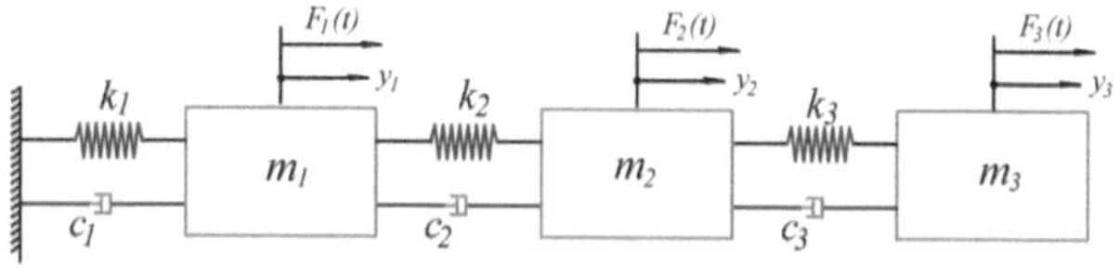

FIGURE 7.12 A three-degree-of-freedom spring-mass-damper system.

SOLUTION

Let y_1, y_2, and y_3 represent the displacements that masses m_1, m_2 and m_3 have moved, respectively. We set $n = 3$ along with $k_{n+1} = 0$ and $c_{n+1} = 0$ for $i = 1, 2, \ldots, n, n+1$ in Figure 7.1. Then, the equations of motion of the system shown in Figure 7.12, for forced vibration, can be obtained from equation (7.5) as

$$\mathbf{M}\ddot{\mathbf{y}} + \mathbf{C}\dot{\mathbf{y}} + \mathbf{K}\mathbf{y} = \mathbf{F}(t) \tag{7.77}$$

in which **M**, **C**, and **K** are referred to as the mass, damping and stiffness matrices, respectively, and are given by

$$\mathbf{M} = \begin{bmatrix} m_1 & 0 & 0 \\ 0 & m_2 & 0 \\ 0 & 0 & m_3 \end{bmatrix} = \begin{bmatrix} m & 0 & 0 \\ 0 & m & 0 \\ 0 & 0 & m \end{bmatrix} \tag{7.78}$$

$$\mathbf{C} = \begin{bmatrix} c_1 + c_2 & -c_2 & 0 \\ -c_2 & c_3 + c_2 & -c_3 \\ 0 & -c_3 & c_3 \end{bmatrix} = \begin{bmatrix} 2c & -c & 0 \\ -c & 2c & -c \\ 0 & -c & c \end{bmatrix} \tag{7.79}$$

$$\mathbf{K} = \begin{bmatrix} k_1 + k_2 & -k_2 & 0 \\ -k_2 & k_3 + k_2 & -k_3 \\ 0 & -k_3 & k_3 \end{bmatrix} = \begin{bmatrix} 2k & -k & 0 \\ -k & 2k & -k \\ 0 & -k & k \end{bmatrix} \tag{7.80}$$

In equation (7.77), $\mathbf{y}$, $\dot{\mathbf{y}}$, $\ddot{\mathbf{y}}$ and $\mathbf{F}(t)$ are referred to as the displacement, velocity, acceleration and force vectors, respectively and are given by

$$\mathbf{y} = \begin{Bmatrix} y_1(t) \\ y_2(t) \\ y_3(t) \end{Bmatrix}, \qquad \dot{\mathbf{y}} = \begin{Bmatrix} \dot{y}_1(t) \\ \dot{y}_2(t) \\ \dot{y}_3(t) \end{Bmatrix}, \qquad \ddot{\mathbf{y}} = \begin{Bmatrix} \ddot{y}_1(t) \\ \ddot{y}_2(t) \\ \ddot{y}_3(t) \end{Bmatrix} \tag{7.81}$$

$$\mathbf{F}(t) = \begin{Bmatrix} F_0 \cos \omega t \\ F_0 \cos \omega t \\ F_0 \cos \omega t \end{Bmatrix} \tag{7.82}$$

By eliminating the damping matrix and the applied-loads vector from equation (7.77), it is possible to construct the equations of motion for a system that is freely vibrating:

$$\mathbf{M}\ddot{\mathbf{y}} + \mathbf{K}\mathbf{y} = \mathbf{0} \tag{7.83}$$

The characteristic equation corresponding to equation (7.83) is

$$\Delta = \left| \mathbf{K} - \omega^2 \mathbf{M} \right| = 0 \tag{7.84}$$

in which ω is the natural frequency of the system.

Expanding the above determinant will result in the generation of an algebraic equation that represents the third degree in the frequency parameter ω^2. The frequencies of the three modes $\left(\mathbf{Y}^{(1)}, \mathbf{Y}^{(2)} \text{ and } \mathbf{Y}^{(3)}\right)$ of vibration that is possible in the system are represented by the three roots of this frequency equation $\left(\omega_1^2 = 0.1981k/m,\ \omega_2^2 = 1.5550k/m,\ \omega_3^2 = 3.2470k/m\right)$. The obtained mode shapes $\mathbf{Y}^{(1)}$, $\mathbf{Y}^{(2)}$ and $\mathbf{Y}^{(3)}$ are, respectively, expressed as follows:

$$\mathbf{Y}^{(1)} = Y_1^{(1)}\begin{Bmatrix} 1 \\ 1.8018 \\ 2.2470 \end{Bmatrix},\quad \mathbf{Y}^{(2)} = Y_1^{(2)}\begin{Bmatrix} 1 \\ 0.4450 \\ -0.8019 \end{Bmatrix},\quad \mathbf{Y}^{(3)} = Y_1^{(3)}\begin{Bmatrix} 1 \\ -1.2470 \\ 0.5550 \end{Bmatrix} \tag{7.85}$$

in which the value of $Y_1^{(1)}$, $Y_1^{(2)}$and $Y_1^{(3)}$can be chosen arbitrarily.

We next perform orthonormalization on the foregoing eigenvectors with respect to the mass matrix.

An eigenvector $\mathbf{Y}^{(i)}$ is considered to be **M**-orthonormal if it meets the following condition:

$$\mathbf{Y}^{(i)T}\mathbf{M}\mathbf{Y}^{(i)} = 1 \tag{7.86}$$

Therefore, for $i = 1$, equation (7.86) leads to

$$\mathbf{Y}^{(1)T}\mathbf{M}\mathbf{Y}^{(1)} = \begin{Bmatrix} Y_1^{(1)} & 1.8018Y_1^{(1)} & 2.2470Y_1^{(1)} \end{Bmatrix}\begin{bmatrix} m & 0 & 0 \\ 0 & m & 0 \\ 0 & 0 & m \end{bmatrix}\begin{Bmatrix} 1Y_1^{(1)} \\ 1.8018Y_1^{(1)} \\ 2.2470Y_1^{(1)} \end{Bmatrix} = 1 \tag{7.87}$$

Solving equation (7.86) leads to

$$Y_1^{(1)} = \frac{1}{\sqrt{m(9.2955)}} = \frac{0.3280}{\sqrt{m}} \tag{7.88}$$

In a similar manner, equation (7.86) yields the following for $i = 2$ and $i = 3$:

$$\mathbf{Y}^{(2)T}\mathbf{M}\mathbf{Y}^{(2)} = \begin{Bmatrix} Y_1^{(2)} & 0.4450Y_1^{(2)} & -0.8019Y_1^{(2)} \end{Bmatrix}\begin{bmatrix} m & 0 & 0 \\ 0 & m & 0 \\ 0 & 0 & m \end{bmatrix}\begin{Bmatrix} 1Y_1^{(2)} \\ 0.4450Y_1^{(2)} \\ -0.8019Y_1^{(2)} \end{Bmatrix} = 1 \tag{7.89}$$

$$\mathbf{Y}^{(3)T}\mathbf{M}\mathbf{Y}^{(3)} = \begin{Bmatrix} Y_1^{(3)} & -1.2470Y_1^{(3)} & 0.5550Y_1^{(3)} \end{Bmatrix}\begin{bmatrix} m & 0 & 0 \\ 0 & m & 0 \\ 0 & 0 & m \end{bmatrix}\begin{Bmatrix} 1Y_1^{(3)} \\ -1.2470Y_1^{(3)} \\ 0.5550Y_1^{(3)} \end{Bmatrix} = 1 \tag{7.90}$$

or

$$Y_1^{(2)} = \frac{0.7370}{\sqrt{m}} Y_1^{(3)} = \frac{0.5910}{\sqrt{m}} \tag{7.91}$$

Therefore, the corresponding **M**-orthonormal mode shapes are given by

$$\mathbf{Y}^{(1)} = \frac{0.3280}{\sqrt{m}}\begin{Bmatrix} 1 \\ 1.8018 \\ 2.2470 \end{Bmatrix}, \quad \mathbf{Y}^{(2)} = \frac{0.7370}{\sqrt{m}}\begin{Bmatrix} 1 \\ 0.4450 \\ -0.8019 \end{Bmatrix}, \quad \mathbf{Y}^{(3)} = \frac{0.591}{\sqrt{m}}\begin{Bmatrix} 1 \\ -1.2470 \\ 0.5550 \end{Bmatrix} \tag{7.92}$$

Therefore, the modal vector can be written as

$$\mathbf{Y} = \left\{\mathbf{Y}^{(1)} \quad \mathbf{Y}^{(2)} \quad \mathbf{Y}^{(3)}\right\} = \frac{1}{\sqrt{m}}\begin{bmatrix} 0.328 & 0.737 & 0.5910 \\ 0.5910 & 0.328 & -0.737 \\ 0.737 & -0.5910 & 0.328 \end{bmatrix} \tag{7.93}$$

By expressing the solution vector **y** as a linear combination of the natural modes of the system

$$\mathbf{y} = \mathbf{Y}\mathbf{q} = \left\{\mathbf{Y}^{(1)} \quad \mathbf{Y}^{(2)} \quad \mathbf{Y}^{(3)}\right\}\begin{Bmatrix} q_1 \\ q_2 \\ q_3 \end{Bmatrix} = \mathbf{Y}^{(1)}q_1 + \mathbf{Y}^{(2)}q_2 + \cdots + \mathbf{Y}^{(3)}q_n \tag{7.94}$$

Equation (7.77) can be rewritten as

$$\mathbf{MY\ddot{q}} + \mathbf{CY\dot{q}} + \mathbf{KYq} = \mathbf{F}(t) \tag{7.95}$$

Pre-multiplication of equation (7.95) by $\mathbf{Y}^T$ leads to

$$\mathbf{Y}^T\mathbf{MY\ddot{q}} + \mathbf{Y}^T\mathbf{CY\dot{q}} + \mathbf{Y}^T\mathbf{KYq} = \mathbf{Y}^T\mathbf{F}(t) \tag{7.96}$$

Because the eigenvector $\mathbf{Y}^{(i)}$ is **M**- orthonormal, equation (7.96) is reduced to

$$\begin{bmatrix} 1 & 0 & 0 \\ 0 & 1 & 0 \\ 0 & 0 & 1 \end{bmatrix}\mathbf{\ddot{q}} + \begin{bmatrix} C_1 & 0 & 0 \\ 0 & C_2 & 0 \\ 0 & 0 & C_3 \end{bmatrix}\mathbf{\dot{q}} + \begin{bmatrix} \omega_1^2 & 0 & 0 \\ 0 & \omega_2^2 & 0 \\ 0 & 0 & \omega_3^2 \end{bmatrix}\mathbf{q} = \mathbf{Y}^T\mathbf{F}(t) = \mathbf{G}(t) \tag{7.97}$$

$$\mathbf{G}(t)=\begin{Bmatrix} G_1(t)\\ G_2(t)\\ G_3(t)\end{Bmatrix}=\mathbf{Y}^T\mathbf{F}(t)=\frac{1}{\sqrt{m}}\begin{bmatrix} 0.328 & 0.5910 & 0.737\\ 0.737 & 0.328 & -0.5910\\ 0.5910 & -0.737 & 0.328\end{bmatrix}\begin{Bmatrix} F_0\cos\omega t\\ F_0\cos\omega t\\ F_0\cos\omega t\end{Bmatrix}$$

$$=\frac{1}{\sqrt{m}}\begin{Bmatrix} 1.656F_0\cos\omega t\\ 0.474F_0\cos\omega t\\ 0.182F_0\cos\omega t\end{Bmatrix}=\begin{Bmatrix} G_{10}\\ G_{20}\\ G_{30}\end{Bmatrix}\cos\omega t=\begin{Bmatrix} \dfrac{1.656F_0}{\sqrt{m}}\\ \dfrac{0.474F_0}{\sqrt{m}}\\ \dfrac{0.182F_0}{\sqrt{m}}\end{Bmatrix}\cos\omega t \tag{7.98}$$

In the event when the normal coordinates or the modal participation factors for the three principal modes are represented by the symbols $q_1(t), q_2(t)$ and $q_3(t)$, then the equations of motion can be expressed as follows:

$$\ddot{q}_i(t)+2\zeta_i\omega_i\dot{q}_i(t)+\omega_i^2 q_i(t)=G_i(t), \qquad i=1,\,2,\,3 \tag{7.99}$$

in which ζ_i is the modal viscous damping ratio defined as follows:

$$\zeta_i=\frac{C_i}{2\omega_i M_i} \tag{7.100}$$

The steady-state solution of equation (7.99) can be written as follows:

$$q_i(t)=q_{i0}\cos(\omega t-\phi_i), \quad i=1,\,2,\,3 \tag{7.101}$$

in which q_{i0} and ϕ are constants to be determined. q_{i0} and ϕ denote the amplitude and phase angle of the response, respectively. By substituting equation (7.101) into equation (7.99), we obtain

$$q_{i0}\left[\left(\omega_i^2-\omega^2\right)\cos(\omega t-\phi_i)-2\zeta_i\omega_i\omega\sin(\omega t-\phi_i)\right]=G_{i0}\cos\omega t \tag{7.102}$$

Utilizing the trigonometric relations

$$\cos(\omega t-\phi_i)=\cos\omega t\cos\phi_i+\sin\omega t\sin\phi_i$$
$$\sin(\omega t-\phi_i)=\sin\omega t\cos\phi_i-\cos\omega t\sin\phi_i$$

in equation (7.102), and by equating the coefficients of cos ωt and sin ωt on both sides of the equation that is produced, we are able to obtain the following:

$$\begin{cases} q_{i0}\left[\left(\omega_i^2-\omega^2\right)\cos\phi_i+2\zeta_i\omega_i\omega\sin\phi_i\right]=G_{i0}\\ q_{i0}\left[\left(\omega_i^2-\omega^2\right)\sin\phi_i-2\zeta_i\omega_i\omega\cos\phi_i\right]=0\end{cases} \tag{7.103}$$

The solution of equation (7.103) yields

$$q_{i0} = \frac{G_{i0}}{\sqrt{\left[\left(\omega_i^2 - \omega^2\right)^2 + \left(2\zeta_i\omega_i\omega\right)^2\right]}} \tag{7.104}$$

$$\phi_i = \tan^{-1}\left(\frac{2\zeta_i\omega_i\omega}{\omega_i^2 - \omega^2}\right) \tag{7.105}$$

Through the process of substituting the G_{i0} values that are provided in equations (7.98) and the ω_i^2 values $\left(\omega_1^2 = 0.1981k/m,\ \omega_2^2 = 1.5550k/m,\ \omega_3^2 = 3.2470k/m\right)$ into equations (7.104) and (7.105), we are able to obtain

$$\begin{aligned} q_{10} &= 44.278\frac{F_0\sqrt{m}}{k}, & \phi_1 &= \tan^{-1}(0.187) \\ q_{20} &= 0.3398\frac{F_0\sqrt{m}}{k}, & \phi_2 &= \tan^{-1}(0.0143) \\ q_{30} &= 0.0191\frac{F_0\sqrt{m}}{k}, & \phi_3 &= \tan^{-1}(0.00934) \end{aligned} \tag{7.106}$$

Finally, the steady-state response can be determined by applying equation (7.94).

7.2 RANDOM VIBRATION OF MULTI-DEGREE-OF-FREEDOM LINEAR SYSTEMS

7.2.1 Stationary Random Vibration: Damped Response Autocorrelation Function

Let us consider a viscously damped MDOF spring-mass-damper system depicted in Figure 7.13. The motion of this MDOF system is completely described by the coordinates $y_1(t), y_2(t),\ldots, y_i(t),\ldots, y_j(t),\ldots, y_n(t)$, which determine the positions of the masses $m_1, m_2,\ldots, m_i,\ldots, m_j,\ldots, m_n$ at any time t from their respective equilibrium positions. The masses $m_1, m_2,\ldots, m_i,\ldots, m_j,\ldots, m_n$ are subject to the external forces $F_1(t), F_2(t),\ldots, F_i(t),\ldots, F_j(t),\ldots, F_n(t)$, respectively.

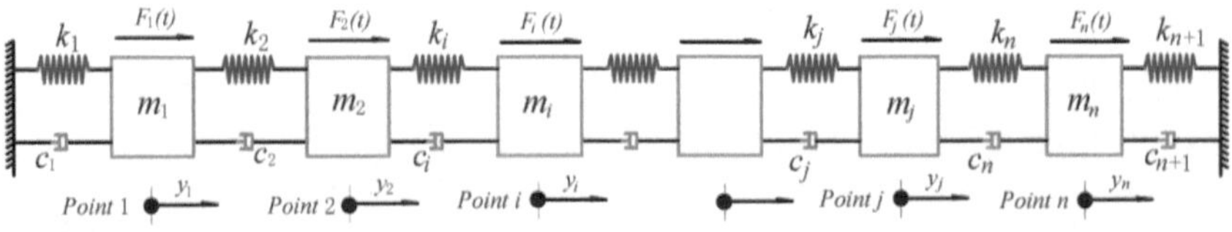

FIGURE 7.13 A spring-mass-damper system with many degrees of freedom.

7.2.1.1 Damped Response Autocorrelation Function

The autocorrelation function for the stationary response displacement $y_1(t)$ of a damped MDOF system is, according to the definition of and derived from equation (7.43) and equation (7.54),

$$\begin{aligned}
R_{y_1}(\tau) &= E\left[y_1(t)y_1(t+\tau)\right] \\
&= E\left\{\left[\psi_{11}q_1(t)+\psi_{12}q_2(t)+\cdots+\psi_{1n}q_n(t)\right]\left[\psi_{11}q_1(t+\tau)+\psi_{12}q_2(t+\tau)\right.\right. \\
&\quad \left.\left.+\cdots+\psi_{1n}q_n(t+\tau)\right]\right\} \\
&= E\left[\psi_{11}^2q_1(t)q_1(t+\tau)+\psi_{11}\psi_{12}q_1(t)q_2(t+\tau)+\cdots+\psi_{11}\psi_{1n}q_1(t)q_n(t+\tau)\right. \\
&\quad +\psi_{12}\psi_{11}q_2(t)q_1(t+\tau)+\psi_{12}^2q_2(t)q_2(t+\tau)+\cdots+\psi_{12}\psi_{1n}q_2(t)q_n(t+\tau)+\cdots \\
&\quad \left.+\psi_{1n}\psi_{11}q_n(t)q_1(t+\tau)+\psi_{1n}\psi_{12}q_n(t)q_2(t+\tau)+\cdots+\psi_{1n}^2q_n(t)q_n(t+\tau)\right] \\
&= \sum_{j=1}^{n}\sum_{k=1}^{n}\psi_{1j}\psi_{1k}E\left[q_j(t)q_k(t+\tau)\right] \qquad (7.107)
\end{aligned}$$

in which ψ_{1j} and ψ_{1k} represent the first elements of modal vectors, respectively.

By employing equation (7.69) to calculate the normal coordinates $q_j(t)$, the cross-correlation can be determined as follows:

$$E\left[q_j(t)q_k(t+\tau)\right]=E\left[\int_{-\infty}^{\infty}G_j(t-\theta_1)g_j(\theta_1)d\theta_1\int_{-\infty}^{\infty}G_k(t+\tau-\theta_2)g_k(\theta_2)d\theta_2\right] \qquad (7.108)$$

in which the generalized force is

$$G_j(t)=\frac{F_j}{M_j}=\frac{\boldsymbol{\psi}_j^T\mathbf{F}(t)}{M_j}=\frac{\boldsymbol{\psi}_j^T\mathbf{F}(t)}{\boldsymbol{\psi}_j^T\mathbf{M}\boldsymbol{\psi}_j} \qquad (7.109)$$

By substituting equation (7.109) into equation (7.108), we derive the cross-correlation relation as follows:

$$\begin{aligned}
&E\left[G_j(t-\theta_1)G_k(t+\tau-\theta_2)\right] \\
&= \frac{1}{M_jM_k}E\left[\boldsymbol{\psi}_j^T\mathrm{F}(t-\theta_1)\boldsymbol{\psi}_k^T\mathbf{F}(t+\tau-\theta_2)\right] \\
&= \frac{1}{M_jM_k}E\left\{\left[\psi_{1j}F_1(t-\theta_1)+\psi_{2j}F_2(t-\theta_1)+\cdots+\psi_{nj}F_n(t-\theta_1)\right]\right. \\
&\quad \left.\times\left[\psi_{1k}F_1(t+\tau-\theta_2)+\psi_{2k}F_2(t+\tau-\theta_2)+\cdots+\psi_{nk}F_n(t+\tau-\theta_2)\right]\right\}
\end{aligned}$$

$$
\begin{aligned}
&= \frac{1}{M_j M_k}\{\psi_{1j}\psi_{1k}R_1(\tau-\theta_2+\theta_1)+\psi_{1j}\psi_{2k}R_{12}(\tau-\theta_2+\theta_1) \\
&\quad +\ldots+\psi_{1j}\psi_{nk}R_{1n}(\tau-\theta_2+\theta_1) \\
&\quad +\psi_{2j}\psi_{1k}R_{21}(\tau-\theta_2+\theta_1)+\psi_{2j}\psi_{2k}R_2(\tau-\theta_2+\theta_1) \\
&\quad +\cdots+\psi_{2j}\psi_{nk}R_{2n}(\tau-\theta_2+\theta_1) \\
&\quad +\cdots\cdots\cdots \\
&\quad +\psi_{nj}\psi_{1k}R_{n1}(\tau-\theta_2+\theta_1)+\psi_{nj}\psi_{2k}R_{n2}(\tau-\theta_2+\theta_1) \\
&\quad +\cdots+\psi_{nj}\psi_{nk}R_n(\tau-\theta_2+\theta_1)\} \\
&= \frac{1}{M_j M_k}\sum_{h=1}^{n}\sum_{m=1}^{n}\psi_{hj}\psi_{mk}R_{hm}(\tau-\theta_2+\theta_1)
\end{aligned}
\tag{7.110}
$$

When there is only one excitation, for example, $F_1(t)$, it is important to note that

$$
E\left[G_j(t-\theta_1)G_k(t+\tau-\theta_2)\right]=\frac{1}{M_j M_k}\psi_{1j}\psi_{1k}R_1(\tau-\theta_2+\theta_1) \tag{7.111}
$$

By first substituting equation (7.110) into equation (7.108) and then substituting the obtained results into equation (7.107), we are able to construct the input-output autocorrelation relation in the following manner:

$$
\begin{aligned}
R_{y1}(\tau) &= \sum_{j=1}^{n}\sum_{k=1}^{n}\psi_{1j}\psi_{1k}\int_{-\infty}^{\infty}\int_{-\infty}^{\infty}\frac{1}{M_j M_k} \\
&\quad \sum_{h=1}^{n}\sum_{m=1}^{n}\psi_{hj}\psi_{mk}R_{hm}(\tau-\theta_2+\theta_1)g_j(\theta_1)g_k(\theta_2)d\theta_1 d\theta_2
\end{aligned}
\tag{7.112}
$$

in which $R_{hm}(\tau)$ are the cross-correlation functions of the excitations $\mathbf{F}(t)$ and are defined as follows:

$$
R_{hm}(\tau)=E\left[F_h(t)F_m(t+\tau)\right] \tag{7.113}
$$

For autocorrelation functions $R_{yi}(\tau)$ corresponding to the response displacement $y_i(t)$, the input-output relationship is identical to equation (7.112), except that the coordinate 1 for the terms $R_{y1}(\tau), \psi_{1j}$ and ψ_{1k} should be replaced with the appropriate coordinate $R_{yi}(\tau)$, ψ_{ij} and ψ_{ik}, respectively.

7.2.2 Stationary Random Vibration: Damped Response Spectral Density

The input-output spectral density relations can be derived from the definition

$$S_{y1}(\omega) = \frac{1}{2\pi}\int_{-\infty}^{\infty} R_{y1}(\tau)e^{-i\omega\tau}\,d\tau \tag{7.114}$$

Following the substitution of the result for $R_{y1}(\tau)$ from equation (7.112) into equation (7.114), we are able to obtain

$$\begin{aligned} S_{y1}(\omega) = {} & \frac{1}{2\pi}\int_{-\infty}^{\infty} e^{-i\omega\tau}\,d\tau \sum_{j=1}^{n}\sum_{k=1}^{n}\psi_{1j}\psi_{1k} \\ & \int_{-\infty}^{\infty}\int_{-\infty}^{\infty}\frac{1}{M_jM_k}\sum_{h=1}^{n}\sum_{m=1}^{n}\psi_{hj}\psi_{mk}R_{hm}(\tau-\theta_2+\theta_1)g_j(\theta_1)g_k(\theta_2)d\theta_1\,d\theta_2 \end{aligned} \tag{7.115}$$

In the following step, the order of integration in equation (7.115) is inverted, and the exponential factors $e^{i\omega\theta_1}$, $e^{-i\omega\theta_2}$ and $e^{-i\omega(\theta_1-\theta_2)}$ are inserted. The product of these factors is identical to unity.

$$\begin{aligned} S_{y1}(\omega) = {} & \sum_{j=1}^{n}\sum_{k=1}^{n}\psi_{1j}\psi_{1k}\frac{1}{M_jM_k}\sum_{h=1}^{n}\sum_{m=1}^{n}\psi_{hj}\psi_{mk} \\ & \int_{-\infty}^{\infty} g_j(\theta_1)e^{i\omega\theta_1}\,d\theta_1 \int_{-\infty}^{\infty} g_k(\theta_2)e^{-i\omega\theta_2}\,d\theta_2\,\frac{1}{2\pi} \\ & \int_{-\infty}^{\infty}\left[R_{hm}(\tau+\theta_1-\theta_2)\right]e^{-i\omega(\tau+\theta_1-\theta_2)}\,d\tau \end{aligned} \tag{7.116}$$

Following the utilization of the definitions provided by equations (6.45), (6.83) and (6.84), we are able to obtain

$$S_{y1}(\omega) = \sum_{j=1}^{n}\sum_{k=1}^{n}\psi_{1j}\psi_{1k}\frac{1}{M_jM_k}\sum_{h=1}^{n}\sum_{m=1}^{n}\psi_{hj}\psi_{mk}S_{hm}(\omega)H_j(-\omega)H_k(\omega) \tag{7.117}$$

in which $S_{hm}(\omega)$ are the cross spectral density functions of the excitations $\mathbf{F}(t)$ and are defined as follows:

$$S_{hm}(\omega) = \frac{1}{2\pi}\int_{-\infty}^{\infty} R_{hm}(\tau)e^{-i\omega\tau}\,d\tau \tag{7.118}$$

Again, in the particular scenario when there is just one random excitation $F_1(t)$, two summing operations with regard to the indexes h and m reduce to a single term, and we obtain the following:

$$S_{y_1}(\omega) = \sum_{j=1}^{n}\sum_{k=1}^{n} \frac{\psi_{1j}^2 \psi_{1k}^2}{M_j M_k} S_1(\omega) H_j(-\omega) H_k(\omega) \tag{7.119}$$

which is the same as equation (6.139) for the two-degree-of-freedom system when the upper limits of the summations in equation (6.139) are replaced by n.

Please take note of the fact that when the single excitation is $F_2(t)$ instead of $F_1(t)$,

$$S_{y_1}(\omega) = \sum_{j=1}^{n}\sum_{k=1}^{n} \frac{\psi_{1j}\psi_{1k}\psi_{2j}\psi_{2k}}{M_j M_k} S_2(\omega) H_j(-\omega) H_k(\omega) \tag{7.120}$$

7.2.3 Response of Base-Excited System to White Noise Process

Let us take into consideration a viscously damped spring-mass-damper system that is under a base excitation, as depicted in Figure 7.14.

This system is subjected to a white-noise stationary ground acceleration. The base motion is denoted as $y(t)$.

Applying Newton's second law to m_3 results in the equation of motion, which appears as follows:

$$m_3 \ddot{x}_3 = -k_3(x_3 - x_2) - c_3(\dot{x}_3 - \dot{x}_2) \tag{7.121}$$

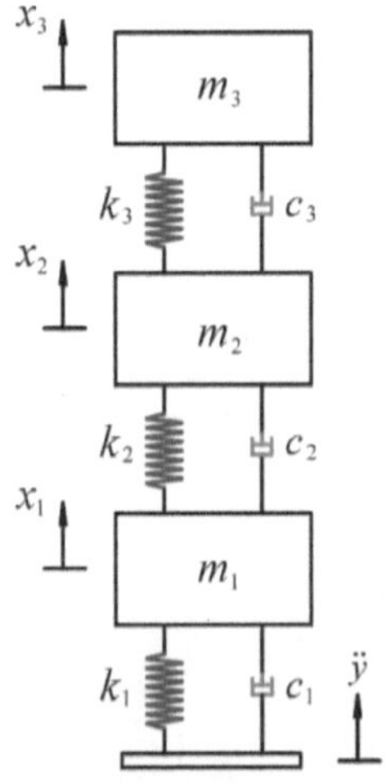

FIGURE 7.14 A base-excited damped three-degree-of-freedom system.

Similarly, applying Newton's second law to m_2 and m_1 results in the following equations of motion

$$m_2\ddot{x}_2 = k_3(x_3 - x_2) + c_3(\dot{x}_3 - \dot{x}_2) - k_2(x_2 - x_1) - c_2(\dot{x}_2 - \dot{x}_1) \quad (7.122)$$

$$m_1\ddot{x}_1 = k_2(x_2 - x_1) + c_2(\dot{x}_2 - \dot{x}_1) - k_1(x_1 - y) - c_1(\dot{x}_1 - \dot{y}) \quad (7.123)$$

Example 7.4

▸ Suppose the system depicted in Figure 7.14 is subjected to a white-noise stationary ground acceleration, $S = 0.01273 m^2 rad/s^3$. Given the base motion $y(t)$, please determine the root mean square value of the relative response of m_1 to the ground. The system parameters are given as follows:

$$m_1 = 40\,\text{kg} \qquad m_2 = 20\,\text{kg} \qquad m_3 = 10\,\text{kg}$$
$$k_1 = 9000\,\text{N/m} \qquad k_2 = 4000\,\text{N/m} \qquad k_3 = 2000\,\text{N/m}$$

The assumed damping factors for each mode are given as follows:

$$\zeta_1 = 0.022 \quad \text{(first mode)}$$
$$\zeta_2 = 0.044 \quad \text{(second mode)}$$
$$\zeta_3 = 0.088 \quad \text{(third mode)}$$

SOLUTION

Let $z_i = x_i - y$, then we have: $z_1 = x_1 - y$, $z_2 = x_2 - y$ and $z_3 = x_3 - y$. Substituting into the original differential equations of motion leads to

$$m_3(\ddot{z}_3 + \ddot{y}) = -k_3((z_3 + y) - (z_2 + y)) - c_3((\dot{z}_3 + \dot{y}) - (\dot{z}_2 + \dot{y})) \quad (7.124)$$

$$\begin{aligned} m_2(\ddot{z}_2 + \ddot{y}) &= k_3((z_3 + y) - (z_2 + y)) + c_3((\dot{z}_3 + \dot{y}) - (\dot{z}_2 + \dot{y})) \\ &\quad - k_2((z_2 + y) - (z_1 + y)) - c_2((\dot{z}_2 + \dot{y}) - (\dot{z}_1 + \dot{y})) \end{aligned} \quad (7.125)$$

$$m_1(\ddot{z}_1 + \ddot{y}) = k_2((z_2 + y) - (z_1 + y)) + c_2((\dot{z}_2 + \dot{y}) - (\dot{z}_1 + \dot{y})) - k_1 z_1 - c_1\dot{z}_1 \quad (7.126)$$

When damping is not considered, the above three differential equations of motion are simplified to

$$m_1\ddot{z}_1 + (k_1 + k_2)z_1 - k_2 z_2 = -m_1\ddot{y} \quad (7.127)$$

$$m_2\ddot{z}_2 - k_2 z_1 + (k_3 + k_2)z_2 - k_3 z_3 = -m_2\ddot{y} \quad (7.128)$$

$$m_3\ddot{z}_3 - k_3 z_2 + k_3 z_3 = -m_3\ddot{y} \quad (7.129)$$

The corresponding three free undamped differential equations of motion are

$$m_1\ddot{z}_1 + (k_1 + k_2)z_1 - k_2 z_2 = 0 \tag{7.130}$$

$$m_2\ddot{z}_2 - k_2 z_1 + (k_3 + k_2)z_2 - k_3 z_3 = 0 \tag{7.131}$$

$$m_3\ddot{z}_3 - k_3 z_2 + k_3 z_3 = 0 \tag{7.132}$$

Then, the free undamped equations of motion of the system can be written as

$$\mathbf{M}\ddot{\mathbf{z}} + \mathbf{K}\mathbf{z} = \mathbf{0} \tag{7.133}$$

in which **M** and **K** are called the mass and stiffness matrices, respectively, and are given by

$$\mathbf{M} = \begin{bmatrix} m_1 & 0 & 0 \\ 0 & m_2 & 0 \\ 0 & 0 & m_3 \end{bmatrix} = \begin{bmatrix} 40 & 0 & 0 \\ 0 & 20 & 0 \\ 0 & 0 & 10 \end{bmatrix} kg \tag{7.134}$$

$$\mathbf{K} = \begin{bmatrix} k_1 + k_2 & -k_2 & 0 \\ -k_2 & k_3 + k_2 & -k_3 \\ 0 & -k_3 & k_3 \end{bmatrix} = \begin{bmatrix} 13000 & -4000 & 0 \\ -4000 & 6000 & -2000 \\ 0 & -2000 & 2000 \end{bmatrix} N/m \tag{7.135}$$

In equation (7.133), $\mathbf{z}$ and $\ddot{\mathbf{z}}$ are called the displacement and acceleration vectors, respectively, and are given by

$$\mathbf{z} = \begin{Bmatrix} z_1(t) \\ z_2(t) \\ z_3(t) \end{Bmatrix}, \qquad \ddot{\mathbf{z}} = \begin{Bmatrix} \ddot{z}_1(t) \\ \ddot{z}_2(t) \\ \ddot{z}_3(t) \end{Bmatrix} \tag{7.136}$$

The three roots of the characteristic equation are found to be

$$\begin{aligned} \omega_1^2 &= 69.0822 \quad &\text{(first mode)} \\ \omega_2^2 &= 265.8277 \quad &\text{(second mode)} \\ \omega_3^2 &= 490.0901 \quad &\text{(third mode)} \end{aligned}$$

The corresponding natural frequencies are as follows:

$$\begin{aligned} \omega_1 &= 8.312 \text{ rad/s} \quad &\text{(first mode)} \\ \omega_2 &= 16.304 \text{ rad/s} \quad &\text{(second mode)} \\ \omega_3 &= 22.138 \text{ rad/s} \quad &\text{(third mode)} \end{aligned}$$

The first mode shape is found to be

$$\phi_1 = \begin{Bmatrix} \phi_{11} \\ \phi_{21} \\ \phi_{31} \end{Bmatrix} = \begin{Bmatrix} -0.0556 \\ -0.1422 \\ -0.2173 \end{Bmatrix}$$

Here, ϕ_{ij} represents the amplitude of displacement of the ith mass in the jth mode. In the literature, the vector ϕ_1 is often called an eigenvector; and the corresponding natural frequency ω_1 is called an eigenvalue.

ϕ_1 defines the first mode shape of the system, or the configuration of the system when it is vibrating with frequency ω_1. Only the ratio of ϕ_{21} to ϕ_{11} and ratio of ϕ_{31} to ϕ_{11} are unique. It is equally valid to write ϕ_1 as

$$\phi_1 = \begin{Bmatrix} -0.1112 \\ -0.2844 \\ -0.4346 \end{Bmatrix} \text{ or } \phi_1 = \begin{Bmatrix} 0.0556 \\ 0.1422 \\ 0.2173 \end{Bmatrix}$$

It is only possible for us to obtain the ratios of the responses in free vibration. For the second mode, $\omega_2 = 16.304\,\text{rad/s}$. For the third mode, $\omega_3 = 22.138\,\text{rad/s}$. Then the second and third mode shapes are, respectively, found to be

$$\phi_2 = \begin{Bmatrix} \phi_{12} \\ \phi_{22} \\ \phi_{32} \end{Bmatrix} = \begin{Bmatrix} 0.1123 \\ 0.0664 \\ -0.2019 \end{Bmatrix} \text{ and } \phi_3 = \begin{Bmatrix} \phi_{13} \\ \phi_{23} \\ \phi_{33} \end{Bmatrix} = \begin{Bmatrix} -0.0965 \\ 0.1593 \\ -0.1098 \end{Bmatrix}$$

ϕ_2 defines the second mode shape of the system, or the configuration of the system, when it is vibrating with frequency ω_2. Similarly, ϕ_3 defines the third mode shape of the system, or the configuration of the system, when it is vibrating with frequency ω_3. The terms ϕ_1, ϕ_2 and ϕ_3 are called eigenvectors. The modal matrix $\mathbf{\Phi}$ is just the collection of the eigenvectors ϕ_1, ϕ_2 and ϕ_3,

$$[\mathbf{\Phi}] = \{\phi_1 \quad \phi_2 \quad \phi_3\} = \begin{bmatrix} -0.0556 & 0.1123 & -0.0965 \\ -0.1422 & 0.0664 & 0.1593 \\ -0.2173 & -0.2019 & -0.1098 \end{bmatrix}$$

It should be noted that, as a matter of convention, ω and ϕ are ranked in ascending order. Therefore, ω_1 represents the lowest or fundamental frequency, while ω_2 represents the second-lowest frequency and so forth.

As a result of the orthogonality conditions, the modal mass and modal stiffness matrices can be calculated as follows:

$$[M] = [\Phi]^T \mathbf{M}[\Phi] = \begin{bmatrix} M_1 & & \\ & M_2 & \\ & & M_3 \end{bmatrix}$$

$$= \begin{bmatrix} -0.0556 & -0.1422 & -0.2173 \\ 0.1123 & 0.0664 & -0.2019 \\ -0.0965 & 0.1593 & -0.1098 \end{bmatrix} \begin{bmatrix} 40 & 0 & 0 \\ 0 & 20 & 0 \\ 0 & 0 & 10 \end{bmatrix} \begin{bmatrix} -0.0556 & 0.1123 & -0.0965 \\ -0.1422 & 0.0664 & 0.1593 \\ -0.2173 & -0.2019 & -0.1098 \end{bmatrix}$$

$$= \begin{bmatrix} 1 & 0 & 0 \\ 0 & 1 & 0 \\ 0 & 0 & 1 \end{bmatrix} \tag{7.137}$$

$$[K] = [\Phi]^T \mathbf{K}[\Phi] = \begin{bmatrix} K_1 & & \\ & K_2 & \\ & & K_3 \end{bmatrix}$$

$$= \begin{bmatrix} -0.0556 & -0.1422 & -0.2173 \\ 0.1123 & 0.0664 & -0.2019 \\ -0.0965 & 0.1593 & -0.1098 \end{bmatrix} \begin{bmatrix} 13000 & -4000 & 0 \\ -4000 & 6000 & -2000 \\ 0 & -2000 & 2000 \end{bmatrix}$$

$$\begin{bmatrix} -0.0556 & 0.1123 & -0.0965 \\ -0.1422 & 0.0664 & 0.1593 \\ -0.2173 & -0.2019 & -0.1098 \end{bmatrix} \tag{7.138}$$

$$= \begin{bmatrix} 69.0822 & 0 & 0 \\ 0 & 265.8277 & 0 \\ 0 & 0 & 490.0901 \end{bmatrix} = \begin{bmatrix} \omega_1^2 & 0 & 0 \\ 0 & \omega_2^2 & 0 \\ 0 & 0 & \omega_2^2 \end{bmatrix}$$

Now let us consider the case of forced vibration with damping,

$$\mathbf{M}\ddot{\mathbf{z}} + \mathbf{C}\dot{\mathbf{z}} + \mathbf{K}\mathbf{z} = \mathbf{F} \tag{7.139}$$

These differential equations in question are coupled. One of the most important aspects of the modal displacement superposition method (or the normal mode method) is that it allows us to convert coupled equations, which are difficult to solve, into uncoupled equations, which are simple to solve.

Make a change of response variables as follows:

$$\mathbf{z} = [\Phi]\mathbf{q} \tag{7.140}$$

in which $\mathbf{q}$ denotes "new" variables.

The transformation matrix is represented by the matrix $[\Phi]$, which consists of the eigenvectors. Substitute equation (7.140) into equation (7.139) and pre-multiply by $[\Phi]^T$

$$[\Phi]^T \mathbf{M}[\Phi]\ddot{\mathbf{q}} + [\Phi]^T \mathbf{C}[\Phi]\dot{\mathbf{q}} + [\Phi]^T \mathbf{K}[\Phi]\mathbf{q} = [\Phi]^T \mathbf{F} \tag{7.141}$$

Equation (7.141) can be re-written as

$$[M]\ddot{\mathbf{q}} + [C]\dot{\mathbf{q}} + [K]\mathbf{q} = [\Phi]^T \mathbf{F} = \mathbf{G} \tag{7.142}$$

The modal force $\mathbf{G}$ can be calculated as follows:

$$[\Phi]^T \mathbf{F} = \begin{bmatrix} -0.0556 & -0.1422 & -0.2173 \\ 0.1123 & 0.0664 & -0.2019 \\ -0.0965 & 0.1593 & -0.1098 \end{bmatrix} \begin{Bmatrix} -m_1\ddot{y} \\ -m_2\ddot{y} \\ -m_3\ddot{y} \end{Bmatrix} = \begin{Bmatrix} 7.2396\ddot{y} \\ -3.8014\ddot{y} \\ 1.7715\ddot{y} \end{Bmatrix} \tag{7.143}$$

We assume that the modal damping matrix $[C]$ is diagonal. Then, equation (7.142) can be re-written into the following three uncoupled equations of motion:

$$\ddot{q}_1 + \frac{C_1}{M_1}\dot{q}_1 + \omega_1{}^2 q_1 = \ddot{q}_1 + 2\zeta_1\omega_1\dot{q}_1 + \omega_1{}^2 q_1 = 7.2396\ddot{y} \tag{7.144}$$

$$\ddot{q}_1 + \frac{C_1}{M_1}\dot{q}_1 + \omega_1{}^2 q_1 = \ddot{q}_1 + 2\zeta_1\omega_1\dot{q}_1 + \omega_1{}^2 q_1 = -3.8014\ddot{y} \tag{7.145}$$

$$\ddot{q}_1 + \frac{C_1}{M_1}\dot{q}_1 + \omega_1{}^2 q_1 = \ddot{q}_1 + 2\zeta_1\omega_1\dot{q}_1 + \omega_1{}^2 q_1 = 1.7715\ddot{y} \tag{7.146}$$

Considering that y is white noise, we are able to apply the formula (i.e., equation (5.137)) for white noise

$$\sigma_z^2 = \frac{\pi S_{\ddot{b}\ddot{b}}(\omega)}{2\omega_n{}^3\xi} \tag{7.147}$$

In the process of utilizing this particular equation, it is important to take into consideration the coefficients of the $\ddot{y}$ term. Simply based on dimensional considerations, these terms would be squared, and we multiply $S_{\ddot{b}\ddot{b}}(\omega)$ in the above expression. Therefore,

$$\sigma_1^2 = \frac{7.2396^2 \times \pi S_{\ddot{b}\ddot{b}}(\omega)}{2\omega_1{}^3\xi_1} = \frac{7.2396^2 \times \pi \times 0.01273}{2 \times 8.312^3 \times 0.022} = 0.09125 \tag{7.148}$$

$$\sigma_2^2 = \frac{(-3.8014)^2 \times \pi S_{\ddot{b}\ddot{b}}(\omega)}{2\omega_2{}^3\xi_2} = \frac{(-3.8014)^2 \times \pi \times 0.01273}{2 \times 16.304^3 \times 0.044} = 0.001515 \tag{7.149}$$

$$\sigma_3^2 = \frac{1.7715^2 \times \pi S_{\ddot{b}\ddot{b}}(\omega)}{2\omega_3{}^3\xi_3} = \frac{1.7715^2 \times \pi \times 0.01273}{2 \times 22.138^3 \times 0.088} = 0.00006573 \tag{7.150}$$

We are solely concerned with $z_1(t)$ at this time. Based on the process of normal mode transformation,

$$z_1 = \phi_{11}q_1 + \phi_{12}q_2 + \phi_{13}q_3 \tag{7.151}$$

Assuming that $q_1(t)$, $q_2(t)$ and $q_3(t)$ are independent processes, the variance of the sum is equal to the sum of the variances. Therefore, the root mean square of $z_1(t)$ can be calculated as follows:

$$\sigma_{z_1}^2 = \phi_{11}^2\sigma_1^2 + \phi_{12}^2\sigma_2^2 + \phi_{13}^2\sigma_3^2$$

$$\begin{aligned}\sigma_{z_1}^2 &= \phi_{11}^2\sigma_1^2 + \phi_{12}^2\sigma_2^2 + \phi_{13}^2\sigma_3^2\\ &= (-0.0556)^2 \times 0.09125 + (0.1123)^2 \times 0.001515 + (-0.0965)^2 \times 0.00006573\\ &= 0.000282 + 0.0000191 + 0.000000612 = 0.000302\end{aligned}$$

$$\sigma_{z_1} = 0.0174\,\text{m} = 17.4\,\text{mm}$$

Therefore, the root mean square value of the relative response of m_1 to the ground is 17.4*mm*.

7.3 EXERCISES

EXERCISE 7.1

Please determine the roots of the following polynomial.

$$f(y) = y^3 - 6.1y^2 + 11.2y - 6.3 = 0$$

EXERCISE 7.2

Please determine the eigenvalues and eigenvectors of the matrix that is presented below.

$$\mathbf{B} = \begin{bmatrix} 2 & 2 & 2 \\ 2 & 4 & 4 \\ 2 & 4 & 6 \end{bmatrix}$$

EXERCISE 7.3

One of the eigenvectors of a vibrating system is known to be

$$\begin{Bmatrix} 0.2755 \\ 0.3995 \\ 0.4491 \end{Bmatrix}$$

Prove that this eigenvector is orthogonal with respect to the mass matrix

$$\mathbf{M} = \begin{bmatrix} 100 & 0 & 0 \\ 0 & 200 & 0 \\ 0 & 0 & 300 \end{bmatrix}$$

EXERCISE 7.4

Determine the modal matrix $[\Phi]$ that corresponds to the system

$$\begin{bmatrix} 5 & 0 & 0 \\ 0 & 5 & 0 \\ 0 & 0 & 5 \end{bmatrix} \begin{Bmatrix} \ddot{y}_1 \\ \ddot{y}_2 \\ \ddot{y}_3 \end{Bmatrix} + \begin{bmatrix} 4000 & -900 & 0 \\ -900 & 6000 & -500 \\ 0 & -500 & 4000 \end{bmatrix} \begin{Bmatrix} y_1 \\ y_2 \\ y_3 \end{Bmatrix} = \begin{Bmatrix} 0 \\ 0 \\ 0 \end{Bmatrix}$$

Show that $[\Phi]^T = \alpha[\Phi]^{-1}$, in which α is a specific constant.

EXERCISE 7.5

Considering the initial conditions to be zero, please determine the response of a three-degree-of-freedom system with viscous damping under forced vibration using the given equations of motion as shown below.

$$\begin{bmatrix} 10 & 0 & 0 \\ 0 & 20 & 0 \\ 0 & 0 & 30 \end{bmatrix} \ddot{\mathbf{y}}(t) + \begin{bmatrix} 15 & -5 & 0 \\ -5 & 20 & -15 \\ 0 & -15 & 15 \end{bmatrix} \dot{\mathbf{y}}(t) + \begin{bmatrix} 140 & -60 & 0 \\ -60 & 100 & -40 \\ 0 & -40 & 40 \end{bmatrix} \mathbf{y}(t) = \begin{Bmatrix} 5\cos 2t \\ 0 \\ 0 \end{Bmatrix}$$

EXERCISE 7.6

A vibrating system's mass and stiffness matrices are provided by

$$\mathbf{M} = \begin{bmatrix} 5 & 0 & 0 & 0 \\ 0 & 4 & 0 & 0 \\ 0 & 0 & 3 & 0 \\ 0 & 0 & 0 & 2 \end{bmatrix} \quad \mathbf{K} = \begin{bmatrix} 6 & -4 & 0 & 0 \\ -4 & 6 & -2 & 0 \\ 0 & -2 & 4 & -2 \\ 0 & 0 & -2 & 2 \end{bmatrix}$$

Please determine the fundamental frequency and the mode shape of this vibrating system.

EXERCISE 7.7

The flexibility matrix is defined as the inverse of the stiffness matrix. For a system with three degrees of freedom, the mass and flexibility matrices are provided by

$$\mathbf{M} = \begin{bmatrix} 2 & 0 & 0 \\ 0 & 4 & 0 \\ 0 & 0 & 2 \end{bmatrix} \quad \mathbf{A} = \mathbf{K}^{-1} = \begin{bmatrix} 2 & 2 & 2 \\ 2 & 4 & 4 \\ 2 & 4 & 6 \end{bmatrix}$$

1. Please determine the lowest natural frequency of vibration of this vibrating system.
2. Please determine the modal matrix of this vibrating system.

EXERCISE 7.8

Let us take into consideration a viscously damped spring-mass-damper system that is under a base excitation, as depicted in Figure 7.15.

Suppose the system depicted in Figure 7.15 is subjected to a white-noise stationary ground acceleration, $S = 0.015\,\text{m}^2\text{rad/s}^3$. Given the base motion $y(t)$, please determine the root mean square value of the relative response of m_1 to the ground. The system parameters are given as follows:

$m_1 = 60$ kg $\quad m_2 = 40$ kg $\quad m_3 = 20$ kg $\quad m_4 = 10$ kg

$k_1 = 8000$ N/m $\quad k_2 = 6000$ N/m $\quad k_3 = 4000$ N/m $\quad k_4 = 2000$ N/m

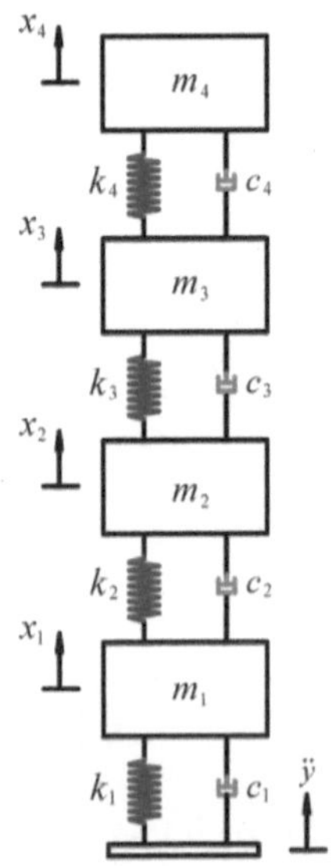

FIGURE 7.15 A base-excited damped four-degree-of-freedom system.

The assumed damping factors for each mode are given as follows:

$$\zeta_1 = 0.015 \quad \text{(first mode)}$$
$$\zeta_2 = 0.030 \quad \text{(second mode)}$$
$$\zeta_3 = 0.060 \quad \text{(third mode)}$$
$$\zeta_4 = 0.090 \quad \text{(fourth mode)}$$

8 Random Vibration of Nonlinear Systems

8.1 INTRODUCTION

The previous three chapters have explained the theories and applications of the random vibration of linear systems. When it comes to engineering applications, the utilization of a linear model for the system that is being considered results in outcomes that are frequently helpful and quite straightforward. In accordance with the linear theory, we can deduce that the response processes are also Gaussian if the excitation processes have a Gaussian distribution. Because of this, it is possible to compute a variety of response statistics that are relevant to reliability, such as level up-crossing rates, using only a few statistical parameters. In the event that the amplitude of motion is relatively low, the vast majority of engineering systems can be modeled, at least to a first approximation, in terms of linear differential equations of motion. Since the treatment of nonlinear equations is a great deal more complicated than the treatment of linear equations, it is evident that the adoption of a linear model is a step that is highly desired in the analysis of any given system that is being investigated. Nevertheless, there are situations in which a linearized mathematical model of the physical problem does not fully account for the quantitative or even the qualitative behavior of the problem. The vibration of elastic systems with large amplitudes is a typical example of the former case. An example of the latter case involves the existence of sub-harmonics and ultra-harmonics in the response of nonlinear systems that are excited periodically.

Furthermore, it is of the utmost importance to keep in mind that there is no real-world system that is exactly linear. Nonlinearities can occur in mechanical and structural systems in a variety of ways, and they often become progressively more significant as vibration amplitude increases. Due to the fact that systems are typically required to be designed in such a way that they are able to withstand with a specified probability, the severest possible levels of excitation that they can encounter over their operational lifetime cycle, it is frequently of vital importance to appropriately account for the effect of nonlinearities. Failure to do so can result in, at best, an overly conservative design or, at worst, a system prone to catastrophic failure.

In the next section, a description of several representative nonlinearities that are encountered in vibration problems will be provided. The objective is to expose the reader in a natural way to the considerations that can lead to the formulation of a nonlinear mathematical model in analyzing a specific engineering problem.

 DOI: 10.1201/9781003431503-8

8.2 THE REPRESENTATION OF NONLINEARITIES

This section will provide a concise discussion on various sources of nonlinearity that can occur in engineering vibration problems.

8.2.1 Motion in a Gravitational Field

In the majority of mechanical and structural systems, the forces that tend to return the system to its static equilibrium position arise from one or more sources, including the gravitational field, hydrostatic pressures acting on floating objects and internal stresses produced in deformed solids, etc.

It will be demonstrated in this subsection, with the help of a straightforward example, how nonlinearity can arise in the restoring force that is derived from the aforementioned source of a gravitational field. In this instance, the force that is involved is conservative in nature.

Take into consideration the motion of a typical pendulum that is depicted in Figure 8.1. At the lower end of a light, rigid rod that is l in length, there is a mass m that is attached to the pendulum. This mass is approximated here as a point mass. The upper end of the rod is able to swivel freely around a fixed point, which is located at O.

A system with one degree of freedom is said to exist when one's attention is limited to the oscillation of the pendulum in a vertical plane. The angle θ that the rod makes with the vertical line through O is an appropriate coordinate to define the motion. The position of static equilibrium is, of course, the position of the rod when it is coincident with the vertical line through O.

The free body diagram in Figure 8.2. provides the basis for deriving the differential equation that governs the free vibration of the pendulum, which is as follows:

$$ml^2\ddot{\theta} + mgl\sin\theta = 0 \tag{8.1}$$

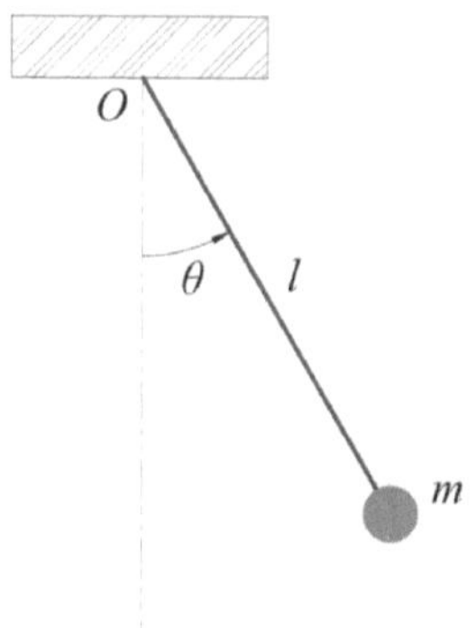

FIGURE 8.1 A simple pendulum.

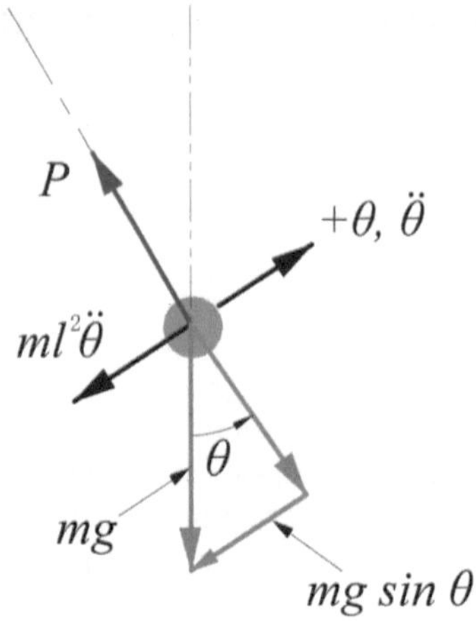

FIGURE 8.2 The free body diagram of the pendulum in Figure 8.1.

In cases where the angle is very small, the sin θ can be approximated by θ, thus equation (8.1) can be reduced to a linear equation:

$$\ddot{\theta} + \left(\sqrt{g/l}\right)^2 \theta = \ddot{\theta} + \omega_n^2 \theta = 0 \tag{8.2}$$

The solution to equation (8.2) can be represented as

$$\theta(t) = A_n \sin\left(\omega_n t + \phi\right) \tag{8.3}$$

in which A_n represents the magnitude of oscillation, ϕ represents the phase angle and ω_n represents the angular frequency. The values of A_n and ϕ are determined by the initial conditions, but the angular frequency ω_n is independent of the amplitude A_n. Equation (8.3) represents an estimated solution for the basic pendulum. At this time, recall that we can use the theory of Taylor series to show that the following identity holds for all real numbers θ (where θ is the angle in radians):

$$\sin(\theta) = \theta - \frac{\theta^3}{3!} + \frac{\theta^5}{5!} - \frac{\theta^7}{7!} + \cdots = \sum_{n=0}^{\infty} \frac{(-1)^n}{(2n+1)!} \theta^{2n+1} \tag{8.4}$$

Considering the relationship in equation (8.4), a more accurate approximation can be achieved by utilizing a two-term approximation for the sin θ near $\theta = 0$ as $\theta - \theta^3/6$ in equation (8.1):

$$ml^2\ddot{\theta} + mgl\left(\theta - \frac{\theta^3}{6}\right) = 0 \tag{8.5}$$

or

$$\ddot{\theta}+\left(\sqrt{g/l}\right)^2\left(\theta-\frac{\theta^3}{6}\right)=\ddot{\theta}+\omega_n^2\left(\theta-\frac{\theta^3}{6}\right)=0 \tag{8.6}$$

It can be noted that equation (8.6) is nonlinear due to the presence of the term involving θ^3, which is a result of geometric nonlinearity. Equation (8.6) has similarities to the equation of motion governing a spring-mass system featuring a nonlinear spring. If the spring exhibits nonlinearity, namely due to material nonlinearity, the restoring force can be mathematically represented as $r(y)$, where y represents the deformation of the spring. Consequently, the equation of motion for the spring-mass system can be described as

$$m\ddot{y}+r(y)=0 \tag{8.7}$$

The spring is linear if the value of $dr/dy(y)=k$ is constant. When the function dr/dy is a strictly increasing function with respect to y, the spring is referred to as a hard spring, as depicted in Figure 8.3.

On the other hand, when the function dr/dy is strictly decreasing with respect to y, the spring is referred to as a soft spring, as depicted in Figure 8.4. In a loose sense, a pendulum with large amplitudes is regarded as a system with a nonlinear elastic (spring) component because of the similarities between equations (8.6) and (8.7).

Figure 8.5 depicts a comparison between the exact restoration characteristic $r(\theta)=gl\sin\theta$, the linear approximation, which can be found in equation (8.2), and the linear-plus-cubic approximation, which can be found in equation (8.6). In this case, the nonlinearity is clearly of the "softening" type, that is, the slope of

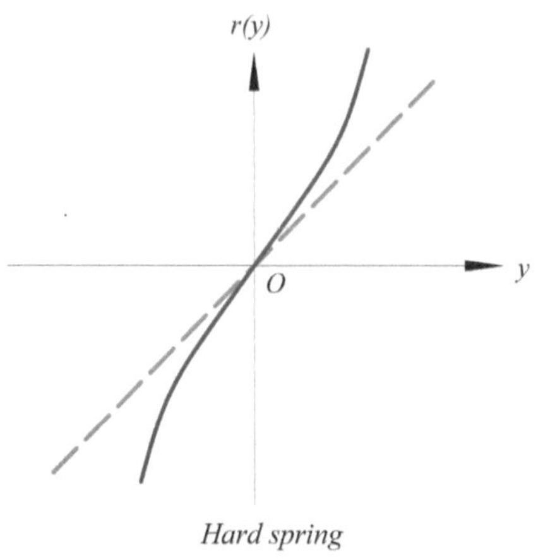

FIGURE 8.3 A hard spring.

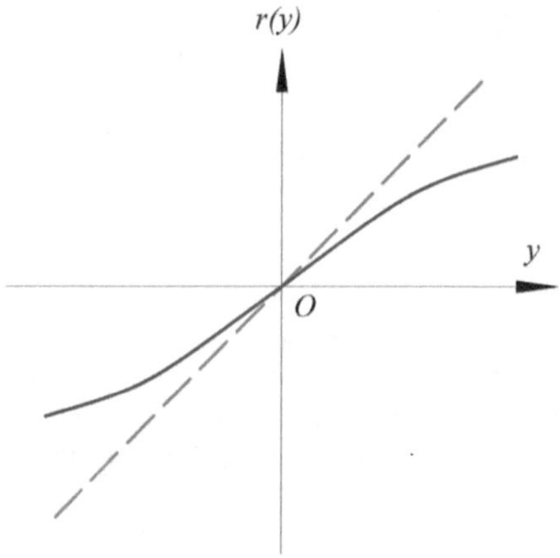

FIGURE 8.4 A soft spring.

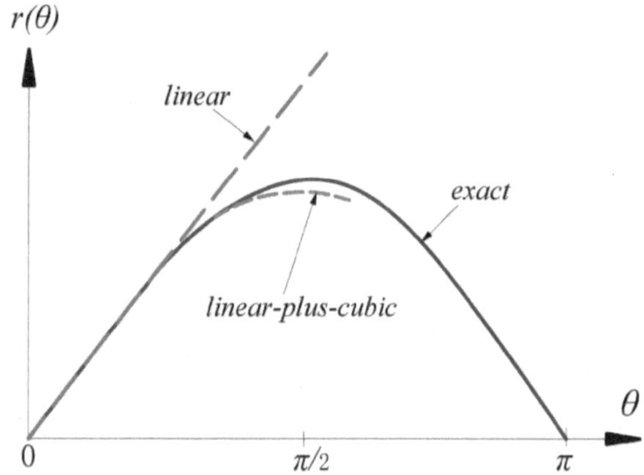

FIGURE 8.5 The restoring force characteristics of the simple pendulum.

the characteristic, which is a measure of the "stiffness" associated with the restoration force, tends to decrease as the amplitude increases.

8.2.2 Hydrostatic Pressures Acting on Floating Objects

Let's take into consideration a floating body that is effectively rigid and symmetrical about a vertical plane. A ship, an offshore drilling rig, or a floating production, storage and offloading system (FPSO) are three examples of what could fall under this category. The body will be in the position depicted in the sectional view of Figure 8.6 when it is placed in still water and is in the static equilibrium position. In this scenario, the buoyancy force, denoted as *B*, is equivalent to the weight of

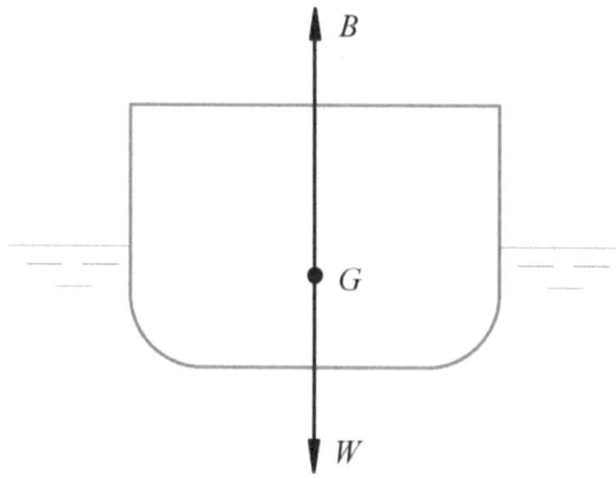

FIGURE 8.6 Forces acting on a floating body when it is in the static equilibrium position.

the body, represented as W. Furthermore, the lines along which these two forces act are aligned, lying in the plane of symmetry. Both forces pass through the body's center of gravity, located at point G.

Assuming that the body is provided with a rotating angle, denoted by the symbol Φ, as depicted in Figure 8.7. It is assumed that the mass of liquid that is displaced by the body does not undergo any changes. In the displaced position, the lines of action of B and W are no longer coinciding; yet, the buoyant force continues to be equal to B. The buoyancy force will intersect the plane of symmetry at a specific point called the "transverse metacenter", denoted as M. For small displacements, M is effectively a fixed point. Point M will remain practically stationary with respect to the ship as long as there is no abrupt change in the ship's shape in the vicinity of the waterline. This is true even when the ship is tilted at modest angles, often up to around 7 or, occasionally, 10 degrees.

Based on Figure 8.7, the calculation of the righting arm for small angles of heel may be easily determined if the positions of G and M are known. This can

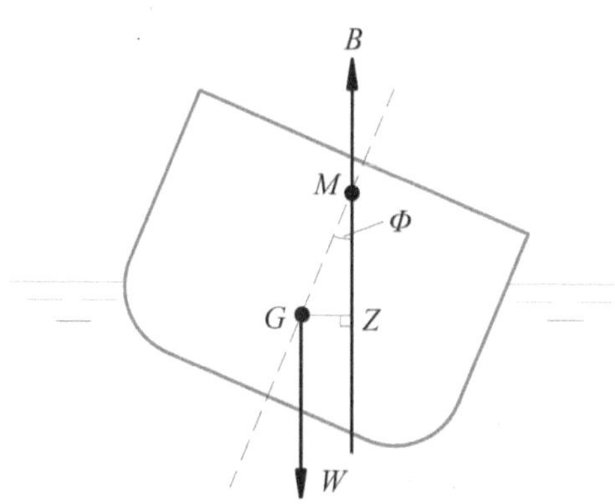

FIGURE 8.7 Forces acting on a floating body with a rotating angle Φ.

be done accurately enough for all practical purposes using the following provided formula

$$GZ \approx GM \sin \Phi \tag{8.8}$$

Consequently, the distance GM, which is referred to as the transverse metacentric height, is a crucial indicator of transverse stability at small angles of the heel. Because GZ is deemed positive when the ship's moment of buoyancy and weight tends to rotate it toward the upright position, GM is positive when the point M is above the point G, and negative when the point M is below the point G.

Obviously, in order to maintain stability, it is necessary for the transverse metacenter to be located above the center of gravity. In the design of floating bodies, the distance between G and M, GM, which is commonly referred to as the "transverse metacentric height", is obviously a parameter that is of the utmost importance.

The distance between the lines of action of forces W and B, represented by GZ in Figure 8.7, is referred to as the "righting lever", and the magnitude of the restoring moment is directly proportional to this distance. For small heeling angles, the relationship between the righting lever (GZ) and the heeling angle (Φ) can be approximated as $GZ \approx GM\Phi$, indicating a linear relationship. However, when there are larger heeling angles, the GZ value changes in a clearly nonlinear way with Φ, and this behavior is influenced by the shape of the body.

Figure 8.8 displays a typical GZ-Φ curve for a small ship. It is evident that a linear approximation is reasonably accurate for small angles of heeling ($\Phi < 10^\circ$).

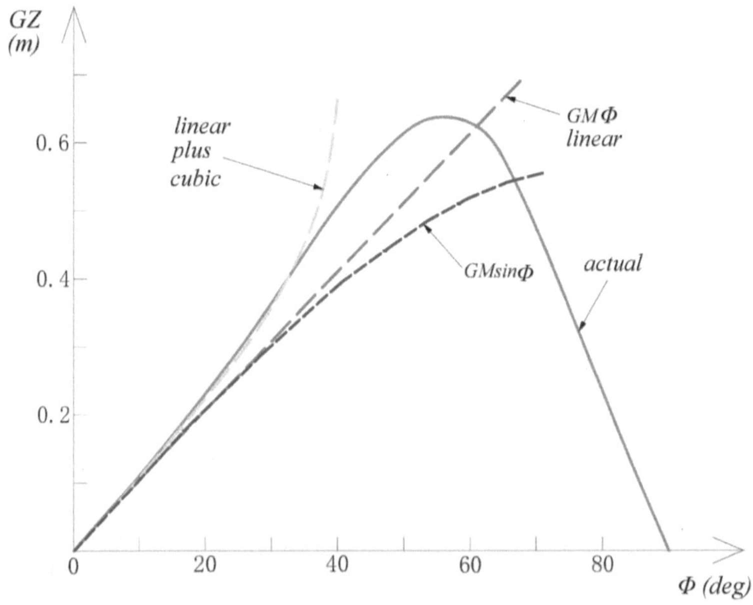

FIGURE 8.8 A typical GZ curve for a small ship.

Figure 8.8 demonstrates that a "hardening" linear-plus-cubic representation is more precise for heeling angles up to 30°. When the heeling angle is extremely great, the *GZ* characteristic exhibits a "softening kind", with *GZ* decreasing to zero at an angle of 90 degrees. When $GZ = 0$, it means that the body is on the verge of capsizing. Because there is no effective restoring moment, even a small disturbance will cause the ship to roll over.

8.2.3 Elastic Restoration Forces

There are many components in mechanical and structural systems that, as a result of undergoing some degree of elastic deformation, generate a restoring force. An example of a component that typically possesses this "spring-like" characteristic is illustrated in Figure 8.9.

The constitutive properties of the material that the component is composed of are, of course, the primary factor that determines the variation of the applied static force, which is equal and opposite to the restoring force, with the resulting deformation. In many instances, especially if the deformation is small, the force-displacement relationship obeys Hooke's law, that is, it is linear. Nevertheless, deviations from the linear characteristics frequently take place when the displacements are large. It is possible for these deviations to be of either the softening or hardening type.

In the context of this discussion, the word "elastic" means that the material does not undergo permanent deformation when it is subjected to loads from the outside. Therefore, if a material exhibits elastic behavior, it will go back to its original state once the external loads are removed. It is clear that elastic restoring forces are of the conservative type.

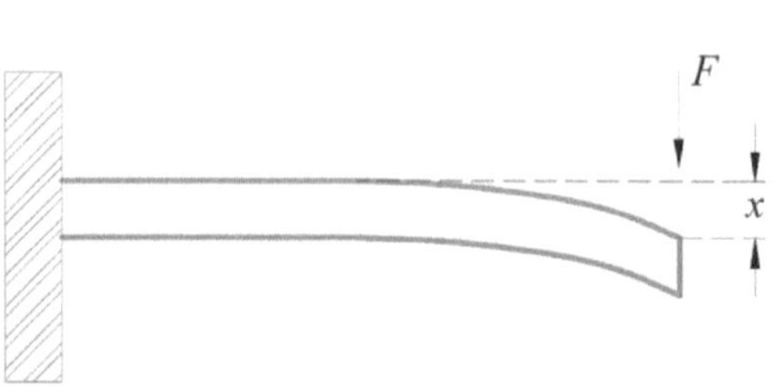

FIGURE 8.9 A typical deformable element: A cantilevered beam.

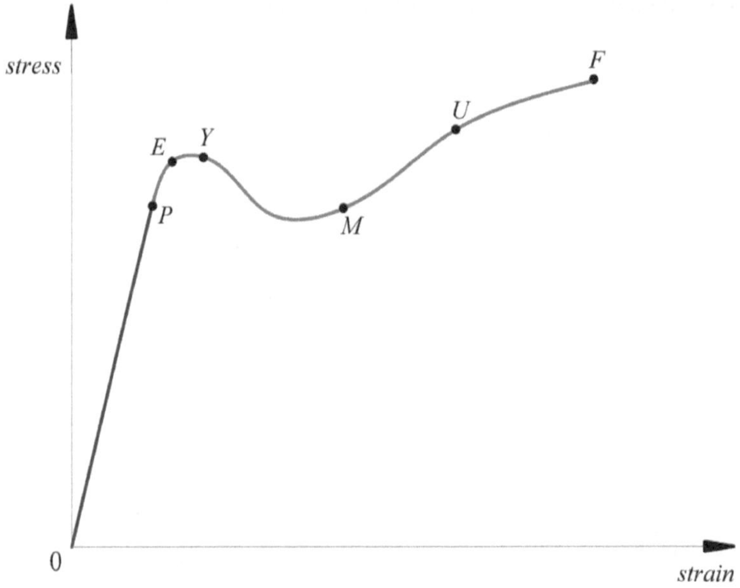

FIGURE 8.10 Stress-strain curve for mild steel.

In many metallic materials, the loading range for which nonlinear elastic behavior occurs is fairly limited. Beyond this range, permanent deformation occurs, which is associated with plastic behavior. An illustration of the stress-strain characteristic of mild steel can be found in Figure 8.10.

In Figure 8.10, to a very close approximation, the material exhibits linear and elastic behavior in the range O to P. As a result of the fact that the curve is essentially a straight line for the majority of this region, the stress is proportional to the strain. The material that is found in this area is considered to be linearly elastic. This linear relationship's upper stress limit is known as the proportional limit, denoted by the symbol P. In the event that the stress marginally exceeds the proportional limit, the curve has a tendency to bend, as demonstrated. This continues until the stress reaches the elastic limit, i.e., point E. If the load is removed from the material once it has reached this point, the material will still return back to its original shape. Nevertheless, the elastic limit is rarely determined for steel in normal circumstances. This is due to the fact that it is so close to the proportional limit, making it rather difficult to detect. Elastic behavior is still predominant from P to E, but the material now exhibits nonlinear behavior. Beyond the "elastic limit", which is denoted by the letter E, some degree of permanent, plastic deformation takes place. In the event that the material experiences a slight increase in stress that is greater than the elastic limit, the material will undergo permanent deformation. This pattern of behavior is referred to as yielding. The stress responsible for causing yielding is referred to as the yield stress or yield point, denoted as Y. The resulting deformation is known as plastic deformation. As you can see in Figure 8.10, once the yield point has been reached, the material will continue to elongate (strain) without any increase in load. This is something that you should take note of. It is common practice to refer to the material as being perfectly plastic when it is in this state. After the yielding has ended in point *M*, the curve rises continuously until it reaches a stress known as the ultimate stress, denoted by the letter *U*. The rise in the curve in this manner is referred to as strain hardening. In Figure 8.10, the material finally breaks at the fracture stress, denoted by the letter *F*.

8.2.4 Flow Induced Forces

In many engineering systems, the forces that are exerted by the fluids that surround the system are not negligible. Due to the fact that it serves as the foundation for either an analytical or numerical determination of the dynamic behavior of the system, the mathematical modeling of the fluid-induced load becomes extremely important in these situations. For instance, Morison's equation is frequently utilized in order to model the wave forces that are experienced by members of offshore structures in situations when the effects of diffraction are secondary. If it is determined that the velocity of the structural member is negligible, then the wave force per unit length of the member can be computed using the following Morison's equation

$$F(t)=\frac{1}{2}\rho\pi D^2 C_M\dot{v}+\frac{1}{2}\rho D C_D v|v| \tag{8.9}$$

in which $v(t)$ represents the wave-induced water particle velocity, relative to the structural member. $\dot{v}$ represents the corresponding acceleration. ρ represents the density of the fluid, and D represents the equivalent diameter of the member. In addition, the coefficients C_M and C_D are the coefficients that represent the inertia and drag, respectively. The non-dissipative inertia force is represented by the first term on the right-hand side of equation (8.9), which is linear with respect to acceleration. The second term in equation (8.9), which is the dissipative drag force, is nonlinear with respect to velocity.

In a number of applications, the dissipative fluid loading can be well represented by a linear-plus-quadratic model, which can be expressed in the following form:

$$F(t)=\alpha v+\beta v|v| \tag{8.10}$$

For example, the field of ship stochastic dynamics is an example of a situation in which the modeling that is represented by equation (8.10) is appropriate. The damping moment for a ship rolling in random waves is found to depend on the roll angular velocity, $\dot{\phi}$, as per the linear-plus-quadratic relationship $a\dot{\phi}+b\dot{\phi}|\dot{\phi}|$, where a and b are constants. Typically, the magnitude of the second quadratic term is significantly larger than that of the first linear term.

8.3 METHODS FOR SOLVING NONLINEAR RANDOM VIBRATION PROBLEMS

Solving nonlinear random vibration equations is significantly more challenging compared to solving linear ones. This situation is unsurprising, given the same circumstances in deterministic vibration theory. There, a comprehensive linear vibration theory exists. However, there is no corresponding general theoretical framework for the treatment of nonlinear deterministic vibration problems.

Here, a brief discussion will be conducted on the various methods that have been developed over the years, beginning with the method that will serve as the primary focus of this section. When dealing with nonlinear problems, one natural method of attacking is to replace the governing set of nonlinear differential equations with an equivalent set of linear equations, with the goal of minimizing the difference between the sets in some appropriate sense. In the field of electrical engineering, this method is frequently referred to as the "describing function method". This method has been utilized widely for the analysis of deterministic nonlinear problems for many years. An adaptation of this approach to deal with stochastic problems had led to the development of the *equivalent linearization technique.*

If the nonlinearities in the system are of negligible magnitude, the *perturbation technique* of solution can be employed. This *technique* has been widely recognized in deterministic vibration theory for many years and was extended to the context of stochastic excitation in the 1960s. The fundamental concept involves expanding the solution to the nonlinear set of equations by utilizing a small scaling parameter, such as ϵ, which is used to characterize the magnitude of the nonlinear terms. The first term in the expansion is just the linear response, which is the response that is obtained after all of the nonlinearities in the system have been removed. The subsequent terms in the expansion express the influence of nonlinearity. Similar to perturbation techniques in general, the computations are typically time-consuming and get increasingly tedious as the order of ϵ increases. In actual practice, results are typically only attained to the first order in ϵ.

In the following, the detailed theories regarding the *equivalent linearization technique* and the *perturbation technique* will be elucidated.

8.3.1 The Equivalent Linearization Technique

In the process of analyzing the structural response or system response to random excitations, one sort of nonlinear problem is represented by nonlinear differential equations. A system with a single degree of freedom is an example of a problem that falls within this category. This sort of problem is characterized by the differential equation that follows:

$$\ddot{y}+g(y,\dot{y})=F(t) \tag{8.11}$$

in which g is a nonlinear function of the displacement y and the velocity $\dot{y}$. $F(t)$ in equation (8.11) is a random excitation force.

There is the possibility of further specialization of the general nonlinear systems that are described by equation (8.11). The case in which g is nonlinear only with respect to y is called a nonlinear stiffness case, which is mathematically expressed as

$$g(y,\dot{y}) = c\dot{y} + f_1(y) \tag{8.12}$$

Similarly, the case in which g is nonlinear only with respect to $\dot{y}$ is called a nonlinear damping case, which is mathematically expressed as

$$g(y,\dot{y}) = f_2(\dot{y}) + ky \tag{8.13}$$

In equations (8.12) and (8.13), f denotes a nonlinear function.

For the purpose of demonstrating how the equivalent linearization technique can be adapted to a system with a single degree of freedom, such as the one that is represented by equation (8.11), we make the assumption that an approximate solution to equation (8.11) may be derived from the linearized equation

$$\ddot{y} + \beta_e\dot{y} + k_e y = F(t) \tag{8.14}$$

The linearization error, which is a stochastic process, is as follows:

$$\varepsilon = g(y,\dot{y}) - \beta_e\dot{y} - k_e y \tag{8.15}$$

ε is the difference between equations (8.11) and (8.14). The method of attack is to minimize this error in an appropriate manner, such as by making the value of $E\left[\varepsilon^2\right]$ or $E\left[|\varepsilon|\right]$ as small as possible. It is common practice to pursue the goal of minimizing the mean-square error, which can be achieved by requiring that

$$\begin{aligned} \frac{\partial}{\partial\beta_e}E\left[\varepsilon^2\right] &= 0 \\ \frac{\partial}{\partial k_e}E\left[\varepsilon^2\right] &= 0 \end{aligned} \tag{8.16}$$

By substituting equation (8.15) into equation (8.16) and interchanging the order of differentiation and expectation, we are able to derive the following:

$$\begin{aligned} \frac{\partial}{\partial\beta_e}E\left[\varepsilon^2\right] &= \frac{\partial}{\partial\beta_e}E\left[\left(g(y,\dot{y}) - \beta_e\dot{y} - k_e y\right)^2\right] \\ &= \frac{\partial}{\partial\beta_e}E\left[\left(g(y,\dot{y})\right)^2 + \beta_e^2\dot{y}^2 + k_e^2y^2 - 2\beta_e\dot{y}g(y,\dot{y}) - 2k_e yg(y,\dot{y}) + 2\beta_e k_e\dot{y}y\right] \\ &= E\left[2\beta_e\dot{y}^2 - 2\dot{y}g(y,\dot{y}) + 2k_e\dot{y}y\right] = 0 \\ &= 2\beta_e E\left[\dot{y}^2\right] - 2E\left[\dot{y}g(y,\dot{y})\right] + 2k_e E\left[\dot{y}y\right] = 0 \end{aligned} \tag{8.17}$$

$$
\begin{aligned}
&\frac{\partial}{\partial k_e} E\left[\varepsilon^2\right] \\
&\quad = \frac{\partial}{\partial k_e} E\left[\left(g(y,\, \dot{y}) - \beta_e \dot{y} - k_e y\right)^2\right] \\
&\quad = \frac{\partial}{\partial k_e} E\left[\left(g(y,\, \dot{y})\right)^2 + \beta_e^{\,2} \dot{y}^2 + k_e^{\,2} y^2 - 2\beta_e \dot{y} g(y,\, \dot{y}) - 2k_e y g(y,\, \dot{y}) + 2\beta_e k_e \dot{y} y\right] \\
&\quad = E\left[2k_e y^2 - 2yg(y,\, \dot{y}) + 2\beta_e \dot{y} y\right] = 0 \\
&\quad = 2k_e E\left[y^2\right] - 2E\left[yg(y,\, \dot{y})\right] + 2\beta_e E\left[\dot{y} y\right] = 0
\end{aligned} \tag{8.18}
$$

Upon solving for the equivalent damping and stiffness constants, denoted as β_e and k_e, we have arrived at the following:

$$
\begin{aligned}
\beta_e &= \frac{E\left[y^2\right] E[\dot{y} g(y, \dot{y})] - E[y\dot{y}] E[yg(y, \dot{y})]}{E\left[y^2\right] E\left[\dot{y}^2\right] - (E[y\dot{y}])^2} \\
k_e &= \frac{E\left[\dot{y}^2\right] E[yg(y, \dot{y})] - E[y\dot{y}] E[\dot{y} g(y, \dot{y})]}{E\left[y^2\right] E\left[\dot{y}^2\right] - (E[y\dot{y}])^2}
\end{aligned} \tag{8.19}
$$

It is important to take note that equations (8.19) do not contain explicit expressions for β_e and k_e. This is because the expectations that are displayed on the right-hand side are dependent on β_e and k_e.

For the purpose of keeping the following discussion to a reasonable length while also preserving the essence of the equivalent-linearization approach, we will make the assumption that the excitation $F(t)$ is stationary, is Gaussian, and has a zero expectation. After that, the approximate displacement and velocity, denoted as y and $\dot{y}$, which are computed from the linearized equation (8.14), are also Gaussian, and they approach stationarity if the dynamical system is stable. Because equation (8.14) is a linear equation, the response y will also has a zero expectation. In the remaining portion of this section, we will be focusing solely on the response that is stationary.

It is possible to define a stationary random process in a variety of different ways. The option that is the simplest and most suited to our needs is the one that we select. In other words, if $E\left[y(t)\right]$ and $E\left[y(t)y(t+\tau)\right]$ are both independent of t, then a random process is said to be (weakly) stationary. This indicates that $E\left[y(t)\right]$ is a constant (in this particular instance, $E\left[y(t)\right]$ equals 0), and that the autocorrelation function $E\left[y(t)y(t+\tau)\right]$ is only dependent on τ.

As per the classical random process theory, a random process $y(t)$ that is weakly stationary is orthogonal to its derivative $\dot{y}(t)$ when assessed at the same time, and the term "orthogonality" is defined as $E[y\dot{y}]=0$. Therefore, the mathematical expectation in equation (8.19) can be reduced to

$$\beta_e = \frac{E[\dot{y}g(y,\dot{y})]}{E[\dot{y}^2]}$$
$$k_e = \frac{E[yg(y,\dot{y})]}{E[y^2]} \tag{8.20}$$

a. The nonlinear damping case
A demonstration of the equivalent linearization technique's application is now presented. Only in the damping term, which is described as $\alpha\dot{y}+\beta\dot{y}|\dot{y}|$, does the system that we are considering exhibit nonlinearities. The vibration system's differential equation of motion can thus be expressed as follows:

$$\ddot{y}+g(y,\dot{y})=\ddot{y}+\alpha\dot{y}+\beta\dot{y}|\dot{y}|+\omega_n^2 y=F(t) \tag{8.21}$$

The equation of motion for a linear system, which is equivalent to the nonlinear system described in equation (8.21), can be expressed as:

$$\ddot{y}+\beta_e\dot{y}+k_e y=F(t) \tag{8.22}$$

Given that the excitation term $F(t)$ in equation (8.22) is a Gaussian random process with zero expectation, the response displacement y will also be normally distributed with zero expectation. The response velocity $\dot{y}$ will also be normally distributed with zero expectation. In the following, we denote the unknown variance of $\dot{y}$ as $\sigma_{\dot{y}}^2$

$$E[\dot{y}^2]=\sigma_{\dot{y}}^2 \tag{8.23}$$

Based on equation (8.20), the equivalent stiffness coefficient k_e can be calculated as

$$k_e = \frac{E[yg(y,\dot{y})]}{E[y^2]} = \frac{E\left[y\left(\alpha\dot{y}+\beta\dot{y}|\dot{y}|+\omega_n^2 y\right)\right]}{E[y^2]}$$
$$= \frac{\alpha E[y\dot{y}]+\beta E[y\dot{y}|\dot{y}|]+\omega_n^2 E[yy]}{E[y^2]} \tag{8.24}$$

Based on the definition of the autocorrelation function, we are aware that the relationship $E[y\dot{y}]=0$ indicates that the random process y and its derivative $\dot{y}$ are uncorrelated. As a result of the fact that being uncorrelated for a Gaussian process means being independent, it follows that the Gaussian process y and its derivative $\dot{y}$ are also independent. It should be noted that the following relationship exists due to the fact that y and its derivative $\dot{y}$ are independent:

$$E\left[y\dot{y}|\dot{y}|\right]=E[y]E\left[\dot{y}|\dot{y}|\right]=0\times E\left[\dot{y}|\dot{y}|\right]=0 \tag{8.25}$$

The equivalent stiffness coefficient k_e in equation (8.24) can then be calculated as

$$k_e=\frac{E[yg(y,\dot{y})]}{E\left[y^2\right]}=\frac{\alpha\times 0+\beta\times 0+\omega_n{}^2E\left[y^2\right]}{E\left[y^2\right]}=\omega_n{}^2 \tag{8.26}$$

Based on equation (8.20), the equivalent damping coefficient β_e can be calculated as

$$\begin{aligned}\beta_e&=\frac{E[\dot{y}g(y,\dot{y})]}{E\left[\dot{y}^2\right]}=\frac{E\left[\dot{y}\left(\alpha\dot{y}+\beta\dot{y}|\dot{y}|+\omega_n{}^2y\right)\right]}{E\left[\dot{y}^2\right]}\\&=\frac{\alpha E[\dot{y}\dot{y}]+\beta E\left[\dot{y}\left(\dot{y}|\dot{y}|\right)\right]+\omega_n{}^2E[y\dot{y}]}{E\left[\dot{y}^2\right]}=\frac{\alpha E\left[\dot{y}^2\right]+\beta E\left[\dot{y}^2|\dot{y}|\right]}{E\left[\dot{y}^2\right]}\end{aligned} \tag{8.27}$$

Because the response velocity $\dot{y}$ is normally distributed with zero expectation, we have

$$\beta E\left[\dot{y}^2|\dot{y}|\right]=\int_{-\infty}^{\infty}\beta\dot{y}^2|\dot{y}|\frac{1}{\sqrt{2\pi}\sigma_{\dot{y}}}exp\left[-\dot{y}^2/\left(2\sigma_{\dot{y}}{}^2\right)\right]d\dot{y} \tag{8.28}$$

Based on equation (3.25), we have

$$\int_0^{\infty}\dot{y}^3\exp\left(-\frac{\dot{y}^2}{2\sigma^2}\right)d\dot{y}=2\sigma^4 \tag{8.29}$$

Combining equations (8.28) and (8.29) leads to

$$\beta E\left[\dot{y}^2|\dot{y}|\right]=\frac{\beta\times 2\times 2\sigma_{\dot{y}}{}^4}{\sqrt{2\pi}\sigma_{\dot{y}}}=\frac{2\sqrt{2}}{\sqrt{\pi}}\beta\sigma_{\dot{y}}{}^3 \tag{8.30}$$

Combining equations (8.27) and (8.30) leads to

$$\beta_e=\frac{\alpha E\left[\dot{y}^2\right]+\beta E\left[\dot{y}^2|\dot{y}|\right]}{E\left[\dot{y}^2\right]}=\frac{\alpha\sigma_{\dot{y}}{}^2+\frac{2\sqrt{2}}{\sqrt{\pi}}\beta\sigma_{\dot{y}}{}^3}{\sigma_{\dot{y}}{}^2}=\alpha+\frac{2\sqrt{2}}{\sqrt{\pi}}\beta\sigma_{\dot{y}} \tag{8.31}$$

The frequency response function of the equivalent linear system (8.22) is

$$H_e(\omega) = \frac{1}{(\omega_n{}^2 - \omega^2) + i\omega\beta_e} \tag{8.32}$$

The magnitude of $H_e(\omega)$ can therefore be expressed in the following way:

$$|H_e(\omega)| = \sqrt{\frac{1}{(\omega_n{}^2 - \omega^2)^2 + (\omega\beta_e)^2}} \tag{8.33}$$

Accordingly, the square of the magnitude of $H_e(\omega)$ can be expressed in the following way:

$$|H_e(\omega)|^2 = \frac{1}{(\omega_n{}^2 - \omega^2)^2 + (\omega\beta_e)^2} \tag{8.34}$$

In the event that $\beta_e = \alpha$, we have the relationship described below:

$$|H(\omega)|^2 = \frac{1}{(\omega_n{}^2 - \omega^2)^2 + (\alpha\omega)^2} \tag{8.35}$$

Applying the relationship described in equation (8.31), we proceed to expand the function $|H_e(\omega)|^2$ into a Taylor series in the vicinity of α:

$$|H_e(\omega)|^2 = |H(\omega)|^2 + \frac{2\sqrt{2}}{\sqrt{\pi}}\beta\sigma_{\dot{y}}\frac{\partial}{\partial\alpha}|H(\omega)|^2 \tag{8.36}$$

Equation (8.36) can be further written as follows:

$$\begin{aligned} |H_e(\omega)|^2 &= |H(\omega)|^2 + \frac{2\sqrt{2}}{\sqrt{\pi}}\beta\sigma_{\dot{y}}\frac{\partial}{\partial\alpha}|H(\omega)|^2 \\ &= |H(\omega)|^2 + \frac{2\sqrt{2}}{\sqrt{\pi}}\beta\sigma_{\dot{y}}\left(-2\alpha\omega^2|H(\omega)|^4\right) \\ &= |H(\omega)|^2\left\{1 - \frac{4\sqrt{2}}{\sqrt{\pi}}\alpha\beta\omega^2\sigma_{\dot{y}}|H(\omega)|^2\right\} \end{aligned} \tag{8.37}$$

After applying equation (8.37), we are able to derive the following relationship between the output and input spectral density functions of the equivalent linear system (8.22):

$$S_{yy}(\omega) = S_{FF}(\omega)|H(\omega)|^2\left\{1 - \frac{4\sqrt{2}}{\sqrt{\pi}}\alpha\beta\omega^2\sigma_{\dot{y}}|H(\omega)|^2\right\} \tag{8.38}$$

$\sigma_{\dot{y}}$, the response velocity standard deviation, is not known in equation (8.38). However, it is possible to determine it in an iterative manner as follows:

1. First, we assign a value to $\sigma_{\dot{y}}$ and subsequently compute the output spectral density function $S_{yy}(\omega)$ in equation (8.38) based on a given input spectral density function $S_{FF}(\omega)$.
2. Compute the second moment of $S_{yy}(\omega)$, which is equal to $\sigma_{\dot{y}}^2$ (this is because $\sigma_{\dot{y}}^2 = \int_{-\infty}^{\infty} S_{\dot{y}\dot{y}}(\omega)d\omega = \int_{-\infty}^{\infty} \omega^2 S_{yy}(\omega)d\omega$). Compare the $\sigma_{\dot{x}}$ value thus obtained with the assigned value in the first step. Compare the value of $\sigma_{\dot{y}}$ that was obtained in this way with the $\sigma_{\dot{y}}$ value that was assigned in the first step.
3. The above procedure should be repeated until the value of the evaluated standard deviation ($\sigma_{\dot{y}}$) is in agreement with the value that was assigned.

b. The nonlinear stiffness case

Next, let us take into consideration the scenario in which nonlinearities are only present in the restoring term, which can be mathematically expressed as

$$\ddot{y} + g(y, \dot{y}) = \ddot{y} + \alpha\dot{y} + \omega_n^2\left(y + \lambda y^3\right) = F(t) \tag{8.39}$$

This kind of nonlinear system is referred to as a Duffing system in the literature. The equation of motion for a linear system, which is mathematically equivalent to the nonlinear system stated in equation (8.39), can be expressed as follows:

$$\ddot{y} + \beta_e\dot{y} + k_e y = F(t) \tag{8.40}$$

Considering that the excitation term $F(t)$ in equation (8.40) follows a Gaussian random process with an expected value of zero, the resulting displacement of the response will also have a Gaussian distribution with an expected value of zero. In the subsequent discussion, we represent the unknown variance of y as σ_y^2.

$$E\left[y^2\right] = \sigma_y^2 \tag{8.41}$$

In accordance with equation (8.20), the equivalent damping coefficient β_e can be computed as follows:

$$\begin{aligned}\beta_e &= \frac{E[\dot{y}g(y,\dot{y})]}{E\left[\dot{y}^2\right]} = \frac{E\left[\dot{y}\left\{\alpha\dot{y} + \omega_n^2\left(y + \lambda y^3\right)\right\}\right]}{E\left[\dot{y}^2\right]} \\ &= \frac{\alpha E\left[\dot{y}\dot{y}\right] + \omega_n^2 E\left[\dot{y}y\right] + \omega_n^2\lambda E\left[\dot{y}y^3\right]}{E\left[\dot{y}^2\right]}\end{aligned} \tag{8.42}$$

In light of the fact that y and its derivative $\dot{y}$ are independent, it is important to point out that the following relationship exists:

$$E\left[\dot{y}y^3\right]=E[\dot{y}]E\left[y^3\right]=E\left[\frac{d}{dt}y\right]E\left[y^3\right]=\frac{d}{dt}E[y]E\left[y^3\right]=0 \quad (8.43)$$

Combining equations (8.42) and (8.43) leads to

$$\beta_e=\frac{\alpha E[\dot{y}\dot{y}]+\omega_n{}^2\times 0+\omega_n{}^2\lambda\times 0}{E\left[\dot{y}^2\right]}=\alpha \quad (8.44)$$

According to equation (8.20), the value of the equivalent stiffness coefficient k_e can be computed as follows:

$$\begin{aligned} k_e &= \frac{E[yg(y,\dot{y})]}{E\left[y^2\right]}=\frac{E\left[y\left\{\alpha\dot{y}+\omega_n{}^2\left(y+\lambda y^3\right)\right\}\right]}{E\left[y^2\right]} \\ &= \frac{\alpha E[y\dot{y}]+\omega_n{}^2E[yy]+\omega_n{}^2\lambda E\left[yy^3\right]}{E\left[y^2\right]} \\ &= \omega_n{}^2+\frac{\omega_n{}^2\lambda E\left[y^4\right]}{E\left[y^2\right]} \end{aligned} \quad (8.45)$$

Because the response displacement y is Gaussian distributed with zero mean, we have

$$\omega_n{}^2\lambda E\left[y^4\right]=\omega_n{}^2\lambda\int_{-\infty}^{\infty}y^4\frac{1}{\sqrt{2\pi}\sigma_y}\exp\left[-y^2/\left(2\sigma_y{}^2\right)\right]dy \quad (8.46)$$

Based on equation (3.25), we have

$$\int_0^{\infty}y^4\exp\left(-\frac{y^2}{2\sigma^2}\right)dy=\frac{3\sqrt{2\pi}\sigma_y{}^5}{2} \quad (8.47)$$

Combining equations (8.45), (8.46) and (8.47) leads to

$$\begin{aligned} k_e &= \omega_n{}^2+\frac{\omega_n{}^2\lambda E\left[y^4\right]}{E\left[y^2\right]}=\omega_n{}^2+\frac{\omega_n{}^2\lambda\frac{1}{\sqrt{2\pi}\sigma_y}\frac{3\sqrt{2\pi}\sigma_y{}^5}{2}\times 2}{\sigma_y{}^2} \\ &= \omega_n{}^2\left(1+3\lambda\sigma_y{}^2\right) \end{aligned} \quad (8.48)$$

In the following, we write $k_e = \omega_e^2$. With regard to the equivalent linear system in equation (8.22), the frequency response function is as follows:

$$H_e(\omega) = \frac{1}{(\omega_e^2 - \omega^2) + i\omega\beta_e} = \frac{1}{(\omega_e^2 - \omega^2) + i\omega\alpha} \tag{8.49}$$

In light of this, the magnitude of $H_e(\omega)$ can be calculated as

$$|H_e(\omega)| = \sqrt{\frac{1}{(\omega_e^2 - \omega^2)^2 + (\alpha\omega)^2}} \tag{8.50}$$

In light of this, the square of the magnitude of $H_e(\omega)$ can be calculated in the following way:

$$|H_e(\omega)|^2 = \frac{1}{(\omega_e^2 - \omega^2)^2 + (\alpha\omega)^2} \tag{8.51}$$

In the case that $\omega_n^2 = \omega_e^2$, we have the relationship that is given below

$$|H(\omega)|^2 = \frac{1}{(\omega_n^2 - \omega^2)^2 + (\alpha\omega)^2} \tag{8.52}$$

Through the utilization of the relationship that is outlined in equation (8.48), we proceed to expand the function $|H_e(\omega)|^2$ into a Taylor series in the neighborhood of ω_n^2:

$$|H_e(\omega)|^2 = |H(\omega)|^2 + 3\lambda\omega_n^2\sigma_y^2 \frac{\partial}{\partial(\omega_n^2)}|H(\omega)|^2 \tag{8.53}$$

It can be readily verified that

$$\frac{\partial}{\partial(\omega_n^2)}|H(\omega)|^2 = -2(\omega_n^2 - \omega^2)|H(\omega)|^4 \tag{8.54}$$

Combining equations (8.53) and (8.54) leads to

$$\begin{aligned} |H_e(\omega)|^2 &= |H(\omega)|^2 + 3\lambda\omega_n^2\sigma_y^2\left\{-2(\omega_n^2 - \omega^2)|H(\omega)|^4\right\} \\ &= |H(\omega)|^2\left\{1 - 6\lambda\omega_n^2\sigma_y^2(\omega_n^2 - \omega^2)|H(\omega)|^2\right\} \end{aligned} \tag{8.55}$$

Through the use of equation (8.55), we are able to deduce the following relationship between the output and input spectral density functions of the equivalent linear system in equation (8.22):

$$\begin{aligned} S_{yy}(\omega) &= S_{FF}(\omega)|H(\omega)|^2\left\{1-6\lambda\omega_n^{\ 2}\sigma_y^{\ 2}\left(\omega_n^{\ 2}-\omega^2\right)|H(\omega)|^2\right\} \\ &= \frac{S_{FF}(\omega)}{\left(\omega_n^{\ 2}-\omega^2\right)^2+(\alpha\omega)^2}\left\{1-\frac{6\lambda\omega_n^{\ 2}\sigma_y^{\ 2}\left(\omega_n^{\ 2}-\omega^2\right)}{\left(\omega_n^{\ 2}-\omega^2\right)^2+(\alpha\omega)^2}\right\} \end{aligned} \tag{8.56}$$

In equation (8.38), the standard deviation of the response displacement, denoted by σ_y, is unknown. Nevertheless, it is feasible to determine it by an iterative approach as outlined below:

1. In the first step, we assign a value to σ_y.
2. We next calculate the spectral density function $S_{yy}(\omega)$ of output displacement in equation (8.56) based on a given input force spectral density function $S_{FF}(\omega)$.
3. We then calculate the zeroth moment of $S_{yy}(\omega)$, which is equal to $\sigma_y^{\ 2}$ (i.e., $\sigma_y^{\ 2} = \int_{-\infty}^{\infty} S_{yy}(\omega)d\omega$).
4. We next compare the σ_y value thus obtained with the assigned value in the first step.
5. Repeating the procedures described above should be done until the value of the standard deviation (σ_y) that was evaluated is in agreement with the value that was assigned.

Example 8.1

▶ Let us consider a system in which nonlinearities are only present in the restoring term, which can be expressed as

$$\ddot{y}+g(y,\dot{y})=\ddot{y}+\alpha\dot{y}+\omega_n^{\ 2}\left(y+\lambda y^5\right)=F(t) \tag{8.57}$$

If the given input force spectral density function is $S_{FF}(\omega)$, please determine the spectral density function $S_{yy}(\omega)$ of output displacement.

SOLUTION

The equation of motion for a linear system, which is equivalent to the nonlinear system in equation (8.57), can be expressed as

$$\ddot{y}+\beta_e\dot{y}+k_e y=F(t) \tag{8.58}$$

Utilizing similar derivation procedures as those from equations (8.42) to (8.44), we can derive

$$\beta_e=\alpha \tag{8.59}$$

Based on equation (8.20), the equivalent stiffness coefficient k_e can be calculated as

$$k_e = \frac{E[yg(y,\dot{y})]}{E\left[y^2\right]} = \frac{E\left[y\left\{\alpha\dot{y} + \omega_n^2\left(y + \lambda y^5\right)\right\}\right]}{E\left[y^2\right]}$$
$$= \frac{\alpha E\left[y\dot{y}\right] + \omega_n^2 E\left[yy\right] + \omega_n^2 \lambda E\left[yy^5\right]}{E\left[y^2\right]} = \omega_n^2 + \frac{\omega_n^2 \lambda E\left[y^6\right]}{E\left[y^2\right]} \tag{8.60}$$

Since the response displacement y is normally distributed with zero expectation, we have

$$\omega_n^2 \lambda E\left[y^6\right] = \omega_n^2 \lambda \int_{-\infty}^{\infty} y^6 \frac{1}{\sqrt{2\pi}\sigma_y} \exp\left[-y^2/\left(2\sigma_y^2\right)\right] dy \tag{8.61}$$

Based on equation (3.24), we can obtain

$$\int_0^{\infty} y^6 \exp\left(-\frac{y^2}{2\sigma^2}\right) dy = \frac{(\sqrt{2}\sigma_y)^{6+1}}{2} \Gamma\left(\frac{6+1}{2}\right) = (4\sqrt{2})\sigma_y^7 \times 3.3234 = 18.797\sigma_y^7 \tag{8.62}$$

Combining equations (8.60), (8.61) and (8.62) leads to

$$k_e = \omega_n^2 + \frac{\omega_n^2 \lambda E\left[y^6\right]}{E\left[y^2\right]} = \omega_n^2 + \frac{\omega_n^2 \lambda \frac{1}{\sqrt{2\pi}\sigma_y} 18.797\sigma_y^7 \times 2}{\sigma_y^2} = \omega_n^2\left(1 + 22.35\lambda\sigma_y^4\right) \tag{8.63}$$

In the following, we write $k_e = \omega_e^2$. With regard to the equivalent linear system in equation (8.22), the frequency response function is as follows:

$$H_e(\omega) = \frac{1}{\left(\omega_e^2 - \omega^2\right) + i\omega\beta_e} = \frac{1}{\left(\omega_e^2 - \omega^2\right) + i\omega\alpha} \tag{8.64}$$

Utilizing similar derivation procedures as those from equations (8.50) to (8.56), we can derive

$$S_{yy}(\omega) = \frac{S_{FF}(\omega)}{\left(\omega_n^2 - \omega^2\right)^2 + (\alpha\omega)^2} \left\{1 - \frac{44.7\lambda\omega_n^2\sigma_y^4\left(\omega_n^2 - \omega^2\right)}{\left(\omega_n^2 - \omega^2\right)^2 + (\alpha\omega)^2}\right\} \tag{8.65}$$

The unknown variance of the displacement response σ_y^2 can be evaluated through an iteration procedure similar to that discussed in regard to the foregoing Duffing system.

The equivalent linearization technique has been found to be a valuable analytical tool in a wide range of engineering applications when it comes to treating physical problems that can be addressed. Examples of research studies applying the equivalent linearization technique include the examination of the sliding motion of a structure on a surface that moves randomly, the investigation of how ships and offshore structures respond to random ocean waves, the study of the large amplitude response of clamped plates and skin structures to acoustic loading, the analysis of the sloshing of liquids in tanks when subjected to earthquake excitation, the examination of the motion of vehicles moving over rough terrain and the investigation of the impact of nonlinear soil-structure interaction on the dynamic response of buildings.

Nevertheless, the equivalent linearization technique has two primary drawbacks: First, the response statistics of linear systems that are subject to Gaussian random excitations are always Gaussian, whereas the response statistics of nonlinear systems that are subject to Gaussian random excitations are always non-Gaussian. As a consequence of this, the motion response statistics that are acquired by the equivalent linearization technique are unable to accurately capture the non-Gaussian properties of the original system response. To make matters even worse, the non-Gaussian properties of the system response have a significant influence on the system fatigue and reliability analysis. While the external forces acting on some engineering systems can be approximated as Gaussian, the differential equations of motion of these engineering systems are often nonlinear. Therefore, it is not suitable to conduct a system analysis of these engineering systems based on the motion responses obtained from the equivalent linearization technique. Furthermore, the equivalent linearization technique lacks the ability to forecast responses in the frequency domain outside the range of excitation frequencies. Consequently, certain response statistics may be underestimated. The equivalent linearization technique is unable to forecast the subharmonic and super-harmonic resonance phenomena exhibited by nonlinear systems when they are subjected to external excitations. In the following, we will discuss the perturbation technique, which is another often-used approach to dealing with nonlinear random vibration problems.

8.3.2 The Perturbation Technique

Another method for assessing the response of a nonlinear vibration system under the action of random external loads is to make the assumption that the response may be described in the form of a power series of the parameter that is associated with the nonlinear term. The perturbation technique is the name given to this method. The perturbation technique is applicable for addressing scenarios where the nonlinear terms in the governing equations are of small magnitude. Let us take into consideration the equation that follows for a system with a single degree of freedom.

$$\ddot{y}+\alpha\dot{y}+\omega_n^{2}y+\varepsilon\eta\left(y,\dot{y}\right)=F(t) \tag{8.66}$$

In the above nonlinear system, denoted by equation (8.66), we have expressed the nonlinear term as "$\varepsilon\eta(y, \dot{y})$", with the parameter "$\varepsilon$" assumed to have a sufficiently small value that the perturbation technique is applicable. The perturbation technique assumes that the solution of equation (8.66) may be expressed as a power series in ε; that is,

$$y(t) = y_0(t) + \varepsilon y_1(t) + \varepsilon^2 y_2(t) + \cdots \tag{8.67}$$

in which $y_0(t)$ is the solution to the linear equation. By substituting equation (8.67) into equation (8.66) and by equating terms containing the same power of ε, we obtain the following set of linear equations for y_0, y_1, y_2, etc.

$$\ddot{y}_0 + \alpha\dot{y}_0 + \omega_n^2 y_0 = F(t) \tag{8.68}$$

$$\ddot{y}_1 + \alpha\dot{y}_1 + \omega_n^2 y_1 = -\eta(y_0, \dot{y}_0) \tag{8.69}$$

$$\ddot{y}_2 + \alpha\dot{y}_2 + \omega_n^2 y_2 = -y_1\frac{\partial}{\partial y}\eta(y_0, \dot{y}_0) - \dot{y}_1\frac{\partial}{\partial \dot{y}}\eta(y_0, \dot{y}_0)\cdots \tag{8.70}$$

Equation (8.70) is obtained by conducting a Taylor series expansion of $\varepsilon\eta(y, \dot{y})$. That is,

$$\begin{aligned}\varepsilon\eta(y, \dot{y}) = \varepsilon\eta(y_0, \dot{y}_0) + \varepsilon(\varepsilon y_1 + \varepsilon^2 y_2 + \cdots)\frac{\partial}{\partial y}\eta(y_0, \dot{y}_0)\\ + \varepsilon(\varepsilon\dot{y}_1 + \varepsilon^2\dot{y}_2 + \cdots)\frac{\partial}{\partial \dot{y}}\eta(y_0, \dot{y}_0) + \cdots\end{aligned} \tag{8.71}$$

It is evident from equations (8.68), (8.69), and (8.70) that $y_{i+1}(t)$ can be derived as the linear response to an excitation that is a nonlinear function of the previously determined $y_i(t)$. The evaluation of $y_i(t)$ for $i \geq 2$ is, however, an extraordinarily difficult task; hence, in practice, only the first two terms are usually taken into account. Therefore, we will limit our following discussion to this particular instance.

If we make the assumption that the excitation process $F(t)$ is weakly stationary and that after the excitation has been operating for a sufficient amount of time, there exists a weakly stationary solution to equation (8.66), then this solution can be formed from the weakly stationary solutions to equations (8.68) and (8.69). The solution $y_0(t)$ from equation (8.68) can be written as a convolution integral in the following way:

$$y_0(t) = \int_{-\infty}^{\infty} F(t-\theta)g(\theta)d\theta \tag{8.72}$$

where $g(\theta)$ represents the linear system's impulse response function, which equals the inverse Fourier transform of the frequency response function $H(\omega)$. In a similar manner, from equation (8.69), $y_1(t)$ can also be expressed as a convolution integral as follows:

$$y_1(t) = -\int_{-\infty}^{\infty} \eta\left(y_0(t-\theta), \dot{y}_0(t-\theta)\right) g(\theta) d\theta \tag{8.73}$$

Therefore, the approximate time-domain response of a nonlinear system represented by equation (8.66) with an impulse function $g(\theta)$ can be computed as follows:

$$\begin{aligned} y(t) = y_0(t) + \varepsilon y_1(t) &= \int_{-\infty}^{\infty} F(t-\theta) g(\theta) d\theta \\ &- \varepsilon \int_{-\infty}^{\infty} \eta\left(y_0(t-\theta), \dot{y}_0(t-\theta)\right) g(\theta) d\theta \end{aligned} \tag{8.74}$$

Equations (8.73) and (8.74) can be utilized to compute a variety of statistical properties of the response $y(t)$, such as the mean function and the correlation function. The expectation of $y(t)$ is merely the following:

$$E[y(t)] = E\left[y_0(t)\right] + \varepsilon E\left[y_1(t)\right] \tag{8.75}$$

Substitution of equations (8.73) and (8.74) into equation (8.75) yields

$$\begin{aligned} E[y(t)] &= E\left[\int_{-\infty}^{\infty} F(t-\theta) g(\theta) d\theta\right] - \varepsilon E\left[\int_{-\infty}^{\infty} \eta\left(y_0(t-\theta), \dot{y}_0(t-\theta)\right) g(\theta) d\theta\right] \\ &= \int_{-\infty}^{\infty} E\left[F(t-\theta)\right] g(\theta) d\theta - \varepsilon \int_{-\infty}^{\infty} E\left[\eta\left(y_0(t-\theta), \dot{y}_0(t-\theta)\right)\right] g(\theta) d\theta \end{aligned} \tag{8.76}$$

Typically, when using a perturbation technique, the results are calculated up to the first order of the perturbation parameter ε. The autocorrelation function of the response $y(t)$, computed to the first order of ε, is

$$\begin{aligned} R_{yy}(\theta) &= E\left[y(t) y(t+\theta)\right] = E\left[\left(y_0(t) + \varepsilon y_1(t)\right)\left(y_0(t+\theta) + \varepsilon y_1(t+\theta)\right)\right] \\ &\approx E\left[y_0(t) y_0(t+\theta)\right] + \varepsilon\left\{E\left[y_0(t) y_1(t+\theta)\right] + E\left[y_1(t) y_0(t+\theta)\right]\right\} \\ &= R_{y_0 y_0}(\theta) + \varepsilon\left\{R_{y_0 y_1}(\theta) + R_{y_1 y_0}(\theta)\right\} \end{aligned} \tag{8.77}$$

In equation (8.77), the autocorrelation function of the linear part of the response, denoted by $R_{y_0y_0}(\theta)$, can be computed as follows:

$$
\begin{aligned}
R_{y_0y_0}(\theta) &= E\left[y_0(t)y_0(t+\theta)\right] \\
&= \int_{-\infty}^{\infty}\int_{-\infty}^{\infty} E\left[F(t-\theta_1)F(t+\theta-\theta_2)\right]g(\theta_1)g(\theta_2)d\theta_1\,d\theta_2 \\
&= \int_{-\infty}^{\infty}\int_{-\infty}^{\infty} R_{FF}(\theta+\theta_1-\theta_2)g(\theta_1)g(\theta_2)d\theta_1\,d\theta_2
\end{aligned}
\tag{8.78}
$$

in which $R_{FF}(\theta+\theta_1-\theta_2)$ = input autocorrelation function.

By applying a formula similar to the one used for the autocorrelation function, the cross-correlation functions $R_{y_0y_1}(\theta)$ and $R_{y_1y_0}(\theta)$ in equation (8.77) can be calculated as

$$
\begin{aligned}
R_{y_0y_1}(\theta) &= E\left[y_0(t)y_1(t+\theta)\right] \\
&= -\int_{-\infty}^{\infty}\int_{-\infty}^{\infty} E\left[F(t-\theta_1)\eta(t+\theta-\theta_2)\right]g(\theta_1)g(\theta_2)d\theta_1\,d\theta_2
\end{aligned}
\tag{8.79}
$$

$$
\begin{aligned}
R_{y_1y_0}(\theta) &= E\left[y_1(t)y_0(t+\theta)\right] \\
&= -\int_{-\infty}^{\infty}\int_{-\infty}^{\infty} E\left[F(t+\theta-\theta_1)\eta(t-\theta_2)\right]g(\theta_1)g(\theta_2)d\theta_1\,d\theta_2
\end{aligned}
\tag{8.80}
$$

In equations (8.79) and (8.80), $\eta(t-\theta)$ represents the abbreviation of $\eta\left(y_0(t-\theta),\dot{y}_0(t-\theta)\right)$. Finally, the response autocorrelation function, denoted as $R_{yy}(\theta)$ can be determined using equations (8.77), (8.78), (8.79) and (8.80).

After obtaining the autocorrelation function, the response spectral density function $S_{yy}(\omega)$ can then be obtained by performing the Fourier transform of the autocorrelation function. It can be written as follows:

$$
S_{yy}(\omega) = S_{y_0y_0}(\omega) + \varepsilon\left\{S_{y_0y_1}(\omega) + S_{y_1y_0}(\omega)\right\} \tag{8.81}
$$

In this subsection, we will revisit a few characteristics of the cross-correlation function and the cross-spectral density function in order to further derive a more succinct expression for $S_{yy}(\omega)$, which is the spectral density of the response. According to classical random process theory, there exists the following relationship between the cross-correlation functions of two random processes, $x(t)$ and $y(t)$:

$$
R_{yx}(\theta) = R_{xy}(-\theta) \tag{8.82}
$$

For weakly stationary ergodic random processes $y_0(t)$ and $y_1(t)$, the cross-correlation function $R_{y_0y_1}(\theta)$ and the cross-spectral density function $S_{y_0y_1}(\omega)$ constitute a pair of Fourier transforms, that is

$$R_{y_0y_1}(\theta) = \frac{1}{2}\int_{-\infty}^{\infty} S_{y_0y_1}(\omega)e^{i\omega\theta}\,d\omega \tag{8.83}$$

$$S_{y_0y_1}(\omega) = \frac{1}{\pi}\int_{-\infty}^{\infty} R_{y_0y_1}(\theta)e^{-i\omega\theta}\,d\theta \tag{8.84}$$

It is important to take note that the cross-spectral density function is a complex-valued function, in contrast to the auto-spectral density function, which is a real-valued function. In the following step, we will conduct a detailed evaluation of the cross-spectral density function $S_{y_0y_1}(\omega)$ in order to demonstrate how the imaginary part of the function is incorporated into $S_{y_0y_1}(\omega)$:

$$\begin{aligned} S_{y_0y_1}(\omega) &= \frac{1}{\pi}\int_{-\infty}^{\infty} R_{y_0y_1}(\theta)e^{-i\omega\theta}\,d\theta \\ &= \frac{1}{\pi}\left\{\int_{-\infty}^{0} R_{y_0y_1}(\theta)e^{-i\omega\theta}\,d\theta + \int_{0}^{\infty} R_{y_0y_1}(\theta)e^{-i\omega\theta}\,d\theta\right\} \end{aligned} \tag{8.85}$$

Through the utilization of the relationship specified in equation (8.82), we are able to derive the following relationship:

$$\begin{aligned} \int_{-\infty}^{0} R_{y_0y_1}(\theta)e^{-i\omega\theta}\,d\theta &= \int_{-\infty}^{0} R_{y_1y_0}(-\theta)e^{-i\omega\theta}\,d\theta \\ &= \int_{0}^{\infty} R_{y_1y_0}(-\theta)e^{i\omega(-\theta)}d(-\theta) = \int_{0}^{\infty} R_{y_1y_0}(\theta)e^{i\omega\theta}\,d\theta \end{aligned} \tag{8.86}$$

Consequently, equation (8.85) can be re-written as follows:

$$\begin{aligned} S_{y_0y_1}(\omega) &= \frac{1}{\pi}\left\{\int_{-\infty}^{0} R_{y_0y_1}(\theta)e^{-i\omega\theta}d\theta + \int_{0}^{\infty} R_{y_0y_1}(\theta)e^{-i\omega\theta}d\theta\right\} \\ &= \frac{1}{\pi}\left\{\int_{0}^{\infty} R_{y_1y_0}(\theta)e^{i\omega\theta}d\theta + \int_{0}^{\infty} R_{y_0y_1}(\theta)e^{-i\omega\theta}\,d\theta\right\} \end{aligned}$$

$$
\begin{aligned}
&= \frac{1}{\pi}\int_0^\infty \left\{R_{y_1y_0}(\theta)+R_{y_0y_1}(\theta)\right\}\cos\omega\theta d\theta \\
&\quad + i\frac{1}{\pi}\int_0^\infty \left\{R_{y_1y_0}(\theta)-R_{y_0y_1}(\theta)\right\}\sin\omega\theta d\theta \\
&= \mathrm{Re}\left[S_{y_0y_1}(\omega)\right]+i\,\mathrm{Im}\,g\left[S_{y_0y_1}(\omega)\right]
\end{aligned}
\tag{8.87}
$$

in which

$$
\mathrm{Re}\left[S_{y_0y_1}(\omega)\right]=\frac{1}{\pi}\int_0^\infty \left\{R_{y_1y_0}(\theta)+R_{y_0y_1}(\theta)\right\}\cos\omega\theta d\theta \tag{8.88}
$$

$$
\mathrm{Im}\,g\left[S_{y_0y_1}(\omega)\right]=\frac{1}{\pi}\int_0^\infty \left\{R_{y_1y_0}(\theta)-R_{y_0y_1}(\theta)\right\}\sin\omega\theta d\theta \tag{8.89}
$$

In a similar manner, $S_{y_1y_0}(\omega)$ in equation (8.81) can be calculated as:

$$
\begin{aligned}
S_{y_1y_0}(\omega) &= \frac{1}{\pi}\left\{\int_{-\infty}^0 R_{y_1y_0}(\theta)e^{-i\omega\theta}d\theta+\int_0^\infty R_{y_1y_0}(\theta)e^{-i\omega\theta}d\theta\right\} \\
&= \frac{1}{\pi}\left\{\int_0^\infty R_{y_0y_1}(\theta)e^{i\omega\theta}d\theta+\int_0^\infty R_{y_1y_0}(\theta)e^{-i\omega\theta}d\theta\right\} \\
&= \frac{1}{\pi}\int_0^\infty \left\{R_{y_1y_0}(\theta)+R_{y_0y_1}(\theta)\right\}\cos\omega\theta d\theta \\
&\quad + i\frac{1}{\pi}\int_0^\infty \left\{R_{y_0y_1}(\theta)-R_{y_1y_0}(\theta)\right\}\sin\omega\theta d\theta \\
&= \mathrm{Re}\left[S_{y_1y_0}(\omega)\right]+i\,\mathrm{Im}\,g\left[S_{y_1y_0}(\omega)\right]
\end{aligned}
\tag{8.90}
$$

By thorough examination of equations (8.87) and (8.90), we can determine:

$$
\mathrm{Re}\left[S_{y_0y_1}(\omega)\right]=\mathrm{Re}\left[S_{y_1y_0}(\omega)\right] \tag{8.91}
$$

$$
\mathrm{Im}\,g\left[S_{y_0y_1}(\omega)\right]=-\mathrm{Im}\,g\left[S_{y_1y_0}(\omega)\right] \tag{8.92}
$$

Combining equations (8.81), (8.87), (8.90), (8.91) and (8.92) leads to:

$$
S_{yy}(\omega)=S_{y_0y_0}(\omega)+2\varepsilon\ \mathrm{Re}\left[S_{y_0y_1}(\omega)\right] \tag{8.93}
$$

a. A special case with only a nonlinear stiffness term
As an implementation of the perturbation technique described above, we will consider replacing the nonlinear term $\eta(y, \dot{y})$ in equation (8.66) with y^3. Therefore, equation (8.66) can be re-written as

$$\ddot{y}+\alpha\dot{y}+\omega_n{}^2 y+\varepsilon y^3 = F(t) \tag{8.94}$$

The aforementioned equation (8.94) can sometimes be used to approximate the rolling motion of a ship in random seas in the field of Ocean Engineering. In the subsequent analysis, we make the assumption that $F(t)$ in equation (8.94) is a steady state Gaussian random process with a mean value of zero. In addition, we define that the spectral density function of $F(t)$ is denoted by $S_{FF}(\omega)$. Given that the frequency response function $H(\omega)$ and the impulse response function $g(\theta)$ of the linear system ($\varepsilon = 0$) are both known, we are able to derive the linear system's response autocorrelation function $R_{y_0 y_0}(\theta)$ by using the following equation:

$$R_{y_0 y_0}(\theta) = \frac{1}{2}\int_{-\infty}^{\infty} S_{y_0 y_0}(\omega)e^{i\omega\theta}\, d\omega = \frac{1}{2}\int_{-\infty}^{\infty} S_{FF}(\omega)\left|H(\omega)\right|^2 e^{i\omega\theta}\, d\omega \tag{8.95}$$

in which

$$H(\omega) = \int_{-\infty}^{\infty} g(\theta)e^{-i\omega\theta}\, d\theta \tag{8.96}$$

Clearly, the response spectral density of the linear system can be calculated by using the following formula:

$$S_{y_0 y_0}(\omega) = S_{FF}(\omega)\left|H(\omega)\right|^2 \tag{8.97}$$

In this particular instance, the cross-correlation function $R_{y_0 y_1}(\theta)$ can be computed as follows:

$$\begin{aligned} R_{y_0 y_1}(\theta) &= E\left[y_0(t)\, y_1(t+\theta)\right] \\ &= -\int_{-\infty}^{\infty}\int_{-\infty}^{\infty} E\left[F(t-\theta_1)\eta(t+\theta-\theta_2)\right] g(\theta_1)g(\theta_2)d\theta_1\, d\theta_2 \\ &= -\int_{-\infty}^{\infty}\int_{-\infty}^{\infty} E\left[F(t-\theta_1)\, y_0^3(t+\theta-\theta_2)\right] g(\theta_1)g(\theta_2)d\theta_1\, d\theta_2 \end{aligned} \tag{8.98}$$

Since we have made the assumption that $F(t)$ in equation (8.94) is a steady-state Gaussian random process with a mean value of zero, then the linear system response $y_0(t)$ will also be a Gaussian random process with a mean value of zero. According to the classical random process theory, for a Gaussian random process with a mean value of zero, we have the following relationship:

$$\begin{aligned} &E\left[F(t-\theta_1)y_0^3(t+\theta-\theta_2)\right] \\ &= E\left[F(t-\theta_1)y_0(t+\theta-\theta_2)\right]E\left[y_0(t+\theta-\theta_2)y_0(t+\theta-\theta_2)\right] \\ &\quad + E\left[y_0(t+\theta-\theta_2)y_0(t+\theta-\theta_2)\right]E\left[F(t-\theta_1)y_0(t+\theta-\theta_2)\right] \\ &\quad + E\left[F(t-\theta_1)y_0(t+\theta-\theta_2)\right]E\left[y_0(t+\theta-\theta_2)y_0(t+\theta-\theta_2)\right] \\ &= 3E\left[F(t-\theta_1)y_0(t+\theta-\theta_2)\right]E\left[y_0(t+\theta-\theta_2)y_0(t+\theta-\theta_2)\right] \end{aligned} \tag{8.99}$$

Furthermore, by utilizing equation (8.72), we can deduce

$$\begin{aligned} &E\left[F(t-\theta_1)y_0(t+\theta-\theta_2)\right] \\ &= \int_{-\infty}^{\infty} E\left[F(t-\theta_1)F(t+\theta-\theta_2-\theta_3)\right]g(\theta_3)d\theta_3 \\ &= \int_{-\infty}^{\infty} R_{FF}(\theta+\theta_1-\theta_2-\theta_3)g(\theta_3)d\theta_3 \end{aligned} \tag{8.100}$$

and

$$E\left[y_0(t+\theta-\theta_2)y_0(t+\theta-\theta_2)\right] = \sigma_{y0}^2 \tag{8.101}$$

in which σ_{y0}^2 is the variance of the linear system's response. Combing equations (8.99), (8.100) and (8.101) leads to

$$\begin{aligned} &E\left[F(t-\theta_1)y_0^3(t+\theta-\theta_2)\right] \\ &= 3\sigma_{y0}^2 \int_{-\infty}^{\infty} R_{FF}(\theta+\theta_1-\theta_2-\theta_3)g(\theta_3)d\theta_3 \\ &= \frac{3}{2}\sigma_{y0}^2 \int_{-\infty}^{\infty}\int_{-\infty}^{\infty} S_{FF}(\omega)e^{i\omega(\theta+\theta_1-\theta_2-\theta_3)}\, g(\theta_3)d\theta_3\, d\omega \end{aligned} \tag{8.102}$$

The cross-correlation function, denoted as $R_{y0y1}(\theta)$ in equation (8.98), can then be computed as

$$\begin{aligned} &R_{y0y1}(\theta) \\ &= -\int_{-\infty}^{\infty}\int_{-\infty}^{\infty} E\left[F(t-\theta_1)y_0^3(t+\theta-\theta_2)\right]g(\theta_1)g(\theta_2)d\theta_1 d\theta_2 \\ &= -\frac{3}{2}\sigma_{y0}^2 \int_{-\infty}^{\infty}\int_{-\infty}^{\infty}\int_{-\infty}^{\infty}\int_{-\infty}^{\infty} S_{FF}(\omega)e^{i\omega(\theta+\theta_1-\theta_2-\theta_3)}\, g(\theta_1)g(\theta_2)g(\theta_3)d\theta_1 d\theta_2 d\theta_3 d\omega \end{aligned}$$

$$
\begin{aligned}
&= -\frac{3}{2}\sigma_{y0}{}^2 \int_{-\infty}^{\infty} S_{FF}(\omega) \int_{-\infty}^{\infty} g(\theta_2)e^{-i\omega\theta_2}\, d\theta_2 \int_{-\infty}^{\infty} g(\theta_3)e^{-i\omega\theta_3}\, d\theta_3 \\
&\quad \int_{-\infty}^{\infty} g(\theta_1)e^{-i(-\omega)\theta_1}\, d\theta_1 e^{i\omega\theta}\, d\omega \\
&= -\frac{3}{2}\sigma_{y0}{}^2 \int_{-\infty}^{\infty} S_{FF}(\omega)H(\omega)H(\omega)H(-\omega)e^{i\omega\theta}\, d\omega \\
&= -\frac{3}{2}\sigma_{y0}{}^2 \int_{-\infty}^{\infty} S_{FF}(\omega)H(\omega)H(-\omega)H(\omega)e^{i\omega\theta}\, d\omega \\
&= -\frac{3}{2}\sigma_{y0}{}^2 \int_{-\infty}^{\infty} S_{FF}(\omega)\left|H(\omega)\right|^2 H(\omega)e^{i\omega\theta}\, d\omega
\end{aligned}
\tag{8.103}
$$

Next, we can derive the cross-spectral density function $S_{y0y1}(\omega)$ in equation (8.93) as

$$
S_{y0y1}(\omega) = -3\sigma_{y0}{}^2 S_{FF}(\omega)\left|H(\omega)\right|^2 H(\omega) \tag{8.104}
$$

Combining equations (8.93), (8.97) and (8.104) leads to:

$$
\begin{aligned}
S_{yy}(\omega) &= S_{FF}(\omega)\left|H(\omega)\right|^2 - 6\varepsilon\sigma_{y0}{}^2 S_{FF}(\omega)\left|H(\omega)\right|^2 \mathrm{Re}\left[H(\omega)\right] \\
&= S_{FF}(\omega)\left|H(\omega)\right|^2 \left\{1 - 6\varepsilon\sigma_{y0}{}^2 \mathrm{Re}\left[H(\omega)\right]\right\}
\end{aligned}
\tag{8.105}
$$

In this case, we have $\varepsilon = \omega_n{}^2\lambda$ for the Duffing system, as shown in equation (8.39). Therefore, equation (8.105) agrees with the outcome derived in equation (8.56) using the equivalent linearization technique because we have the following relationships:

$$
\begin{aligned}
H(\omega) &= \frac{1}{\left(\omega_n{}^2 - \omega^2\right) + i\omega\alpha} \\
&= \frac{\left(\omega_n{}^2 - \omega^2\right) - i\omega\alpha}{\left\{\left(\omega_n{}^2 - \omega^2\right) + i\omega\alpha\right\}\left\{\left(\omega_n{}^2 - \omega^2\right) - i\omega\alpha\right\}} \\
&= \frac{\left(\omega_n{}^2 - \omega^2\right) - i\omega\alpha}{\left(\omega_n{}^2 - \omega^2\right)^2 + (\omega\alpha)^2}
\end{aligned}
\tag{8.106}
$$

$$
\mathrm{Re}\left[H(\omega)\right] = \frac{\left(\omega_n{}^2 - \omega^2\right)}{\left(\omega_n{}^2 - \omega^2\right)^2 + (\omega\alpha)^2} \tag{8.107}
$$

$$|H(\omega)| = \sqrt{\frac{1}{\left(\omega_n^2 - \omega^2\right)^2 + (\alpha\omega)^2}} \tag{8.108}$$

$$|H(\omega)|^2 = \frac{1}{\left(\omega_n^2 - \omega^2\right)^2 + (\alpha\omega)^2} \tag{8.109}$$

$$\left(\omega_n^2 - \omega^2\right)|H(\omega)|^2 = \frac{\left(\omega_n^2 - \omega^2\right)}{\left(\omega_n^2 - \omega^2\right)^2 + (\alpha\omega)^2} = \text{Re}\left[H(\omega)\right] \tag{8.110}$$

Let us now proceed to assess the variance of the response of the Duffing system that is taken into consideration in equation (8.94). The cross-correlation function $R_{y_1 y_0}(\theta)$ can be calculated by replacing θ by $-\theta$ in the equation (8.103), i.e.,

$$\begin{aligned} R_{y_1 y_0}(\theta) &= R_{y_0 y_1}(-\theta) \\ &= -\int_{-\infty}^{\infty}\int_{-\infty}^{\infty} E\left[F(t-\theta_1)y_0^3(t-\theta-\theta_2)\right]g(\theta_1)g(\theta_2)d\theta_1\,d\theta_2 \end{aligned} \tag{8.111}$$

Because the linear system response $y_0(t)$ is a Gaussian random process with a zero mean, we have the following relationship according to the classical random process theory:

$$\begin{aligned} &E\left[F(t-\theta_1)y_0^3(t-\theta-\theta_2)\right] \\ &= E\left[F(t-\theta_1)y_0(t-\theta-\theta_2)\right]E\left[y_0(t-\theta-\theta_2)y_0(t-\theta-\theta_2)\right] \\ &\quad + E\left[y_0(t-\theta-\theta_2)y_0(t-\theta-\theta_2)\right]E\left[F(t-\theta_1)y_0(t-\theta-\theta_2)\right] \\ &\quad + E\left[F(t-\theta_1)y_0(t-\theta-\theta_2)\right]E\left[y_0(t-\theta-\theta_2)y_0(t-\theta-\theta_2)\right] \\ &= 3E\left[F(t-\theta_1)y_0(t-\theta-\theta_2)\right]E\left[y_0(t-\theta-\theta_2)y_0(t-\theta-\theta_2)\right] \end{aligned} \tag{8.112}$$

In addition, we have the following relationship

$$\begin{aligned} &E\left[F(t-\theta_1)y_0(t-\theta-\theta_2)\right] \\ &= \int_{-\infty}^{\infty} E\left[F(t-\theta_1)F(t-\theta-\theta_2-\theta_3)\right]g(\theta_3)d\theta_3 \\ &= \int_{-\infty}^{\infty} R_{FF}(-\theta+\theta_1-\theta_2-\theta_3)g(\theta_3)d\theta_3 \end{aligned} \tag{8.113}$$

and

$$E\left[y_0(t-\theta-\theta_2)y_0(t-\theta-\theta_2)\right]=\sigma_{y0}{}^2 \tag{8.114}$$

where $\sigma_{y0}{}^2$ is the linear system's response variance. Combing equations (8.112), (8.113) and (8.114) leads to

$$\begin{aligned}&E\left[F(t-\theta_1)y_0{}^3(t-\theta-\theta_2)\right]\\&=3\sigma_{y0}{}^2\int_{-\infty}^{\infty}R_{FF}(-\theta+\theta_1-\theta_2-\theta_3)g(\theta_3)d\theta_3\\&=\frac{3}{2}\sigma_{y0}{}^2\int_{-\infty}^{\infty}\int_{-\infty}^{\infty}S_{FF}(\omega)e^{i\omega(-\theta+\theta_1-\theta_2-\theta_3)}g(\theta_3)d\theta_3\,d\omega\end{aligned} \tag{8.115}$$

The cross-correlation function $R_{y1y0}(\theta)$ in equation (8.111), can thus be calculated as

$$\begin{aligned}&R_{y1y0}(\theta)\\&=-\int_{-\infty}^{\infty}\int_{-\infty}^{\infty}E\left[F(t-\theta_1)y_0{}^3(t-\theta-\theta_2)\right]g(\theta_1)g(\theta_2)d\theta_1\,d\theta_2\\&=-\frac{3}{2}\sigma_{y0}{}^2\int_{-\infty}^{\infty}\int_{-\infty}^{\infty}\int_{-\infty}^{\infty}\int_{-\infty}^{\infty}S_{FF}(\omega)e^{i\omega(-\theta+\theta_1-\theta_2-\theta_3)}g(\theta_1)\,g(\theta_2)g(\theta_3)d\theta_1\,d\theta_2 d\theta_3 d\omega\\&=-\frac{3}{2}\sigma_{y0}{}^2\int_{-\infty}^{\infty}S_{FF}(\omega)\int_{-\infty}^{\infty}g(\theta_2)e^{-i\omega\theta_2}\,d\theta_2\int_{-\infty}^{\infty}g(\theta_3)e^{-i\omega\theta_3}\,d\theta_3\\&\int_{-\infty}^{\infty}g(\theta_1)e^{-i(-\omega)\theta_1}\,d\theta_1 e^{-i\omega\theta}d\omega\\&=-\frac{3}{2}\sigma_{y0}{}^2\int_{-\infty}^{\infty}S_{FF}(\omega)H(\omega)H(\omega)H(-\omega)e^{-i\omega\theta}d\omega\\&=-\frac{3}{2}\sigma_{y0}{}^2\int_{-\infty}^{\infty}S_{FF}(\omega)H(\omega)H(-\omega)H(\omega)e^{-i\omega\theta}\,d\omega\\&=-\frac{3}{2}\sigma_{y0}{}^2\int_{-\infty}^{\infty}S_{FF}(\omega)\left|H(\omega)\right|^2 H(\omega)e^{-i\omega\theta}\,d\omega\end{aligned} \tag{8.116}$$

As a result, the autocorrelation function of the nonlinear response, denoted by $R_{yy}(\theta)$, can be determined by utilizing equations (8.77), (8.95), (8.103) and (8.116) as follows:

$$\begin{aligned}
&R_{yy}(\theta)\\
&= R_{y_0y_0}(\theta)+\varepsilon\left\{R_{y_0y_1}(\theta)+R_{y_1y_0}(\theta)\right\}\\
&=\frac{1}{2}\int_{-\infty}^{\infty} S_{FF}(\omega)\left|H(\omega)\right|^2 e^{i\omega\theta}d\omega-\varepsilon\frac{3}{2}\sigma_{y_0}{}^2\left\{\int_{-\infty}^{\infty} S_{FF}(\omega)\left|H(\omega)\right|^2 H(\omega)e^{i\omega\theta}\,d\omega\right.\\
&\left.+\int_{-\infty}^{\infty} S_{FF}(\omega)\left|H(\omega)\right|^2 H(\omega)e^{-i\omega\theta}\,d\omega\right\}\\
&=\frac{1}{2}\int_{-\infty}^{\infty} S_{FF}(\omega)\left|H(\omega)\right|^2 e^{i\omega\theta}d\omega-3\varepsilon\sigma_{y_0}{}^2\left\{\int_{-\infty}^{\infty} S_{FF}(\omega)\left|H(\omega)\right|^2 H(\omega)\cos\omega\theta\,d\omega\right\}
\end{aligned} \tag{8.117}$$

The variance of the response of the Duffing system can be calculated from equation (8.117) by setting $\theta = 0$, i.e.,

$$\sigma_y{}^2=\frac{1}{2}\int_{-\infty}^{\infty} S_{FF}(\omega)\left|H(\omega)\right|^2 d\omega-3\varepsilon\sigma_{y_0}{}^2\left\{\int_{-\infty}^{\infty} S_{FF}(\omega)\left|H(\omega)\right|^2 H(\omega)d\omega\right\} \tag{8.118}$$

In spite of the fact that the variance may be calculated from equation (8.118), it might be an appropriate choice to demonstrate the computation technique directly from equation (8.67). Due to the fact that the mean response, denoted by $E[y]$, is equal to zero, the varince is equal to the second moment, denoted by $E[y^2]$, which may be approximately given from equation (8.67) as follows:

$$E\left[y^2\right]\approx E\left[y_0^2\right]+2\varepsilon E\left[y_0y_1\right] \tag{8.119}$$

The equation (8.78) can be used to obtain the value of the first term. Specifically, by letting $\theta = 0$, we can derive

$$\begin{aligned}
E\left[y_0^2\right]&=R_{y_0y_0}(0)=E\left[y_0(t)\,y_0(t+0)\right]\\
&=\int_{-\infty}^{\infty}\int_{-\infty}^{\infty} E\left[F(t-\theta_1)F(t+0-\theta_2)\right]g(\theta_1)g(\theta_2)d\theta_1\,d\theta_2\\
&=\int_{-\infty}^{\infty}\int_{-\infty}^{\infty} R_{FF}(\theta_1-\theta_2)g(\theta_1)g(\theta_2)d\theta_1\,d\theta_2
\end{aligned} \tag{8.120}$$

For the Duffing system, the expression $E\left[y_0 y_1\right]$ in the second term of equation (8.119) can be represented as follows, derived from equation (8.73):

$$E\left[y_0 y_1\right] = -\int_{-\infty}^{\infty} g(\theta) E\left[y_0(t) y_0^3(t-\theta)\right] d\theta \tag{8.121}$$

Moreover, by utilizing equation (8.72), the abovementioned equation (8.121) can be expressed as

$$\begin{aligned} E\left[y_0 y_1\right] &= -\int_{-\infty}^{\infty} g(\theta) E\left[y_0(t) y_0^3(t-\theta)\right] d\theta \\ &= -\int_{-\infty}^{\infty} g(\theta) d\theta \int_{-\infty}^{\infty}\int_{-\infty}^{\infty}\int_{-\infty}^{\infty}\int_{-\infty}^{\infty} g(\theta_1) g(\theta_2) g(\theta_3) g(\theta_4) \times E\left[F(t-\theta_1)\right. \\ &\quad \left.\times F(t-\theta-\theta_2) F(t-\theta-\theta_3) F(t-\theta-\theta_4)\right] d\theta_1 d\theta_2 d\theta_3 d\theta_4 \end{aligned} \tag{8.122}$$

In light of the fact that we have assumed that $F(t)$ in equation (8.122) represents a steady-state Gaussian random process with a mean value of zero, then, according to the classical random process theory, the expected value that is involved in equation (8.122) can be written as follows:

$$\begin{aligned} & E\left[F(t-\theta_1) F(t-\theta-\theta_2) F(t-\theta-\theta_3) F(t-\theta-\theta_4)\right] \\ &\quad = E\left[F(t-\theta_1) F(t-\theta-\theta_2)\right] E\left[F(t-\theta-\theta_3) F(t-\theta-\theta_4)\right] \\ &\quad\quad + E\left[F(t-\theta_1) F(t-\theta-\theta_4)\right] E\left[F(t-\theta-\theta_3) F(t-\theta-\theta_2)\right] \\ &\quad\quad + E\left[F(t-\theta_1) F(t-\theta-\theta_3)\right] E\left[F(t-\theta-\theta_4) F(t-\theta-\theta_2)\right] \\ &\quad = R_{FF}(\theta-\theta_1+\theta_2) R_{FF}(\theta_4-\theta_3) + R_{FF}(\theta-\theta_1+\theta_4) R_{FF}(\theta_2-\theta_3) \\ &\quad\quad + R_{FF}(\theta-\theta_1+\theta_3) R_{FF}(\theta_2-\theta_4) \end{aligned} \tag{8.123}$$

Therefore, we can write

$$\begin{aligned} E\left[y_0 y_1\right] &= -\int_{-\infty}^{\infty} g(\theta) d\theta \int_{-\infty}^{\infty}\int_{-\infty}^{\infty}\int_{-\infty}^{\infty}\int_{-\infty}^{\infty} g(\theta_1) g(\theta_2) g(\theta_3) g(\theta_4) \\ &\quad \times E\left[F(t-\theta_1) F(t-\theta-\theta_2) F(t-\theta-\theta_3) F(t-\theta-\theta_4)\right] d\theta_1 d\theta_2 d\theta_3 d\theta_4 \\ &= -\int_{-\infty}^{\infty} g(\theta) d\theta \int_{-\infty}^{\infty}\int_{-\infty}^{\infty}\int_{-\infty}^{\infty}\int_{-\infty}^{\infty} g(\theta_1) g(\theta_2) g(\theta_3) g(\theta_4) \\ &\quad \times \left[R_{FF}(\theta-\theta_1+\theta_2) R_{FF}(\theta_4-\theta_3) + R_{FF}(\theta-\theta_1+\theta_4) R_{FF}(\theta_2-\theta_3)\right. \\ &\quad \left. + R_{FF}(\theta-\theta_1+\theta_3) R_{FF}(\theta_2-\theta_4)\right] d\theta_1 d\theta_2 d\theta_3 d\theta_4 \end{aligned} \tag{8.124a}$$

Equation (8.124a) can be further written as

$$E[y_0 y_1] = -\int_{-\infty}^{\infty} g(\theta)d\theta \int_{-\infty}^{\infty}\int_{-\infty}^{\infty}\int_{-\infty}^{\infty}\int_{-\infty}^{\infty} g(\theta_1)g(\theta_2)g(\theta_3)g(\theta_4)$$
$$\times\left[R_{FF}(\theta-\theta_1+\theta_2)R_{FF}(\theta_4-\theta_3)+R_{FF}(\theta-\theta_1+\theta_4)R_{FF}(\theta_2-\theta_3)\right.$$
$$\left.+R_{FF}(\theta-\theta_1+\theta_3)R_{FF}(\theta_2-\theta_4)\right]d\theta_1 d\theta_2 d\theta_3 d\theta_4$$
$$= -\int_{-\infty}^{\infty} g(\theta)d\theta \int_{-\infty}^{\infty}\int_{-\infty}^{\infty} R_{FF}(\theta_4-\theta_3)g(\theta_3)g(\theta_4)d\theta_3\, d\theta_4$$
$$\times \int_{-\infty}^{\infty}\int_{-\infty}^{\infty}\left[R_{FF}(\theta-\theta_1+\theta_2)\right]g(\theta_1)g(\theta_2)d\theta_1 d\theta_2$$
$$-\int_{-\infty}^{\infty} g(\theta)d\theta \int_{-\infty}^{\infty}\int_{-\infty}^{\infty} R_{FF}(\theta_2-\theta_3)g(\theta_2)g(\theta_3)d\theta_2\, d\theta_3$$
$$\times \int_{-\infty}^{\infty}\int_{-\infty}^{\infty}\left[R_{FF}(\theta-\theta_1+\theta_4)\right]g(\theta_1)g(\theta_4)d\theta_1 d\theta_4$$
$$-\int_{-\infty}^{\infty} g(\theta)d\theta \int_{-\infty}^{\infty}\int_{-\infty}^{\infty} R_{FF}(\theta_2-\theta_4)g(\theta_2)g(\theta_4)d\theta_2\, d\theta_4$$
$$\times \int_{-\infty}^{\infty}\int_{-\infty}^{\infty}\left[R_{FF}(\theta-\theta_1+\theta_3)\right]g(\theta_1)g(\theta_3)d\theta_1 d\theta_3 \qquad (8.124b)$$

On the other hand, we can deduce from equation (8.78) that

$$R_{y_0 y_0}(0) = \int_{-\infty}^{\infty}\int_{-\infty}^{\infty} R_{FF}(\theta_1-\theta_2)g(\theta_1)g(\theta_2)d\theta_1\, d\theta_2 \qquad (8.125)$$

Through the utilization of the equation presented above, equation (8.125) can be written as

$$E[y_0 y_1] = -\int_{-\infty}^{\infty} g(\theta)d\theta R_{y_0 y_0}(0)\int_{-\infty}^{\infty}\int_{-\infty}^{\infty}\left[R_{FF}(\theta-\theta_1+\theta_2)\right]g(\theta_1)g(\theta_2)d\theta_1 d\theta_2$$
$$-\int_{-\infty}^{\infty} g(\theta)d\theta R_{y_0 y_0}(0)\int_{-\infty}^{\infty}\int_{-\infty}^{\infty}\left[R_{FF}(\theta-\theta_1+\theta_4)\right]g(\theta_1)g(\theta_4)d\theta_1 d\theta_4$$
$$-\int_{-\infty}^{\infty} g(\theta)d\theta R_{y_0 y_0}(0)\int_{-\infty}^{\infty}\int_{-\infty}^{\infty}\left[R_{FF}(\theta-\theta_1+\theta_3)\right]g(\theta_1)g(\theta_3)d\theta_1 d\theta_3$$
$$(8.126)$$

Through the utilization of the equation presented above in conjunction with equation (8.78), the expression $E\left[y_0 y_1\right]$ can be expressed as

$$E\left[y_0 y_1\right] = -3R_{y0y0}(0)\int_{-\infty}^{\infty} R_{y0y0}(\theta)g(\theta)d\theta \tag{8.127}$$

Therefore, by substituting equations (8.120) and (8.127) into equation (8.119), the second moment, which is equivalent to the variance for zero mean, can be evaluated using the formula that is presented below:

$$E\left[y^2\right] = R_{y0y0}(0)\left[1 - 6\varepsilon\int_{-\infty}^{\infty} R_{y0y0}(\theta)g(\theta)d\theta\right] \tag{8.128}$$

b. A generalized case

As another implementation of the perturbation technique described in this chapter, we next consider replacing the nonlinear term $\eta\left(y,\ \dot{y}\right)$ in equation (8.66) with y^n. By doing this, equation (8.66) can be re-written as

$$\ddot{y} + \alpha\dot{y} + \omega_n{}^2 y + \varepsilon y^n = F(t) \tag{8.129}$$

in which n is an odd integer. This equation represents a vibration system with a small stiffness nonlinearity. In the following analysis, we also assume that $F(t)$ in equation (8.129) is a stationary Gaussian random process with a zero mean. In addition, we define that the spectral density of the excitation force $F(t)$ is $S_{FF}(\omega)$.

First, let us calculate $R_{y0y0}(\theta)$; the calculation is identical to that for a linear system. Because the frequency response function $H(\omega)$ and the impulse response function $g(\theta)$ of the linear system ($\varepsilon = 0$) are both known, we can calculate the linear system's response autocorrelation function $R_{y0y0}(\theta)$ as follows:

$$R_{y0y0}(\theta) = \frac{1}{2}\int_{-\infty}^{\infty} S_{y0y0}(\omega)e^{i\omega\theta}\,d\omega = \frac{1}{2}\int_{-\infty}^{\infty} S_{FF}(\omega)\left|H(\omega)\right|^2 e^{i\omega\theta}\,d\omega \tag{8.130}$$

where

$$H(\omega) = \int_{-\infty}^{\infty} g(\theta)e^{-i\omega\theta}\,d\theta \tag{8.131}$$

It is evident that the response spectral density of the linear system can be determined by employing the formula that is presented below:

$$S_{y0y0}(\omega) = S_{FF}(\omega)\left|H(\omega)\right|^2 \tag{8.132}$$

To compute the cross-correlation function $R_{y0y1}(\theta)$ in this generalized case, the following formula can be used:

$$\begin{aligned} R_{y0y1}(\theta) &= E\left[y_0(t)y_1(t+\theta)\right] \\ &= -\int_{-\infty}^{\infty}\int_{-\infty}^{\infty} E\left[F(t-\theta_1)\eta(t+\theta-\theta_2)\right]g(\theta_1)g(\theta_2)d\theta_1\,d\theta_2 \\ &= -\int_{-\infty}^{\infty}\int_{-\infty}^{\infty} E\left[F(t-\theta_1)y_0{}^n(t+\theta-\theta_2)\right]g(\theta_1)g(\theta_2)d\theta_1\,d\theta_2 \end{aligned} \tag{8.133}$$

Considering that we have assumed that the function $F(t)$ in equation (8.129) is a stationary Gaussian random process with a mean value of zero, it follows that the response of the linear system, denoted by $y_0(t)$, will also be a stationary Gaussian random process with a zero mean value. According to the classical theory of random processes, the following relationship is true for a Gaussian random process that has a mean value of zero:

$$\begin{aligned} &E\left[F(t-\theta_1)y_0{}^n(t+\theta-\theta_2)\right] \\ &\quad = mE\left[F(t-\theta_1)y_0(t+\theta-\theta_2)\right]E\left[y_0{}^{n-1}(t+\theta-\theta_2)\right] \end{aligned} \tag{8.134}$$

m, which is located on the right-hand side of equation (8.134), represents the number of different ways in which we can pick a $y_0(t+\theta-\theta_2)$ from the product $y_0{}^{n-1}(t+\theta-\theta_2)$, which means that $m = n$ in this case. Considering that $y_0(t)$ is weakly stationary and since $E\left[y_0(t)\right] = 0$,

$$E\left[y_0{}^{n-1}(t)\right] = \sigma_{y0}{}^{n-1}(n-2)(n-4)\cdots 5\cdot 3\cdot 1 \tag{8.135}$$

At this point, we are reminded that n is an odd integer. More concisely,

$$E\left[y_0{}^{n-1}(t)\right] = \frac{\left(\sqrt{2}\sigma_{y0}\right)^{n-1}}{\sqrt{\pi}}\Gamma\left(\frac{n}{2}\right) \tag{8.136}$$

in which $\Gamma(\cdot)$ denotes the Gamma function and use has been made of the recurrence relation $\Gamma(\nu)=(\nu-1)\Gamma(\nu-1)$ and $\Gamma\left(\frac{1}{2}\right)=\sqrt{\pi}$. Moreover, we have

$$\begin{aligned}
&E\left[F(t-\theta_1)y_0(t+\theta-\theta_2)\right] \\
&=\int_{-\infty}^{\infty} E\left[F(t-\theta_1)F(t+\theta-\theta_2-\theta_3)\right]g(\theta_3)d\theta_3 \\
&=\int_{-\infty}^{\infty} R_{FF}(\theta+\theta_1-\theta_2-\theta_3)g(\theta_3)d\theta_3
\end{aligned} \tag{8.137}$$

and

$$E\left[y_0(t+\theta-\theta_2)y_0(t+\theta-\theta_2)\right]=\sigma_{y0}{}^2 \tag{8.138}$$

in which $\sigma_{y0}{}^2$ is the variance of the linear system's response. Combing equations (8.134) and (8.137) yields

$$\begin{aligned}
&E\left[F\left(t-\theta_1\right)y_0^n\left(t+\theta-\theta_2\right)\right] \\
&=mE\left[F(t-\theta_1)y_0(t+\theta-\theta_2)\right]E\left[y_0^{n-1}(t+\theta-\theta_2)\right] \\
&=\sqrt{\left(\frac{2^{n+1}}{\pi}\right)}\sigma_{y0}{}^{n-1}\Gamma\left(\frac{n+2}{2}\right)\int_{-\infty}^{\infty} R_{FF}(\theta+\theta_1-\theta_2-\theta_3)g(\theta_3)d\theta_3 \\
&=\frac{1}{2}\sqrt{\left(\frac{2^{n+1}}{\pi}\right)}\sigma_{y0}{}^{n-1}\Gamma\left(\frac{n+2}{2}\right)\int_{-\infty}^{\infty}\int_{-\infty}^{\infty} S_{FF}(\omega)\,e^{i\omega(\theta+\theta_1-\theta_2-\theta_3)}\,g(\theta_3)d\theta_3 d\omega
\end{aligned} \tag{8.139}$$

The cross-correlation function, which is represented by $R_{y0y1}(\theta)$ in equation (8.133), can then be determined as follows:

$$\begin{aligned}
&R_{y0y1}(\theta) \\
&=-\int_{-\infty}^{\infty}\int_{-\infty}^{\infty} E\left[F(t-\theta_1)y_0^n(t+\theta-\theta_2)\right]g(\theta_1)g(\theta_2)d\theta_1 d\theta_2 \\
&=-\frac{1}{2}\sqrt{\left(\frac{2^{n+1}}{\pi}\right)}\sigma_{y0}{}^{n-1}\Gamma\left(\frac{n+2}{2}\right) \\
&\int_{-\infty}^{\infty}\int_{-\infty}^{\infty}\int_{-\infty}^{\infty}\int_{-\infty}^{\infty} S_{FF}(\omega)e^{i\omega(\theta+\theta_1-\theta_2-\theta_3)}\,g(\theta_1)g(\theta_2)g(\theta_3)d\theta_1 d\theta_2 d\theta_3 d\omega
\end{aligned}$$

$$=-\frac{1}{2}\sqrt{\left(\frac{2^{n+1}}{\pi}\right)}\sigma_{y0}{}^{n-1}\Gamma\left(\frac{n+2}{2}\right)$$

$$\int_{-\infty}^{\infty} S_{FF}(\omega)\int_{-\infty}^{\infty} g(\theta_2)e^{-i\omega\theta_2}\,d\theta_2\int_{-\infty}^{\infty} g(\theta_3)e^{-i\omega\theta_3}\,d\theta_3\int_{-\infty}^{\infty} g(\theta_1)e^{-i(-\omega)\theta_1}\,d\theta_1\, e^{i\omega\theta}d\omega$$

$$=-\frac{1}{2}\sqrt{\left(\frac{2^{n+1}}{\pi}\right)}\sigma_{y0}{}^{n-1}\Gamma\left(\frac{n+2}{2}\right)\int_{-\infty}^{\infty} S_{FF}(\omega)H(\omega)H(\omega)H(-\omega)e^{i\omega\theta}\,d\omega$$

$$=-\frac{1}{2}\sqrt{\left(\frac{2^{n+1}}{\pi}\right)}\sigma_{y0}{}^{n-1}\Gamma\left(\frac{n+2}{2}\right)\int_{-\infty}^{\infty} S_{FF}(\omega)H(\omega)H(-\omega)H(\omega)e^{i\omega\theta}\,d\omega$$

$$=-\frac{1}{2}\sqrt{\left(\frac{2^{n+1}}{\pi}\right)}\sigma_{y0}{}^{n-1}\Gamma\left(\frac{n+2}{2}\right)\int_{-\infty}^{\infty} S_{FF}(\omega)\left|H(\omega)\right|^2 H(\omega)e^{i\omega\theta}\,d\omega \qquad (8.140)$$

Next, we can get the cross-spectral density function $S_{y0y1}(\omega)$ by using the following formula:

$$S_{y0y1}(\omega)=-\sqrt{\left(\frac{2^{n+1}}{\pi}\right)}\sigma_{y0}{}^{n-1}\Gamma\left(\frac{n+2}{2}\right)S_{FF}(\omega)\left|H(\omega)\right|^2 H(\omega) \qquad (8.141)$$

Combining equations (8.93), (8.132) and (8.141) yields:

$$S_{yy}(\omega)=S_{FF}(\omega)\left|H(\omega)\right|^2-2\varepsilon\sqrt{\left(\frac{2^{n+1}}{\pi}\right)}\sigma_{y0}{}^{n-1}\Gamma\left(\frac{n+2}{2}\right)S_{FF}(\omega)\left|H(\omega)\right|^2\operatorname{Re}\left[H(\omega)\right]$$

$$=S_{FF}(\omega)\left|H(\omega)\right|^2\left\{1-2\varepsilon\sqrt{\left(\frac{2^{n+1}}{\pi}\right)}\sigma_{y0}{}^{n-1}\Gamma\left(\frac{n+2}{2}\right)\operatorname{Re}\left[H(\omega)\right]\right\} \qquad (8.142)$$

These results, represented by equation (8.142), for the system described by equation (8.129) can be somewhat generalized. If the dynamic system is governed by the more general equation

$$\ddot{y}+\alpha\dot{y}+\omega_n{}^2y+\varepsilon\sum_{n=3,5\cdots} g_n y^n=F(t) \qquad (8.143)$$

then the spectral density of the response to the first order in ε is

$$S_{yy}(\omega)=S_{FF}(\omega)\left|H(\omega)\right|^2\left\{1-2\varepsilon\left[\sum_{n=3,5\cdots} g_n\sqrt{\left(\frac{2^{n+1}}{\pi}\right)}\sigma_{y0}{}^{n-1}\Gamma\left(\frac{n+2}{2}\right)\right]\operatorname{Re}\left[H(\omega)\right]\right\}$$

(8.144)

When applied to the scenario of a white-noise excitation, equation (8.144) becomes,

$$S_{yy}(\omega)=\frac{S_0}{\left(\omega_n^{\,2}-\omega^2\right)^2+(\alpha\omega)^2}$$

$$\left\{1-2\varepsilon\left[\sum_{n=3,5\cdots} g_n\sqrt{\left(\frac{2^{n+1}}{\pi}\right)}\sigma_{y0}^{\;n-1}\Gamma\left(\frac{n+2}{2}\right)\right]\frac{\left(\omega_n^{\,2}-\omega^2\right)}{\left(\omega_n^{\,2}-\omega^2\right)^2+(\omega\alpha)^2}\right\} \tag{8.145}$$

in which S_0 is the spectral density of the exciting white noise.

c. A special case with only a nonlinear damping term
As another implementation of the perturbation technique described in this chapter, we now replace the nonlinear term $\eta(y,\,\dot{y})$ in equation (8.66) with $\dot{y}^3$. Therefore, equation (8.66) can be re-written as

$$\ddot{y}+\alpha\dot{y}+\varepsilon\dot{y}^3+\omega_n^{\,2}y=F(t) \tag{8.146}$$

In the following analysis, we assume that the function $F(t)$ in equation (8.146) is a stationary Gaussian random process with a mean value of zero. Furthermore, we designate the spectral density function of $F(t)$ as $S_{FF}(\omega)$. With the knowledge that both the frequency response function $H(\omega)$ and the impulse response function $g(\theta)$ of the linear system ($\varepsilon=0$) are known, we are able to determine the response autocorrelation function $R_{y0y0}(\theta)$ of the linear system by employing the equation that is presented below:

$$R_{y0y0}(\theta)=\frac{1}{2}\int_{-\infty}^{\infty}S_{y0y0}(\omega)e^{i\omega\theta}\,d\omega=\frac{1}{2}\int_{-\infty}^{\infty}S_{FF}(\omega)\left|H(\omega)\right|^2e^{i\omega\theta}\,d\omega \tag{8.147}$$

where

$$H(\omega)=\int_{-\infty}^{\infty}g(\theta)e^{-i\omega\theta}\,d\theta \tag{8.148}$$

Without a doubt, the response spectral density of the linear system can be determined by employing the formula that is presented below:

$$S_{y0y0}(\omega)=S_{FF}(\omega)\left|H(\omega)\right|^2 \tag{8.149}$$

The cross-correlation function $R_{y_0y_1}(\theta)$ can be determined in this particular case using the following formula:

$$\begin{aligned} R_{y_0y_1}(\theta) &= E\left[y_0(t)y_1(t+\theta)\right] \\ &= -\int_{-\infty}^{\infty}\int_{-\infty}^{\infty} E\left[F(t-\theta_1)\eta(t+\theta-\theta_2)\right]g(\theta_1)g(\theta_2)d\theta_1\,d\theta_2 \\ &= -\int_{-\infty}^{\infty}\int_{-\infty}^{\infty} E\left[F(t-\theta_1)\dot{y}_0^{\,3}(t+\theta-\theta_2)\right]g(\theta_1)g(\theta_2)d\theta_1\,d\theta_2 \end{aligned} \tag{8.150}$$

Considering that we have assumed that the function $F(t)$ in equation (8.146) is a Gaussian random process with a mean value of zero and that it is weakly stationary, it follows that the response velocity of the linear system, denoted by $\dot{y}_0(t)$, will also be a Gaussian random process with a zero mean value. According to the classical theory of random processes, the following relationship is true for a Gaussian random process that has a mean value equal to zero:

$$\begin{aligned} &E\left[F(t-\theta_1)\dot{y}_0^{\,3}(t+\theta-\theta_2)\right] \\ &\quad= E\left[F(t-\theta_1)\dot{y}_0(t+\theta-\theta_2)\right]E\left[\dot{y}_0(t+\theta-\theta_2)y_0(t+\theta-\theta_2)\right] \\ &\qquad+ E\left[\dot{y}_0(t+\theta-\theta_2)\dot{y}_0(t+\theta-\theta_2)\right]E\left[F(t-\theta_1)\dot{y}_0(t+\theta-\theta_2)\right] \\ &\qquad+ E\left[F(t-\theta_1)\dot{y}_0(t+\theta-\theta_2)\right]E\left[\dot{y}_0(t+\theta-\theta_2)\dot{y}_0(t+\theta-\theta_2)\right] \\ &\quad= 3E\left[F(t-\theta_1)\dot{y}_0(t+\theta-\theta_2)\right]E\left[\dot{y}_0(t+\theta-\theta_2)\dot{y}_0(t+\theta-\theta_2)\right] \end{aligned} \tag{8.151}$$

In addition, we have

$$\begin{aligned} &E\left[F(t-\theta_1)\dot{y}_0(t+\theta-\theta_2)\right] \\ &\quad= E\left[F(t-\theta_1)\frac{d}{d\theta}\left\{\int_{-\infty}^{\infty}F(t+\theta-\theta_2-\theta_3)g(\theta_3)d\theta_3\right\}\right] \\ &\quad= \int_{-\infty}^{\infty}\frac{d}{d\theta}\left\{E\left[F(t-\theta_1)F(t+\theta-\theta_2-\theta_3)\right]\right\}g(\theta_3)d\theta_3 \\ &\quad= \int_{-\infty}^{\infty}\frac{d}{d\theta}\left[R_{FF}(\theta+\theta_1-\theta_2-\theta_3)\right]g(\theta_3)d\theta_3 \end{aligned} \tag{8.152}$$

and

$$E\left[\dot{y}_0(t+\theta-\theta_2)\dot{y}_0(t+\theta-\theta_2)\right] = \sigma_{\dot{y}_0}{}^2 \tag{8.153}$$

where $\sigma_{\dot{y}0}{}^2$ is the variance of the linear system's response velocity. Combing equations (8.151), (8.152) and (8.153) yields

$$E\left[F(t-\theta_1)\dot{y}_0{}^3(t+\theta-\theta_2)\right]$$
$$=3\sigma_{\dot{y}0}{}^2\int_{-\infty}^{\infty}\frac{d}{d\theta}\left[R_{FF}(\theta+\theta_1-\theta_2-\theta_3)\right]g(\theta_3)d\theta_3 \tag{8.154}$$
$$=\frac{3}{2}\sigma_{\dot{y}0}{}^2\int_{-\infty}^{\infty}\int_{-\infty}^{\infty}\frac{d}{d\theta}\left[S_{FF}(\omega)e^{i\omega(\theta+\theta_1-\theta_2-\theta_3)}\right]g(\theta_3)d\theta_3\,d\omega$$

The cross-correlation function, which is represented by $R_{y0y1}(\theta)$ in equation (8.150), can then be determined as follows:

$$R_{y0y1}(\theta)$$
$$=-\int_{-\infty}^{\infty}\int_{-\infty}^{\infty}E\left[F(t-\theta_1)\dot{y}_0{}^3(t+\theta-\theta_2)\right]g(\theta_1)g(\theta_2)d\theta_1\,d\theta_2$$
$$=-\frac{3}{2}\sigma_{\dot{y}0}{}^2\int_{-\infty}^{\infty}\int_{-\infty}^{\infty}\int_{-\infty}^{\infty}\int_{-\infty}^{\infty}\frac{d}{d\theta}\left[S_{FF}(\omega)e^{i\omega(\theta+\theta_1-\theta_2-\theta_3)}\right]g(\theta_1)g(\theta_2)g(\theta_3)d\theta_1\,d\theta_2d\theta_3\,d\omega$$
$$=-\frac{3}{2}\sigma_{\dot{y}0}{}^2\int_{-\infty}^{\infty}S_{FF}(\omega)\int_{-\infty}^{\infty}g(\theta_2)e^{-i\omega\theta_2}\,d\theta_2\int_{-\infty}^{\infty}g(\theta_3)e^{-i\omega\theta_3}\,d\theta_3$$
$$\times\int_{-\infty}^{\infty}g(\theta_1)e^{-i(-\omega)\theta_1}\,d\theta_1\frac{d}{d\theta}\left(e^{i\omega\theta}\right)d\omega \tag{8.155}$$
$$=-\frac{3}{2}\sigma_{\dot{y}0}{}^2\int_{-\infty}^{\infty}S_{FF}(\omega)H(\omega)H(\omega)H(-\omega)(i\omega)e^{i\omega\theta}\,d\omega$$
$$=-\frac{3}{2}\sigma_{\dot{y}0}{}^2\int_{-\infty}^{\infty}S_{FF}(\omega)H(\omega)H(-\omega)H(\omega)(i\omega)e^{i\omega\theta}\,d\omega$$
$$=-\frac{3}{2}\sigma_{\dot{y}0}{}^2\int_{-\infty}^{\infty}S_{FF}(\omega)\left|H(\omega)\right|^2H(\omega)(i\omega)e^{i\omega\theta}\,d\omega$$

We now can proceed to obtain the cross-spectral density function $S_{y0y1}(\omega)$ as follows:

$$S_{y0y1}(\omega)=-3\sigma_{\dot{y}0}{}^2S_{FF}(\omega)\left|H(\omega)\right|^2H(\omega)(i\omega) \tag{8.156}$$

Combining equations (8.93), (8.149) and (8.156) yields:

$$\begin{aligned} S_{yy}(\omega) &= S_{FF}(\omega)|H(\omega)|^2 - 6\varepsilon\sigma_{\dot{y}0}{}^2 S_{FF}(\omega)|H(\omega)|^2 \operatorname{Re}[(i\omega)H(\omega)] \\ &= S_{FF}(\omega)|H(\omega)|^2\{1-6\varepsilon\sigma_{\dot{y}0}{}^2 \operatorname{Re}[(i\omega)H(\omega)]\} \end{aligned} \tag{8.157}$$

d. A generalized case with only a nonlinear damping term

As the last implementation example of the perturbation technique in this chapter, we will now investigate the results of substituting the nonlinear term $\eta(y, \dot{y})$ in equation (8.66) with $\dot{y}^n$. When this is done, the equation (8.66) can be re-written as

$$\ddot{y}+\alpha\dot{y}+\varepsilon\dot{y}^n+\omega_n{}^2 y = F(t) \tag{8.158}$$

where n is an odd integer. The vibration system that is represented by this equation has a nonlinearity of stiffness that is very small. In the subsequent analysis, we additionally make the assumption that the function $F(t)$ in equation (8.158) is a Gaussian random process that is weakly stationary and has a mean of zero. Additionally, we define the spectral density function of the excitation force $F(t)$ as $S_{FF}(\omega)$. Again, the response autocorrelation function $R_{y_0y_0}(\theta)$ of the linear system ($\varepsilon = 0$) can be calculated as follows:

$$R_{y_0y_0}(\theta) = \frac{1}{2}\int_{-\infty}^{\infty} S_{y_0y_0}(\omega)e^{i\omega\theta}\,d\omega = \frac{1}{2}\int_{-\infty}^{\infty} S_{FF}(\omega)|H(\omega)|^2 e^{i\omega\theta}\,d\omega \tag{8.159}$$

where

$$H(\omega) = \int_{-\infty}^{\infty} g(\theta)e^{-i\omega\theta}\,d\theta \tag{8.160}$$

where $H(\omega)$ is the frequency response function of the linear system ($\varepsilon = 0$) and $g(\theta)$ is the impulse response function $g(\theta)$ of the linear system. For the purpose of computing the cross-correlation function $R_{y_0y_1}(\theta)$ in this generalized scenario, the formula that follows can be utilized:

$$\begin{aligned} R_{y_0y_1}(\theta) &= E[y_0(t)y_1(t+\theta)] \\ &= -\int_{-\infty}^{\infty}\int_{-\infty}^{\infty} E[F(t-\theta_1)\eta(t+\theta-\theta_2)]g(\theta_1)g(\theta_2)d\theta_1\,d\theta_2 \\ &= -\int_{-\infty}^{\infty}\int_{-\infty}^{\infty} E[F(t-\theta_1)\dot{y}_0{}^n(t+\theta-\theta_2)]g(\theta_1)g(\theta_2)d\theta_1\,d\theta_2 \end{aligned} \tag{8.161}$$

In light of the fact that we have made the assumption that the function $F(t)$ in equation (8.158) is a weakly stationary Gaussian random process with a mean value of zero, it follows that the response velocity of the linear system, which is denoted by $\dot{y}_0(t)$, will also be a weakly stationary Gaussian random process for which the mean value is zero. For a Gaussian random process with a mean value of zero, the following relationship holds true according to the classical theory of random processes:

$$\begin{aligned} &E\left[F(t-\theta_1)\dot{y}_0^{\,n}(t+\theta-\theta_2)\right] \\ &\quad = mE\left[F(t-\theta_1)\dot{y}_0(t+\theta-\theta_2)\right]E\left[\dot{y}_0^{\,n-1}(t+\theta-\theta_2)\right] \end{aligned} \tag{8.162}$$

The letter m, which can be found on the right-hand side of equation (8.162), denotes the number of different ways that can be used to pick an $y_0(t+\theta-\theta_2)$ from the product $y_0^{\,n-1}(t+\theta-\theta_2)$. In this particular case, m is equal to n. Taking into account the fact that $y_0(t)$ is weakly stationary and as $E[y_0(t)]$ equals zero,

$$E\left[\dot{y}_0^{\,n-1}(t)\right] = \frac{\left(\sqrt{2}\sigma_{\dot{y}0}\right)^{n-1}}{\sqrt{\pi}}\Gamma\left(\frac{n}{2}\right) \tag{8.163}$$

where in which $\Gamma(\cdot)$ is the Gamma function. Furthermore, we have

$$\begin{aligned} &E\left[F(t-\theta_1)\dot{y}_0(t+\theta-\theta_2)\right] \\ &= E\left[F(t-\theta_1)\frac{d}{d\theta}\left\{\int_{-\infty}^{\infty}F(t+\theta-\theta_2-\theta_3)g(\theta_3)d\theta_3\right\}\right] \\ &= \int_{-\infty}^{\infty}\frac{d}{d\theta}\left\{E\left[F(t-\theta_1)F(t+\theta-\theta_2-\theta_3)\right]\right\}g(\theta_3)d\theta_3 \\ &= \int_{-\infty}^{\infty}\frac{d}{d\theta}\left[R_{FF}(\theta+\theta_1-\theta_2-\theta_3)\right]g(\theta_3)d\theta_3 \end{aligned} \tag{8.164}$$

and

$$E\left[\dot{y}_0(t+\theta-\theta_2)\dot{y}_0(t+\theta-\theta_2)\right] = \sigma_{\dot{y}0}^{\;2} \tag{8.165}$$

in which $\sigma_{\dot{y}0}^{\;2}$ is the variance of the response velocity of the linear system. By combining equations (8.162), (8.163) and (8.164), one obtains the following:

$$\begin{aligned} &E\left[F(t-\theta_1)\dot{y}_0^{\,n}(t+\theta-\theta_2)\right] \\ &= mE\left[F(t-\theta_1)\dot{y}_0(t+\theta-\theta_2)\right]E\left[\dot{y}_0^{\,n-1}(t+\theta-\theta_2)\right] \\ &= \sqrt{\left(\frac{2^{n+1}}{\pi}\right)}\sigma_{\dot{y}0}^{\;n-1}\Gamma\left(\frac{n+2}{2}\right)\int_{-\infty}^{\infty}\frac{d}{d\theta}\left[R_{FF}(\theta+\theta_1-\theta_2-\theta_3)\right]g(\theta_3)d\theta_3 \\ &= \frac{1}{2}\sqrt{\left(\frac{2^{n+1}}{\pi}\right)}\sigma_{\dot{y}0}^{\;n-1}\Gamma\left(\frac{n+2}{2}\right)\int_{-\infty}^{\infty}\int_{-\infty}^{\infty}\frac{d}{d\theta}\left[S_{FF}(\omega)e^{i\omega(\theta+\theta_1-\theta_2-\theta_3)}\right]g(\theta_3)d\theta_3 d\omega \end{aligned} \tag{8.166}$$

The cross-correlation function, which is denoted by $R_{y_0y_1}(\theta)$ in equation (8.161), can then be computed in the following manner:

$$
\begin{aligned}
R_{y_0y_1}(\theta) \\
&= -\int_{-\infty}^{\infty}\int_{-\infty}^{\infty} E\left[F(t-\theta_1)\dot{y}_0^{\,n}(t+\theta-\theta_2)\right] g(\theta_1)g(\theta_2)d\theta_1 d\theta_2 \\
&= -\frac{1}{2}\sqrt{\left(\frac{2^{n+1}}{\pi}\right)}\sigma_{\dot{y}_0}^{\ \ n-1}\Gamma\left(\frac{n+2}{2}\right) \\
&\quad \int_{-\infty}^{\infty}\int_{-\infty}^{\infty}\int_{-\infty}^{\infty}\int_{-\infty}^{\infty}\frac{d}{d\theta}\left[S_{FF}(\omega)e^{i\omega(\theta+\theta_1-\theta_2-\theta_3)}\right] g(\theta_1)g(\theta_2)g(\theta_3)d\theta_1 d\theta_2 d\theta_3 d\omega \\
&= -\frac{1}{2}\sqrt{\left(\frac{2^{n+1}}{\pi}\right)}\sigma_{\dot{y}_0}^{\ \ n-1}\Gamma\left(\frac{n+2}{2}\right) \\
&\quad \int_{-\infty}^{\infty} S_{FF}(\omega)\int_{-\infty}^{\infty} g(\theta_2)e^{-i\omega\theta_2}\,d\theta_2 \int_{-\infty}^{\infty} g(\theta_3)e^{-i\omega\theta_3}\,d\theta_3 \\
&\quad \int_{-\infty}^{\infty} g(\theta_1)e^{-i(-\omega)\theta_1}\,d\theta_1 \frac{d}{d\theta}\left[e^{i\omega\theta}\right]d\omega \\
&= -\frac{1}{2}\sqrt{\left(\frac{2^{n+1}}{\pi}\right)}\sigma_{\dot{y}_0}^{\ \ n-1}\Gamma\left(\frac{n+2}{2}\right)\int_{-\infty}^{\infty} S_{FF}(\omega)H(\omega)H(\omega)H(-\omega)\frac{d}{d\theta}\left[e^{i\omega\theta}\right]d\omega \\
&= -\frac{1}{2}\sqrt{\left(\frac{2^{n+1}}{\pi}\right)}\sigma_{\dot{y}_0}^{\ \ n-1}\Gamma\left(\frac{n+2}{2}\right)\int_{-\infty}^{\infty} S_{FF}(\omega)H(\omega)H(-\omega)H(\omega)\frac{d}{d\theta}\left[e^{i\omega\theta}\right]d\omega \\
&= -\frac{1}{2}\sqrt{\left(\frac{2^{n+1}}{\pi}\right)}\sigma_{\dot{y}_0}^{\ \ n-1}\Gamma\left(\frac{n+2}{2}\right)\int_{-\infty}^{\infty} S_{FF}(\omega)\left|H(\omega)\right|^2 H(\omega)(i\omega)e^{i\omega\theta}\,d\omega
\end{aligned}
\tag{8.167}
$$

Using the following formula, we can then obtain the cross-spectral density function, which is denoted by $S_{y_0y_1}(\omega)$:

$$S_{y_0y_1}(\omega) = -\sqrt{\left(\frac{2^{n+1}}{\pi}\right)}\sigma_{\dot{y}_0}^{\ \ n-1}\Gamma\left(\frac{n+2}{2}\right)S_{FF}(\omega)\left|H(\omega)\right|^2(i\omega)H(\omega) \tag{8.168}$$

Undoubtedly, the response spectral density of the linear system can be calculated using the formula shown below:

$$S_{y_0y_0}(\omega) = S_{FF}(\omega)\left|H(\omega)\right|^2 \tag{8.169}$$

Combining equations (8.93), (8.168) and (8.169) yields:

$$S_{yy}(\omega)=S_{FF}(\omega)|H(\omega)|^2-2\varepsilon\sqrt{\left(\frac{2^{n+1}}{\pi}\right)}\sigma_{\dot{y}0}{}^{n-1}\Gamma\left(\frac{n+2}{2}\right)S_{FF}(\omega)|H(\omega)|^2\,\mathrm{Re}\left[(i\omega)H(\omega)\right]$$

$$=S_{FF}(\omega)|H(\omega)|^2\left\{1-2\varepsilon\sqrt{\left(\frac{2^{n+1}}{\pi}\right)}\sigma_{\dot{y}0}{}^{n-1}\Gamma\left(\frac{n+2}{2}\right)\mathrm{Re}\left[(i\omega)H(\omega)\right]\right\} \quad (8.170)$$

where

$$\sigma_{\dot{y}0}=\sqrt{\int_{-\infty}^{\infty}\omega^2 S_{FF}(\omega)|H(\omega)|^2\,d\omega} \quad (8.171)$$

These results, which are represented by equation (8.170), for the system that is defined by equation (8.158) are capable of being generalized to some extent. That being said, if the dynamic system is governed by the equation that is more general,

$$\ddot{y}+\alpha\dot{y}+\varepsilon\sum_{n=3,5\cdots}\beta_n\dot{y}^n+\omega_n{}^2y=F(t) \quad (8.172)$$

then the spectral density function of the response, when approximated to the first order in ε, is

$$S_{yy}(\omega)=S_{FF}(\omega)|H(\omega)|^2\left\{1-2\varepsilon\left[\sum_{n=3,5\cdots}\beta_n\sqrt{\left(\frac{2^{n+1}}{\pi}\right)}\sigma_{\dot{y}0}{}^{n-1}\Gamma\left(\frac{n+2}{2}\right)\right]\mathrm{Re}\left[(i\omega)H(\omega)\right]\right\} \quad (8.173)$$

If we apply equation (8.173) to the case of a white-noise excitation, we obtain the following:

$$S_{yy}(\omega)=\frac{S_0}{\left(\omega_n{}^2-\omega^2\right)^2+(\alpha\omega)^2}\left\{1-2\varepsilon\left[\sum_{n=3,5\cdots}\beta_n\sqrt{\left(\frac{2^{n+1}}{\pi}\right)}\sigma_{\dot{y}0}{}^{n-1}\Gamma\left(\frac{n+2}{2}\right)\right]\frac{\omega^2\alpha}{\left(\omega_n{}^2-\omega^2\right)^2+(\omega\alpha)^2}\right\} \quad (8.174)$$

where S_0 is the spectral density function of the white noise exciting force.

Using equation (8.105), one may determine the relationship that exists between the power spectral densities of the excitation and the response of the nonlinear system (8.94). After obtaining the response spectrum, it is possible to proceed with the computation of the time history of the nonlinear system responses, response statistics, and motion extrema. The perturbation technique has been used to solve a number of problems in a variety of engineering fields. For example, in the field of ocean engineering, the perturbation technique had once been applied to investigate the stochastic responses of a deep-water jacket platform when it was subjected to the action of Morison-type nonlinear wave forces. However, it is common knowledge that the solutions that are produced by the perturbation technique are only reliable in situations when the nonlinearity is very weak (that is, when the parameter ε in equation (8.94) is very small). In Chapter 9, we will discuss the numerical simulation method, which is yet another frequent approach to dealing with nonlinear random vibration problems.

8.3.3 Comparing the Two Techniques

In this subsection, we are going to demonstrate that when dealing with a nonlinear random vibration problem, the use of the equivalent-linearization technique will yield a result that is in agreement with the outcome obtained from the perturbation technique up to the first order in ε. We will first consider a generalized case with only a nonlinear damping term in the equation of motion of a nonlinear random vibration system.

a. A generalized case with only nonlinear damping terms

Now, we will show a demonstration of the application of the equivalent linearization technique. The system under consideration only exhibits nonlinearities in the damping term, which is denoted as $\alpha\dot{y}+\varepsilon\dot{y}^n$. The differential equation of motion for the random vibration system can therefore be described as follows:

$$\ddot{y}+g(y,\dot{y})=\ddot{y}+\alpha\dot{y}+\varepsilon\dot{y}^n+\omega_n^{\ 2}y=F(t) \tag{8.175}$$

in which n is an odd integer. We can represent the equation of motion for a linear system, which is equivalent to the equation of motion for the nonlinear system specified in equation (8.175), as follows:

$$\ddot{y}+\beta_e\dot{y}+k_e y=F(t) \tag{8.176}$$

Based on the fact that the excitation term $F(t)$ in equation (8.176) is a Gaussian random process with zero expectation, it follows that the response displacement y will also be normally distributed with zero expectation. Additionally, the response velocity, denoted by $\dot{y}$, will also be distributed normally with zero expectation. In the following, we will refer to the variance of $\dot{y}$ that is unknown as $\sigma_{\dot{y}}^{\ 2}$.

$$E\left[\dot{y}^2\right]=\sigma_{\dot{y}}^{\ 2} \tag{8.177}$$

In accordance with equation (8.20), the equivalent stiffness coefficient k_e can be computed as follows:

$$\begin{aligned} k_e &= \frac{E[yg(y,\dot{y})]}{E\left[y^2\right]} = \frac{E\left[y\left(\alpha\dot{y}+\varepsilon\dot{y}^n+\omega_n^{\ 2}y\right)\right]}{E\left[y^2\right]} \\ &= \frac{\alpha E\left[y\dot{y}\right]+\varepsilon E\left[y\dot{y}^n\right]+\omega_n^{\ 2}E\left[yy\right]}{E\left[y^2\right]} \end{aligned} \tag{8.178}$$

We are aware, on the basis of the definition of the autocorrelation function, that the relationship $E\left[y\dot{y}\right]=0$ shows that the random process y and its derivative $\dot{y}$ are uncorrelated. As a consequence of the fact that being uncorrelated for a Gaussian process is synonymous with being independent, it then follows that the Gaussian process y and its derivative $\dot{y}$ are independent as well. It is important to notice that the following relationship exists because y and its derivative are independent.

$$E\left[y\dot{y}^n\right]=E\left[y\right]E\left[\dot{y}^n\right]=0\times E\left[\dot{y}^n\right]=0 \tag{8.179}$$

With the above information, the equivalent stiffness coefficient k_e in equation (8.178) can be computed as follows:

$$\begin{aligned} k_e &= \frac{\alpha E\left[y\dot{y}\right]+\varepsilon E\left[y\dot{y}^n\right]+\omega_n^{\ 2}E\left[yy\right]}{E\left[y^2\right]} \\ &= \frac{\alpha\times 0+\varepsilon\times 0+\omega_n^{\ 2}E\left[y^2\right]}{E\left[y^2\right]} = \omega_n^{\ 2} \end{aligned} \tag{8.180}$$

In accordance with equation (8.20), the equivalent damping coefficient β_e can be computed as follows:

$$\begin{aligned} \beta_e &= \frac{E[\dot{y}g(y,\dot{y})]}{E\left[\dot{y}^2\right]} = \frac{E\left[\dot{y}\left(\alpha\dot{y}+\varepsilon\dot{y}^n+\omega_n^{\ 2}y\right)\right]}{E\left[\dot{y}^2\right]} \\ &= \frac{\alpha E\left[\dot{y}\dot{y}\right]+\varepsilon E\left[\dot{y}\dot{y}^n\right]+\omega_n^{\ 2}E\left[y\dot{y}\right]}{E\left[\dot{y}^2\right]} = \frac{\alpha E\left[\dot{y}^2\right]+\varepsilon E\left[\dot{y}^{n+1}\right]}{E\left[\dot{y}^2\right]} \end{aligned} \tag{8.181}$$

Due to the fact that the response velocity $\dot{y}$ follows a normal distribution with zero mean, we can obtain that

$$\varepsilon E\left[\dot{y}^{n+1}\right]=\int_{-\infty}^{\infty}\varepsilon\dot{y}^{n+1}\frac{1}{\sqrt{2\pi}\sigma_{\dot{y}}}exp\left[-\dot{y}^2/\left(2\sigma_{\dot{y}}^{\ 2}\right)\right]d\dot{y} \tag{8.182}$$

According to equation (3.24), we have the following:

$$\int_0^\infty \dot{y}^{n+1}\exp\left(-\frac{\dot{y}^2}{2\sigma_{\dot{y}}^2}\right)d\dot{y} = \frac{(\sqrt{2}\sigma_{\dot{y}})^{n+2}}{2}\Gamma\left(\frac{n+2}{2}\right) \tag{8.183}$$

When the equations (8.182) and (8.183) are combined, the result is

$$\begin{aligned} \varepsilon E\left[\dot{y}^{n+1}\right] &= \frac{\varepsilon \times 2 \times \dfrac{(\sqrt{2}\sigma_{\dot{y}})^{n+2}}{2}\Gamma\left(\dfrac{n+2}{2}\right)}{\sqrt{2\pi}\sigma_{\dot{y}}} \\ &= \varepsilon\sqrt{\frac{2^{n+1}}{\pi}}\sigma_{\dot{y}}^{n+1}\Gamma\left(\frac{n+2}{2}\right) \end{aligned} \tag{8.184}$$

Combining equations (8.181) and (8.184) yields

$$\begin{aligned} \beta_e &= \frac{\alpha E\left[\dot{y}^2\right] + \varepsilon E\left[\dot{y}^{n+1}\right]}{E\left[\dot{y}^2\right]} = \frac{\alpha\sigma_{\dot{y}}^2 + \varepsilon\sqrt{\dfrac{2^{n+1}}{\pi}}\sigma_{\dot{y}}^{n+1}\Gamma\left(\dfrac{n+2}{2}\right)}{\sigma_{\dot{y}}^2} \\ &= \alpha + \varepsilon\sqrt{\frac{2^{n+1}}{\pi}}\sigma_{\dot{y}}^{n-1}\Gamma\left(\frac{n+2}{2}\right) \end{aligned} \tag{8.185}$$

The equivalent linear system (8.176) has a frequency response function that is as follows:

$$H_e(\omega) = \frac{1}{\left(\omega_n^2 - \omega^2\right) + i\omega\beta_e} \tag{8.186}$$

Consequently, the following formula can be used to calculate the magnitude of $H_e(\omega)$:

$$|H_e(\omega)| = \sqrt{\frac{1}{\left(\omega_n^2 - \omega^2\right)^2 + (\omega\beta_e)^2}} \tag{8.187}$$

In light of this, the square of the magnitude of $H_e(\omega)$ can be calculated in the following way:

$$|H_e(\omega)|^2 = \frac{1}{\left(\omega_n^2 - \omega^2\right)^2 + (\omega\beta_e)^2} \tag{8.188}$$

We have the relationship that is given below in the event that $\beta_e = \alpha$ occurs:

$$|H(\omega)|^2 = \frac{1}{\left(\omega_n^2 - \omega^2\right)^2 + (\alpha\omega)^2} \tag{8.189}$$

Through the utilization of the relationship that is outlined in equation (8.185), we proceed to expand the function $|H_e(\omega)|^2$ into a Taylor series in the vicinity of α:

$$|H_e(\omega)|^2 = |H(\omega)|^2 + \varepsilon\sqrt{\frac{2^{n+1}}{\pi}}\sigma_{\dot{y}}^{\,n-1}\Gamma\left(\frac{n+2}{2}\right)\frac{\partial}{\partial\alpha}|H(\omega)|^2 \tag{8.190}$$

Equation (8.190) can be further expressed in the following way:

$$\begin{aligned} |H_e(\omega)|^2 &= |H(\omega)|^2 + \varepsilon\sqrt{\frac{2^{n+1}}{\pi}}\sigma_{\dot{y}}^{\,n-1}\Gamma\left(\frac{n+2}{2}\right)\frac{\partial}{\partial\alpha}|H(\omega)|^2 \\ &= |H(\omega)|^2 + \varepsilon\sqrt{\frac{2^{n+1}}{\pi}}\sigma_{\dot{y}}^{\,n-1}\Gamma\left(\frac{n+2}{2}\right)\left(-2\alpha\omega^2|H(\omega)|^4\right) \\ &= |H(\omega)|^2\left\{1 - 2\varepsilon\alpha\omega^2\sqrt{\frac{2^{n+1}}{\pi}}\sigma_{\dot{y}}^{\,n-1}\Gamma\left(\frac{n+2}{2}\right)\left(|H(\omega)|^2\right)\right\} \end{aligned} \tag{8.191}$$

By utilizing equation (8.191), we can deduce the subsequent relationship between the output and input spectral density functions of the equivalent linear system (8.176):

$$S_{yy}(\omega) = S_{FF}(\omega)|H(\omega)|^2\left\{1 - 2\varepsilon\alpha\omega^2\sqrt{\frac{2^{n+1}}{\pi}}\sigma_{\dot{y}}^{\,n-1}\Gamma\left(\frac{n+2}{2}\right)\left(|H(\omega)|^2\right)\right\} \tag{8.192}$$

We have the following relationships:

$$\begin{aligned} (i\omega)H(\omega) &= (i\omega)\frac{1}{\left(\omega_n^2 - \omega^2\right) + i\omega\alpha} \\ &= (i\omega)\frac{\left(\omega_n^2 - \omega^2\right) - i\omega\alpha}{\left\{\left(\omega_n^2 - \omega^2\right) + i\omega\alpha\right\}\left\{\left(\omega_n^2 - \omega^2\right) - i\omega\alpha\right\}} \\ &= (i\omega)\frac{\left(\omega_n^2 - \omega^2\right) - i\omega\alpha}{\left(\omega_n^2 - \omega^2\right)^2 + (\omega\alpha)^2} = \frac{\left(\omega_n^2 - \omega^2\right)\omega i + \omega^2\alpha}{\left(\omega_n^2 - \omega^2\right)^2 + (\omega\alpha)^2} \end{aligned} \tag{8.193}$$

$$|H(\omega)|^2 = \frac{1}{\left(\omega_n^2 - \omega^2\right)^2 + (\alpha\omega)^2} \tag{8.194}$$

Based on equations (8.193) and (8.194), we have

$$\operatorname{Re}\left[(i\omega)H(\omega)\right]=\frac{\omega^2\alpha}{\left(\omega_n^{\,2}-\omega^2\right)^2+(\omega\alpha)^2}=\alpha\omega^2\left|H(\omega)\right|^2 \quad (8.195)$$

Consequently, equation (8.191) can be re-written as

$$\begin{aligned}\left|H_e(\omega)\right|^2&=\left|H(\omega)\right|^2\left\{1-2\varepsilon\alpha\omega^2\sqrt{\frac{2^{n+1}}{\pi}}\sigma_{\dot{y}}^{\,n-1}\Gamma\left(\frac{n+2}{2}\right)\left(\left|H(\omega)\right|^2\right)\right\}\\&=\left|H(\omega)\right|^2\left\{1-2\varepsilon\sqrt{\frac{2^{n+1}}{\pi}}\sigma_{\dot{y}}^{\,n-1}\Gamma\left(\frac{n+2}{2}\right)\left(\alpha\omega^2\left|H(\omega)\right|^2\right)\right\}\\&=\left|H(\omega)\right|^2\left\{1-2\varepsilon\sqrt{\frac{2^{n+1}}{\pi}}\sigma_{\dot{y}}^{\,n-1}\Gamma\left(\frac{n+2}{2}\right)\left(\operatorname{Re}\left[(i\omega)H(\omega)\right]\right)\right\}\end{aligned} \quad (8.196)$$

Therefore, the relationship between the output and input spectral density functions of the equivalent linear system in equation (8.176) can also be re-written as:

$$\begin{aligned}S_{yy}(\omega)&=S_{FF}(\omega)\left|H(\omega)\right|^2\\&\left\{1-2\varepsilon\sqrt{\frac{2^{n+1}}{\pi}}\sigma_{\dot{y}}^{\,n-1}\Gamma\left(\frac{n+2}{2}\right)\left(\operatorname{Re}\left[(i\omega)H(\omega)\right]\right)\right\}\end{aligned} \quad (8.197)$$

The result in equation (8.197) agrees with the result as shown in equation (8.170), which has been derived by applying the perturbation technique. We have demonstrated that the equivalent-linearization technique can yield a result that agrees with the outcome obtained from the perturbation technique to the first order in ε.

The above-obtained result, which is represented by equation (8.197), for the nonlinear random vibration system defined by equation (8.175) can be further generalized to some extent. Having stated that, if the nonlinear random vibration system is governed by the equation of a more general form,

$$\ddot{y}+\alpha\dot{y}+\varepsilon\sum_{n=3,5\cdots}\beta_n\dot{y}^n+\omega_n^{\,2}y=F(t) \quad (8.198)$$

then the spectral density function of the response will be

$$\begin{aligned}S_{yy}(\omega)&=S_{FF}(\omega)\left|H(\omega)\right|^2\\&\left\{1-2\varepsilon\left[\sum_{n=3,5\cdots}\beta_n\sqrt{\left(\frac{2^{n+1}}{\pi}\right)}\sigma_{\dot{y}}^{\,n-1}\Gamma\left(\frac{n+2}{2}\right)\right]\operatorname{Re}\left[(i\omega)H(\omega)\right]\right\}\end{aligned} \quad (8.199)$$

When we apply equation (8.199) to the case of a white-noise excitation, we obtain the following result:

$$S_{yy}(\omega) = \frac{S_0}{\left(\omega_n^2 - \omega^2\right)^2 + (\alpha\omega)^2} \left\{1 - 2\varepsilon\left[\sum_{n=3,5\cdots} \beta_n \sqrt{\left(\frac{2^{n+1}}{\pi}\right)} \sigma_{\dot{y}}^{\,n-1} \Gamma\left(\frac{n+2}{2}\right)\right] \frac{\omega^2\alpha}{\left(\omega_n^2 - \omega^2\right)^2 + (\omega\alpha)^2}\right\} \tag{8.200}$$

in which S_0 is the spectral density of the white noise excitation force.

b. A generalized case with only nonlinear stiffness terms
In the following step, let us take into consideration the case in which nonlinearities are only present in the restoring term, which can be expressed mathematically as

$$\ddot{y} + g\left(y, \dot{y}\right) = \ddot{y} + \alpha\dot{y} + \omega_n^2 y + \varepsilon y^n = F(t) \tag{8.201}$$

in which n is an odd integer. The vibration system that is represented by this equation has a nonlinearity of stiffness that is very small. In the subsequent analysis, we additionally make the assumption that the function $F(t)$ in equation (8.201) is a Gaussian random process that is weakly stationary and has a mean of zero. Additionally, we define the spectral density function of the excitation force $F(t)$ as $S_{FF}(\omega)$.

We can express the equation of motion for a linear system, which is mathematically equivalent to the equation for a nonlinear system that is presented in equation (8.201), as follows:

$$\ddot{y} + \beta_e \dot{y} + k_e y = F(t) \tag{8.202}$$

Taking into consideration that the excitation term $F(t)$ in equation (8.202) is a Gaussian random process with a mean value of zero, the response displacement that is produced as a result will also have a Gaussian distribution with a mean value of zero. For the sake of the remaining discussion, we will refer to the unknown variance of y as σ_y^2, that is, $E\left[y^2\right] = \sigma_y^2$.

By utilizing equation (8.20), we can calculate the equivalent damping coefficient β_e as expressed below:

$$\begin{aligned} \beta_e &= \frac{E[\dot{y}g(y,\dot{y})]}{E\left[\dot{y}^2\right]} = \frac{E\left[\dot{y}\left\{\alpha\dot{y} + \omega_n^2 y + \varepsilon y^n\right\}\right]}{E\left[\dot{y}^2\right]} \\ &= \frac{\alpha E\left[\dot{y}\dot{y}\right] + \omega_n^2 E\left[\dot{y}y\right] + \varepsilon E\left[\dot{y}y^n\right]}{E\left[\dot{y}^2\right]} \end{aligned} \tag{8.203}$$

Given that y and its derivative $\dot{y}$ are independent, it is important to emphasize that the subsequent relationship exists:

$$E\left[\dot{y}y^n\right]=E[\dot{y}]E\left[y^n\right]=E\left[\frac{d}{dt}y\right]E\left[y^n\right]=\frac{d}{dt}E[y]E\left[y^n\right]=0 \quad (8.204)$$

Consequently, equation (8.203) can be simplified as

$$\beta_e=\frac{\alpha E[\dot{y}\dot{y}]+\omega_n{}^2\times 0+\varepsilon\times 0}{E\left[\dot{y}^2\right]}=\alpha \quad (8.205)$$

In accordance with equation (8.20), the equivalent stiffness coefficient k_e can be determined as follows:

$$\begin{aligned}k_e&=\frac{E[yg(y,\dot{y})]}{E\left[y^2\right]}=\frac{E\left[y\left\{\alpha\dot{y}+\omega_n{}^2y+\varepsilon y^n\right\}\right]}{E\left[y^2\right]}\\&=\frac{\alpha E[y\dot{y}]+\omega_n{}^2E[yy]+\varepsilon E\left[yy^n\right]}{E\left[y^2\right]}=\omega_n{}^2+\frac{\varepsilon E\left[y^{n+1}\right]}{E\left[y^2\right]}\end{aligned} \quad (8.206)$$

Given that the response displacement y follows a Gaussian distribution with a mean of zero, we can obtain

$$\varepsilon E\left[y^{n+1}\right]=\varepsilon\int_{-\infty}^{\infty}y^{n+1}\frac{1}{\sqrt{2\pi}\sigma_y}\exp\left[-y^2/\left(2\sigma_y{}^2\right)\right]dy \quad (8.207)$$

According to equation (3.24), we have the following because n is an odd integer:

$$\begin{aligned}E\left[y^{n+1}\right]&=\int_{-\infty}^{\infty}y^{n+1}\frac{1}{\sqrt{2\pi}\sigma_y}\exp\left[-y^2/\left(2\sigma_y{}^2\right)\right]dy\\&=2\times\frac{1}{\sqrt{2\pi}\sigma_y}\frac{\left(\sqrt{2}\sigma_y\right)^{n+2}}{2}\Gamma\left(\frac{n+2}{2}\right)\end{aligned} \quad (8.208)$$

Consequently, equation (8.206) can be re-written as

$$\begin{aligned}k_e&=\omega_n{}^2+\frac{\varepsilon E\left[y^{n+1}\right]}{E\left[y^2\right]}=\omega_n{}^2+\frac{2\varepsilon\times\frac{1}{\sqrt{2\pi}\sigma_y}\frac{(\sqrt{2}\sigma_y)^{n+2}}{2}\Gamma\left(\frac{n+2}{2}\right)}{\sigma_y{}^2}\\&=\omega_n{}^2+2\varepsilon\times\frac{1}{\sqrt{2\pi}}\frac{(\sqrt{2})^{n+2}}{2}\sigma_y{}^{n-1}\Gamma\left(\frac{n+2}{2}\right)\\&=\omega_n{}^2+\varepsilon\times\sqrt{\frac{2^{n+1}}{\pi}}\sigma_y{}^{n-1}\Gamma\left(\frac{n+2}{2}\right)\end{aligned} \quad (8.209)$$

We are going to write $k_e = \omega_e^2$ in the following. Regarding the equivalent linear system (8.202), the frequency response function can be expressed as follows:

$$H_e(\omega) = \frac{1}{(\omega_e^2 - \omega^2) + i\omega\beta_e} = \frac{1}{(\omega_e^2 - \omega^2) + i\omega\alpha} \tag{8.210}$$

Then, the magnitude of $H_e(\omega)$ and the square of the magnitude of $H_e(\omega)$ can be calculated as

$$|H_e(\omega)| = \sqrt{\frac{1}{(\omega_e^2 - \omega^2)^2 + (\alpha\omega)^2}} \tag{8.211}$$

$$|H_e(\omega)|^2 = \frac{1}{(\omega_e^2 - \omega^2)^2 + (\alpha\omega)^2} \tag{8.212}$$

We have the relationship that is presented below in the event that $\omega_n^2 = \omega_e^2$:

$$|H(\omega)|^2 = \frac{1}{(\omega_n^2 - \omega^2)^2 + (\alpha\omega)^2} \tag{8.213}$$

By employing the relationship specified in equation (8.209), we proceed to express the function $|H_e(\omega)|^2$ as a Taylor series expansion in the neighborhood of ω_n^2:

$$|H_e(\omega)|^2 = |H(\omega)|^2 + \varepsilon \times \sqrt{\frac{2^{n+1}}{\pi}} \sigma_y^{n-1} \Gamma\left(\frac{n+2}{2}\right) \frac{\partial}{\partial(\omega_n^2)} |H(\omega)|^2 \tag{8.214}$$

Based on equation (8.213), it can be shown that

$$\frac{\partial}{\partial(\omega_n^2)} |H(\omega)|^2 = -2(\omega_n^2 - \omega^2)|H(\omega)|^4 \tag{8.215}$$

By combining equations (8.214) and (8.215), we obtain

$$\begin{aligned}
|H_e(\omega)|^2 &= |H(\omega)|^2 + \varepsilon \times \sqrt{\frac{2^{n+1}}{\pi}} \sigma_y^{n-1} \Gamma\left(\frac{n+2}{2}\right) \left\{-2(\omega_n^2 - \omega^2)|H(\omega)|^4\right\} \\
&= |H(\omega)|^2 - 2\varepsilon \times \sqrt{\frac{2^{n+1}}{\pi}} \sigma_y^{n-1} \Gamma\left(\frac{n+2}{2}\right) (\omega_n^2 - \omega^2) \left\{|H(\omega)|^4\right\} \\
&= |H(\omega)|^2 \left\{1 - 2\varepsilon \times \sqrt{\frac{2^{n+1}}{\pi}} \sigma_y^{n-1} \Gamma\left(\frac{n+2}{2}\right) (\omega_n^2 - \omega^2)|H(\omega)|^2\right\}
\end{aligned} \tag{8.216}$$

We can derive the following relationship between the output and input spectral density functions of the equivalent linear system in equation (8.202) by employing equation (8.216):

$$S_{yy}(\omega) = S_{FF}(\omega)|H(\omega)|^2 \left\{1 - 2\varepsilon \times \sqrt{\frac{2^{n+1}}{\pi}}\sigma_y^{\,n-1}\Gamma\left(\frac{n+2}{2}\right)(\omega_n^{\,2} - \omega^2)|H(\omega)|^2\right\} \tag{8.217}$$

By taking into account the relationship in equation (8.110), equation (8.217) can be re-written as:

$$S_{yy}(\omega) = S_{FF}(\omega)|H(\omega)|^2 \left\{1 - 2\varepsilon \times \sqrt{\frac{2^{n+1}}{\pi}}\sigma_y^{\,n-1}\Gamma\left(\frac{n+2}{2}\right)\mathrm{Re}[H(\omega)]\right\} \tag{8.218}$$

Equation (8.218) yields a result that is consistent with the result that is displayed in equation (8.142), which was obtained through the utilization of the perturbation technique. As a consequence of this, we have proved that the technique of equivalent linearization is capable of producing a result that is in agreement with the outcome that was obtained using the perturbation technique to the first order in ε.

The results that were obtained previously, which are represented by equations (8.217) and (8.218), for the nonlinear random vibration system that is specified by equation (8.201) can be further generalized to a certain extent. This being said, if the nonlinear random vibration system is governed by the equation of a more general form

$$\ddot{y} + g(y, \dot{y}) = \ddot{y} + \alpha\dot{y} + \omega_n^{\,2}y + \varepsilon \sum_{n=3,5\cdots} k_n y^n = F(t) \tag{8.219}$$

the spectral density function of the response will then be calculated as follows:

$$S_{yy}(\omega) = S_{FF}(\omega)|H(\omega)|^2 \left\{1 - 2\varepsilon\left[\sum_{n=3,5\cdots} k_n\sqrt{\frac{2^{n+1}}{\pi}}\sigma_y^{\,n-1}\Gamma\left(\frac{n+2}{2}\right)\right]\mathrm{Re}[H(\omega)]\right\} \tag{8.220}$$

By applying equation (8.220) to the scenario of a white-noise excitation, we can obtain the following result:

$$S_{yy}(\omega)=\frac{S_0}{\left(\omega_n^2-\omega^2\right)^2+(\alpha\omega)^2}$$

$$\left\{1-2\varepsilon\left[\sum_{n=3,5\cdots}k_n\sqrt{\frac{2^{n+1}}{\pi}}\sigma_y^{n-1}\Gamma\left(\frac{n+2}{2}\right)\right]\frac{\left(\omega_n^2-\omega^2\right)}{\left(\omega_n^2-\omega^2\right)^2+(\omega\alpha)^2}\right\} \tag{8.221}$$

where the spectral density of the white noise excitation force is denoted by S_0.

8.4 EXERCISES

EXERCISE 8.1

In this exercise, you are required to utilize the equivalent-linearization technique to solve a nonlinear random vibration problem. The vibration system's differential equation of motion is expressed as follows:

$$\ddot{y}+g(y,\dot{y})=\ddot{y}+\alpha\dot{y}+\beta\dot{y}^3+\omega_n^2y=F(t) \tag{8.222}$$

$F(t)$ in the above equation is a random excitation force with a power spectral density $S_{FF}(\omega)$. It can be seen that only in the damping term, which is described as $\alpha\dot{y}+\beta\dot{y}^3$, does the system that you are considering exhibit nonlinearities. The equation of motion for a linear system, which is equivalent to the nonlinear system described in equation (8.222), can be expressed as:

$$\ddot{y}+\beta_e\dot{y}+k_ey=F(t) \tag{8.223}$$

1. Please determine the equivalent stiffness coefficient k_e and the equivalent damping coefficient β_e.
2. Please derive the relationship between the output spectral density function $S_{yy}(\omega)$ and the input spectral density function $S_{FF}(\omega)$ of the equivalent linear system equation (8.223).

EXERCISE 8.2

In this exercise, you are also required to utilize the equivalent linearization technique to solve a nonlinear random vibration problem. The vibration system's differential equation of motion is expressed as follows:

$$\ddot{y}+g(y,\dot{y})=\ddot{y}+\alpha\dot{y}+\omega_n^2\left(y+\lambda y^7\right)=F(t) \tag{8.224}$$

$F(t)$ in the above equation is a random excitation force with a power spectral density $S_{FF}(\omega)$. It can be seen that only in the restoring term, which is described as $\omega_n^2\left(y+\lambda y^7\right)$, does the system that you are considering exhibit nonlinearities. The equation of motion for a linear system, which is equivalent to the nonlinear system described in equation (8.224), can be expressed as:

$$\ddot{y}+\beta_e\dot{y}+k_e y=F(t) \tag{8.225}$$

1. Please determine the equivalent stiffness coefficient k_e and the equivalent damping coefficient β_e.
2. Please derive the relationship between the output spectral density function $S_{yy}(\omega)$ and the input spectral density function $S_{FF}(\omega)$ of the equivalent linear system in equation (8.225).

EXERCISE 8.3

The following dynamic system only has a system nonlinearity term $h(y,\dot{y})$, but it does not have a nonlinear excitation term or any other nonlinear terms:

$$M\ddot{y}+C\dot{y}+Ky+h(y,\dot{y})=f(t) \tag{8.226}$$

In equation (8.226), $f(t)$ is a random excitation force, and we assume that $f(t)$ is a stationary Gaussian random process with a zero mean. After dividing each term of the equation shown above by the value of M, the equation of motion that results is as follows:

$$\ddot{y}+\alpha\dot{y}+\omega_n^2 y+h_0(y,\dot{y})=f_0(t) \tag{8.227}$$

in which $\alpha=C/M$, $\omega_n=\sqrt{K/M}$, $h_0(y,\dot{y})=h(y,\dot{y})/M$ and $f_0(t)=\dot{f}(t)/M$.

The equation of motion for a linear system that is equivalent to the nonlinear system described in equation (8.227) can be expressed as:

$$\ddot{y}+\alpha_e\dot{y}+\omega_e^2 y=f_0(t)+\varepsilon(t) \tag{8.228}$$

In equation (8.228), $\varepsilon(t)$ represents the error associated with the equivalent linearization, and it is a random process.

1. Show that the mathematical expectation of $(\varepsilon(t))^2$ is given by

$$E\left\{(\varepsilon(t))^2\right\}=(\alpha_e-\alpha)^2E\left[\dot{y}^2\right]+\left(\omega_e^2-\omega_n^2\right)^2E\left[y^2\right]+E\left[\left(h_0(y,\dot{y})\right)^2\right]$$
$$-2(\alpha_e-\alpha)E\left[h_0(y,\dot{y})\dot{y}\right]-2\left(\omega_e^2-\omega_n^2\right)E\left[h_0(y,\dot{y})y\right]$$

2. Show that the parameters α_e and ω_e^2 can be calculated as follows:

$$\alpha_e=\alpha+\frac{E\left[h_0(y,\dot{y})\dot{y}\right]}{E\left[\dot{y}^2\right]}$$

$$\omega_e^2=\omega_n^2+\frac{E\left[h_0(y,\dot{y})y\right]}{E\left[y^2\right]}$$

EXERCISE 8.4

In this exercise, we assume that the nonlinearities $h_0(y,\dot{y})$ in equation (8.227) are separable, that is, $h_0(y,\dot{y}) = h_1(y) + h_2(\dot{y})$. Therefore, the equation of motion of the nonlinear system in this exercise is

$$\ddot{y} + \alpha\dot{y} + \omega_n^2 y + h_0(y,\dot{y}) = \ddot{y} + \alpha\dot{y} + \omega_n^2 y + h_1(y) + h_2(\dot{y}) = f_0(t) \quad (8.229)$$

The equation of motion for a linear system that is equivalent to the nonlinear system described in equation (8.229) can be expressed as:

$$\ddot{y} + \alpha_e\dot{y} + \omega_e^2 y = f_0(t) + \varepsilon(t) \quad (8.230)$$

1. Show that:

$$E\left[h_0(y,\dot{y})\dot{y}\right] = E\left[h_2(\dot{y})\dot{y}\right]$$

$$E\left[h_0(y,\dot{y})y\right] = E\left[h_1(y)y\right]$$

2. Show that in this exercise the parameters α_e and ω_e^2 can be calculated as follows:

$$\alpha_e = \alpha + \frac{E\left[h_2(\dot{y})\dot{y}\right]}{E\left[\dot{y}^2\right]}$$

$$\omega_e^2 = \omega_n^2 + \frac{E\left[h_1(y)y\right]}{E\left[y^2\right]}$$

EXERCISE 8.5

In order to solve the problem in this exercise, the following relationship should be utilized:

$$E\left[Y_1Y_2\cdots Y_{2m}\right] = \sum E\left[Y_jY_k\right]E\left[Y_rY_s\right]\cdots \quad (8.231)$$

In equation (8.231), $Y_1, Y_2, \cdots Y_{2m}$ are jointly distributed Gaussian random variables, and $E[\bullet]$ denotes mathematical expectation. The sum in equation (8.231) is to be taken over all different ways by which you can group $2m$ elements into m pairs. Equation (8.231) has been derived under the assumption that the expected value of each of the jointly distributed random variables $(Y_1, Y_2, \cdots Y_{2m})$ is zero.

In this exercise, as an implementation of the perturbation technique described in this chapter, you should consider replacing the nonlinear term $\eta(y,\dot{y})$ in equation (8.66) with y^5. Therefore, equation (8.66) can be re-written as

$$\ddot{y} + \alpha\dot{y} + \omega_n^2 y + \varepsilon y^5 = F(t) \quad (8.232)$$

$F(t)$ in equation (8.232) is a weakly stationary Gaussian random process with a zero mean. Furthermore, the spectral density function of $F(t)$ is denoted by $S_{FF}(\omega)$.

1. Derive a closed-form expression for the frequency response function $H(\omega)$ of the linear system ($\varepsilon = 0$) corresponding to the nonlinear system in equation (8.232).
2. Derive a closed-form expression for the variance of the above linear system's response $\left(\sigma_{y_0}{}^2\right)$
3. Derive a closed-form expression for the relationship between the nonlinear system response displacement spectral density function $S_{yy}(\omega)$ and the input force spectral density function $S_{FF}(\omega)$.

EXERCISE 8.6

For the purpose of this exercise, you will be required to solve a nonlinear random vibration problem by employing the equivalent linearization technique. Specifically, the differential equation of motion for the vibration system is expressed as follows:

$$\ddot{y} + g(y, \dot{y}) = \ddot{y} + \alpha\dot{y} + \varepsilon\left(3\dot{y}^3 + 5\dot{y}^5\right) + \omega_n{}^2 y = F(t) \tag{8.233}$$

In the equation that was presented above, $F(t)$ represents a random excitation force that has a power spectral density function of $S_{FF}(\omega)$. Upon closer inspection, it becomes evident that the system under consideration only displays nonlinearities in the damping term, which is denoted by $\alpha\dot{y} + \varepsilon\left(3\dot{y}^3 + 5\dot{y}^5\right)$. One way to express the equation of motion for a linear system, which is equivalent to the equation (8.233) that describes a nonlinear system, is as follows:

$$\ddot{y} + \beta_e\dot{y} + k_e y = F(t) \tag{8.234}$$

1. Please calculate the equivalent stiffness coefficient k_e as well as the equivalent damping coefficient β_e.
2. In the case of a white-noise excitation, i.e., $S_{FF}(\omega) = S_0$, please derive the relationship between the output spectral density function $S_{yy}(\omega)$ and the input spectral density function S_0 of the equivalent linear system in equation (8.234).

9 Simulation Methods for Random Vibration Analysis

9.1 INTRODUCTION

In Chapters 5, 6, 7 and 8, we have examined analytical techniques for obtaining statistical properties of the response of both linear and nonlinear systems when they are exposed to random excitations. In this chapter, we will demonstrate that numerical simulation offers still another powerful method for obtaining the response statistics of random vibration systems. The goal of this chapter is to elucidate some of the fundamental methods of simulation and their applications in analyzing random vibration problems.

Figure 9.1 depicts the major stages of response and performance analysis of random vibration systems through numerical simulation (Figure 9.1). They are as follows:

1. A sample of random excitation time histories is generated.
2. The equations of motion are numerically solved using a time-stepping method for each random excitation time history, resulting in a corresponding sample of response time histories.
3. Statistical techniques are employed to calculate estimates of the requested statistics based on the sample of response time histories. The required sample size will be determined by the desired accuracy of these estimates.

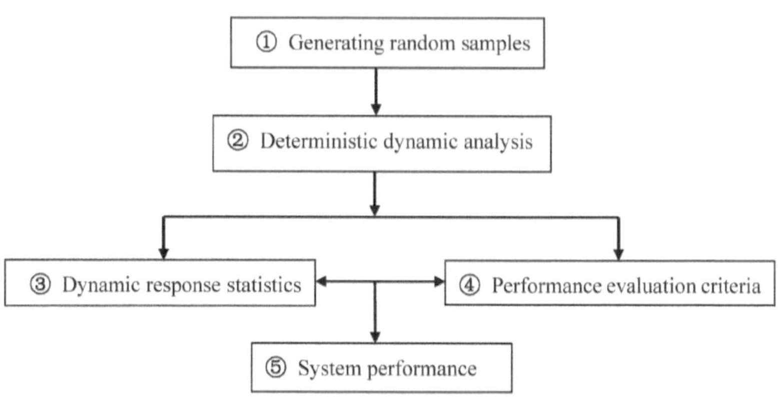

FIGURE 9.1 The numerical simulation method.

DOI: 10.1201/9781003431503-9

4. The obtained results of the dynamic response statistics will be compared with the requirements specified in the performance evaluation criteria.
5. After the comparison in stage ④, the system performance will be determined.

One of the most important characteristics of numerical simulation methods is their generality. It is possible to apply these methods to the response and performance analyses of any system that is subjected to random excitations, provided that computer algorithms are available to determine the response to deterministic excitations. The computing time, which can become excessive when dealing with very complex systems, is a serious limitation of numerical simulation. The development of high-speed computers, on the other hand, has made this limitation less serious. In the past 20 years, there has been a significant surge in the utilization of numerical simulation techniques for the purpose of solving random vibration problems. There are primarily two factors that have contributed to this growth. First, the amount of computational power that is currently accessible is enormous, even for a laptop computer, and it is always growing. Second, numerical simulation methods are particularly appealing as a means of obtaining solutions to random vibration problems because of their versatility. The drawback of numerical simulation methods for a range of problems has been that the necessary numerical calculations can be time-consuming, taking days, weeks or even months to complete. This situation, however, is undergoing changes. A decade ago, the solution to some numerical problems would have required several days of computer time; however, these problems can today be solved in a matter of minutes or hours. This has really opened the door to the use of numerical simulation methods for solving a wide array of random vibration problems.

We next provide an example of applying the numerical simulation method to solve a random vibration problem. The governing equation of motion for a nonlinear system with a single degree of freedom, including both system nonlinearity and external force nonlinearity, can be expressed as:

$$m\ddot{y}+c\dot{y}+ky+g(y,\dot{y})=F_L(v)+F_N(v,\dot{y}) \tag{9.1}$$

in which the constants m, c and k denote the mass coefficient, damping coefficient and stiffness coefficient of the nonlinear system. y, $\dot{y}$ and $\ddot{y}$ represent the system response displacement, response velocity and response acceleration, respectively. In equation (9.1), $F_L(v)$ is a linear external excitation force term. $g(y,\dot{y})$ is a system nonlinearity term, and $F_N(v,\dot{y})$ denotes a nonlinear external excitation force term. If the considered system is an offshore rig excited by Morison forces, then $F_N(v,\dot{y})$ can be considered as a nonlinear Morison drag force, with v representing the velocity of the water particles in the waves. Considering that the wave water particle velocity is a random process, the term $F_N(v,\dot{y})$ can be considered a random excitation term, thus equation (9.1) can be classified as a nonlinear stochastic differential equation. Directly finding the analytical solution to the problem (9.1) is a very challenging task. Indeed, it is a truth that in the vast majority of real-world scenarios, it is not possible to derive the analytical solution to this equation. Consequently, the numerical simulation method should be considered in order to find approximations of solutions. The main benefit of numerical simulation is that if a deterministic problem has a solution (either analytical or numerical), then the exact solution of the stochastic problem that corresponds to that deterministic problem may also be obtained. One more significant benefit of the numerical simulation method is that it can easily handle various sorts of nonlinear terms.

Professor Masanobu Shinozuka was the pioneer who established the theoretical foundation for the numerical simulation method for solving random vibration problems and provided the simulation formulations for one-dimensional and multi-dimensional Gaussian random processes. The generation of sample functions for the random processes that are present in the problem is an important part of the numerical simulation method. It is necessary for the generated sample functions to accurately express the probabilistic properties of the random processes that correspond to them. In the field of ocean engineering, this means that the sample functions for the wave water particle velocity and acceleration random processes should be generated based on a variety of wave spectral density functions that are already in existence in the literature. In this regard, Professor Masanobu Shinozuka is credited with being a pioneer in the theory of simulating random ocean waves by super-imposing a finite number of trigonometric functions with different amplitudes, frequencies and random phase angles. As an illustration, the following equation can be utilized to generate the sample function of the water particle velocity of a wave

$$v(t)=\sqrt{2}\sum_{n=0}^{N-1}\left\{S_{v0v0}\left[n(\Delta\omega)\right]\Delta\omega\right\}^{1/2}\cos\left[n(\Delta\omega)t+\Phi_n\right] \tag{9.2}$$

in which the frequency spacing is denoted by "$\Delta\omega$", while "$S_{v0v0}\left[n(\Delta\omega)\right]$" represents the one-sided power spectral density function of wave water particle velocity. This function can be obtained from the one-sided power spectral density function of wavewater particle displacement. $\Phi_0, \Phi_1, \Phi_2, \ldots, \Phi_{N-1}$ are independent random phase angles that follow a uniform distribution in the interval $[0, 2\pi]$. Upon transformation by a transfer function, equation (9.2) is subsequently substituted into v in the $F_N(v,\dot{y})$ term in equation (9.1). Following this, the fourth-order Runge–Kutta method can be utilized to numerically integrate the differential equation (9.1) in order to obtain a time history of the displacement response of the offshore rig. Additionally, a time history of the first-order derivative of the displacement response of the offshore rig can be obtained in a similar manner. A probability density curve of the displacement responses of the offshore rig can be constructed by calculating the average values of a sufficiently large number of time histories of the offshore rig's displacement responses. It is possible to construct the probability density curve of the first-order derivative of the displacement responses of the offshore rig by following a similar procedure. By utilizing the information obtained from the probability density curves of these individual random variables, it is possible to determine the joint probability density of the response values and the first-order derivatives of the response values. Subsequently, the level up-crossing rates of the responses of the offshore rig system can be calculated by employing Rice's formula. The above description illustrates the fundamental principles of the numerical simulation method and the associated simulation procedures.

During the course of this chapter, we will be discussing three topics: (1) Some techniques that are commonly used for generating samples of random inputs (such as the $v(t)$ term mentioned above); (2) some time-stepping methods (such as the fourth-order Runge–Kutta method mentioned above) for numerically solving the differential equations of motion of a deterministic vibration system; and (3) factors to take into account when choosing a particular approach for numerical simulation.

9.2 TECHNIQUES FOR GENERATING RANDOM SAMPLES

This section discusses three key techniques that are required when employing the numerical simulation method for solving random vibration problems. These techniques include the generation of uniformly distributed random numbers, the generation of random variables and the generation of sample time histories based on a specific power spectral density function.

9.2.1 Generation of Uniformly Distributed Random Numbers

For the purpose of applying the numerical simulation method to the solution of random vibration problems, it is necessary to generate random numbers for different probability distributions. The random number that follows a uniform distribution in the interval [0, 1] is considered the simplest and most important random number. On the basis of this random number, it is possible to obtain the random number with arbitrary probability distributions by employing a variety of sampling techniques. As a result, the [0, 1] interval uniformly distributed random number serves as the basis for the generation of various distributed random numbers.

Assuming that y is the random variable distributed over the interval [0, 1], its probability density function is given as

$$f(y)=\begin{cases}1, & 0\le y\le 1\\ 0, & otherwise\end{cases} \tag{9.3}$$

and its probability distribution function is given as

$$F(y)=\begin{cases}0, & y<0,\\ y, & 0\le y\le 1\\ 1, & y>1\end{cases} \tag{9.4}$$

The sequence of samples $\hat{s}_1, \hat{s}_2, \ldots, \hat{s}_{\hat{n}}, \ldots$, taken from a uniformly distributed random variable y over the interval [0, 1] is referred to as the random number uniformly distributed within the [0, 1] interval.

The generation of random numbers can be accomplished through a variety of different approaches. The use of an appropriate "pseudo" random number generator, which is available on practically all computers of numbers, is the most popular and practical option. They are considered "pseudo" due to the fact that they generate a sequence of numbers by utilizing a formula. This sequence is reproducible and repeats in a consistent manner following a lengthy cycle interval. The "pseudo" random number is usually generated using the following recurrence formula.

$$y_{\hat{n}+1} = \beta y_{\hat{n}} (\bmod M) \tag{9.5}$$

$$\hat{s}_{\hat{n}+1} = \frac{y_{\hat{n}+1}}{M} \tag{9.6}$$

where β, M and y_0 are the positive integers that have been preselected and for which semi-theoretical and semi-experiential methods are used to determine them.

According to equation (9.5), the value of $y_{\hat{n}+1}$ equals the remainder of the product $\beta y_{\hat{n}}$ divided by M. The value obtained from $y_{\hat{n}+1}$ dividing by M is the $(\hat{n}+1)th$ uniformly distributed random number, denoted as $\hat{s}_{\hat{n}+1}$. By applying the process repeatedly, a sequence of random numbers can be generated.

9.2.2 Simulation of Continuous Random Variables

The inverse-function method can be utilized to effectively simulate a random variable that follows a continuous distribution. Assume that the cumulative distribution function of the random variable y_i is denoted by the notation $F_Y(y_i)$. The numerical value of $F_Y(y_i)$ is a value that falls inside the interval of [0, 1] according to the definition. Assuming that ξ_i is the random number that is generated with a uniform distribution ($0 \le \xi_i \le 1$), the inverse-function method is utilized to equate the ξ_i to $F_Y(y_i)$ in the following manner:

$$F_Y(y_i) = \xi_i \quad \text{or} \quad y_i = F_Y^{-1}(\xi_i) \tag{9.7}$$

This approach is applicable to variables for which a cumulative distribution function has been determined through direct observation or an analytic expression for the inverse cumulative function, $F^{-1}(.)$, is available.

As an illustration, given a random variable with a Rayleigh distribution, we have

$$f_Y(y) = \frac{y}{a^2} \exp\left(-\frac{y^2}{2a^2}\right), \quad y \geqslant 0 \tag{9.8}$$

$$F_Y(y) = 1 - \exp\left(-\frac{y^2}{2a^2}\right) = \xi, \quad y \geqslant 0 \tag{9.9}$$

in which a is the distribution parameter and $0 \le \xi \le 1$. This random variable can be obtained from the generated random number ξ_i (with a uniform distribution in the interval [0, 1]) as

$$y_i = a\left[-2\ln\left(1-\xi_i\right)\right]^{1/2} \tag{9.10}$$

9.2.3 Generation of Sample Time Histories Based on a Specific Spectrum

In the field of mechanics, there are currently a variety of different approaches that can be utilized to solve a large number of problems that involve uncertain quantities that are described by random processes. At this time, the numerical simulation method, however, appears to be the only universal method that can provide accurate solutions for certain problems in stochastic mechanics involving nonlinearity, system stochasticity, large variations of uncertain parameters and so on. Additionally, the numerical simulation method is able to assess the accuracy of other approximate methods, such as perturbation, equivalent linearization, stochastic averaging and so on.

When it comes to the numerical simulation method, the generation of sample functions of the random processes that are involved in the problem is one of the most important parts of the method. The sample functions that are generated must accurately describe the probabilistic characteristics of the corresponding random processes, which may be either stationary or non-stationary, one-dimensional or multidimensional, univariate or multivariate, Gaussian or non-Gaussian. In this section, the following representative methods are identified from the extensive bibliography that pertains to the diverse methodologies that are currently available for the generation of such sample functions: the spectral representation method; the covariance decomposition method; simulation using fast Hartley transform; simulation using a fast Fourier transform (FFT) based approach; etc.

Among the various methods that have been stated above for the purpose of generating sample functions of random processes, *the spectral representation method* is one of the most extensively used methods in the present day due to its versatility and robustness. Using this method, sample functions of random processes are generated in accordance with the power spectral density information that has been given. One significant benefit of the spectral representation method is that there are currently spectral representation-based algorithms that can be used to simulate random processes that can be either stationary or non-stationary, one-dimensional or multi-dimensional, univariate or multivariate, Gaussian or non-Gaussian. Furthermore, these algorithms provide a high level of simplicity and robustness, and it is not difficult to develop computer programs based on them. We next present the basic simulation formulas of the spectral representation method.

Let's examine a 1D-1V stationary random process $f_0(t)$ that has a mean of zero. This random process is characterized by an autocorrelation function $R_{f_0 f_0}(\tau)$ and a one-sided power spectral density function $S_{f_0 f_0}(\omega)$. As N approaches infinity, the random process $f_0(t)$ can be simulated using the following series:

$$f(t) = \sqrt{2}\sum_{n=0}^{N-1} A_n \cos(\omega_n t + \Phi_n), \tag{9.11}$$

in which

$$A_n = \left(S_{f_0 f_0}(\omega_n)\Delta\omega\right)^{1/2}, \quad n = 0, 1, 2, \ldots N-1 \tag{9.12}$$

$$\omega_n = n\Delta\omega, \qquad n = 0, 1, 2, \ldots N-1, \tag{9.13}$$

$$\Delta\omega = \omega_u / N \tag{9.14}$$

In equation (9.14), ω_u denotes an upper cutoff frequency beyond which the power spectral density function $S_{f_0 f_0}(\omega)$ can be assumed to be zero for reasons that are either mathematical or physical in nature. It is common practice to estimate the value of ω_u by applying the criteria that are stated below:

$$\int_0^{\omega_u} S_{f_0 f_0}(\omega)d\omega = (1-\epsilon)\int_0^{\infty} S_{f_0 f_0}(\omega)d\omega \tag{9.15}$$

where $\epsilon \le 1$ (for example, $\epsilon = 0.0005$).

In equation (9.11), Φ_0, Φ_1, $\Phi_2, \ldots$, Φ_{N-1} represent independent random phase angles uniformly distributed in the interval $[0, 2\pi]$. The simulated random process $f(t)$ given by equation (9.11) is periodic with period T_0:

$$T_0 = \frac{2\pi}{\Delta\omega} \tag{9.16}$$

One other feature that is of great significance is that the random process $f(t)$ that is simulated is asymptotically Gaussian as $N \to \infty$. It should also be mentioned here that when the sample functions for the simulated random process are generated in accordance with equation (9.11), the time steps that separate the generated $f(t)$ values in the time domain should satisfy the following requirements

$$\Delta t \le 2\pi / 2\omega_u \tag{9.17}$$

The procedures from equation (9.11) to equation (9.17) delineate the mathematical theories involved in utilizing the spectral representation method to simulate a time history of a random process.

Example 9.1

▶ In this example, we will utilize the spectral representation method to simulate random ocean waves for a sea state characterized by a JOint North Sea WAve Project (JONSWAP) wave spectrum. A multinational initiative was undertaken to characterize standardized wave spectra for the Southeast region of the North Sea. The result of this research is the JONSWAP spectrum. In the case of sea states that are not yet fully developed, the JONSWAP spectrum is valid. However, it is also used to represent sea states that are fully developed. This spectrum is especially suitable for characterizing wind-generated waves when $3.6\sqrt{H_{m0}} < T_p < 5\sqrt{H_{m0}}$, in which T_p represents the peak spectral period and H_{m0} represents the significant wave height. The one-sided JONSWAP spectrum is expressed mathematically as follows:

$$S(\omega) = \beta \frac{g^2}{\omega^M} \exp\left(-\frac{M}{N}\left(\frac{2\pi}{\omega T_p}\right)^N\right) \gamma^{\exp\left(-\frac{\left(\frac{\omega T_p}{2\pi}-1\right)^2}{2\delta^2}\right)} \tag{9.18}$$

where $M = 5$ and $N = 4$, and

$$\delta = 0.07 \quad if \quad \omega < 2\pi/T_p$$

$$\delta = 0.09 \quad if \quad \omega \geq 2\pi/T_p$$

$$\beta = 5.061 \frac{H_{m0}^2}{T_p^4}\{1 - 0.287\ln(\gamma)\}$$

$$\gamma = \exp\left(3.484\left(1 - 0.1975\left(0.036 - 0.0056T_p/\sqrt{H_{m0}}\right)T_p^4/H_{m0}^2\right)\right)$$

In equation (9.18), ω denotes the wave angular frequency in rad/s, H_{m0} represents the significant wave height of a specific sea state, T_p denotes the peak spectral period of this sea state and γ represents the spectral peakedness factor. In this example, we will select H_{m0} = 7m and T_p = 11s and then we will utilize the spectral representation method to simulate a wave elevation time history record with a duration of 10s based on this JONSWAP wave spectrum. We will select $\Delta t = 0.05$s as the time spacing between the two neighboring wave elevation points that we will simulate.

SOLUTION

The one-sided JONSWAP wave spectrum with H_{m0} = 7m and T_p = 11s is shown in the following Figure 9.2.

Utilizing the procedures from equations (9.11)–(9.17) of the spectral representation method and N = 128, we have simulated a wave elevation time history record with a duration of 100s based on the JONSWAP wave spectrum as shown in Figure 9.2. Our simulated results are displayed in Figure 9.3. ◀

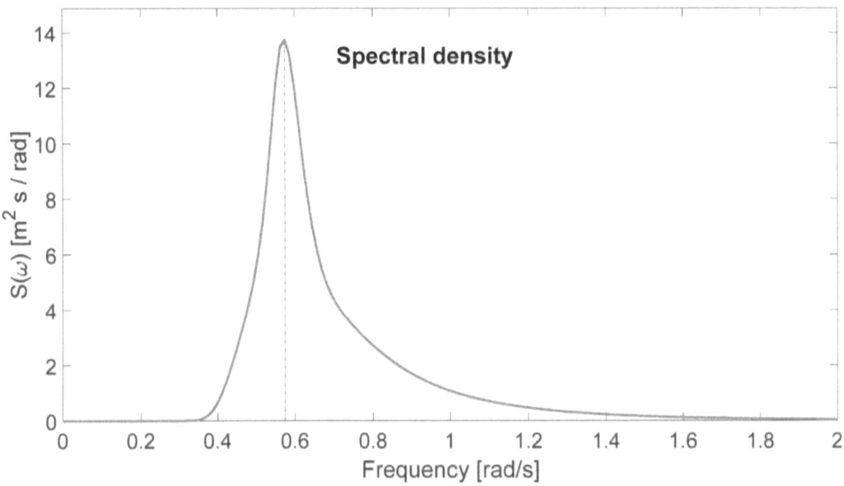

FIGURE 9.2 The JONSWAP wave spectrum with $H_{m0} = 7\text{m}$ and $T_p = 11\text{s}$.

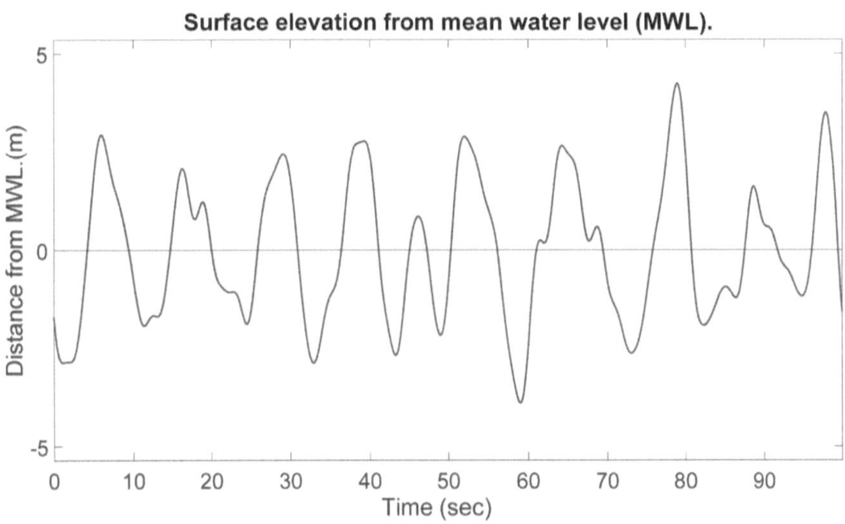

FIGURE 9.3 The simulated wave elevation time record with a duration of 100s.

9.3 NUMERICAL INTEGRATION METHODS IN RANDOM VIBRATION ANALYSIS

We have elucidated in the Introduction section in this chapter the major stages of response and performance analysis of random vibration systems through numerical simulation. We have explained that after a sample of random excitation time histories is generated, the equations of motion should then be numerically solved using a time-stepping method for each random excitation time history for obtaining a corresponding sample of response time histories. In this section, we will discuss some time-stepping methods for numerically integrating the differential equations of motion in the random vibration analysis.

When it is not possible to integrate the differential equation of motion of a vibrating system in closed form, it is necessary to employ a numerical approach. There are a number of numerical integration methods (numerical time-stepping methods) that can be utilized in order to solve vibrational problems. There are two fundamental properties that are shared by numerical integration methods (numerical time-stepping methods). First, it is important to note that they are not designed to satisfy the governing differential equation(s) of the vibrating system at all time t, but rather only at discrete time intervals Δt apart. Second, it is assumed that there is an appropriate sort of variation of the displacement y, velocity $\dot{y}$ and acceleration $\ddot{y}$ within each time interval Δt. It is possible to obtain different numerical integration methods (numerical time-stepping methods), depending on the type of variation assumed for the displacement y, velocity $\dot{y}$ and acceleration $\ddot{y}$, within each time interval Δt. We are going to make the assumption that the values of y and $\dot{y}$ are known to be y_0 and $\dot{y}_0$, respectively, at time $t = 0$, and that the solution to the problem is required from the time $t = 0$ to the time $t = T$. In the following, we will divide the time duration T into n equal steps Δt so that $\Delta t = T/n$. We will then seek the solution at $t_0 = 0$, $t_1 = \Delta t$, $t_2 = 2\Delta t, \ldots$, $t_n = n\Delta t = T$. For the purpose of determining the solution at $t_i = i\Delta t$ from the known solution at $t_{i-1} = (i-1)\Delta t$, we will derive formulas according to five different numerical integration schemes, namely the finite difference method, the Runge–Kutta method, the Houbolt method, the Wilson method and the Newmark method. In the Runge–Kutta method and the finite difference method, the current displacement (solution) is expressed in terms of the values of displacement, velocity and acceleration that were determined previously. The resulting equations are then solved in order to obtain the current displacement. These methods are classified as explicit integration methods. When using the Houbolt method, the Wilson method and the Newmark method, the current equations of motion are combined with the temporal difference equations, and then the resulting equations are solved in order to obtain the current displacement. These methods are classified as implicit integration methods.

9.3.1 The Finite Difference Method

Employing approximations to derivatives is the core concept behind the finite difference method. Therefore, the differential equation that governs the system's motion and the associated boundary conditions, if applicable, are replaced by the corresponding finite difference equations. Deriving the finite difference equations can be accomplished by the utilization of three different types of formulas: forward, backward and center difference formulas. Because the central difference formulas are the most accurate, we will only consider them in this chapter.

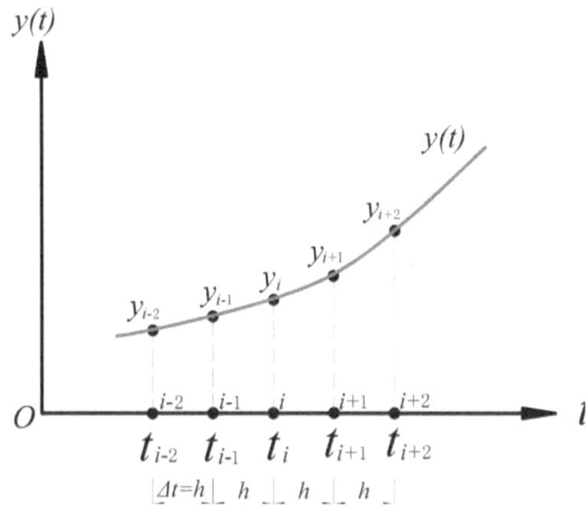

FIGURE 9.4 The grid points in the finite difference method.

The finite difference method involves the replacement of the solution domain (over which the solution of the given differential equation is required) with a finite number of points, known as mesh or grid points, and the subsequent determination of the values of the desired solution at these points. It is common practice to consider the grid points to be equally spaced along each of the independent coordinates (see Figure 9.4).

With the use of Taylor's series expansion, the values of y_{i+1} and y_{i-1} can be expressed about the grid point i as

$$y_{i+1} = y_i + \dot{y}_i \Delta t + \ddot{y}_i \frac{(\Delta t)^2}{2!} + \dddot{y}_i \frac{(\Delta t)^3}{3!} + \ddddot{y}_i \frac{(\Delta t)^4}{4!} + \cdots \tag{9.19}$$

$$y_{i-1} = y_i - \dot{y}_i \Delta t + \ddot{y}_i \frac{(\Delta t)^2}{2!} - \dddot{y}_i \frac{(\Delta t)^3}{3!} + \ddddot{y}_i \frac{(\Delta t)^4}{4!} + \cdots \tag{9.20}$$

in which $y_i = y(t = t_i)$. The central difference approximation to the first derivative of y at $t = t_i$ can be obtained by taking only two terms and subtracting equation (9.20) from equation (9.19):

$$\dot{y}_i = \left.\frac{dy}{dt}\right|_{t_i} = \frac{1}{2\Delta t}\left(y_{i+1} - y_{i-1}\right) \tag{9.21}$$

The central difference approximation to the second derivative of y at $t = t_i$ can be obtained by taking terms up to the second derivative and adding equation (9.20) and equation (9.19):

$$\ddot{y}_i = \left.\frac{d^2 y}{dt^2}\right|_{t_i} = \frac{1}{(\Delta t)^2}\left(y_{i+1} - 2y_i + y_{i-1}\right) \tag{9.22}$$

The governing equation of motion for a viscously damped vibration system with a single degree of freedom can be expressed as:

$$m\ddot{y}+c\dot{y}+ky=m\frac{d^2y}{dt^2}+c\frac{dy}{dt}+ky=F(t) \tag{9.23}$$

where the constants m, c and k denote the mass coefficient, damping coefficient and stiffness coefficient of the vibration system. y, $\dot{y}$ and $\ddot{y}$ denote the vibration system response displacement, response velocity and response acceleration, respectively. In equation (9.23), $F(t)$ denotes an external excitation force term. Let us divide the time period over which the solution to equation (9.23) is required into n equal parts, each of which is an interval with a value of $h=\Delta t$. Let the initial conditions be represented by $y(t=0)=y_0$ and $\dot{y}(t=0)=\dot{y}_0$. By substituting the derivatives with the central differences and writing equation (9.23) at the grid point i, we can obtain the following:

$$\frac{m}{(\Delta t)^2}\{y_{i+1}-2y_i+y_{i-1}\}+\frac{c}{2\Delta t}\{y_{i+1}-y_{i-1}\}+k\,y_i=F_i \tag{9.24}$$

in which $y_i=y(t_i)$ and $F_i=F(t_i)$. When equation (9.24) is solved for y_{i+1}, the result is

$$y_{i+1}=\left[\frac{1}{m/(\Delta t)^2+c/(2\Delta t)}\right]\left\{\left[\frac{2m}{(\Delta t)^2}-k\right]y_i+\left[\frac{c}{2\Delta t}-\frac{m}{(\Delta t)^2}\right]y_{i-1}+F_i\right\} \tag{9.25}$$

This is what is known as the recurrence formula. It permits us to compute the displacement of the mass (y_{i+1}) if we are aware of the previous history of displacements at t_i and t_{i-1}, as well as the present external force F_i. By repeatedly applying equation (9.25), one can obtain the complete time history of the behavior of the system. It is important to keep in mind that the solution of y_{i+1} is arrived at by employing the equilibrium equation at time t_i – that is, equation (9.24). For this reason, this integration procedure is referred to as an explicit integration method. Explicit numerical integration methods are only conditionally stable. In other words, they offer bounded numerical solutions only for (very) small time steps, for instance, $\Delta t\le\tau_{\text{crit}}=T_n/\pi$, in which T_n is the undamped natural period of the vibration system.

Applying equation (9.25) to the case when $i=0$ requires a certain amount of care to be exercised. Given that the calculation of y_1 requires both y_0 and y_{-1}, and the initial conditions only offer the values of y_0 and $\dot{y}_0$, it is necessary to determine the value of y_{-1}. On account of this, the method is not self-starting. Nevertheless, we can calculate the value of y_{-1} by utilizing equations (9.21) and (9.22) in the following manner. By plugging in the given values of y_0 and $\dot{y}_0$ into equation (9.23), the value of $\ddot{y}_0$ can be determined:

$$\ddot{y}_0=\frac{1}{m}\left[F(t=0)-c\,\dot{y}_0-k\,y_0\right] \tag{9.26}$$

Using equations (9.21) and (9.22) at $i = 0$, the following value of y_{-1} is obtained:

$$y_{-1} = y_0 - \dot{y}_0 \Delta t + \ddot{y}_0 \frac{(\Delta t)^2}{2} \tag{9.27}$$

We now extend the foregoing central difference method for solving vibration problems of multi-degree-of-freedom systems. Let us consider a viscously damped spring-mass-damper system that has many degrees of freedom. The equation of motion of this system can be expressed as

$$\mathbf{M}\ddot{\mathbf{y}} + \mathbf{C}\dot{\mathbf{y}} + \mathbf{K}\mathbf{y} = \mathbf{F}(t) \tag{9.28}$$

in which $\mathbf{M}$, $\mathbf{C}$ and $\mathbf{K}$ are referred to as the mass, damping and stiffness matrices, respectively. In equation (9.28), $\mathbf{y}$, $\dot{\mathbf{y}}$, $\ddot{\mathbf{y}}$ and $\mathbf{F}(t)$ are referred to as the displacement, velocity, acceleration and force vectors, respectively. One can directly apply the central difference method that was described for the scenario of a system with a single degree of freedom to this case of a multi-degree-of-freedom system. The central difference formulas for the velocity and acceleration vectors at a particular time $t_i = i\Delta t$ ($\dot{\mathbf{y}}$ and $\ddot{\mathbf{y}}$) are given by

$$\dot{\mathbf{y}}_i = \frac{1}{2\Delta t}\left(\mathbf{y}_{i+1} - \mathbf{y}_{i-1}\right) \tag{9.29}$$

$$\ddot{\mathbf{y}}_i = \frac{1}{(\Delta t)^2}\left(\mathbf{y}_{i+1} - 2\mathbf{y}_i + \mathbf{y}_{i-1}\right) \tag{9.30}$$

It is therefore possible to write the equation of motion in equation (9.28) at the time t_i as

$$\frac{\mathbf{M}}{(\Delta t)^2}\left(\mathbf{y}_{i+1} - 2\mathbf{y}_i + \mathbf{y}_{i-1}\right) + \frac{\mathbf{C}}{2\Delta t}\left(\mathbf{y}_{i+1} - \mathbf{y}_{i-1}\right) + \mathbf{K}\mathbf{y}_i = \mathbf{F}_i \tag{9.31}$$

in which $\mathbf{y}_{i+1} = \mathbf{y}(t = t_{i+1})$, $\mathbf{y}_i = \mathbf{y}(t = t_i)$, $\mathbf{y}_{i-1} = \mathbf{y}(t = t_{i-1})$, $\mathbf{F}_i = \mathbf{F}(t = t_i)$ and $t_i = i\Delta t$. After rearranging the equation (9.31), one can obtain

$$\left(\frac{\mathbf{M}}{(\Delta t)^2} + \frac{\mathbf{C}}{2\Delta t}\right)\mathbf{y}_{i+1} = \mathbf{F}_i - \left(\mathbf{K} - \frac{2\mathbf{M}}{(\Delta t)^2}\right)\mathbf{y}_i + \left(\frac{\mathbf{C}}{2\Delta t} - \frac{\mathbf{M}}{(\Delta t)^2}\right)\mathbf{y}_{i-1} \tag{9.32}$$

As a result, equation (9.32) provides the solution vector $\mathbf{y}_{i+1}$ once both $\mathbf{y}_i$ and $\mathbf{y}_{i-1}$ are known. Given that equation (9.32) is to be utilized for $i = 0, 1, 2, \ldots, n$, the evaluation of $\mathbf{y}_1$ necessitates the utilization of $\mathbf{y}_0$ and $\mathbf{y}_{-1}$. Therefore, a special starting procedure is needed to determine $\mathbf{y}_{-1} = \mathbf{y}(t = -\Delta t)$. To accomplish this, equations (9.28) to (9.30) are evaluated at $i = 0$ to obtain

$$\mathbf{M}\ddot{\mathbf{y}}_0 + \mathbf{C}\dot{\mathbf{y}}_0 + \mathbf{K}\mathbf{y}_0 = \mathbf{F}_0 = \mathbf{F}(t = 0) \tag{9.33}$$

$$\dot{\mathbf{y}}_0 = \frac{1}{2\Delta t}(\mathbf{y}_1 - \mathbf{y}_{-1}) \tag{9.34}$$

$$\ddot{\mathbf{y}}_0 = \frac{1}{(\Delta t)^2}(\mathbf{y}_1 - 2\mathbf{y}_0 + \mathbf{y}_{-1}) \tag{9.35}$$

According to equation (9.33), the initial acceleration vector is as follows:

$$\ddot{\mathbf{y}}_0 = \mathbf{M}^{-1}(\mathbf{F}_0 - \mathbf{C}\dot{\mathbf{y}}_0 - \mathbf{K}\mathbf{y}_0) \tag{9.36}$$

Equation (9.34) provides the displacement vector at the time t_1 as follows:

$$\mathbf{y}_1 = 2\Delta t\dot{\mathbf{y}}_0 + \mathbf{y}_{-1} \tag{9.37}$$

Substituting equation (9.37) for $\mathbf{y}_1$, equation (9.35) leads to

$$\ddot{\mathbf{y}}_0 = \frac{2}{(\Delta t)^2}(\Delta t\dot{\mathbf{y}}_0 + \mathbf{y}_{-1} - \mathbf{y}_0) \tag{9.38}$$

or

$$\mathbf{y}_{-1} = \mathbf{y}_0 - \Delta t\dot{\mathbf{y}}_0 + \frac{(\Delta t)^2}{2}\ddot{\mathbf{y}}_0 \tag{9.39}$$

9.3.2 The Runge–Kutta Method

Let us take into consideration the first-order differential equation and the initial condition that follows:

$$\frac{dy}{dt} = f(t, y), \quad y(t_0) = y_0 \tag{9.40}$$

The theory behind Euler's approach is based on an expansion of the solution using the Taylor series around y_0, or

$$y(t_0 + h) = y(t_0) + hy'(t_0) + \frac{1}{2}y''(\xi)h^2 \quad t_0 < \xi < t_0 + h \tag{9.41}$$

in which h represents the size of the step. Euler's approach involves selecting a value for h that is sufficiently small to ensure that the first two terms of this Taylor expansion are the only ones that possess any significance.

Let us now substitute $f(t_0, y_0)$ for $y'(t_0)$ in the above equation. We obtain

$$y(t_0+h) = y(t_0) + hf(t_0, y_0) + \frac{1}{2}y''(\xi)h^2 \quad t_0 < \xi < t_0 + h \tag{9.42}$$

By using the notation of subscripts, we have that

$$y_{i+1} = y_i + hf(t_i, y_i) + O(h^2) \tag{9.43}$$

According to equation (9.43), if we are aware of the values of y_i and $f(t_i, y_i)$ at the position t_i, then it is possible to derive the solution at t_{i+1} with an error $O(h^2)$.

The problem with Euler's approach is that it is not very accurate, and it frequently calls for an extremely small time step. How can we make this method more efficient while requiring only a little amount of more work? One approach that could be taken would be to keep the first three terms of the Taylor expansion rather than the first two terms. This scheme, known as the modified Euler method, is mathematically expressed as

$$y_{i+1} = y_i + hy'(t_i) + \frac{1}{2}h^2 y_i'' + O(h^3) \tag{9.44}$$

It is very evident that this is a more accurate model than equation (9.43). Due to the fact that we do not possess any information regarding the value of y_i'', the question that arises is quite obvious: how do we evaluate it? Through the utilization of the forward derivative approximation, we find that

$$y_i'' = \frac{y_{i+1}' - y_i'}{h} \tag{9.45}$$

By substituting equation (9.45) into equation (9.44) and simplifying the equation,

$$y_{i+1} = y_i + \frac{h}{2}(y_{i+1}' + y_i') + O(h^3) = y_i + \frac{h}{2}\left(f(t_{i+1}, y_{i+1}) + f(t_i, y_i)\right) + O(h^3) \tag{9.46}$$

In spite of the fact that $f(t_i, y_i)$ at (t_i, y_i) can be easily computed, the question remains: how can we compute $f(t_{i+1}, y_{i+1})$ at (t_{i+1}, y_{i+1})? In order to accomplish this, we first compute a first guess using the Euler approach, which is equation (9.43). Equation (9.46) then offers a refinement on the value of y_{i+1}.

Therefore, to sum up, the simple Euler scheme is

$$y_{i+1} = y_i + k_1 + O(h^2), \qquad k_1 = hf(t_i, y_i) \tag{9.47}$$

while the modified Euler scheme is

$$y_{i+1} = y_i + \frac{1}{2}(k_1 + k_2) + O(h^3), \quad k_1 = hf(t_i, y_i), \quad k_2 = hf(t_i + h, y_i + k_1) \tag{9.48}$$

The accuracy of numerical solutions of ordinary differential equations can be enhanced by an increase in the number of terms that are included in the Taylor expansion, as we have just demonstrated. The Runge–Kutta method builds upon this idea, similar to how the modified Euler scheme did. Let us make the assumption that the numerical solution can be approximated by

$$y_{i+1} = y_i + ak_1 + bk_2 \tag{9.49}$$

in which

$$k_1 = hf(t_i, y_i), \quad k_2 = hf(t_i + A_1 h, y_i + B_1 k_1) \tag{9.50}$$

In this case, there are four unknowns: a, b, A_1 and B_1. Equation (9.49) was suggested by the modified Euler scheme that we have just introduced. In that case, the error in the truncated Taylor series was $O(h^3)$. We anticipate that such an error will occur in the present case.

Since the Taylor series expansion of $f(t+h, y+k)$ about (t, y) is

$$\begin{aligned} f(t+h, y+k) = f(t,y) &+ \left[f_t(t,y)h + f_y(t,y)k\right] \\ &+ \frac{1}{2!}\left[(h)^2 f_{tt}(t,y) + 2hkf_{ty}(t,y) + (k)^2 f_{yy}(t,y)\right] \\ &+ \frac{1}{3!}\left[h^3 f_{ttt}(t,y) + 3h^2 k f_{tty}(t,y) + 3hk^2 f_{tyy}(t,y) + k^3 f_{yyy}(t,y)\right] + \cdots. \end{aligned} \tag{9.51}$$

k_2 in equation (9.50) can be re-written as

$$\begin{aligned} k_2 &= hf\left(t_i + A_1 h, y_i + B_1 hf(t_i, y_i)\right) \\ &= h\left\{f(t_i, y_i) + \left[A_1 hf_{t_i}(t_i, y_i) + B_1 hf(t_i, y_i) f_{y_i}(t_i, y_i)\right]\right\} \\ &= hf(t_i, y_i) + A_1 h^2 f_{t_i}(t_i, y_i) + B_1 h^2 f(t_i, y_i) f_{y_i}(t_i, y_i) \end{aligned} \tag{9.52}$$

in which we have retained only the terms up to $O(h^2)$ and neglected all the higher-order terms. Finally, the substitution of equation (9.52) into equation (9.49) results in

$$y_{i+1} = y_i + (a+b)hf(t_i, y_i) + \left[A_1 bf_{t_i}(t_i, y_i) + B_1 bf(t_i, y_i) f_{y_i}(t_i, y_i)\right]h^2 \tag{9.53}$$

Equation (9.53) corresponds to the second-order Taylor series expansion:

$$y_{i+1} = y_i + hy_i' + \frac{1}{2}h^2 y_i'' \tag{9.54}$$

Consequently, if we want to find a solution to the differential equation $y' = f(t, y)$, then

$$y_i'' = f_{t_i}(t_i, y_i) + f_{y_i}(t_i, y_i) y_i' = f_{t_i}(t_i, y_i) + f_{y_i}(t_i, y_i) f_{t_i}(t_i, y_i) \tag{9.55}$$

We obtain the following result by substituting equation (9.55) into equation (9.54):

$$y_{i+1} = y_i + hf_{t_i}(t_i, y_i)_i + \frac{1}{2}h^2\left[f_{t_i}(t_i, y_i) + f_{t_i}(t_i, y_i) f_{y_i}(t_i, y_i)\right] \tag{9.56}$$

A direct comparison of equations (9.53) and (9.56) leads to

$$(a+b) = 1, \qquad A_1 b = \frac{1}{2}, \qquad B_1 b = \frac{1}{2} \tag{9.57}$$

In the above equations, there are four variables that are unknown. If we select $a = \frac{1}{2}$, we immediately obtain $b = \frac{1}{2}$ and $A_1 = B_1 = 1$. Therefore, the Runge–Kutta scheme of the second order is

$$y_{i+1} = y_i + \frac{1}{2}(k_1 + k_2) \tag{9.58}$$

in which $k_1 = hf(t_i, y_i)$ and $k_2 = hf(t_i + h, y_i + k_1)$. Therefore, the second-order Runge–Kutta scheme is the same as the modified Euler scheme.

While the derivation of the second-order Runge–Kutta scheme results in the modified Euler scheme, it does provide a framework for computing higher-order and more accurate Runge–Kutta schemes. One that is particularly popular is the Runge–Kutta scheme of the fourth order:

$$y_{i+1} = y_i + h(ak_1 + bk_2 + ck_3 + dk_4) + O(h^5) \tag{9.59}$$

in which

$$k_1 = f(t_i, y_i) \tag{9.60}$$

$$k_2 = f\left(t_i + \frac{1}{2}h, y_i + \frac{h}{2}k_1\right) \tag{9.61}$$

$$k_3 = f\left(t_i + \frac{1}{2}h, y_i + \frac{h}{2}k_2\right) \tag{9.62}$$

$$k_4 = f(t_i + h, y_i + hk_3) \tag{9.63}$$

We next provide the detailed derivation process for obtaining this fourth-order Runge–Kutta scheme. Let $f = (f)_{t=t_i} = f(t_i, y_i) = k_i$. We recall that a Taylor series of a real function in two variables $f(x, y)$ is given by

$$\begin{aligned} f(x+\Delta x, y+\Delta y) &= f(x,y) + \left[f_x(x,y)\Delta x + f_y(x,y)\Delta y \right] \\ &+ \frac{1}{2!}\left[(\Delta x)^2 f_{xx}(x,y) + 2\Delta x \Delta y f_{xy}(x,y) + (\Delta y)^2 f_{yy}(x,y) \right] \\ &+ \frac{1}{3!}\left[(\Delta x)^3 f_{xxx}(x,y) + 3(\Delta x)^2 \Delta y f_{xxy}(x,y) + 3\Delta x (\Delta y)^2 f_{xyy}(x,y) \right. \\ &\left. + (\Delta y)^3 f_{yyy}(x,y) \right] + \cdots. \end{aligned} \tag{9.64}$$

Equation (9.64) can be written in another form:

$$f(x_1 + a_1,\ x_2 + a_2) = \sum_{j=0}^{\infty} \left\{ \frac{1}{j!} \left(\sum_{k=1}^{2} a_k \frac{\partial}{\partial x'_k} \right)^j f(x'_1, \ldots, x'_n) \right\}_{x'_1 = x_1,\ x'_2 = x_2} \tag{9.65}$$

Substituting the two-dimensional Taylor's series expansion of the function $f\left(t_i + \frac{1}{2}h, y_i + \frac{h}{2}(f)_{t=t_i} \right)$ into equation (9.61) leads to

$$k_2 = f\left(t_i + \frac{1}{2}h, y_i + \frac{1}{2}hf(t_i, y_i) \right) = f(t_i, y_i) + \frac{1}{2}h\left[\frac{\partial f}{\partial t} + f(t_i, y_i)\frac{\partial f}{\partial y} \right]_{t=t_i} + O(h^2) \tag{9.66}$$

For $dy(t)/dt = f[t, y(t)]$, let $f = (f)_{t=t_i} = f(t_i,\ y_i)$, then the total derivative of $f(t_i,\ y_i)$ with respect to time is

$$\frac{d}{dt} f(t_i, y_i) = \frac{\partial f(t_i, y_i)}{\partial t} + f\frac{\partial f(t_i, y_i)}{\partial y} = \frac{\partial f(t_i, y_i)}{\partial t} + \dot{y}_i \frac{\partial f(t_i, y_i)}{\partial y} = \ddot{y}_i \tag{9.67}$$

Considering equation (9.67), equation (9.66) can be re-written as

$$k_2 = f\left(t_i + \frac{1}{2}h, y_i + \frac{1}{2}hf(t_i, y_i) \right) = f(t_i, y_i) + \frac{h}{2}\frac{d}{dt} f(t_i, y_i) \tag{9.68}$$

Substituting the two-dimensional Taylor's series expansion of the function $f\left(t_i + \frac{1}{2}h, y_i + \frac{h}{2}k_2 \right)$ into equation (9.62) leads to

$$k_3 = f\left(t_i + \frac{1}{2}h, y_i + \frac{h}{2}k_2 \right) = f(t_i, y_i) + \frac{1}{2}h\left[\frac{\partial f}{\partial t} + k_2\frac{\partial f}{\partial y} \right]_{t=t_i} + O(h^2) \tag{9.69}$$

Considering equations (9.66) and (9.68), equation (9.69) can be re-written as

$$\begin{aligned} k_3 &= f\left(t_i + \frac{1}{2}h,\ y_i + \frac{h}{2}k_2\right) = f\left(t_i + \frac{1}{2}h,\ y_i + \frac{h}{2}f\left(t_i + \frac{1}{2}h,\ y_i + \frac{1}{2}hk_1\right)\right) \\ &= f\left(t_i + \frac{1}{2}h,\ y_i + \frac{h}{2}f\left(t_i + \frac{1}{2}h,\ y_i + \frac{1}{2}hf\left(t_i,\ y_i\right)\right)\right) \\ &= f\left(t_i,\ y_i\right) + \frac{h}{2}\frac{d}{dt}f\left(t_i + \frac{1}{2}h,\ y_i + \frac{1}{2}hf\left(t_i,\ y_i\right)\right) \\ &= f\left(t_i,\ y_i\right) + \frac{h}{2}\frac{d}{dt}\left[f\left(t_i,\ y_i\right) + \frac{h}{2}\frac{d}{dt}f\left(t_i,\ y_i\right)\right] \end{aligned} \tag{9.70}$$

Substituting the two-dimensional Taylor's series expansion of the function $f\left(t_i + h, y_i + hk_3\right)$ into equation (9.63) leads to

$$k_4 = f\left(t_i + h, y_i + hk_3\right) = f\left(t_i, y_i\right) + h\left[\frac{\partial f}{\partial t} + k_3\frac{\partial f}{\partial y}\right]_{t=t_i} + O\left(h^2\right) \tag{9.71}$$

Considering equations (9.66), (9.68) and (9.70), equation (9.71) can be re-written as

$$\begin{aligned} k_4 &= f\left(t_i + h,\ y_i + hk_3\right) = f\left(t_i + h,\ y_i + hf\left(t_i + \frac{1}{2}h,\ y_i + \frac{h}{2}k_2\right)\right) \\ &= f\left(t_i + h,\ y_i + hf\left(t_i + \frac{1}{2}h,\ y_i + \frac{h}{2}f\left(t_i + \frac{1}{2}h,\ y_i + \frac{1}{2}hf\left(t_i,\ y_i\right)\right)\right)\right) \\ &= f\left(t_i,\ y_i\right) + h\frac{d}{dt}f\left(t_i + \frac{1}{2}h,\ y_i + \frac{h}{2}f\left(t_i + \frac{1}{2}h,\ y_i + \frac{1}{2}hf\left(t_i,\ y_i\right)\right)\right) \\ &= f\left(t_i,\ y_i\right) + h\frac{d}{dt}\left\{f\left(t_i,\ y_i\right) + \frac{h}{2}\frac{d}{dt}\left[f\left(t_i,\ y_i\right) + \frac{h}{2}\frac{d}{dt}f\left(t_i,\ y_i\right)\right]\right\} \end{aligned} \tag{9.72}$$

When we now express the general formula (i.e., equation (9.59)) by making use of what we have just derived, we obtain the following:

$$\begin{aligned} y_{i+1} &= y_i + h\left(ak_1 + bk_2 + ck_3 + dk_4\right) + O\left(h^5\right) \\ &= y_i + h\left\{a \cdot f\left(t_i,\ y_i\right) + b \cdot \left[f\left(t_i,\ y_i\right) + \frac{h}{2}\frac{d}{dt}f\left(t_i,\ y_i\right)\right]\right. \\ &\quad + c \cdot \left[f\left(t_i,\ y_i\right) + \frac{h}{2}\frac{d}{dt}\left[f\left(t_i,\ y_i\right) + \frac{h}{2}\frac{d}{dt}f\left(t_i,\ y_i\right)\right]\right] \\ &\quad + d \cdot \left[f\left(t_i,\ y_i\right) + h\frac{d}{dt}\left[f\left(t_i,\ y_i\right) + \frac{h}{2}\frac{d}{dt}\left[f\left(t_i,\ y_i\right) + \frac{h}{2}\frac{d}{dt}f\left(t_i,\ y_i\right)\right]\right]\right] + O\left(h^5\right) \end{aligned}$$

$$
\begin{aligned}
&= y_i + a\cdot hf(t_i,\ y_i) + b\cdot hf(t_i,\ y_i) + b\cdot\frac{h^2}{2}\frac{d}{dt}f(t_i,\ y_i) \\
&+ c\cdot hf(t_i,\ y_i) + c\cdot\frac{h^2}{2}\frac{d}{dt}f(t_i,\ y_i) + c\cdot\frac{h^3}{4}\frac{d^2f(t_i,\ y_i)}{dt^2} \\
&+ d\cdot hf(t_i,\ y_i) + d\cdot h^2\frac{d}{dt}f(t_i,\ y_i) + d\cdot\frac{h^3}{2}\frac{d^2f(t_i,\ y_i)}{dt^2} \\
&+ d\cdot\frac{h^4}{4}\frac{d^3f(t_i,\ y_i)}{dt^3} + O(h^5)
\end{aligned}
\tag{9.73}
$$

For $\dot{y}_t = dy_t/dt = f(y_t, t)$, the Taylor series expansion of y_{t+h} around t is

$$
\begin{aligned}
y_{t+h} &= y_t + h\dot{y}_t + \frac{h^2}{2}\ddot{y}_t + \frac{h^3}{6}y_t^{(3)} + \frac{h^4}{24}y_t^{(4)} + O(h^5) \\
&= y_t + hf(y_t,\ t) + \frac{h^2}{2}\frac{d}{dt}f(y_t,\ t) + \frac{h^3}{6}\frac{d^2}{dt^2}f(y_t,\ t) + \frac{h^4}{24}\frac{d^3}{dt^3}f(y_t,\ t)
\end{aligned}
\tag{9.74}
$$

Equation (9.74) can be re-written as

$$
y_{i+1} = y_i + hf(t_i,\ y_i) + \frac{h^2}{2}\frac{d}{dt}f(t_i,\ y_i) + \frac{h^3}{6}\frac{d^2f(t_i,\ y_i)}{dt^2} + \frac{h^4}{24}\frac{d^3f(t_i,\ y_i)}{dt^3} \tag{9.75}
$$

If equation (9.73) is equal to equation (9.75), then the coefficients in equations (9.73) and (9.75) ought to be identical to each other. To be specific,

$$
a + b + c + d = 1 \tag{9.76}
$$

$$
\frac{b}{2} + \frac{c}{2} + d = \frac{1}{2} \tag{9.77}
$$

$$
\frac{c}{4} + \frac{d}{2} = \frac{1}{6} \tag{9.78}
$$

$$
\frac{d}{4} = \frac{1}{24} \tag{9.79}
$$

By solving equations (9.76) through (9.79), we are able to find a set of solutions consisting of $a = d = 1/6$ and $b = c = 1/3$. Consequently, we have been able to obtain the expression of the fourth-order Runge–Kutta method as follows:

$$
y_{i+1} = y_i + h\left(\frac{1}{6}k_1 + \frac{1}{3}k_2 + \frac{1}{3}k_3 + \frac{1}{6}k_4\right) + O(h^5) \tag{9.80}
$$

in which

$$k_1 = f(t_i, y_i) \tag{9.81}$$

$$k_2 = f\left(t_i + \frac{1}{2}h, y_i + \frac{h}{2}k_1\right) \tag{9.82}$$

$$k_3 = f\left(t_i + \frac{1}{2}h, y_i + \frac{h}{2}k_2\right) \tag{9.83}$$

$$k_4 = f(t_i + h, y_i + hk_3) \tag{9.84}$$

9.3.3 Houbolt Method for Multi-Degree-of-Freedom Systems

The Houbolt method will be consider ed in the context of a multi-degree-of-freedom system. The subsequent finite difference expansions are employed in this method (Figure 9.5):

$$\dot{\mathbf{y}}_{i+1} = \frac{1}{6\Delta t}(11\mathbf{y}_{i+1} - 18\mathbf{y}_i + 9\mathbf{y}_{i-1} - 2\mathbf{y}_{i-2}) \tag{9.85}$$

$$\ddot{\mathbf{y}}_{i+1} = \frac{1}{(\Delta t)^2}(2\mathbf{y}_{i+1} - 5\mathbf{y}_i + 4\mathbf{y}_{i-1} - \mathbf{y}_{i-2}) \tag{9.86}$$

For the purpose of deriving equations (9.85) and (9.86), the function $y(t)$ should be considered. As illustrated in Figure 11.8, let the values of y at the equally spaced grid points $t_{i-2} = t_i - 2\Delta t$, $t_{i-1} = t_i - \Delta t$, t_i and $t_{i+1} = t_i + \Delta t$ be given by y_{i-2}, y_{i-1}, y_i and y_{i+1}, respectively.

The Taylor series expansion, utilizing a backward step, offers several possibilities.

- Using a step size of Δt:

$$y(t) = y(t+\Delta t) - \Delta t\dot{y}(t+\Delta t) + \frac{(\Delta t)^2}{2!}\ddot{y}(t+\Delta t) - \frac{(\Delta t)^3}{3!}\dddot{y}(t+\Delta t) \tag{9.87}$$

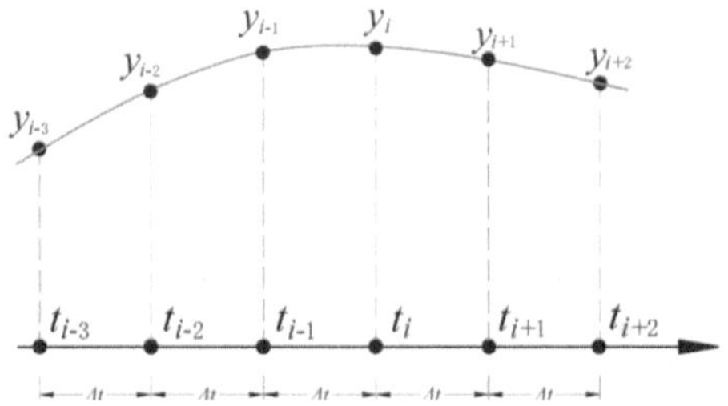

FIGURE 9.5 Equally spaced grid points in the Houbolt method.

Equation (9.87) can be re-written as

$$y_i = y_{i+1} - \Delta t\dot{y}_{i+1} + \frac{(\Delta t)^2}{2!}\ddot{y}_{i+1} - \frac{(\Delta t)^3}{3!}\dddot{y}_{i+1} + \ldots \tag{9.88}$$

- Using a step size of $2\Delta t$:

$$\begin{aligned} y(t-\Delta t) &= y(t+\Delta t) - (2\Delta t)\dot{y}(t+\Delta t) \\ &+ \frac{(2\Delta t)^2}{2!}\ddot{y}(t+\Delta t) - \frac{(2\Delta t)^3}{3!}\dddot{y}(t+\Delta t) + \cdots \end{aligned} \tag{9.89}$$

Equation (9.89) can be re-written as

$$y_{i-1} = y_{i+1} - 2\Delta t\dot{y}_{i+1} + 2(\Delta t)^2\ddot{y}_{i+1} - \frac{4(\Delta t)^3}{3}\dddot{y}_{i+1} + \cdots \tag{9.90}$$

- Using a step size of $3\Delta t$:

$$\begin{aligned} y(t-2\Delta t) &= y(t+\Delta t) - (3\Delta t)\dot{y}(t+\Delta t) \\ &+ \frac{(3\Delta t)^2}{2!}\ddot{y}(t+\Delta t) - \frac{(3\Delta t)^3}{3!}\dddot{y}(t+\Delta t) + \cdots \end{aligned} \tag{9.91}$$

Equation (9.91) can be re-written as

$$y_{i-2} = y_{i+1} - 3\Delta t\dot{y}_{i+1} + \frac{9}{2}(\Delta t)^2\ddot{y}_{i+1} - \frac{9}{2}(\Delta t)^3\dddot{y}_{i+1} + \ldots \tag{9.92}$$

By focusing solely on terms up to $(\Delta t)^3$, equations (9.88), (9.90) and (9.92) can be solved to express $\dot{y}_{i+1}, \ddot{y}_{i+1}$ and $\dddot{y}_{i+1}$ in terms of y_{i-2}, y_{i-1}, y_i and y_{i+1}. This yields $\dot{y}_{i+1}$ and $\ddot{y}_{i+1}$ as

$$\dot{y}_{i+1} = \frac{1}{6(\Delta t)}\left(11y_{i+1} - 18y_i + 9y_{i-1} - 2y_{i-2}\right) \tag{9.93}$$

$$\ddot{y}_{i+1} = \frac{1}{(\Delta t)^2}\left(2y_{i+1} - 5y_i + 4y_{i-1} - y_{i-2}\right) \tag{9.94}$$

In their vector form, these equations are represented by the equations (9.85) and (9.86). In order to find the solution at step i+1 ($\mathbf{y}_{i+1}$), we take into consideration equation (9.28) at time t_{i+1}, so that

$$\mathbf{M}\ddot{\mathbf{y}}_{i+1} + \mathbf{C}\dot{\mathbf{y}}_{i+1} + \mathbf{K}\mathbf{y}_{i+1} = \mathbf{F}_{i+1} \equiv \mathbf{F}(t = t_{i+1}) \tag{9.95}$$

When we plug equations (9.85) and (9.86) into equation (9.95), we obtain the following result:

$$\frac{\mathbf{M}}{(\Delta t)^2}\left(2\mathbf{y}_{i+1}-5\mathbf{y}_i+4\mathbf{y}_{i-1}-\mathbf{y}_{i-2}\right)+\frac{\mathbf{C}}{6\Delta t}\left(11\mathbf{y}_{i+1}-18\mathbf{y}_i+9\mathbf{y}_{i-1}-2\mathbf{y}_{i-2}\right) + \mathbf{K}\mathbf{y}_{i+1}=\mathbf{F}_{i+1} \tag{9.96}$$

Rearranging equation (9.96) leads to

$$\begin{aligned}\left(\frac{2\mathbf{M}}{(\Delta t)^2}+\frac{11\mathbf{C}}{6\Delta t}+\mathbf{K}\right)\mathbf{y}_{i+1}&=\mathbf{F}_{i+1}+\left(\frac{5\mathbf{M}}{(\Delta t)^2}+\frac{3\mathbf{C}}{\Delta t}\right)\mathbf{y}_i\\&-\left(\frac{4\mathbf{M}}{(\Delta t)^2}+\frac{3\mathbf{C}}{2\Delta t}\right)\mathbf{y}_{i-1}+\left(\frac{\mathbf{M}}{(\Delta t)^2}+\frac{\mathbf{C}}{3\Delta t}\right)\mathbf{y}_{i-2}\end{aligned} \tag{9.97}$$

It should be noted that the equation of equilibrium at time t_{i+1}, which is denoted by equation (9.95), is utilized in the process of determining the solution $\mathbf{y}_{i+1}$ by means of equation (9.97). This also applies to the Wilson and Newmark methods. Because of this, these methodologies are referred to as implicit integration methods.

As can be observed from equation (9.97), in order to find the solution $\mathbf{y}_{i+1}$, it is necessary to have knowledge about $\mathbf{y}_i$, $\mathbf{y}_{i-1}$ and $\mathbf{y}_{i-2}$. Consequently, it is necessary to find the values of $\mathbf{y}_{-1}$ and $\mathbf{y}_{-2}$ prior to making an attempt to find the vector $\mathbf{y}_1$ by utilizing equation (9.97). It is not possible to utilize equation (9.97) to determine $\mathbf{y}_1$ and $\mathbf{y}_2$ because there is no direct method that can be used to calculate $\mathbf{y}_{-1}$ and $\mathbf{y}_{-2}$. This renders the method non-self-starting. In order to start the method, we can use the finite difference method to find $\mathbf{y}_1$ and $\mathbf{y}_2$. Once $\mathbf{y}_0$ is known from the given initial conditions of the problem and $\mathbf{y}_1$ and $\mathbf{y}_2$ are known from the finite difference method, the subsequent solutions $\mathbf{y}_3, \mathbf{y}_4, \ldots$ can be found by using equation (9.97).

9.3.4 Wilson θ Method

According to the Wilson method, the acceleration of the system is assumed to change in a linear fashion between two different instants of time. More specifically, the two instants of time are taken in the manner that is depicted in Figure 9.6.

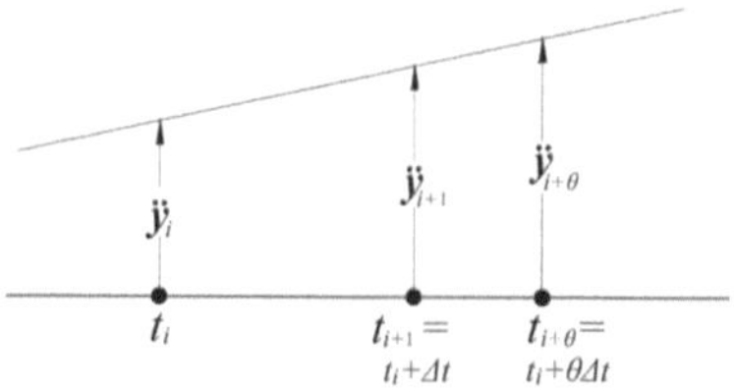

FIGURE 9.6 Linear acceleration assumption of the Wilson θ method.

Therefore, it is assumed that the acceleration is linear from the time $t_i = i\Delta t$ to the time $t_{i+\theta} = t_i + \theta\Delta t$, where $\theta \geq 1.0$. Due to this particular reason, this method is also known as the Wilson θ method. If $\theta = 1.0$, then this method is reduced to the linear acceleration scheme.

Based on the results of a stability analysis, it can be shown that the Wilson method is unconditionally stable, given that the value of θ satisfies $\theta \geq 1.37$. Consideration of the Wilson method for a multi-degree-of-freedom system is going to be given in this section.

Because $\ddot{\mathbf{y}}(t)$ is assumed to change in a linear fashion between t_i and $t_{i+\theta}$, we can predict the value of $\ddot{\mathbf{y}}$ at any time $t_i + \tau$, $0 \leq \tau \leq \theta\Delta t$:

$$\ddot{\mathbf{y}}(t_i + \tau) = \ddot{\mathbf{y}}_i + \frac{\tau}{\theta\Delta t}(\ddot{\mathbf{y}}_{i+\theta} - \ddot{\mathbf{y}}_i) \tag{9.98}$$

When we integrate equation (9.98), we obtain the following:

$$\dot{\mathbf{y}}(t_i + \tau) = \dot{\mathbf{y}}_i + \ddot{\mathbf{y}}_i\tau + \frac{\tau^2}{2\theta\Delta t}(\ddot{\mathbf{y}}_{i+\theta} - \ddot{\mathbf{y}}_i) \tag{9.99}$$

and

$$\mathbf{y}(t_i + \tau) = \mathbf{y}_i + \dot{\mathbf{y}}_i\tau + \frac{1}{2}\ddot{\mathbf{y}}_i\tau^2 + \frac{\tau^3}{6\theta\Delta t}(\ddot{\mathbf{y}}_{i+\theta} - \ddot{\mathbf{y}}_i) \tag{9.100}$$

Through the process of substituting $\tau = \theta\Delta t$ into equations (9.99) and (9.100), we can derive

$$\dot{\mathbf{y}}_{i+\theta} = \dot{\mathbf{y}}(t_i + \theta\Delta t) = \dot{\mathbf{y}}_i + \frac{\theta\Delta t}{2}(\ddot{\mathbf{y}}_{i+\theta} + \ddot{\mathbf{y}}_i) \tag{9.101}$$

$$\mathbf{y}_{i+\theta} = \mathbf{y}(t_i + \theta\Delta t) = \mathbf{y}_i + \theta\Delta t\dot{\mathbf{y}}_i + \frac{(\theta\Delta t)^2}{6}(\ddot{\mathbf{y}}_{i+\theta} + 2\ddot{\mathbf{y}}_i) \tag{9.102}$$

We can solve equation (9.102) to obtain

$$\ddot{\mathbf{y}}_{i+\theta} = \frac{6}{(\theta\Delta t)^2}(\mathbf{y}_{i+\theta} - \mathbf{y}_i) - \frac{6}{\theta\Delta t}\dot{\mathbf{y}}_i - 2\ddot{\mathbf{y}}_i \tag{9.103}$$

Equation (9.103) is substituted into equation (9.101), which results in the following:

$$\dot{\mathbf{y}}_{i+\theta} = \frac{3}{(\theta\Delta t)}(\mathbf{y}_{i+\theta} - \mathbf{y}_i) - 2\dot{\mathbf{y}}_i - \frac{\theta\Delta t}{2}\ddot{\mathbf{y}}_i \tag{9.104}$$

In order to obtain the value of $y_{i+\theta}$, we take into consideration the equilibrium equation (9.28) at the time $t_{i+\theta} = t_i + \theta\Delta t$ and write

$$\mathbf{M}\ddot{\mathbf{y}}_{i+\theta} + \mathbf{C}\dot{\mathbf{y}}_{i+\theta} + \mathbf{K}\mathbf{y}_{i+\theta} = \mathbf{F}_{i+\theta} \tag{9.105}$$

in which the force vector $\mathbf{F}_{i+\theta}$ is also derived by employing the linear assumption:

$$\mathbf{F}_{i+\theta} = \mathbf{F}_i + \theta\left(\mathbf{F}_{i+1} - \mathbf{F}_i\right) \tag{9.106}$$

When we substitute equations (9.103), (9.104) and (9.106) for $\ddot{\mathbf{y}}_{i+\theta}$, $\dot{\mathbf{y}}_{i+\theta}$ and $\mathbf{F}_{i+\theta}$, equation (9.105) gives

$$\begin{aligned}&\mathbf{M}\left[\frac{6}{(\theta\Delta t)^2}\left(\mathbf{y}_{i+\theta} - \mathbf{y}_i\right) - \frac{6}{\theta\Delta t}\dot{\mathbf{y}}_i - 2\ddot{\mathbf{y}}_i\right]\\ &+\mathbf{C}\left[\frac{3}{(\theta\Delta t)}\left(\mathbf{y}_{i+\theta} - \mathbf{y}_i\right) - 2\dot{\mathbf{y}}_i - \frac{\theta\Delta t}{2}\ddot{\mathbf{y}}_i\right] + \mathbf{K}\mathbf{y}_{i+\theta}\\ &= \mathbf{F}_i + \theta\left(\mathbf{F}_{i+1} - \mathbf{F}_i\right)\end{aligned} \tag{9.107}$$

Rearranging equation (9.107) leads to

$$\begin{aligned}&\left[\frac{6\mathbf{M}}{(\theta\Delta t)^2} + \frac{3\mathbf{C}}{(\theta\Delta t)} + \mathbf{K}\right]\mathbf{y}_{i+\theta}\\ &= \mathbf{F}_i + \theta\left(\mathbf{F}_{i+1} - \mathbf{F}_i\right) + \left[\frac{6\mathbf{M}}{(\theta\Delta t)^2} + \frac{3\mathbf{C}}{(\theta\Delta t)}\right]\mathbf{y}_i\\ &+\left(\frac{6\mathbf{M}}{\theta\Delta t} + 2\mathbf{C}\right)\dot{\mathbf{y}}_i + \left(2\mathbf{M} + \frac{\theta\Delta t}{2}\mathbf{C}\right)\ddot{\mathbf{y}}_i\end{aligned} \tag{9.108a}$$

the solution to which can be found for $\mathbf{y}_{i+\theta}$.

The steps that make up the Wilson method are as follows:

1. With the knowledge of the initial conditions $\mathbf{y}_0$ and $\dot{\mathbf{y}}_0$, determine $\ddot{\mathbf{y}}_0$ by using equation (9.26);
2. Choose an appropriate time step, denoted as Δt, and an acceptable value for the variable θ which is typically assumed to be 1.4.
3. Compute the effective load vector $\tilde{\mathbf{F}}_{i+\theta}$ starting with $i = 0$:

$$\begin{aligned}\tilde{\mathbf{F}}_{i+\theta} &= \mathbf{F}_i + \theta\left(\mathbf{F}_{i+1} - \mathbf{F}_i\right) + \left[\frac{6\mathbf{M}}{(\theta\Delta t)^2} + \frac{3\mathbf{C}}{(\theta\Delta t)}\right]\mathbf{y}_i\\ &+\left(\frac{6\mathbf{M}}{\theta\Delta t} + 2\mathbf{C}\right)\dot{\mathbf{y}}_i + \left(2\mathbf{M} + \frac{\theta\Delta t}{2}\mathbf{C}\right)\ddot{\mathbf{y}}_i\end{aligned} \tag{9.108b}$$

4. Find the displacement vector $\mathbf{y}_{i+\theta}$ at time $t_{i+\theta}$:

$$\mathbf{y}_{i+\theta} = \left[\frac{6\mathbf{M}}{(\theta\Delta t)^2} + \frac{3\mathbf{C}}{(\theta\Delta t)} + \mathbf{K}\right]^{-1}\tilde{\mathbf{F}}_{i+\theta} \tag{9.109}$$

5. Next, substitute the displacement vector $\mathbf{y}_{i+\theta}$ into equations (9.104) and (9.103) to obtain $\dot{\mathbf{y}}_{i+\theta}$ and $\ddot{\mathbf{y}}_{i+\theta}$;

$$\dot{\mathbf{y}}_{i+\theta} = \frac{3}{(\theta\Delta t)}(\mathbf{y}_{i+\theta} - \mathbf{y}_i) - 2\dot{\mathbf{y}}_i - \frac{\theta\Delta t}{2}\ddot{\mathbf{y}}_i \tag{9.110}$$

$$\ddot{\mathbf{y}}_{i+\theta} = \frac{6}{(\theta\Delta t)^2}(\mathbf{y}_{i+\theta} - \mathbf{y}_i) - \frac{6}{\theta\Delta t}\dot{\mathbf{y}}_i - 2\ddot{\mathbf{y}}_i \tag{9.111}$$

6. Lastly, use these vectors to determine in equations (9.100), (9.99) and (9.98) the displacement, velocity and acceleration vectors at $\tau = \Delta t$, i.e., $\dot{\mathbf{y}}(t_i + \Delta t)$ and $\mathbf{y}(t_i + \Delta t)$.

$$\mathbf{y}(t_i + \Delta t) = \mathbf{y}_{i+1} = \mathbf{y}_i + \dot{\mathbf{y}}_i\Delta t + \frac{1}{2}\ddot{\mathbf{y}}_i(\Delta t)^2 + \frac{(\Delta t)^2}{6\theta}(\ddot{\mathbf{y}}_{i+\theta} - \ddot{\mathbf{y}}_i) \tag{9.112}$$

$$\dot{\mathbf{y}}(t_i + \Delta t) = \dot{\mathbf{y}}_{i+1} = \dot{\mathbf{y}}_i + \ddot{\mathbf{y}}_i\Delta t + \frac{\Delta t}{2\theta}(\ddot{\mathbf{y}}_{i+\theta} - \ddot{\mathbf{y}}_i) \tag{9.113}$$

$$\ddot{\mathbf{y}}(t_i + \Delta t) = \ddot{\mathbf{y}}_{i+1} = \ddot{\mathbf{y}}_i + \frac{1}{\theta}(\ddot{\mathbf{y}}_{i+\theta} - \ddot{\mathbf{y}}_i) \tag{9.114}$$

9.3.5 Newmark β Method

The Newmark integration method also relies on the assumption that the acceleration changes in a linear manner between two specific moments in time. More specifically, the two instants of time are taken in the manner that is depicted in Figure 9.7. Therefore, it is assumed that the acceleration is linear from the time $t_i = i\Delta t$ to the time $t_{i+\theta} = t_i + \theta\Delta t$, where $\theta \geq 1.0$.

In matrix notation, the equation of motion for a system with many degrees of freedom that is evaluated at time t_i should be written as

$$\mathbf{M}\ddot{\mathbf{y}}_i + \mathbf{C}\dot{\mathbf{y}}_i + \mathbf{K}\mathbf{y}_i = \mathbf{F}_i(t) \tag{9.115}$$

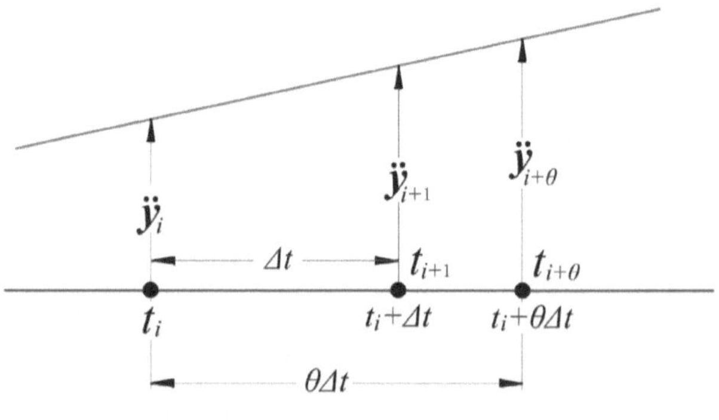

FIGURE 9.7 Linear acceleration assumption in the extended time interval.

By calculating the difference between the dynamic equilibrium conditions that are defined at the time t_i and at $t_i + \delta$, where $\delta = \theta\Delta t$, we are able to determine the incremental equations of motion:

$$\mathbf{M}\hat{\Delta}\ddot{\mathbf{y}}_i + \mathbf{C}\hat{\Delta}\dot{\mathbf{y}}_i + \mathbf{K}\hat{\Delta}\mathbf{y}_i = \hat{\Delta}\mathbf{F}_i(t) \tag{9.116}$$

where the circumflex above Δ denotes that the increments are associated with the extended time step $\delta = \theta\Delta t$. Therefore,

$$\hat{\Delta}\mathbf{y}_i = \mathbf{y}(t_i + \delta) - \mathbf{y}(t_i) \tag{9.117}$$

$$\hat{\Delta}\dot{\mathbf{y}}_i = \dot{\mathbf{y}}(t_i + \delta) - \dot{\mathbf{y}}(t_i) \tag{9.118}$$

$$\hat{\Delta}\ddot{\mathbf{y}}_i = \ddot{\mathbf{y}}(t_i + \delta) - \ddot{\mathbf{y}}(t_i) \tag{9.119}$$

$$\hat{\Delta}\mathbf{F}_i = \mathbf{F}(t_i + \delta) - \mathbf{F}(t_i) \tag{9.120}$$

The integration of the dynamic equations of motion by the step-by-step linear acceleration method with the extended time step is based on the assumption that the acceleration can be expressed as a linear function during the time step $\delta = \theta\Delta t$ as shown in Figure 9.7. Based on this figure, we can write the linear expression for the acceleration during the extended time step $\theta\Delta t$ as

$$\ddot{\mathbf{y}}_i(t) = \ddot{\mathbf{y}}_i + \frac{\hat{\Delta}\ddot{\mathbf{y}}_i}{\theta\Delta t}(t - t_i) = \ddot{\mathbf{y}}_i + \frac{(t - t_i)}{\theta\Delta t}\left[\ddot{\mathbf{y}}(t_i + \delta) - \ddot{\mathbf{y}}(t_i)\right] \tag{9.121}$$

When we integrate equation (9.121) twice between limits t_i and t, we obtain the following:

$$\dot{\mathbf{y}}(t) = \dot{\mathbf{y}}_i + \ddot{\mathbf{y}}_i(t - t_i) + \frac{1}{2}\frac{\hat{\Delta}\ddot{\mathbf{y}}_i}{\theta\Delta t}(t - t_i)^2 \tag{9.122}$$

$$\mathbf{y}(t) = \mathbf{y}_i + \dot{\mathbf{y}}_i(t - t_i) + \frac{1}{2}\ddot{\mathbf{y}}_i(t - t_i)^2 + \frac{1}{6}\frac{\hat{\Delta}\ddot{\mathbf{y}}_i}{\theta\Delta t}(t - t_i)^3 \tag{9.123}$$

The evaluation of equations (9.122) and (9.123) at the end of the extended interval $t = t_i + \theta\Delta t$ yields the following results:

$$\hat{\Delta}\dot{\mathbf{y}}_i = \dot{\mathbf{y}}(t_i + \theta\Delta t) - \dot{\mathbf{y}}_i = \ddot{\mathbf{y}}_i(\theta\Delta t) + \frac{1}{2}(\theta\Delta t)\hat{\Delta}\ddot{\mathbf{y}}_i \tag{9.124}$$

$$\hat{\Delta}\mathbf{y}_i = \dot{\mathbf{y}}_i(\theta\Delta t) + \frac{1}{2}\ddot{\mathbf{y}}_i(\theta\Delta t)^2 + \frac{1}{6}\hat{\Delta}\ddot{\mathbf{y}}_i(\theta\Delta t)^2 \tag{9.125}$$

Solving equation (9.125) for the incremental acceleration $\hat{\Delta}\ddot{\mathbf{y}}$ leads to:

$$\hat{\Delta}\ddot{\mathbf{y}}_i = \frac{6}{(\theta\Delta t)^2}\hat{\Delta}\mathbf{y}_i - 3\ddot{\mathbf{y}}_i - \frac{6}{\theta\Delta t}\dot{\mathbf{y}}_i \tag{9.126}$$

The solution in equation (9.126) is then substituted into equation (9.124) to yield the following:

$$\hat{\Delta}\dot{\mathbf{y}}_i = \frac{1}{2}(\theta\Delta t)\left[\frac{6}{(\theta\Delta t)^2}\hat{\Delta}\mathbf{y}_i - 1\ddot{\mathbf{y}}_i - \frac{6}{\theta\Delta t}\dot{\mathbf{y}}_i\right] = \frac{3}{(\theta\Delta t)}\hat{\Delta}\mathbf{y}_i - \frac{1}{2}(\theta\Delta t)\ddot{\mathbf{y}}_i - 3\dot{\mathbf{y}}_i \tag{9.127}$$

Lastly, an equation for the incremental displacement $\hat{\Delta}y_i$ can be obtained by substituting equations (9.126) and (9.127) into the incremental equation of motion, that is, equation (9.116). This equation can be written in a convenient manner as

$$\begin{aligned}&\mathbf{M}\left[\frac{6}{(\theta\Delta t)^2}\hat{\Delta}\mathbf{y}_i - 3\ddot{\mathbf{y}}_i - \frac{6}{\theta\Delta t}\dot{\mathbf{y}}_i\right]\\ &\quad+\mathbf{C}\left[\frac{3}{(\theta\Delta t)}\hat{\Delta}\mathbf{y}_i - \frac{1}{2}(\theta\Delta t)\ddot{\mathbf{y}}_i - 3\dot{\mathbf{y}}_i\right] + \mathbf{K}\hat{\Delta}\mathbf{y}_i = \hat{\Delta}\mathbf{F}_i(t)\end{aligned} \tag{9.128}$$

Similar in form to the static incremental equilibrium equation, the matrix equation (9.128) can be solved for the incremental displacements $\hat{\Delta}\mathbf{y}_i$ by merely solving a system of linear equations.

In order to obtain the incremental accelerations $\hat{\Delta}\ddot{\mathbf{y}}_i$ for the extended time interval, the value of $\hat{\Delta}\mathbf{y}_i$ that is obtained from the solution of equation (9.128) is substituted into equation (9.126). After that, the incremental acceleration $\Delta\ddot{\mathbf{y}}_i$ for the normal time interval Δt is determined using a straightforward linear interpolation. Therefore,

$$\Delta\ddot{\mathbf{y}}_i = \frac{\hat{\Delta}\ddot{\mathbf{y}}_i}{\theta} \tag{9.129}$$

In order to determine the incremental velocity $\Delta\dot{\mathbf{y}}_i$ and the incremental displacement $\Delta\mathbf{y}_i$, which correspond to the normal interval Δt, equations (9.124) and (9.125) are utilized. The extended time interval parameter $\theta\Delta t$ is substituted by $\theta\Delta t$ in order to compute the incremental velocity and displacement, respectively.

$$\Delta\dot{\mathbf{y}}_i = \ddot{\mathbf{y}}_i(\Delta t) + \frac{1}{2}(\Delta t)\Delta\ddot{\mathbf{y}}_i \tag{9.130}$$

$$\Delta\mathbf{y}_i = \dot{\mathbf{y}}_i(\Delta t) + \frac{1}{2}\ddot{\mathbf{y}}_i(\Delta t)^2 + \frac{1}{6}\Delta\ddot{\mathbf{y}}_i(\Delta t)^2 \tag{9.131}$$

Eventually, at the end of the normal time interval Δt, the displacement $\mathbf{y}_{i+1}$ and velocity $\dot{\mathbf{y}}_{i+1}$ are computed by

$$\mathbf{y}_{i+1} = \mathbf{y}_i + \Delta\mathbf{y}_i \tag{9.132}$$

$$\dot{\mathbf{y}}_{i+1} = \dot{\mathbf{y}}_i + \Delta\dot{\mathbf{y}}_i \tag{9.133}$$

The initial acceleration for the next step is computed from the condition of dynamic equilibrium at the time $t + \Delta t$; therefore,

$$\ddot{\mathbf{y}}_{i+1} = \mathbf{M}^{-1}\left[\mathbf{F}_{i+1} - \mathbf{F}_D\left(\dot{\mathbf{y}}_{i+1}\right) - \mathbf{F}_S\left(\mathbf{y}_{i+1}\right)\right] \tag{9.134}$$

where $\mathbf{F}_D\left(\dot{\mathbf{y}}_{i+1}\right)$ and $\mathbf{F}_S\left(\mathbf{y}_{i+1}\right)$ are the vectors that represent the damping force and stiffness force, respectively, that are evaluated at the end of the time step $t_{i+1} = t_i + \Delta t$. Following the determination of the vectors representing the displacement, velocity and acceleration at $t_{i+1} = t_i + \Delta t$, the procedure that has been explained is repeated in order to calculate these quantities at the next time step $t_{i+2} = t_{i+1} + \Delta t$. This process is continued until reaching the desired final time.

The Newmark β method can be regarded as a generalization of the linear acceleration method. It utilizes a numerical parameter denoted as β. The method contained in addition to β, a second parameter γ. In equations (9.130) and (9.131), respectively, these parameters take the place of the numerical coefficients 1/2 and 1/6 that are found in the terms that contain the incremental acceleration $\Delta\ddot{\mathbf{y}}_i$. As a result, by substituting the coefficient 1/2 of $\Delta\ddot{\mathbf{y}}_i$ with γ in equation (9.130) and the coefficient 1/6 also of $\Delta\ddot{\mathbf{y}}_i$ with β in equation (9.131), we are able to obtain the following:

$$\Delta\dot{\mathbf{y}}_i = \ddot{\mathbf{y}}_i(\Delta t) + \gamma(\Delta t)\Delta\ddot{\mathbf{y}}_i \tag{9.135}$$

$$\Delta\mathbf{y}_i = \dot{\mathbf{y}}_i(\Delta t) + \frac{1}{2}\ddot{\mathbf{y}}_i(\Delta t)^2 + \beta\Delta\ddot{\mathbf{y}}_i(\Delta t)^2 \tag{9.136}$$

It has been discovered that when the value of γ is not equal to 1/2, the method introduces a superfluous damping into the system. Due to this particular reason, this parameter is generally set as $\gamma = 1/2$. Following the setting of $\gamma = 1/2$, the solution of equation (9.136) for $\Delta\ddot{\mathbf{y}}_i$ and its subsequent substitution into equation (9.135) yield

$$\Delta\ddot{\mathbf{y}}_i = \frac{\Delta\mathbf{y}_i}{\beta(\Delta t)^2} - \frac{\dot{\mathbf{y}}_i}{\beta(\Delta t)} - \frac{1}{2\beta}\ddot{\mathbf{y}}_i \tag{9.137}$$

$$\begin{aligned}\Delta\dot{\mathbf{y}}_i &= \ddot{\mathbf{y}}_i(\Delta t) + \frac{1}{2}(\Delta t)\left[\frac{\Delta\mathbf{y}_i}{\beta(\Delta t)^2} - \frac{\dot{\mathbf{y}}_i}{\beta(\Delta t)} - \frac{1}{2\beta}\ddot{\mathbf{y}}_i\right] \\ &= \frac{\Delta\mathbf{y}_i}{2\beta(\Delta t)} - \frac{\dot{\mathbf{y}}_i}{2\beta} + \left(1 - \frac{1}{4\beta}\right)(\Delta t)\ddot{\mathbf{y}}_i\end{aligned} \tag{9.138}$$

After that, equations (9.137) and (9.138) are substituted into the incremental equation of motion

$$\mathbf{M}\Delta\ddot{\mathbf{y}}_i + \mathbf{C}\Delta\dot{\mathbf{y}}_i + \mathbf{K}\Delta\mathbf{y}_i = \Delta\mathbf{F}_i(t) \tag{9.139}$$

to obtain an equation to compute the incremental displacement $\Delta\mathbf{y}_i$, that is

$$\begin{aligned} &\mathbf{M}\left[\frac{\Delta\mathbf{y}_i}{\beta(\Delta t)^2} - \frac{\dot{\mathbf{y}}_i}{\beta(\Delta t)} - \frac{1}{2\beta}\ddot{\mathbf{y}}_i\right] \\ &+\mathbf{C}\left[\frac{\Delta\mathbf{y}_i}{2\beta(\Delta t)} - \frac{\dot{\mathbf{y}}_i}{2\beta} + \left(1 - \frac{1}{4\beta}\right)(\Delta t)\ddot{\mathbf{y}}_i\right] + \mathbf{K}\Delta\mathbf{y}_i = \Delta\mathbf{F}_i(t) \end{aligned} \tag{9.140}$$

Within these formulas, **C** and **K** denote the damping and stiffness matrices, respectively, whose coefficients are evaluated at the initial time t_i of the time step $\Delta t = t_{i+1} - t_i$. After the incremental displacement $\Delta\mathbf{y}_i$ has been obtained, similar procedures as those in equations (9.129)–(9.134) can be used to obtain the displacement $\mathbf{y}_{i+1}$, velocity $\dot{\mathbf{y}}_{i+1}$ and acceleration $\ddot{\mathbf{y}}_{i+1}$ at the end of the normal time interval Δt. Once the vectors for acceleration, displacement and velocity have been determined at $t_{i+1} = t_i + \Delta t$, the previously described technique is repeated to compute these quantities at the subsequent time step, $t_{i+2} = t_{i+1} + \Delta t$. This process is carried out repeatedly until the desired final time is reached.

In the process of implementing the Newmark beta method, the initial step involves the selection of a numerical value for the parameter β. The value that Newmark proposed was within the range of $\frac{1}{6} \le \beta \le \frac{1}{2}$. The method is conditionally stable and is exactly equivalent to the linear acceleration method when $\beta = \frac{1}{6}$. In the case where $\beta = \frac{1}{4}$, the method is equivalent to making the assumption that the velocity varied linearly during the time step, which would require that the mean acceleration is maintained for the interval. In the final scenario, where the value of β is $\beta = \frac{1}{2}$, the Newmark beta method is unconditionally stable and this method provides satisfactory accuracy.

9.4 APPLICATION EXAMPLES OF THE NUMERICAL SIMULATION METHODS

In this section, we provide several calculation examples of applying the numerical simulation methods explained in the foregoing sections. The random vibration system studied in this section is described by a nonlinear Duffing equation.

9.4.1 Numerically Simulating a Duffing Oscillator Involving a Wiener Process

The equation of motion of a Duffing oscillator with a random excitation term involving a Wiener process is as follows:

$$\frac{dy(t)}{dt} + 2zy(t) + by(t)dt + a\left[y(t)\right]^3 dt = 2\sqrt{z}dW(t) \tag{9.141}$$

in which $W(t)$ denotes the Wiener process. Equation (9.141) can be re-written as

$$\frac{d^2y(t)}{dt^2} + 2z\frac{dy(t)}{dt} + by(t) + a\left[y(t)\right]^3 = 2\sqrt{z}\frac{dW(t)}{dt} \tag{9.142}$$

The Duffing equation (9.142), also known as the Duffing oscillator, is a nonlinear differential equation of the second order that is used to model specific damped and driven oscillators. It was named after Georg Duffing (1861–1944). In equation (9.142), the (unknown) function $y = y(t)$ is the displacement at time t, $\frac{dy(t)}{dt}$ is the first derivative of $y(t)$ with respect to time, i.e., velocity and $\frac{d^2y(t)}{dt^2}$ is the second time-derivative of $y(t)$, i.e., acceleration. The numbers z, b and a are given constants. In terms of physical modeling, equation (9.142) can be used to model, for instance, an elastic pendulum whose spring's stiffness does not perfectly obey Hooke's law.

The parameters that are included in the equation (9.142) are as follows:

- z is responsible for controlling the amount of damping.
- b is responsible for controlling the linear stiffness.
- a is responsible for controlling the amount of nonlinearity in the restoring force; if $a = 0$. equation (9.142) describes a damped and forced simple harmonic oscillator.

One interpretation of the Duffing equation is that it describes the oscillations of a mass that is attached to a linear damper and a nonlinear spring. Following this, the restoring force that is provided by the nonlinear spring is $by(t) + a\left[y(t)\right]^3$. When $b > 0$ and $a > 0$, the spring is referred to as a hardening spring. In contrast, when $a < 0$, the spring is referred to as a softening spring (still with $b > 0$).

In equation (9.141), $W(t)$ is a Wiener process. The Wiener process is a random process that is considered to be of utmost importance in the theory of stochastic processes. Not only is this process valuable as a model for a wide variety of important phenomena, but it is also valuable as the building block that can be used to construct other useful random processes. In order to provide a definition for the Wiener process, it is necessary to introduce the notion of a random process with independent increments.

A continuous parameter random process $\{W(t), 0 \le t < \infty\}$ is said to have independent increments if $W(0) = 0$ and, for all choices of indices $t_0 < t_1 < \cdots < t_n$, the n random variables

$$W(t_1) - W(t_0), \cdots, W(t_n) - W(t_{n-1}) \tag{9.143}$$

are independent. The process is considered to have stationary independent increments if, in addition, $W(t_n + h) - W(t_{n-1} + h)$ has the same distribution as $W(t_n) - W(t_{n-1})$ for all choice of indices t_{n-1} and t_n, and every $h > 0$.

In the theory of random processes and its applications, the Wiener process plays a fundamental role. The Wiener process serves as a model for Brownian motion and thermal noise in electric circuits, among other applications. When a particle with a diameter of 10^{-4} cm is submerged in a liquid or gas, it displays ceaseless erratic motions that can be observed through the use of a microscope. The motion of such a particle is referred to as Brownian motion, named after the English botanist Robert Brown, who first observed this phenomenon in 1827 (shortly after the introduction of microscopes with achromatic objectives). The same phenomenon is also displayed in a dramatic manner by smoke particles that are suspended in the air.

Within the realms of statistical mechanics and kinetic theory, one of the major successes was the explanation of the phenomenon of Brownian motion. In the year 1905, Einstein demonstrated that Brownian motion could be explained by making the assumption that the particles are subject to the continual bombardment of the molecules of the surrounding medium. His pioneering work was generalized, extended and experimentally verified by a number of different physicists.

Let us indicate the displacement of a particle in Brownian motion after time t by the symbol $W(t)$. This displacement is measured from the particle's starting point. By definition, $W(0)=0$. As a result of the continual "impacts" that are exerted upon it by the force field of the medium that surrounds it, a Brownian particle is in a state of perpetual motion. At a time interval (s, t) that is long in comparison to the amount of time that passes between impacts, the displacement of the particle can be thought of as the sum of a large number of small displacements. As a consequence of this, according to the central limit theorem, it seems reasonable to make the assumption that $W(t)-W(s)$ follows a normal distribution. Moving forward, it makes sense to make the assumption that, for any $h>0$, the probability distribution of the displacement $W(t)-W(s)$ should be identical to that of $W(t+h)-W(s+h)$. This is because it is assumed that the medium is in equilibrium, and the probability law of a particle's displacement over an interval should depend only on the length $t-s$ of the interval, and not on the time at which we start observation.

Next, we make the assumption that the motion of the particle is totally caused by highly frequent irregular molecular impacts. The mathematical interpretation of this is that the random process $W(t)$ has independent increments; the displacements over non-overlapping intervals are independent because the number and magnitude of impacts in non-overlapping intervals are independent.

Therefore, we are led to define the notion of a Wiener process (named after Norbert Wiener, who was one of the pioneers in studying mathematical Brownian motion). The Wiener process is alternatively referred to as the Wiener–Lévy process and the Brownian motion process.

A random process $\{W(t), t \geq 0\}$ is said to be a Wiener process if

i. $\{W(t), t \geq 0\}$ has stationary independent increments
ii. For every $t>0$, $W(t)$ has a normal distribution
iii. For all $t>0$, the expectation of $W(t)$ is 0, that is: $E[X(t)]=0$
iv. $W(0)=0$

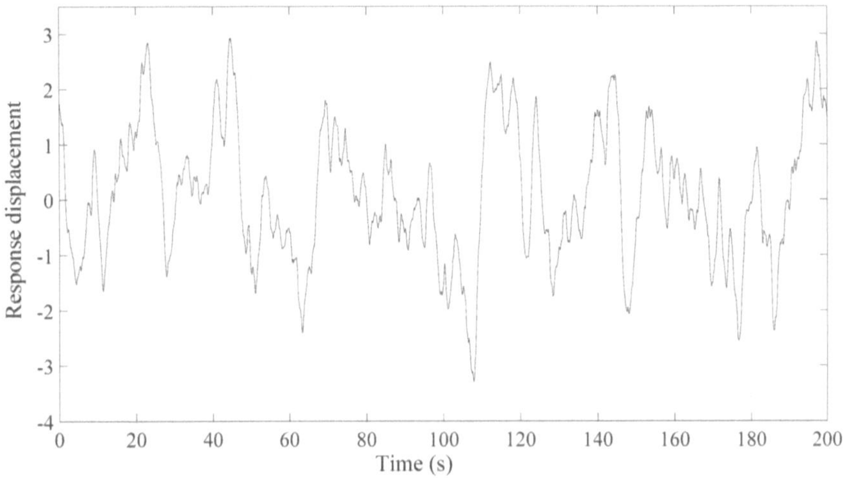

FIGURE 9.8 The simulated 200-s response displacement time history of a Duffing oscillator involving a Wiener process.

We next numerically simulate the Duffing oscillator as represented by equation (9.142) involving a Wiener process $W(t)$. $\frac{dW(t)}{dt}$ in equation (9.142) is a Gaussian white noise process with a constant power spectral density function. Without loss of generality, we choose the following values for the coefficients in equation (9.142): $z = 0.99$; $a = 0.1$; $b = 0.2$. Therefore, in this case, equation (9.142) represents a Duffing oscillator with a hardening spring. The simulation length is chosen to be 200 s, and the simulation time interval is chosen to be dt = 0.1s. The random excitation sample time histories have been generated based on the Gaussian white noise constant power spectrum according to the theories as elucidated in Section 9.2.3. The subsequent deterministic simulation technique utilized is Euler's discretization scheme as explained in Section 9.3.2. The simulated 200-s response displacement time history is depicted in Figure 9.8.

9.4.2 Numerically Simulating a Duffing Oscillator with a Softening Spring

At this point, we proceed to perform another numerical simulation of the Duffing oscillator, which is described by equation (9.142) and includes a Wiener process. In this case, for the coefficients in equation (9.142), we select the following values: $z = 0.99$; $a = -0.1$; $b = 2.2$. These choices are made without any loss of generality. We notice that in the present case $a < 0$ and $b > 0$. Thus, in this particular scenario, equation (9.142) represents a Duffing oscillator characterized by a spring that exhibits softening behavior. The duration of the simulation is set to 200 s, and the time interval for each simulation step is set to dt = 0.1s.

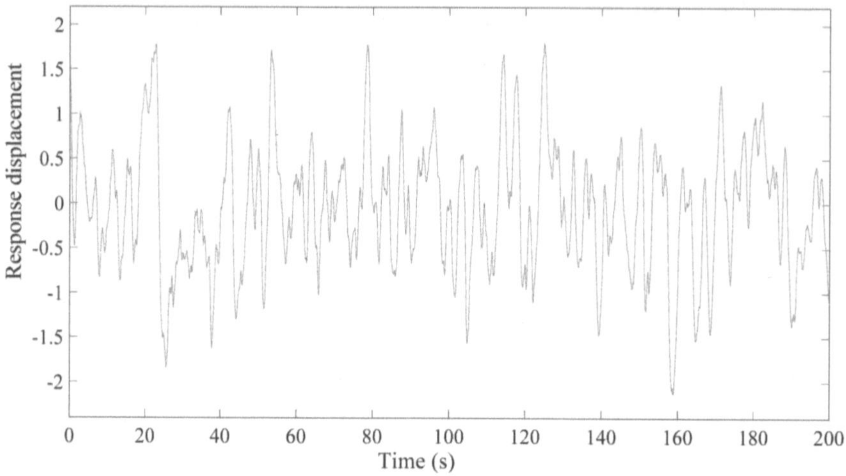

FIGURE 9.9 The simulated 200-s response displacement time history of a Duffing oscillator with a softening spring.

To generate the random excitation sample time histories, the Gaussian white noise constant power spectrum is used as the basis. This has been done in accordance with the theories that were explained in the Section 9.2.3. The subsequent deterministic simulation technique employed is Euler's discretization scheme, as elucidated in Section 9.3.2. Figure 9.9 shows the time history of the simulated displacement response of 200s.

9.4.3 A Response Ornstein–Uhlenbeck Process in Numerical Simulation

In the field of mathematics, the Ornstein–Uhlenbeck process is a random process that has applications in the field of physical sciences. The original application of the Ornstein–Uhlenbeck process in physics was as a model for the velocity of a massive Brownian particle when it was subjected to the action of friction. It is named in honor of Leonard Ornstein and George Eugene Uhlenbeck.

The Ornstein–Uhlenbeck process, denoted by the symbol y_t, is defined by the stochastic differential equation that is presented below:

$$dy_t = -\beta y_t dt + \sigma dW(t) \tag{9.144}$$

in which $\beta > 0$ and $\sigma > 0$ are parameters and $W(t)$ denotes the Wiener process.

We next write $y_t = \dfrac{dy(t)}{dt}$ and set $a = 0$ and $b = 0$ in equation (9.142), we thus obtain

$$\frac{d}{dt}(y_t) + 2zy_t = 2\sqrt{z}\frac{dW(t)}{dt} \tag{9.145}$$

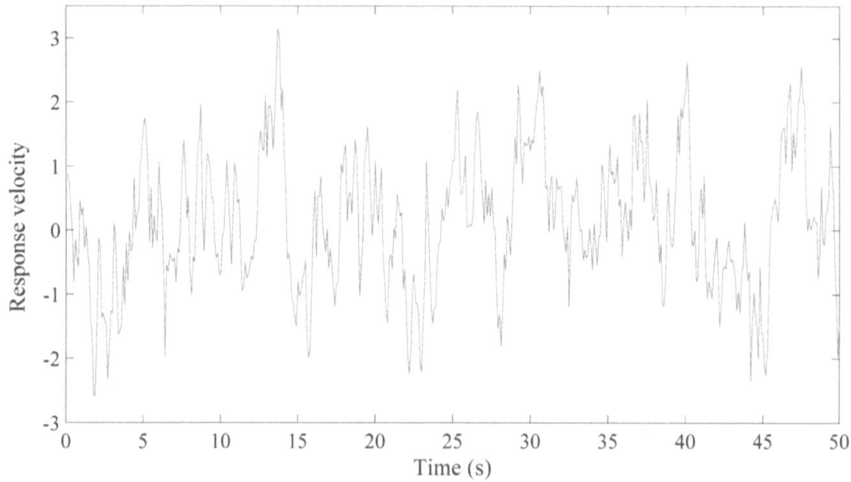

FIGURE 9.10 A simulated 50 s response velocity time history that is a sample record of an Ornstein–Uhlenbeck process.

Equation (9.145) can be re-written as

$$dy_t = -2zy_t dt + 2\sqrt{z}dW(t) \tag{9.146}$$

After comparing equations (9.144) and (9.146), we find that the simulated response velocity of a Duffing oscillator (when $a = 0$ and $b = 0$) is an Ornstein–Uhlenbeck process.

In the following, we will proceed to carry out yet another numerical simulation of the Duffing oscillator, which is characterized by equation (9.142) and incorporates a Wiener process. In this particular instance, we use the following values for the coefficients in equation (9.142): z = 0.99; $a = 0$; and $b = 0$. The numerical simulation is going to run for a total of 50 s, and the time interval of each step of the numerical simulation is going to be set at 0.1 s. Gaussian white noise constant power spectrum is utilized as the foundation to generate the random excitation sample time histories. This has been carried out in compliance with the theories elucidated in Section 9.2.3. Euler's discretization scheme, which is covered in detail in Section 9.3.2, is the subsequent deterministic simulation technique that is utilized. In Figure 9.10, the time history of the simulated displacement velocity over a period of 50 s (which is a sample record of an Ornstein–Uhlenbeck process) is displayed.

9.5 EXERCISES

EXERCISE 9.1

Please utilize the MATLAB software to generate a 7-by-7 matrix of uniformly distributed random numbers between 0 and 1.

EXERCISE 9.2

Please utilize the MATLAB software to generate a column vector containing 20,000 elements of uniformly distributed random numbers between 0 and 1. Plot a histogram based on these 20,000 numbers. Plot a probability density curve based on these 20,000 numbers.

EXERCISE 9.3

Please utilize the MATLAB software to generate a row vector containing 1000 elements of Poisson-distributed random numbers based on a Poisson distribution with the parameter $\mu = 5$. Calculate the mean value of these generated 1000 Poisson distributed random numbers.

EXERCISE 9.4

Please utilize the MATLAB software to generate a column vector containing 100,000 elements of Gaussian-distributed random numbers based on a Gaussian distribution with the following parameters: mean $\mu = 1$; variance $\sigma^2 = 2$. Plot a histogram based on these generated 100,000 Gaussian distributed numbers. Plot a probability density curve based on these generated 100,000 Gaussian distributed numbers.

EXERCISE 9.5

Assume that both of the random variables X_1 and X_2 follow a normal distribution, with the mean being zero and the variance being σ^2, denoted as $N(0, \sigma^2)$. It can be concluded that the random variable $\sqrt{X_1^2 + X_2^2}$ follows a Rayleigh distribution as follows:

$$f_X(x) = \frac{x}{\sigma^2} \exp\left[-\frac{x^2}{2\sigma^2}\right], \qquad for \qquad x \geq 0$$

Please utilize the MATLAB software to generate a column vector containing 30,000 elements of Rayleigh-distributed random numbers based on a Rayleigh distribution with the following parameters: $\sigma = 200$. Plot a histogram based on these generated 30,000 Rayleigh-distributed numbers. Plot a probability density curve based on these generated 30,000 Rayleigh-distributed numbers.

EXERCISE 9.6

In this exercise, you are required to generate sample time histories based on a standard McCormick wave spectrum. The following is the mathematical

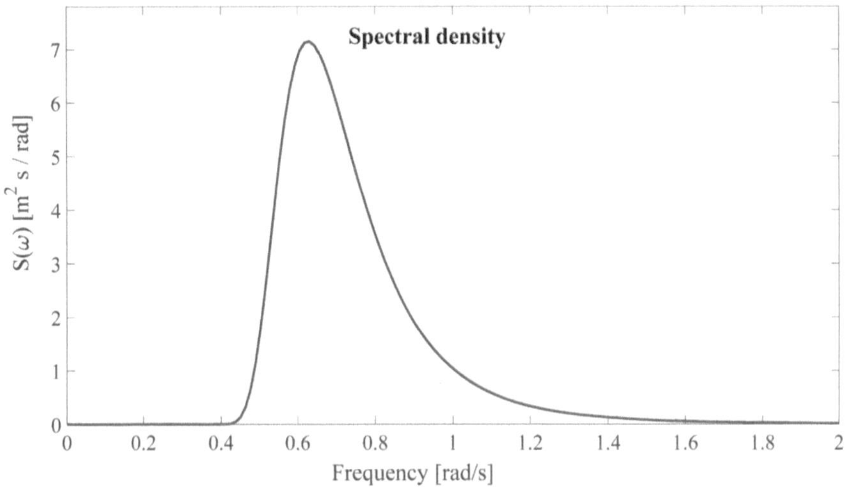

FIGURE 9.11 A McCormick wave spectrum with $H_s = 6.0\ m, T_p = 10.0\ s$ and $T_z = 8.143\ s$.

formulation of the McCormick frequency spectrum, which has a significant wave height H_s, a spectral peak period T_p and a zero up-crossing time T_z:

$$S(\omega) = (M+1)\frac{(0.25\times H_s)^2}{\omega_p}\left(\frac{\omega_p}{\omega}\right)^{(M+1)}\exp\left(-\frac{M+1}{M}\left(\frac{\omega_p}{\omega}\right)^M\right)$$

in which

$$\omega_p = \frac{2\pi}{T_p}$$

$$\frac{T_p}{T_z} = \frac{\left(\frac{M+1}{M}\right)^{1/M}}{\Gamma\left(\frac{M+1}{M}\right)}$$

in which $\Gamma\ ()$ is a Gamma function. In Figure 9.11, the blue curve shows a McCormick wave spectrum with $H_s = 6.0\ m$, $T_p = 10.0\ s$ and $T_z = 8.143\ s$.

Please generate a 100 s wave elevation time record based on the McCormick wave spectrum as shown in Figure 9.11 using a simulation time step of 0.1 s.

EXERCISE 9.7

The equation of motion of a viscously damped single-degree-of-freedom vibration system subjected to an external force is as follows:

$$m\ddot{y} + 0.4\dot{y} + ky = F(t) = 3\left(1 - sin\frac{t}{2}\right)$$

In the above equation the mass value is chosen to be $m = 2$ and the stiffness coefficient value is chosen to be $k = 2$. The undamped natural frequency and the natural period of this vibration system can be respectively determined by the following equations:

$$\omega_n = \left(\frac{k}{m}\right)^{1/2} = \left(\frac{2}{2}\right)^{1/2} = 1$$

$$T_n = \frac{2\pi}{\omega_n} = 2\pi$$

Explicit numerical schemes are only conditionally stable. In other words, they offer numerical solutions that are bounded only for (very) small time steps. For instance,

$$\Delta t \le \tau_{crit} = \frac{T_n}{\pi}$$

Therefore, when using explicit numerical schemes for finding the response of the vibration system in this exercise, the time step Δt must be less than $T_n/\pi = 2.0$.

Assume that the displacement and velocity values of the vibrating mass in this exercise at $t = 0$ are both equal to zero, utilizing the finite difference numerical integration method (an explicit numerical scheme) and a simulation time step $\Delta t < 2.0$ to find the response of this vibration system.

EXERCISE 9.8

One expression of the second-order Runge–Kutta scheme is as follows:

$$y_{i+1} = y_i + h(ak_1 + bk_2) + O(h^2) \tag{9.147}$$

in which

$$k_1 = f(t_i, y_i) \tag{9.148}$$

$$k_2 = f\left(t_i + \frac{1}{2}h, y_i + \frac{h}{2}k_1\right) \tag{9.149}$$

Utilizing two-dimensional Taylor's series expansion, show that the constant values in equation (9.147) should be $a = 0$ and $b = 1$, that is, one possible expression of the second-order Runge-Kutta scheme is

$$\begin{aligned} y_{i+1} &= y_i + h(ak_1 + bk_2) + O(h^2) \\ &= y_i + h\left\{f\left(t_i + \frac{1}{2}h, y_i + \frac{h}{2}f(t_i, y_i)\right)\right\} + O(h^2) \end{aligned} \tag{9.150}$$

EXERCISE 9.9

One expression of the third-order Runge–Kutta scheme is as follows:

$$y_{i+1} = y_i + h(ak_1 + bk_2 + ck_3) + O(h^4) \tag{9.151}$$

in which

$$k_1 = f(t_i, y_i) \tag{9.152}$$

$$k_2 = f\left(t_i + \frac{1}{p}h, y_i + \frac{h}{p}k_1\right) \tag{9.153}$$

$$k_3 = f\left(t_i + \frac{1}{q}h, y_i + \frac{h}{q}k_2\right) \tag{9.154}$$

Utilizing two-dimensional Taylor's series expansion, show that the constant values in equation (9.151) should be $a = \frac{1}{4}$, $b = 0$, $c = \frac{3}{4}$, $p = 3$ and $q = 3/2$, that is, one possible expression of the third-order Runge–Kutta scheme is

$$y_{i+1} = y_i + h\left(\frac{1}{4}k_1 + \frac{3}{4}k_3\right) + O(h^4) \tag{9.155}$$

$$k_1 = f(t_i, y_i) \tag{9.156}$$

$$k_2 = f\left(t_i + \frac{1}{3}h, y_i + \frac{h}{3}k_1\right) \tag{9.157}$$

$$k_3 = f\left(t_i + \frac{2}{3}h, y_i + \frac{2h}{3}k_2\right) \tag{9.158}$$

EXERCISE 9.10

The equation of motion of a Duffing oscillator with a random excitation term involving a Wiener process is as follows:

$$\frac{d^2y(t)}{dt^2} + 2z\frac{dy(t)}{dt} + by(t) + a[y(t)]^3 = 2\sqrt{z}\frac{dW(t)}{dt} \tag{9.159}$$

in which $W(t)$ denotes the Wiener process. $\frac{dW(t)}{dt}$ in equation (9.159) is a Gaussian white noise process with a constant power spectral density function. In this exercise, you are required to numerically simulate the Duffing oscillator as represented by equation (9.159). Please choose the following values for the coefficients in equation (9.159): $z = 0.5$; $a = 1$; $b = 2$. Therefore, in this case, equation (9.159) represents a Duffing oscillator with a hardening spring. The simulation length should be chosen to be 300 s, and the simulation time step should be chosen to be $dt = 0.1$ s. Please plot your simulated 300 s response displacement time history in a figure. Please also plot your simulated 300 s response velocity time history in another figure.

EXERCISE 9.11

Please choose the following values for the coefficients in equation (9.159): $z = 0.5$; $a = 0$; $b = 0$. Therefore, in this case, the response velocity of equation (9.159) will be an Ornstein–Uhlenbeck process. In this exercise, your simulation length should be chosen to be 500 s, and the simulation time step should be chosen to be $dt = 0.1$ s. Please plot your simulated 500 s response velocity time history in a figure.

10 Random Vibration Induced Failures

10.1 INTRODUCTION

The ultimate goal of utilizing random-process theory in a dynamic system response analysis is to be able to evaluate the reliability of a system that has been designed to endure random excitations. In this context, reliability refers to the probability of success (one minus the probability of failure). In this context, we will only be concerned with failures that are the result of the dynamic response of a stable structure. This means that we will not take into account failures that occur under static loads (even when they are random), failures that are attributable to static or dynamic instability, failures that are caused by corrosion, abrasion and other similar factors.

Let us assume that the dynamic system we are considering is an engineering structure. For the purpose of this discussion, let $Y(t)$ represent the dynamic response (either a deflection, a strain or a stress) at a critical point in the engineering structure. We postulate that the engineering structure will fail in the case that any of the two scenarios that are listed below take place:

1. For the very first time, $Y(t)$ reaches either an upper bound level U or a lower bound level $-L$, where both U and L are defined as large positive numbers.
2. Damage to the engineering structure accumulated as a result of $Y(t)$ fluctuating at minor or moderate excursions that are not substantial enough to trigger a failure of the first type, and failure happens when the overall amount of damage that has been accumulated reaches a fixed total.

We refer to failures that are caused by the first reason as first-excursion failures and failures that are caused by the second reason as fatigue failures. It will be demonstrated that in order to have a better understanding of these two types of failure, it is beneficial to obtain the probabilistic characteristics of the following two items: (1) within a given time interval, the number of times that the random process $Y(t)$ crosses a certain threshold; and (2) within a given time interval, the number of peaks (or troughs) in $Y(t)$ that are above (or below) a certain level. The studies of items (1) and (2) are referred to as the threshold-crossing problem and the peak (or amplitude) distribution problem, respectively. Thus, before we go on to a discussion of structural failures, we will first take a closer look at these problems in greater depth.

DOI: 10.1201/9781003431503-10

10.2 THE THRESHOLD-CROSSING PROBLEM

To analyze the first-excursion type failure, it is useful to obtain the probabilistic characteristics of the number of times with which the random process $Y(t)$ crosses a specific threshold ξ within a designated time interval.

Let's say that $Y(t)$ is a random process with continuous values, and $Y(t)$ is also continuous with regard to its parameter t. We will conduct an investigation into the statistical properties of an associated counting process known as $\mathbb{N}(\xi, t_1, t_2)$, which is a process that counts the random number of times that $Y(t)$ crosses a threshold ξ from either above or below during the time period $\left(t_1, t_2\right]$. An example of a sample function for the random process $Y(t)$ is depicted in Figure 10.1, along with a specified threshold ξ.

For the purpose of facilitating the study of the counting process $\mathbb{N}(\xi, t_1, t_2)$, let us construct a three-valued random process in the following manner:

$$Z(t) = \mathrm{H}\left[Y(t) - \xi\right] \tag{10.1}$$

in which H[ ] represents the Heaviside step function, i.e.,

$$Z(t) = \begin{cases} 1 & Y(t) > \xi \\ 0.5 & Y(t) = \xi \\ 0 & Y(t) < \xi \end{cases} \tag{10.2}$$

Given that the Dirac δ function is the derivative of the Heaviside step function, carrying out formal differentiation of $Z(t)$ with respect to t results in the following:

$$\dot{Z}(t) = \dot{Y}(t)\delta\left[Y(t) - \xi\right] \tag{10.3}$$

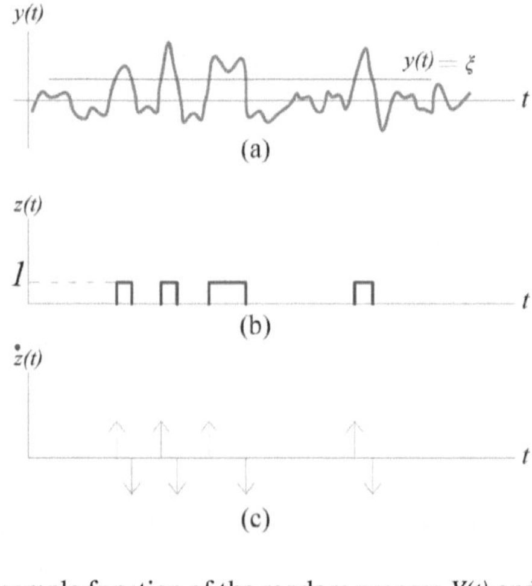

FIGURE 10.1 The sample function of the random process $Y(t)$ and a specified threshold value ξ.

in which $\delta[\ ]$ is the Dirac δ function. In Figure 10.1, the associated typical sample functions of $Y(t)$, $Z(t)$ and $\dot{Z}(t)$ are shown. As can be seen, the impulses in Figure 10.1c are unit impulses with alternative signs each of which is positioned under a threshold crossing in Figure 10.1a. The counting process $\mathbb{N}(\xi, t_1, t_2)$ provides the number of impulses of this type within the time interval $(t_i, t_2]$, without taking into account the direction of the impulses. On account of this, this counting process can be formulated as:

$$\mathbb{N}(\xi, t_1, t_2) = \int_{t_1+}^{t_2+} \left|\dot{Y}(t)\right| \delta\left[Y(t) - \xi\right] dt \tag{10.4}$$

For the purpose of obtaining the statistical properties of $\mathbb{N}$, the equation (10.4) can be utilized. For instance, the mathematical expectation of $\mathbb{N}$ can be expressed as follows:

$$\begin{aligned} E\left[\mathbb{N}(\xi, t_1, t_2)\right] &= \int_{t_1}^{t_2} E\left\{\left|\dot{Y}(t)\right| \delta\left[Y(t) - \xi\right]\right\} dt \\ &= \int_{t_1}^{t_2}\int_{-\infty}^{\infty}\int_{-\infty}^{\infty} |\dot{y}|\ \delta(y - \xi)\, p_{\{Y\}\{\dot{Y}\}}(y, \dot{y}, t)\, dy d\dot{y} dt \\ &= \int_{t_1}^{t_2}\int_{-\infty}^{\infty} |\dot{y}|\, p_{\{Y\}\{\dot{Y}\}}(\xi, \dot{y}, t)\, d\dot{y}\, dt \end{aligned} \tag{10.5}$$

In deriving the above equation, an assumption has been made that $E\left\{\left|\dot{Y}(t)\right| \delta\left[Y(t) - \xi\right]\right\}$ is bounded. Thus, no distinction is made between t_1 and $t_1 +$. Similarly, no distinction is made between t_2 and $t_2 +$.

The rate of threshold crossings per unit of time is frequently a convenient metric to work with. Let us write

$$\mathbb{N}(\xi, t_1, t_2) = \int_{t_1+}^{t_2+} n(\xi, t) dt \tag{10.6}$$

in order that

$$n(\xi, t) = \left|\dot{Y}(t)\right| \delta\left[Y(t) - \xi\right] \tag{10.7}$$

Calculating the mathematical expectation of $n(\xi, t)$, assuming that it exists, can be done as follows:

$$E\left[n(\xi, t)\right] = \int_{-\infty}^{\infty} |\dot{y}|\, p_{\{Y\}\{\dot{Y}\}}(\xi, \dot{y}, t) d\dot{y} \tag{10.8}$$

The mathematical expectation of $\mathbb{N}(\xi, t_1, t_2)$ can be computed using the following formula:

$$E\left[\mathbb{N}(\xi, t_1, t_2)\right] = \int_{t_1}^{t_2} E\left[n(\xi, t)\right]dt \tag{10.9}$$

In the event that the random process $Y(t)$ is weakly stationary, the joint probability density function of $Y(t)$ and $\dot{Y}(t)$ is independent of t. Thus, for a given threshold, the crossing rate remains constant. Similarly, one can use the following formula to find the correlation function for $n(\xi,t)$:

$$\begin{aligned}\phi_{nn}(\xi, t_1, t_2) &= E\left[n(\xi, t_1)\right] E\left[n(\xi, t_2)\right] \\ &= \int_{-\infty}^{\infty}\int_{-\infty}^{\infty} |\dot{y}_1||\dot{y}_2| \, p_{\{Y\}\{\dot{Y}\}}(\xi, \dot{y}_1, t_1; \xi, \dot{y}_2, t_2)\,d\dot{y}_1\, d\dot{y}_2\end{aligned} \tag{10.10}$$

The joint probability density function of $Y(t)$ and $\dot{Y}(t)$ that is required in this context is $p_{\{Y\}\{\dot{Y}\}}(y_1, \dot{y}_1, t_1; y_2, \dot{y}_2, t_2)$, which has two distinct times: t_1 and t_2. When being used in equation (10.10), y_1 and y_2 in the foregoing joint probability density function are both replaced by the given threshold value ξ. In certain applications, only the information related to the crossing of the threshold by $Y(t)$ from below is needed. In light of the fact that slopes must be positive at such crossings, the expected rate of threshold up-crossings from below can be computed as follows:

$$E\left[n_+(\xi, t)\right] = \int_0^{\infty} \dot{y}\, p_{\{Y\}\{\dot{Y}\}}(\xi, \dot{y}, t)\,d\dot{y} \tag{10.11}$$

This particular crossing rate is characterized by the correlation function that is as follows:

$$\begin{aligned}\phi_{n_+n_+}(\xi, t_1, t_2) &= E\left[n_+(\xi, t_1)\right] E\left[n_+(\xi, t_2)\right] \\ &= \int_0^{\infty}\int_0^{\infty} \dot{y}_1\dot{y}_2\, p_{\{Y\}\{\dot{Y}\}}(\xi, \dot{y}_1, t_1; \xi, \dot{y}_2, t_2)\,d\dot{y}_1\, d\dot{y}_2\end{aligned} \tag{10.12}$$

Equations (10.11) and (10.12) can be regarded as equivalent to equations (10.9) and (10.11), with the only difference being that the average is now calculated exclusively over the positive sample space of the derivative process $\dot{Y}(t)$. In the case of a weakly stationary random process, the occurrence of a positive slope crossing at a certain time must be accompanied by a negative slope crossing at another time. The aforementioned physical argument allows us to write in the case of weakly stationary random process:

$$E\left[n_+(\xi,t)\right]=\frac{1}{2}E\left[n(\xi,t)\right] \tag{10.13}$$

For instance, suppose that $Y(t)$ is stationary, Gaussian and has a 0 mean. Then, $Y(t)$ and $\dot{Y}(t)$ are independent, and their joint probability density is

$$p_{Y\dot{Y}}(y,\dot{y})=\frac{1}{2\pi\sigma_Y\sigma_{\dot{Y}}}\exp\left(-\frac{y^2}{2\sigma_Y^2}-\frac{\dot{y}^2}{2\sigma_{\dot{Y}}^2}\right) \tag{10.14}$$

The result of substituting equation (10.14) into equation (10.11) is

$$E\left[n_+(\xi,t)\right]=\frac{1}{2\pi}\frac{\sigma_{\dot{Y}}}{\sigma_Y}\exp\left(-\frac{\xi^2}{2\sigma_Y^2}\right) \tag{10.15}$$

Please note that the following relationship has been utilized for obtaining the result in equation (10.15).

$$\int_0^\infty x\exp\left(-\frac{x^2}{2\sigma^2}\right)dx=\sigma^2 \tag{10.16}$$

10.3 AMPLITUDE DISTRIBUTION OF A NARROW BAND GAUSSIAN PROCESS

Let us assume that a random process $Y(t)$ is a Gaussian random process with a narrow-band spectrum. The term "narrow-band" is derived from the fact that the spectral density function of the random process is sharply concentrated in the vicinity of the frequency ω_0. A random process being "narrow-band" means that the amplitude and phase of the process change slowly and randomly over time, while the frequency remains constant as shown in Figure 10.2.

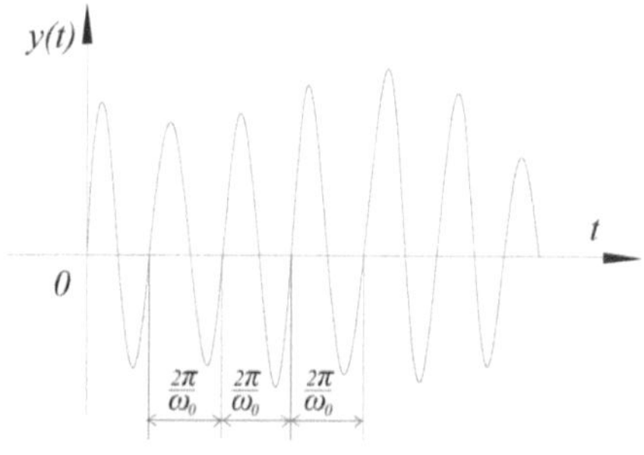

FIGURE 10.2 A sample record of a narrow-band process.

We can then write the sample function $y(t)$ of $Y(t)$ as

$$\begin{aligned} y(t) &= A(t)\cos\{\omega_0 t + \varepsilon(t)\} \\ &= A(t)\{\cos\varepsilon(t)\cos\omega_0 t - \sin\varepsilon(t)\sin\omega_0 t\} \end{aligned} \quad (10.17)$$

in which the frequency ω_0 is a constant. In addition, if we assume that $y(t)$ is a Gaussian random process with a mean of 0 and a variance of σ^2 by expanding $y(t)$ in a Fourier series with a period $T = 2\pi/\omega$, we can express $y(t)$ in the following manner:

$$y(t) = \sum_{n=1}^{\infty} (a_n \cos n\omega t + b_n \sin n\omega t) \quad (10.18)$$

in which

$$a_n = \frac{2}{T}\int_0^T y(t)\cos n\omega t \, dt \quad (10.19)$$

$$b_n = \frac{2}{T}\int_0^T y(t)\sin n\omega t \, dt \quad (10.20)$$

and the coefficients a_n and b_n in equation (10.18) are normally distributed, with a mean of 0 and a variance of σ^2. By writing $n\omega t$ in equation (10.18) as $(n\omega - \omega_0)t + \omega_0 t$, we can formulate equation (10.18) as follows:

$$y(t) = y_c(t)\cos\omega_0 t - y_s(t)\sin\omega_0 t \quad (10.21)$$

in which

$$y_c(t) = \sum_{n=1}^{\infty} \{a_n \cos(n\omega - \omega_0)t + b_n \sin(n\omega - \omega_0)t\} \quad (10.22)$$

$$y_s(t) = \sum_{n=1}^{\infty} \{a_n \sin(n\omega - \omega_0)t - b_n \cos(n\omega - \omega_0)t\} \quad (10.23)$$

By comparing equations (10.17) and (10.21), we find

$$y_c(t) = A(t)\cos\varepsilon(t) \quad (10.24)$$

$$y_s(t) = A(t)\sin\varepsilon(t) \quad (10.25)$$

At this point in time, let us define the random variables Y_c and Y_s, which will refer to the values of $y_c(t)$ and $y_s(t)$, respectively at time t. Y_c and Y_s are both normally distributed, as demonstrated by equations (10.22) and (10.23). To acquire the joint probability distribution of Y_c and Y_s, it is necessary to determine the mean and variance of Y_c and Y_s, in addition to the correlation between them. Due to the fact that the random variables a_n and b_n all have a mean of 0, the mean of both Y_c and Y_s is also 0. As a result, we have,

$$E(Y_c) = E(Y_s) = 0 \tag{10.26}$$

Next, let us take into consideration the covariance of the two random variables Y_c and Y_s. Given that the mean values of Y_c and Y_s are both 0, the covariance can be expressed as follows from equations (10.22) and (10.23):

$$\begin{aligned} Cov\left[Y_C, Y_S\right] &= E\left[Y_C, Y_S\right] \\ &= E\left[\sum_{n=1}^{\infty}\left\{a_n \cos\left(n\omega - \omega_0\right)t + b_n \sin\left(n\omega - \omega_0\right)t\right\}\right. \\ &\quad \left.\times \sum_{m=1}^{\infty}\left\{a_m \sin\left(m\omega - \omega_0\right)t - b_m \cos\left(m\omega - \omega_0\right)t\right\}\right] \\ &= \sum_{n=1}^{\infty}\sum_{m=1}^{\infty}\left\{E\left[a_n a_m\right]\cos\left(n\omega - \omega_0\right)t \times \sin\left(m\omega - \omega_0\right)t\right. \\ &\quad + E\left[b_n a_m\right]\sin\left(n\omega - \omega_0\right)t \times \sin\left(m\omega - \omega_0\right)t \\ &\quad - E\left[a_n b_m\right]\cos\left(n\omega - \omega_0\right)t \times \cos\left(m\omega - \omega_0\right)t \\ &\quad - E\left[b_n b_m\right]\sin\left(n\omega - \omega_0\right)t \times \cos\left(m\omega - \omega_0\right)t \end{aligned} \tag{10.27}$$

$E\left[a_n a_m\right]$ in the first term of the aforementioned equation can be evaluated using equations (10.19) and (10.20) in the following manner:

$$\begin{aligned} E\left[a_n a_m\right] &= \frac{4}{T^2}\int_0^T\int_0^T E\left[y(t)y(s)\right]\cos n\omega t \times \cos m\omega s\, dt\, ds \\ &= \frac{4}{T^2}\int_0^T\int_0^T R(t-s)\cos n\omega t \times \cos m\omega s\, dt\, ds \end{aligned} \tag{10.28}$$

in which $R(t-s)$ is the autocorrelation function of $y(t)$ and $y(s)$. Writing $t - s = \tau$ gives us

$$E\left[a_n a_m\right] = \frac{4}{T^2}\int_0^T\int_{-s}^{T-s} R(\tau)\cos n\omega(s+\tau) \times \cos m\omega s\, d\tau\, ds \tag{10.29}$$

Moreover, if we allow $u = s/T$ and $\omega T = 2\pi$, we obtain

$$
\begin{aligned}
E[a_n a_m] &= \frac{4}{T}\int_{-Tu}^{T(1-u)}\int_{0}^{1} R(\tau)\cos n\omega(Tu+\tau)\times\cos(m\omega Tu)\,du\,d\tau \\
&= \frac{4}{T}\int_{-Tu}^{T(1-u)}\int_{0}^{1} R(\tau)\{(\cos 2\pi nu\times\cos 2\pi mu)\cos n\omega\tau \\
&\quad -(\sin 2\pi nu\times\cos 2\pi mu)\sin n\omega\tau\}\,du\,d\tau
\end{aligned}
\tag{10.30}
$$

As the results of the characteristics of the orthogonal function, we obtain

$$
\int_{0}^{1}\cos(2\pi nu)\times\cos(2\pi mu)\,du = \begin{cases} 1/2 & \text{if } m = n \\ 0 & \text{otherwise} \end{cases}
\tag{10.31}
$$

as well as

$$
\int_{0}^{1}\sin(2\pi nu)\times\cos(2\pi mu)\,du = 0 \text{ for all } m,\, n
\tag{10.32}
$$

Consequently, equation (10.30) can be simplified to

$$
E[a_n a_m] = \frac{2}{T}\int_{-Tu}^{T(1-u)} R(\tau)\cos(n\omega\tau)\,d\tau \text{ if } m = n
\tag{10.33}
$$

where $n \to \infty$ as $T \to \infty$ such that $n\omega = \omega_n$ remains constant. Therefore, we can write

$$
\lim_{T\to\infty} TE[a_n a_m] = 2\lim_{T\to\infty}\int_{-Tu}^{T(1-u)} R(\tau)\cos\omega_n\tau\,d\tau
\tag{10.34}
$$

On the other hand, according to the Wiener–Khintchine theorem, the integration in equation (10.34) for $T \to \infty$ is equal to $\pi S(\omega_n)$, where $S(\omega_n)$ represents the spectral density function. Consequently, equation (10.34) becomes

$$
\lim_{T\to\infty} TE[a_n a_m] = \begin{cases} 2\pi S(\omega_n) & \text{for } n = m \\ 0 & \text{otherwise} \end{cases}
\tag{10.35}
$$

Applying the same procedure employed in the derivation of the preceding equation to the other three terms of equation (10.27), we can show that

$$\lim_{T\to\infty} TE\left[a_n b_m\right] = 0 \tag{10.36}$$

$$\lim_{T\to\infty} TE\left[b_n a_n\right] = 0 \tag{10.37}$$

$$\lim_{T\to\infty} TE\left[b_n b_m\right] = \begin{cases} 2\pi S(\omega_n) & \text{for } n = m \\ 0 & \text{otherwise} \end{cases} \tag{10.38}$$

Combining equations (10.27), (10.35), (10.36), (10.37) and (10.38) leads to

$$Cov\left[Y_c, Y_s\right] = E\left[Y_c, Y_s\right] = 0 \tag{10.39}$$

This suggests that the two random variables Y_c and Y_s are uncorrelated. Additionally, the second moments of the random variables Y_c and Y_s can be evaluated in the same manner as shown in the derivation of equation (10.39):

$$E\left[Y_c^2\right] = E\left[Y_s^2\right] = \sigma = \lim_{T\to\infty} \sum_{n=1}^{\infty} \frac{2\pi}{T} S(\omega_n) = \int_0^{\infty} S(\omega) d\omega \tag{10.40}$$

Therefore, based on the findings derived from equations (10.39) and (10.40), it can be concluded that the two random variables, Y_c and Y_s, are statistically independent. More specifically, each of these variables follows a normal probability distribution with a mean of 0 and a variance of σ^2. Therefore, the formulation of the joint probability density function of Y_c and Y_s can be expressed as

$$f(Y_c, Y_s) = \frac{1}{2\pi\sigma^2} \exp\left[-\frac{1}{2\sigma^2}\left(Y_c^2 + Y_s^2\right)\right], \quad \begin{matrix} -\infty < Y_c < \infty, \\ -\infty < Y_s < \infty \end{matrix} \tag{10.41}$$

Following this, the joint probability density, which is represented by the equation (10.41), will be converted into the joint probability density function of the amplitude A and the phase ε by utilizing the relationships that are presented in equations (10.24) and (10.25). As for the transform technique, please refer to Chapter 2. The transformation results in the following:

$$\begin{aligned} f(A,\varepsilon) &= \left[f(Y_c, Y_s)\right]_{\substack{Y_c = A\cos\varepsilon \\ Y_s = A\sin\varepsilon}} \cdot \begin{vmatrix} \dfrac{\partial Y_c}{\partial A} & \dfrac{\partial Y_c}{\partial \varepsilon} \\ \dfrac{\partial Y_s}{\partial A} & \dfrac{\partial Y_s}{\partial \varepsilon} \end{vmatrix} \\ &= \frac{A}{2\pi\sigma^2} \exp\left(-\frac{A^2}{2\sigma^2}\right), \quad 0 \le A < \infty, 0 \le \varepsilon \le 2\pi \end{aligned} \tag{10.42}$$

Following that, the marginal probability density function of the amplitude A can be expressed as described below:

$$f(A)=\int_0^{2\pi} f(A,\varepsilon)d\varepsilon=\frac{A}{\sigma^2}\exp\left(-\frac{A^2}{2\sigma^2}\right),0\le A<\infty \quad (10.43)$$

This probability density function, which can be found above, is the Rayleigh probability distribution with the parameter $2\sigma^2$. Therefore, we can draw the conclusion that, under the assumption of a narrow-band Gaussian random process with a mean of 0 and a variance of σ^2, the amplitude follows a Rayleigh distribution with a parameter of $2\sigma^2$. It is important to take note that the variance σ^2 is equal to the area under the spectrum of the random process. As a result, we can analyze the statistical characteristics of the amplitude of a narrow-band Gaussian random process provided that we have knowledge about the spectrum of the process.

10.4 FIRST-EXCURSION FAILURES

We have defined a first-excursion failure as the occurrence of a random process $Y(t)$ passing an upper bound U or a lower bound $-L$ for the first time. In an ideal scenario, we would like to determine the probabilistic structure of the random time T when a failure of this nature occurs under appropriate initial conditions concerning $Y(t)$ and, maybe, its derivatives. This challenging problem is referred to as the first-passage problem. The purpose of this section will be restricted to obtaining approximate solutions to first-passage problems.

For the sake of simplicity, we will first investigate the scenario in which $Y(t)$ is stationary and symmetrically distributed in both the positive and negative ranges, and where the upper and lower bounds are also symmetrical, which is denoted by the equation $U=L$. Then the probability of failure at the time t is as follows:

$$\mathbf{P}\left[\{Y(t)\ge U\}\cup\{Y(t)\le -U\}\right]=\mathbf{P}\left[\{|Y(t)|\ge U\}\right]=2\mathbf{P}\left[\{Y(t)\ge U\}\right] \quad (10.44)$$

In spite of the fact that a first-excursion failure of a structure is typically devastating, we shall take into consideration an imaginary scenario in which a structure is able to recover immediately after experiencing a failure. If this is the case, then the number of failures that occur within a time period is just two times the number of U threshold crossings that occur at positive slopes during the interval. An additional arbitrary assumption that we will make is that failures of this nature arrive independently. The number of failures that occurred within the time interval $(0,t]$ is denoted by the symbol $Q(t)$. There is no doubt that $Q(t)$ is a Poisson process. In the event that $Y(t)$ is stationary, $Q(t)$ is a Poisson process with stationary increments. The probability function of $Q(t)$ is

$$P_{\{Q\}}(n,t)=\exp(-\lambda t)\frac{(\lambda t)^n}{n!} \quad (10.45)$$

Through the process of setting n in equation (10.45) to 0, we can compute the probability that there will be no failure during the time period $(0,t]$ as

$$P_{\{Q\}}(0, t) = \exp(-\lambda t) \tag{10.46}$$

with the failure rate

$$\lambda = 2E\left[n_+(U)\right] = E\left[n(U)\right] \tag{10.47}$$

Equation (10.46), together with equation (10.47), can be used to calculate the reliability of a structure against first-excursion failures. In equation (10.47), $E\left[n_+(U)\right]$ are the expected rates of crossings of the upper bound U with positive slopes. $E\left[n(U)\right]$ are the expected rates of crossings of the upper bound U with positive and negative slopes.

The probability of failure during the time period $(0, t]$ is simply given as

$$1 - P_{\{Q\}}(0, t) = 1 - \exp(-\lambda t) \tag{10.48}$$

Nevertheless, equation (10.48) can be regarded as the probability that the first passage time is either equal to t or less than t. This is the probability distribution function of the random time T, which is expressed as,

$$F_T(t) = 1 - \exp(-\lambda t) \tag{10.49}$$

Taking the equation (10.49) and differentiating it yields the probability density of T

$$p_T(t) = \lambda \exp(-\lambda t) \tag{10.50}$$

For the purpose of computing the statistical properties of the first-passage time T, the equation (10.50) can be utilized. In particular, the mean and the standard deviation of the first-passage time can be easily calculated as

$$E[T] = \int_0^\infty t p_T(t) dt = \lambda^{-1} \tag{10.51}$$

$$\sigma_T = \sqrt{\int_0^\infty \left(t - \lambda^{-1}\right)^2 p_T(t) dt} = \sqrt{\lambda^2} = \lambda^{-1} \tag{10.52}$$

10.5 FATIGUE FAILURES

The fatigue failure model postulates that each excursion or "cycle" of the random response $y(t)$ inflicts an increment of damage that is dependent on the peak amplitude of the "cycle". Each subsequent "cycle" inflicts additional damage, and the point at which the total damage that has been accumulated reaches 100% is considered to be the point at which failure has occurred. A model that is reasonably valid for the fatigue failure of metals under random vibration has been developed with the assistance of certain rational hypotheses. The following part provides a description of the background of these hypotheses.

10.5.1 FATIGUE TEST AND THE PALMGREN-MINER HYPOTHESIS

During the process of a metal tensile specimen being subjected to a fluctuating stress history, a great number of small changes take place. In addition to the migration of dislocations, there is also localized plastic deformation. Microscopically small cracks are formed, and these cracks slowly grow over time. If the stress excursions are sufficiently large, one of the cracks will eventually begin to grow rapidly, which will ultimately result in the material rupturing.

Figure 10.3 provides a description of the physical process wherein fatigue occurs. The metal test specimen in Figure 10.3 is subject to the action of oscillatory stresses as shown in Figure 10.4. When oscillatory tensile stresses of sufficient magnitude are applied to the metal test specimen in Figure 10.3, a small crack will begin to form at a spot of the stress concentration. When it comes to

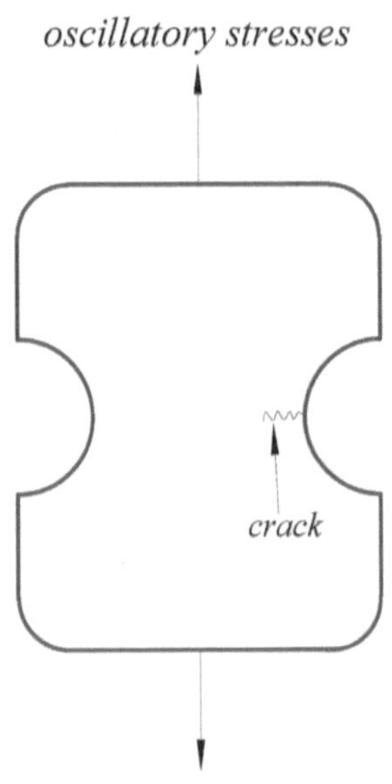

FIGURE 10.3 The description of fatigue process.

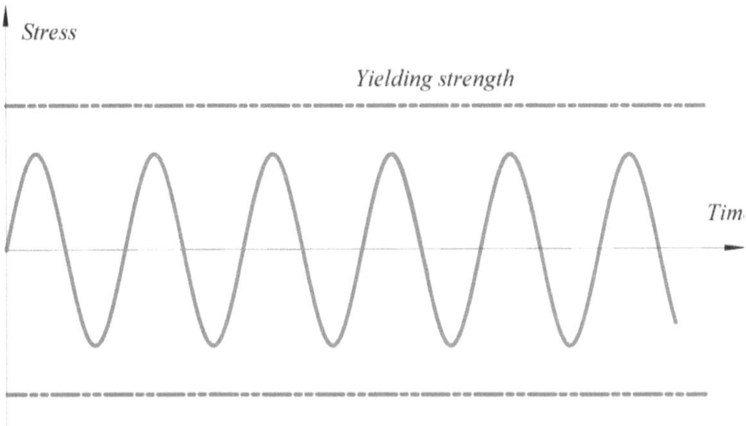

FIGURE 10.4 The oscillatory stresses acting on the specimen in Figure 10.2.

metallic materials, a rough rule of thumb is that the amplitude of the stress at the notch should be greater than 50% of the material's ultimate strength. On the other hand, if there is already a crack or imperfection in the material (for example, a tool mark), the stress amplitude that is required to initiate a growing crack may be considerably lower.

Following the initiation of the crack, it will have a tendency to grow in a direction that is orthogonal to the direction in which the oscillatory tensile loads are acting. Finally, a quasi-static brittle or ductile fracture occurs in the material having a reduced cross-sectional area. Figures 10.3 and 10.4 illustrate that fatigue can occur even though stress amplitudes are significantly below the yielding strength of the metal material.

In the realm of mechanical and structural design, fatigue is perhaps the most important failure mode that must be taken into consideration. Traditionally, fatigue has been associated with metallic materials; however, fatigue in other materials, such as filamentary composites and concrete, is also taken into consideration. It has been reported that fatigue is responsible for eighty to ninety percent of all service failures that have been observed in mechanical and structural systems. Failures caused by fatigue are frequently catastrophic; they occur without any prior warning and have the potential to cause severe property damage in addition to loss of life.

The dominance of this failure mode and the difficulties associated with designing to prevent it can be attributed to a number of factors, including the following: (1) The fatigue process is inherently unpredictable, as demonstrated by the statistical scatter in laboratory data; (2) it is frequently challenging to translate laboratory data of material behavior into field predictions; (3) it is extremely challenging to accurately model the mechanical environments to which the system is exposed throughout its entire design lifetime; and (4) environmental effects produce complex stress states at fatigue-sensitive "hot spots" in the system. In

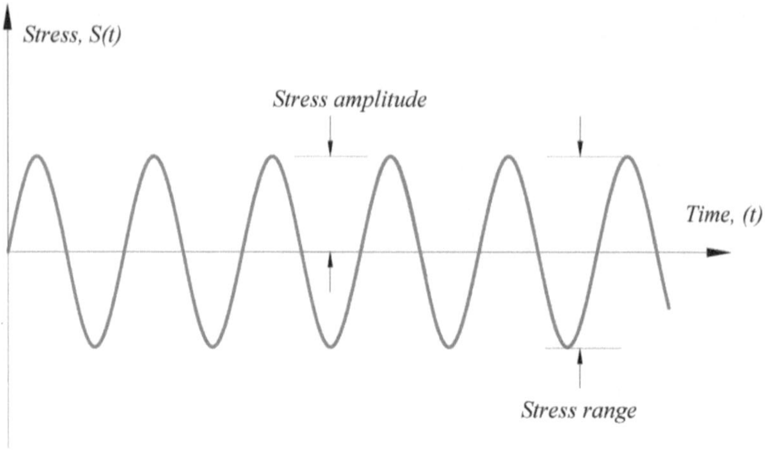

FIGURE 10.5 The oscillatory stresses history during a constant amplitude fatigue test.

conclusion, material fatigue can be caused by the interplay of multiple processes, which might be quite intricate. To sum up, material fatigue may result from a very complicated interaction of several processes.

Even if there is a lot of interest in the fundamental physics of the fatigue process at the moment, it is still not possible to forecast the phenomenological aspects of fatigue in a quantitative manner based on microscopic considerations. In order to obtain fatigue data, it is necessary to conduct tests on the material of interest.

In a classical fatigue test, a constant amplitude stress S fluctuating over time as shown in Figure 10.5 is applied to a specimen. The procedure continues until the specimen is broken. Cycles to failure (N) are quantified.

The majority of the fatigue data has been acquired from tests in which the mean stress is 0; nevertheless, there is some information that may be found regarding the influence of mean stress. In the event that the mean stress is not too high (for example, if it is less than one-third of the fluctuation amplitude), it has been discovered that the amount of mean stress has a relatively insignificant impact on the fatigue life. In the following discussion, we shall restrict ourselves to the scenario in which the mean stress is 0.

As shown in Figure 10.6, the results of testing a large number of identical samples with a variety of stress amplitudes are said to define the $S-N$ curve or "fatigue curve" for the material. The number of cycles till fracture is denoted by the letter N, while the fixed stress amplitude is denoted by the letter S. The fatigue curve is not, strictly speaking, a single curve. It is more precisely the mean of the statistical distribution of fatigue lives that was obtained for the specified stress amplitudes. The "scatter" of the experimental points provides some insight into the breadth of the distribution. For a complete theory of fatigue under random loading, the statistics of this process ought to be appropriately incorporated. However, in the following discussion, for the sake of simplicity, a single

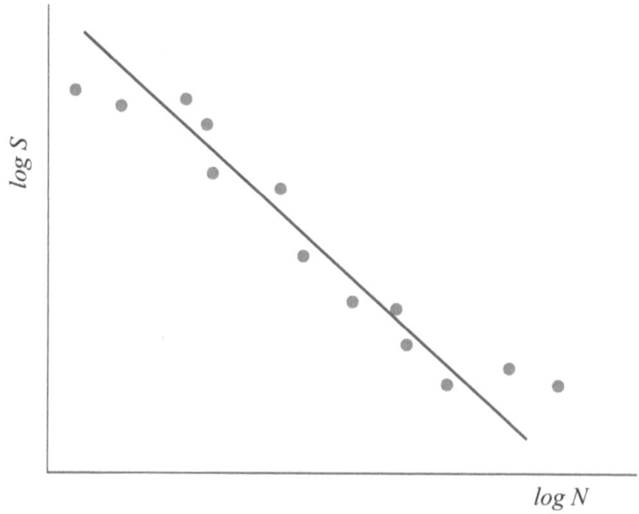

FIGURE 10.6 Typical $S-N$ experimental data.

well-defined $S-N$ curve for the material will be assumed to exist. This curve can be well approximated by a straight line for many different types of materials when log$S$ is plotted against log$N$, which means that the $S-N$ curve can be approximated by the following equation

$$NS^m = c \tag{10.53}$$

in which m and c, determined empirically from fatigue experiments, are the fatigue strength exponent and fatigue strength coefficient, respectively. Often, $1/c$ is considered as a random variable, usually having a lognormal distribution, that is, with $c = E\epsilon^{-1}$ where $\ln E \in \mathrm{N}\left(0, \sigma_E^2\right)$ and ϵ, m are fixed constants. The analytical model that is going to be presented in the following discussions will make use of the relation (10.53)

In the case where the stress in the specimen is a random process instead of the fixed amplitude fluctuation depicted in Figure 10.5, the problem of forecasting fatigue failure becomes far more challenging. As a temporary expedient alternative recourse has been made to a "rational" extrapolation of fixed amplitude fatigue data by utilizing the linear damage accumulation hypothesis that was given by Palmgren and Miner. According to the Palmgren and Miner hypothesis, when n cycles of stress amplitude S have been experienced, the material has "used up" a fraction of its fatigue life equal to n/N, where N is the number of cycles at which failure occurs under uniform amplitude S as determined from the $S-N$ curve. It is presumed that the fractional damage that occurred at this stress level can be added to the corresponding fractional damages of other stress levels in order to obtain the total damage experienced. Therefore, if the specimen experiences n_i cycles of stress amplitude S_i for $i = 1, 2, \cdots M$, the total or cumulative fatigue damage (D) is the linear summation of the individual damage from all the considered stress amplitudes. D is expressed mathematically by the following equation:

$$D = \sum_{i=1}^{M} \frac{n_i}{N_i} \tag{10.54}$$

in which D is the damage ratio, n_i is the number of cycles the material endures at stress amplitude S_i, N_i is the number of cycles to failure at stress amplitude S_i, as determined by using the appropriate $S-N$ curve. M in equation (10.54) is the total number of considered stress amplitudes. In accordance with the Palmgren–Miner hypothesis, the specimen will experience fatigue failure when the value of D exceeds unity. It is important to note that the order in which the various stress levels are applied is not restricted under this hypothesis. Therefore, it can be applied to a stochastic process in which the stress amplitude varies from cycle to cycle.

It is acknowledged that the Palmgren–Miner hypothesis has many deficiencies. This hypothesis glosses over many observed phenomena. However, it offers a model that has at least some similarities with the real situation. Despite the fact that its absolute validity is undoubtedly going to be called into question, its comparative usefulness as a straightforward analytical criterion for comparing alternative modes of operation of the same system may be of significant value.

10.5.2 Fatigue Caused by a Narrow-band Random Stress Process

Let's say that $y(t)$ is a stationary random stress. For the sake of simplicity, let us imagine a simple tension specimen in order that $y(t)$ represents the uniform tensile stress throughout the test section. In the general case, if the system is linear, stress is linearly proportional to strain, and the strain at a specific point is proportional to the element's overall deformation. Consequently, it should be feasible to establish a linear relationship between the stress intensity at a specific point within a structural member and the relative displacements of the member's points of attachment. We assume that the stress response $y(t)$ can be related to the other dynamic response quantities and, as a result, with the excitation.

We are going to make the assumption that $y(t)$ is a narrow-band random process. Figure 10.7 shows an example of what the stress history of a narrow-band response process might look like. The individual "cycles" are discernible, despite the fact that their amplitudes may fluctuate in a random manner. By employing the Palmgren–Miner hypothesis, one can ascribe an incremental "damage" to each cycle. The accumulation of these damages results in a total damage $D(T)$ for a time interval T. As the value of T is increased, the damage $D(T)$ increases in a monotonic manner. At some point in time (T_F), the total damage exceeds the value of unity, and failure occurs according to the Palmgren–Miner criterion.

If we now consider an ensemble of stress histories starting from an arbitrary origin in time, the procedure described above can be utilized to assign a value of

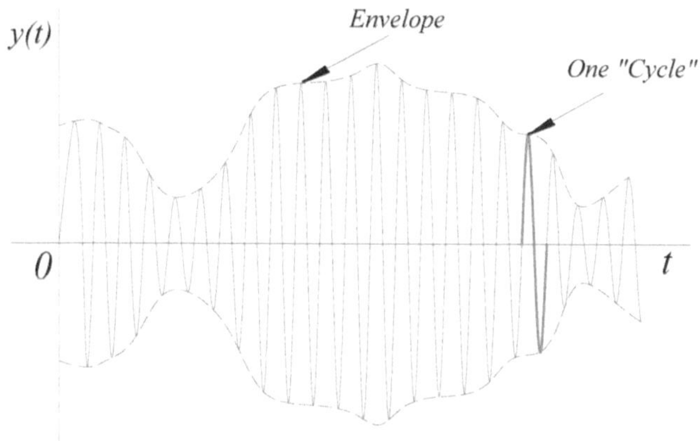

FIGURE 10.7 A sample time history of a narrow-band stationary process.

failure time T_F to each individual sample that is part of the ensemble. It is expected that the failure times, denoted by T_F, will vary randomly from one sample to the next. It would be desirable to have knowledge of the probability density distribution $p(T_F)$ for the failure times in order to obtain a statistical picture that is satisfactory. Unfortunately, this probability density distribution is unknown. We can employ a different approach to determine the mean and variance of the damage D at an arbitrary time T. Even if the damage distribution is still unknown, the central limit theorem can be invoked to demonstrate that the distribution tends toward the normal distribution as $T \to \infty$. At this point, we will merely conduct the analysis for the mean value of the damage.

Let us denote $E[n_+(0, t)]$ to be the expected zero-crossing frequency of the narrow-band random stress history $y(t)$; that is the average number of zero crossings with positive slope per unit time. Then, the expected number of "cycles" is $E[n_+(0, t)]T$ during the time period T. The expected proportion of these cycles whose stress amplitudes fall within the range of A and $A + dA$ is $p(A)dA$, where $p(A)$ represents the probability density of the peaks (see equation (10.43)). The expected number of such peaks, denoted as $n(A)$, is

$$n(A) = E\left[n_+(0, t)\right]T \cdot p(A)dA \tag{10.55}$$

According to the Palmgren–Miner hypothesis, a single peak of amplitude A results in an incremental damage of $1/N(A)$, where $N(A)$ is the number of cycles until failure in a constant-amplitude fatigue test with stress amplitude A. The expected fractional damage resulting from all "cycles" having peaks between A and $A + dA$ can be calculated as

$$\frac{n(A)}{N(A)} = \frac{E\left[n_+(0, t)\right]T \cdot p(A)dA}{N(A)} = E\left[n_+(0,t)\right]T\frac{p(A)}{N(A)}dA \tag{10.56}$$

The total expected damage, denoted by $E[D]$, is the sum of all the contributions like equation (10.56) for all of the stress amplitudes that are encountered. This sum can be calculated by the following integral

$$E[D(T)] = E[n_+(0,t)]T\int_0^\infty \frac{p(A)}{N(A)}dA \tag{10.57}$$

In order to calculate this, it is required to have knowledge of the probability distribution of the peaks, which is denoted by $p(A)$, as well as the fatigue curve in Figure 10.6 that was obtained through experimentation, which gives $N(A)$.

If we adopt the additional assumptions that $y(t)$ is a normal or Gaussian random process and the fatigue curve shown in Figure 10.6 may be represented by an equation of the form of equation (10.53), then it is feasible to evaluate equation (10.57) in closed form. In the event that $y(t)$ is normally distributed, then the peaks of $y(t)$ will have a Rayleigh distribution, as specified by the equation (10.43). Through the substitution of equations (10.43) and (10.53) into equation (10.57), we obtain the result

$$\begin{aligned} E[D(T)] &= E[n_+(0,t)]T\int_0^\infty \frac{p(A)}{N(A)}dA \\ &= \frac{E[n_+(0,t)]T}{c\sigma_y{}^2}\int_0^\infty A^{m+1}\exp\left(-\frac{A}{2\sigma_y{}^2}\right)dA \\ &= \frac{E[n_+(0,t)]T}{c\sigma_y{}^2}\frac{\left(\sqrt{2}\sigma_y\right)^{m+2}}{2}\Gamma\left(\frac{m+2}{2}\right) \\ &= \frac{E[n_+(0,t)]T}{c}\left(\sqrt{2}\sigma_y\right)^m\Gamma\left(1+\frac{m}{2}\right) \end{aligned} \tag{10.58}$$

Please note that in deriving equation (10.58) use has been made of the following equation

$$\int_0^\infty x^n \exp\left(-\frac{x^2}{2\sigma^2}\right)dx = \frac{(\sqrt{2}\sigma)^{n+1}}{2}\Gamma\left(\frac{n+1}{2}\right) \tag{10.59}$$

10.5.3 Fatigue Caused by a Wide-band Random Stress Process

(1). Considering a wide-band process as an equivalent narrow-band process. Let's assume that $S(t)$ is a stationary Gaussian random process with a mean value of 0. Figure 10.8 displays two realizations of $S(t)$. The upper portion of Figure 10.8 shows a realization of a narrow-band Gaussian random process. The lower portion of Figure 10.8 shows a realization of a wide-band Gaussian random

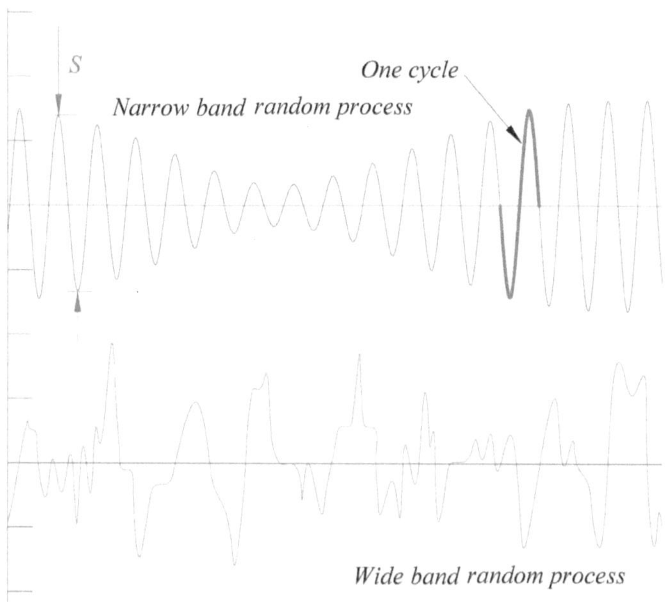

FIGURE 10.8 Two samples of random processes.

process. Each process in Figure 10.8 has the same root-mean-square value and expected rate of zero up-crossings. The stress cycles can be readily identified in the narrow-band random process that has Rayleigh-distributed peaks. A closed-form expression for the damage D has been obtained for the scenario when the $S-N$ curve is linear, $NS^m = c$, as indicated by equation (10.58).

However, taking a closer look at the realization of the wide-band random process shown in the lower portion of Figure 10.8, it is not immediately evident how to calculate stress cycles in order to apply Palmgren–Miner's rule. One method of approximation involves making the assumption that the damage caused by the wide-band random process in Figure 10.8 with an identical zero up-crossing rate $E[n_+(0,t)]$ and an identical root-mean-square value of σ_y will be the same as the damage caused by the corresponding narrow-band random process in Figure 10.8. Therefore, it is possible to estimate the fatigue damage under a wide-band random stress process by considering it as an equivalent narrow-band random process. To do so, compute the $E[n_+(0,t)]$ and σ_y of the wide-band random stress process and then apply the equation (10.58) to calculate the fatigue damage. Fortunately, this technique results in a damage estimate that is on the cautious side.

(2). Rain flow cycle counting algorithm. The identification of stress cycles in a given wide-band record has been the subject of a number of algorithmic proposals over the course of the years. The Fatigue Design Handbook (Society of Automotive Engineers (SAE), 1988) provides a discussion of these methodologies. However, the rain flow method is the one that is currently universally acknowledged as being the one that yields the most accurate results.

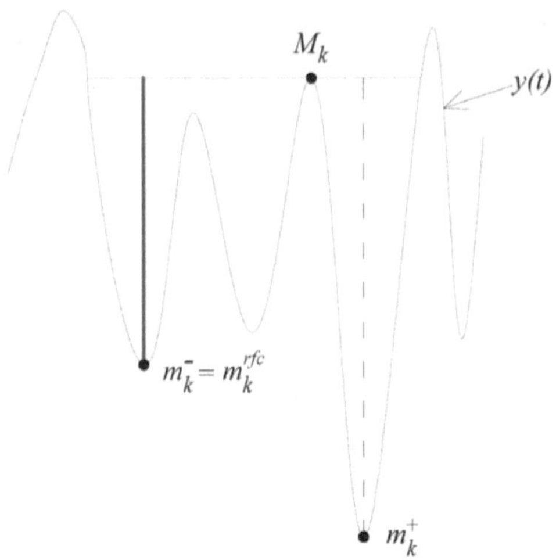

FIGURE 10.9 Definition of the rain flow cycle.

The rain flow cycle counting method is employed to identify the stress cycles, namely the stress range and mean stress for each cycle, based on a given realization of a wide-band random stress process $S(t)$. A comprehensive explanation of the rain flow cycle counting method can be found in the Fatigue Design Handbook (Society of Automotive Engineers (SAE), 1988). The formal definition of the rain flow cycle is illustrated in Figure 10.9.

Definition (Rain flow cycle). Assume that $y(t)$, where t belongs to the interval $[0,T]$, is a variable load function that has a finite number of local maxima. We make the assumption that a local maximum $M_k = y(t_k)$ in $y(t)$ is paired with one particular local minimum m_k^{rfc}. The position of m_k^{rfc} is determined by the following method: starting with the kth local maximum (value M_k), one determines the lowest value in the forward direction between M_k and the nearest point at which $y(t)$ exceeds M_k and designate this obtained lowest value as m_k^+. In addition, starting with the kth local maximum (value M_k), one also determines the lowest value in the backward direction between M_k and the nearest point at which $y(t)$ exceeds M_k and designate this obtained lowest value as m_k^-. The larger (less negative) of these two obtained values, denoted by m_k^{rfc}, is the rain-flow minimum paired with M_k. Therefore, the kth rain-flow pair is $\left(m_k^{rfc}, M_k\right)$, as seen in Figure 10.9.

The definition of the rain flow cycle can also be stated as follows: Starting from each local maximum M_k, one shall try to reach above the same level, in the backward (left) direction as well as in the forward (right) direction, with an as small downward excursion as possible. A minimum m_k^- on the left side is identified. In addition, a minimum m_k^+ on the right side is also identified. The minimum representing the smallest deviation from the maximum M_k is defined to be the corresponding rain flow minimum m_k^{rfc}. The kth rain flow cycle is defined as $\left(m_k^{rfc}, M_k\right)$, as seen in Figure 10.9.

If t_k is the time of the kth local maximum and the corresponding rain flow amplitude is $S_k^{rfc} = \left(M_k - m_k^{rfc}\right)/2$, then the total damage at the time T is

$$D(t) = \sum_{t_k \le T} \frac{1}{N\left(S_k^{rfc}\right)} \tag{10.60}$$

in which the sum contains all rain flow cycles that have been completed up to time T.

We next provide an example of extracting rain flow cycles from a measured dataset. The wave data utilized in this example was measured on December 24, 1989, at the Gullfaks C platform in the North Sea, and this data set contains 39,000 measurements. The sampling rate is 2.5 Hz. The use of this wave dataset here is for educational purposes only and is not for commercial purposes. Figure 10.10 shows the first 200 s of this measured wave dataset.

The vertical blue line in Figure 10.10 represents a rain flow amplitude for a specific local maximum at point A. Figure 10.11 shows the corresponding rain flow cycle plot for the local maximum at point A in Figure 10.10.

We shall now examine the rain flow cycle counts in this measured dataset containing 39,000 measurements. From the sequence of turning points, we find the rain flow cycles in the dataset. Because each cycle is a pair of a local maximum and a local minimum in the measured dataset, a cycle count can be visualized as a set of pairs in the $\mathbb{R}^2$-plane as shown in Figure 10.12.

Figure 10.13 shows the rain flow amplitude distribution for this measured dataset containing 39,000 measurements.

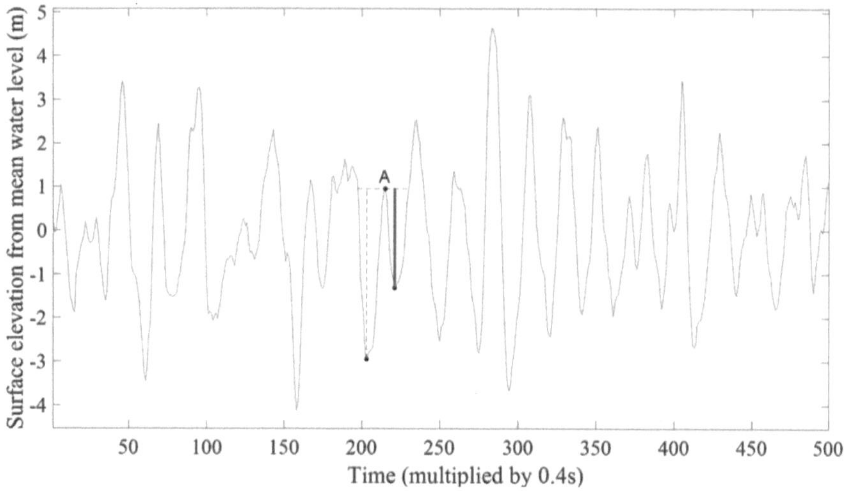

FIGURE 10.10 Part of the measured wave dataset at the Gullfaks C platform.

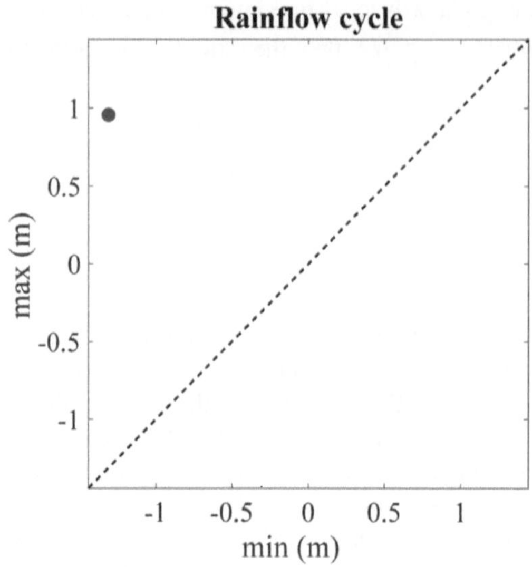

FIGURE 10.11 Rain flow cycle plot for the local maximum at point A in Figure 10.10.

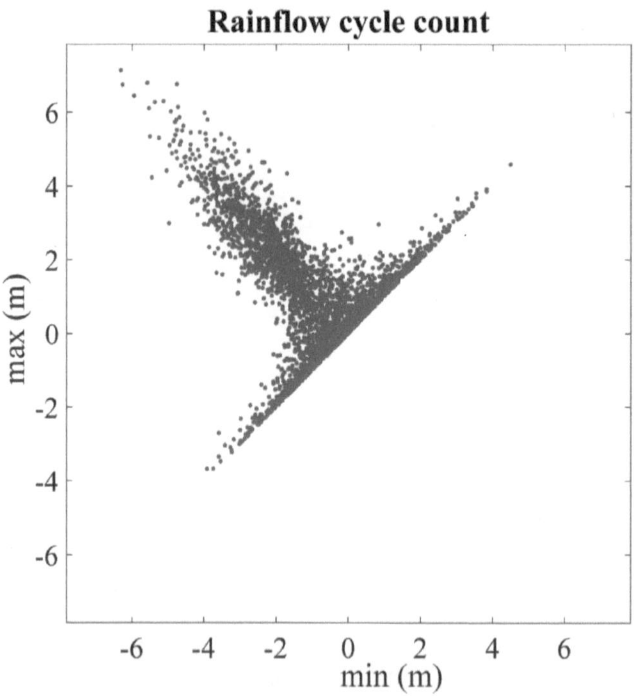

FIGURE 10.12 Rain flow cycle counts in this measured dataset of 39,000 measurements.

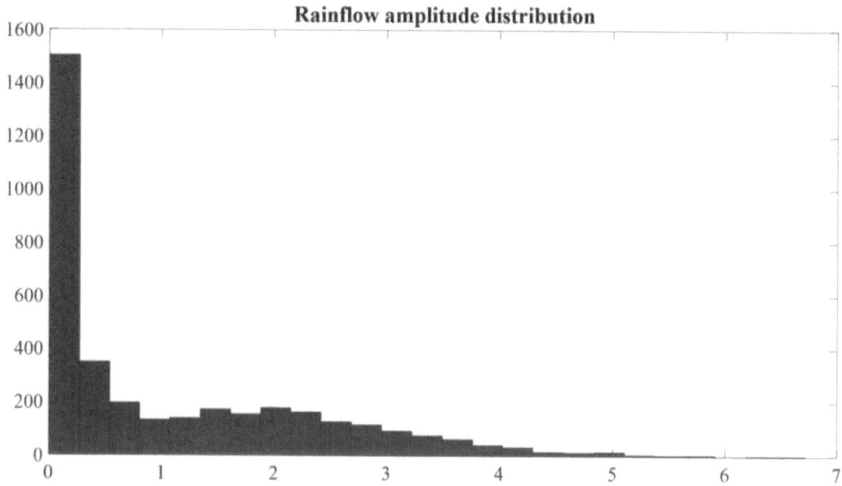

FIGURE 10.13 Rain flow amplitude distribution for this measured dataset of 39,000 measurements.

We will now calculate the fatigue damage induced by this measured dataset of 39,000 measurements. To do this, equation (10.60) can be rewritten as

$$D(t)=\sum_{t_k\le T}\frac{1}{N\left(S_k^{rfc}\right)}=\frac{1}{c}\sum_{t_k\le t}\left(S_k^{rfc}\right)^m,\quad S_k^{rfc}=\frac{\left(M_k-m_k^{rfc}\right)}{2} \tag{10.61}$$

Then, the fatigue failure will occur if the total damage becomes greater than 1, that is, if $D(t)>1$. A fairly straightforward predictor of $D(t)$ can be produced by substituting $1/c=E^{-1}\epsilon$ in equation (10.53) with a constant, such as the median value of $1/c$, which is equal to ϵ, in accordance with the assumption of a lognormal distribution.

The rain flow cycle count of this measured dataset can be directly utilized for the prediction of the expected fatigue damage ratio. The parameters ϵ and m can be estimated from an S–N curve. In this example, we will use $\epsilon=5.5\times10^{-10}$, $m=3.2$. For our measured dataset of 39,000 measurements, the computations go directly from the rain flow cycles based on equation (10.61). The obtained fatigue damage ratio is 2.1369e-05, that is, fatigue failure has not occurred.

Because the fatigue damage ratio is only 2.1369e-05, it is obvious that this load causes very little damage to the material that has the properties that have been stated. Of course, the measured data of the sea waves in this example is not a fatigue load sequence, so this calculation example is meaningless from a fatigue point of view.

10.5.4 Closed-form Expressions for Rain Flow Damage

It has been attempted to develop closed-form formulas for fatigue damage under wide-band loads given a load spectral density function. This would eliminate the need to carry out a simulation. It is assumed that the S–N curve is linear throughout the subsequent discussions on this topic.

(1). Wirsching's wide band rain flow correction factor. Based on the results of extensive Monte Carlo simulations, Wirsching and Light (1980) proposed the following as an empirical form for the general expression for fatigue damage

$$D = \lambda(m,\varepsilon_i)D_{NB} \tag{10.62}$$

in which $\lambda(m,\varepsilon_i)$ represents the rain flow correction factor, whereas ε_i represents the bandwidth parameter. The term D_{NB} refers to damage that has been evaluated by the utilization of an equivalent narrow-band process.

$$D_{NB} = \frac{E\left[n_+(0,t)\right]T}{c}\left(\sqrt{2}\sigma_y\right)^m \Gamma\left(1+\frac{m}{2}\right) \tag{10.63}$$

in which $E\left[n_+(0,t)\right]$ is equivalent frequency (rate of zero up-crossings), T is time of exposure and σ_y is the root-mean-square value. After simulating random processes with a wide range of spectral forms, the empirical rain flow correction factor $\lambda(m,\varepsilon)$ was found by Wirsching and Light (1980) to be as follows:

$$\lambda(m,\varepsilon) = a(m) + \left[1 - a(m)\right]\left[1-\varepsilon\right]^{b(m)} \tag{10.64}$$

$$a(m) = 0.926 - 0.033m,\ b(m) = 1.587m - 2.323 \tag{10.65}$$

$$\varepsilon = \sqrt{1 - \frac{m_2^2}{m_0 m_4}} \tag{10.66}$$

Within the equation (10.66), m_n denotes the nth-order spectral moment calculated based on the load spectral density function $S(\omega)$ as $m_n = \int_0^\infty \omega^n S(\omega)d\omega$. The parameter ε in equation (10.66) represents the bandwidth. The term "bandwith" conveys the notion that it is a representation of the span of frequencies that are encompassed by the load spectral density function. However, this is not the situation at all. Due to the fact that the parameter ε is dependent on the magnitude of the 4th moment of the load spectral density function, the magnitude of ε is highly dependent on the frequency range in which dominating load energy is present. The spectral width parameter v is a parameter that we suggest in order to construct a parameter that represents the degree of spectral energy spreading over the frequency range in a reasonably accurate manner:

$$v = \sqrt{\frac{m_0 m_2}{m_1^2} - 1} \tag{10.67}$$

(2). The Rice formula. In the case of a general wide-banded Gaussian random process, it is possible to find a theoretical approximation of the cycle correction factor. In accordance with the Rice distribution, the cycle counting correction factor can be obtained as follows:

$$\lambda(m,\beta)=\frac{1+\beta}{2}\left[1+\frac{\sqrt{1-\beta^2}}{2\sqrt{\pi}\beta}\left(1-\beta^2\right)^{\frac{m+1}{2}}\frac{\Gamma\left(\frac{m+1}{2}\right)}{\Gamma\left(\frac{m}{2}+1\right)}\right] \tag{10.68}$$

in which the regularity factor β is defined as:

$$\beta=\sqrt{1-\varepsilon^2} \tag{10.69}$$

(3). The Dirlik formula. On the basis of extensive computer simulations, Dirlik proposed a combination of an exponential distribution and two Rayleigh distributions for the probability density function of the stress range, which can be expressed as follows:

$$f_{\Delta\sigma}(S)=\frac{1}{2\sqrt{m_0}}\left[\frac{D_1}{Q}\exp\left(-\frac{S}{2\sqrt{m_0}Q}\right)+\frac{D_2 S}{2\sqrt{m_0}R^2}\exp\left(-\frac{S^2}{8m_0R^2}\right)\right.$$
$$\left.+\frac{D_3 S}{2\sqrt{m_0}}\exp\left(-\frac{S^2}{8m_0}\right)\right] \tag{10.70}$$

in which

$$D_1=\frac{2\left(x_m-\beta^2\right)}{1+\beta^2},\ D_2=\frac{1-\beta-D_1+D_1^2}{1-R},\ D_3=1-D_1-D_2$$
$$Q=\frac{1.25\left(\beta-D_3-D_2R\right)}{D_1},\ R=\frac{\beta-x_m-D_1^2}{1-\beta-D_1+D_1^2},$$
$$x_m=\frac{m_1}{m_0}\sqrt{\frac{m_2}{m_4}},\ \beta=\sqrt{1-\varepsilon^2},\ \varepsilon=\sqrt{1-\frac{m_2^2}{m_0m_4}}$$

Following this, the correction factor can be obtained as follows:

$$\lambda(m,\beta)=D_1\left(\frac{Q}{\sqrt{2}}\right)^m\frac{\Gamma(m+1)}{\Gamma\left(\frac{m}{2}+1\right)}+\left(\frac{R}{\sqrt{2}}\right)^m D_2+D_3 \tag{10.71}$$

Example 10.1

▶ We assume that the spectrum of a wide-band stress process is defined by a JONSWAP wave spectrum. The one-sided JONSWAP spectrum is expressed mathematically as follows:

$$S(\omega)=\beta\frac{g^2}{\omega^M}\exp\left(-\frac{M}{N}\left(\frac{2\pi}{\omega T_p}\right)^N\right)\gamma^{\exp\left(\frac{\left(\frac{\omega T_p}{2\pi}-1\right)^2}{2\delta^2}\right)} \tag{10.72}$$

where $M = 5$ and $N = 4$, and

$$\delta = 0.07 \quad \textit{if} \quad \omega < 2\pi/T_p$$

$$\delta = 0.09 \quad \textit{if} \quad \omega \geq 2\pi/T_p$$

$$\beta = 5.061\frac{H_{m0}^2}{T_p^4}\{1-0.287\ln(\gamma)\}$$

$$\gamma = \exp\left(3.484\left(1-0.1975\left(0.036-0.0056T_p/\sqrt{H_{m0}}\right)T_p^4/H_{m0}^2\right)\right)$$

In equation (10.72), ω represents the wave angular frequency in rad/s. H_{m0} is the significant wave height of a specific sea state. T_p is the peak spectral period of this sea state. γ denotes the spectral peakedness factor. In this example, we will choose $H_{m0} = 6$ m and $T_p = 12$ s.

The slope of the $S-N$ curve is chosen to be $m = 3$, a typical value for welded joints. Compare the rain flow correction factors of (a) Wirsching's wide band rain flow correction factor and (b) The Rice formula.

SOLUTION

The one-sided JONSWAP wave spectrum with $H_{m0} = 6$ m and $T_p = 12$ s is shown in the following Figure 10.14.

For this spectrum, the following parameters are calculated:

$$m_0 = \int_0^\infty \omega^0 S(\omega)d\omega = 2.2485$$

$$m_2 = \int_0^\infty \omega^2 S(\omega)d\omega = 1.1548$$

$$m_4 = \int_0^\infty \omega^4 S(\omega)d\omega = 1.2093$$

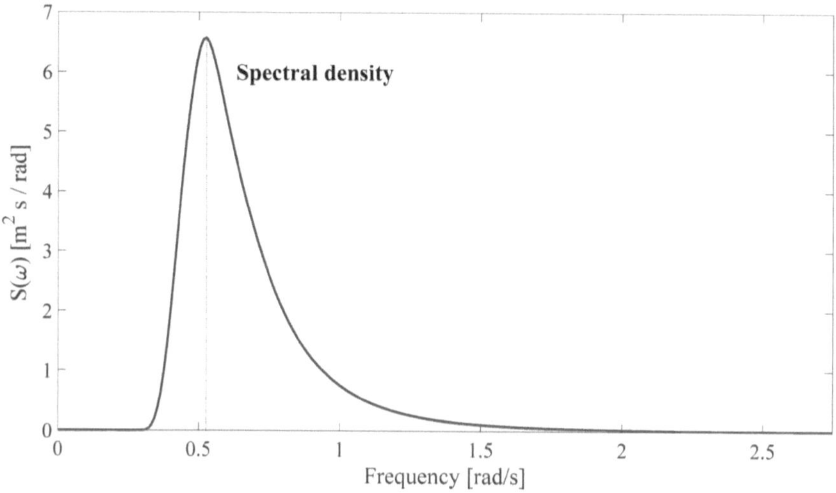

FIGURE 10.14 The JONSWAP wave spectrum with $H_{m0} = 6$ m and $T_{p=12\,s}$.

Then, the empirical rain flow correction factor $\lambda(m,\varepsilon)$ proposed by Wirsching and Light (1980) can be calculated as follows:

$$\lambda(m,\varepsilon) = a(m) + \left[1 - a(m)\right]\left[1-\varepsilon\right]^{b(m)} = 0.8352$$

$$a(m) = 0.926 - 0.033m = 0.827,\ b(m) = 1.587m - 2.323 = 2.438$$

$$\varepsilon = \sqrt{1 - \frac{m_2^2}{m_0 m_4}} = 0.7139$$

The rain flow correction factor $\lambda(m,\beta)$ according to the Rice distribution is:

$$\lambda(m,\beta) = \frac{1+\beta}{2}\left[1 + \frac{\sqrt{1-\beta^2}}{2\sqrt{\pi}\beta}\left(1-\beta^2\right)^{\frac{m+1}{2}} \frac{\Gamma\left(\frac{m+1}{2}\right)}{\Gamma\left(\frac{m}{2}+1\right)}\right]$$
$$= 0.8501\left[1 + \frac{\sqrt{1-0.7002^2}}{2\sqrt{\pi}0.7002}\left(1-0.7002^2\right)^2 \frac{\Gamma(2)}{\Gamma(2.5)}\right]$$
$$= 0.8501\left[1 + \frac{0.7139}{2\times 1.7725\times 0.7002}(0.5097)^2 \frac{1}{1.3293}\right] = 0.8979$$

in which the regularity factor β is:

$$\beta = \sqrt{1-\varepsilon^2} = 0.7002 \blacktriangleleft$$

10.6 EXERCISES

EXERCISE 10.1

Suppose that $Y(t)$ is a stationary Gaussian random process with a zero mean. In addition, $Y(t)$ has a power spectral density function $S(\omega)$. Then, $Y(t)$ and $\dot{Y}(t)$ are independent, and their joint probability density is

$$p_{Y\dot{Y}}(y,\dot{y}) = \frac{1}{2\pi\sigma_Y\sigma_{\dot{Y}}}\exp\left(-\frac{y^2}{2\sigma_Y^2}-\frac{\dot{y}^2}{2\sigma_{\dot{Y}}{}^2}\right)$$

Show that the expected rate of zero crossings of $Y(t)$ from below is

$$E\left[n_+(0,t)\right] = \frac{1}{2\pi}\sqrt{\frac{\int_{-\infty}^{\infty}\omega^2 S(\omega)d\omega}{\int_{-\infty}^{\infty}S(\omega)d\omega}}$$

EXERCISE 10.2

Figure 10.1 displays the sample function of a random process $Y(t)$ as well as a specified threshold value ξ. The corresponding Heaviside step function $\mathrm{H}\left[Y(t)-\xi\right]$ is

$$Z(t) = \mathrm{H}\left[Y(t)-\xi\right] = \begin{cases} 1 & Y(t) > \xi \\ 0.5 & Y(t) = \xi \\ 0 & Y(t) < \xi \end{cases}$$

Please draw another figure to demonstrate that the Heaviside step function $\mathrm{H}\left[Y(t)-\xi\right]$ can be regarded as the integral of the Dirac delta function, that is,

$$Z(t) = \mathrm{H}\left[Y(t)-\xi\right] = \int_{-\infty}^{Y(t)-\xi}\delta(u)du$$

EXERCISE 10.3

In this exercise, you are required to find another way to derive the probability density function of the amplitude of a random process. In this way, the probability distribution of amplitude is assumed to be equal to that of the envelope of the profile of the random process shown in Figure 10.7. You can write the sample function $y(t)$ of a narrow band Gaussian random process $Y(t)$ as

$$y(t) = A(t)\cos\left\{\omega_0 t + \varepsilon(t)\right\}$$

in which the frequency ω_0 is a constant. It was stated in this chapter in connection with a narrow-band random process that the amplitude $A(t)$ and phase $\varepsilon(t)$ both vary slowly with time. Therefore, if their time derivatives, $\dot{A}(t)$ and $\dot{\varepsilon}(t)$, are assumed to be negligibly small, the velocity process $\dot{y}(t)$ can be expressed as

$$\frac{\dot{y}(t)}{\omega_0} = -A(t)\sin\{\omega_0 t + \varepsilon(t)\}$$

Based on the properties of a Gaussian random process, it can be proved that $\dot{y}(t)$ is statistically independent of $y(t)$ and that $\dot{y}(t)$ is also a Gaussian random process with zero mean and variance $\sigma^2\omega_0^2$ (why?). Therefore, for a given time t, $\dot{y}/\omega_0$ has a Gaussian distribution with zero mean and variance σ^2. Then, the joint probability density function of y and $\dot{y}/\omega_0$ can be expressed as

$$f(y,\dot{y}/\omega_0) = \frac{1}{2\pi\sigma^2} e^{-(1/2\sigma^2)\left(y^2+(\dot{y}/\omega_0)^2\right)},$$
$$-\infty < y < \infty, -\infty < \dot{y}/\omega < \infty$$

Based on the above equation and utilizing the polar coordinates, show that the probability density function of the amplitude $A(t)$ is

$$f(A) = \frac{A}{\sigma^2}\exp\left(-\frac{A^2}{2\sigma^2}\right), \ 0 \leqslant A < \infty$$

EXERCISE 10.4

In this exercise, you are required to first generate sample time histories based on a standard McCormick wave spectrum whose mathematical formulation (containing the significant wave height H_s) is as follows:

$$S(\omega) = (M+1)\frac{(0.25\times H_s)^2}{\omega_p}\left(\frac{\omega_p}{\omega}\right)^{(M+1)}\exp\left(-\frac{M+1}{M}\left(\frac{\omega_p}{\omega}\right)^M\right)$$

in which

$$\omega_p = \frac{2\pi}{T_p}$$

$$\frac{T_p}{T_z} = \frac{\left(\dfrac{M+1}{M}\right)^{1/M}}{\Gamma\left(\dfrac{M+1}{M}\right)}$$

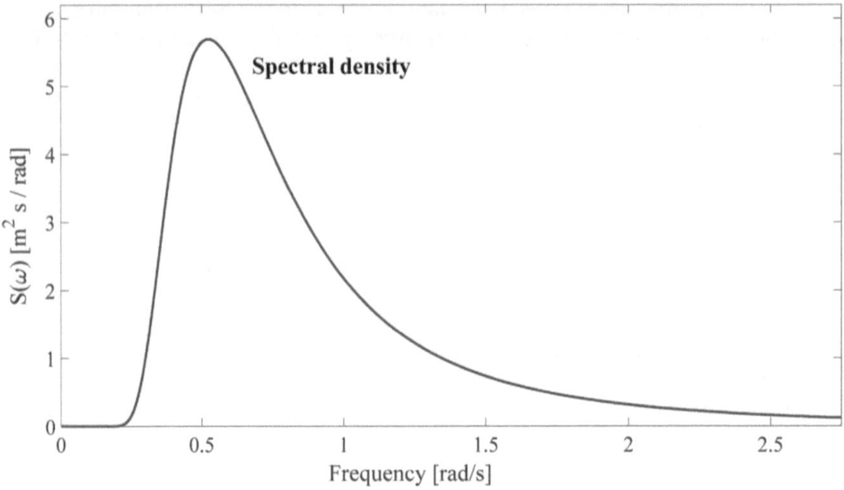

FIGURE 10.15 A McCormick wave spectrum with $H_s = 8.0\ m, T_p = 12.0\ s$ and $T_z = 9.7716\ s$.

in which Γ () is a Gamma function. T_p denotes the spectral peak period, and T_z represents zero up-crossing period. In Figure 10.15, the blue curve shows a McCormick wave spectrum with $H_s = 8.0\ m$, $T_p = 12.0\ s$ and $T_z = 9.7716\ s$.

1. Please generate a 100,000 s wave elevation time record based on the McCormick wave spectrum, as shown in Figure 10.15, using a simulation time step of 0.1s.
2. Please develop a rain flow cycle count plot based on your generated 100,000 s wave elevation time record mentioned above.
3. Please develop a rain flow amplitude distribution plot based on your generated rain flow cycle count plot mentioned before.
4. Please calculate the fatigue damage ratio based on your obtained rain flow cycles of your simulated 100,000 s wave elevation time record. The parameters ϵ and m should be chosen to be: $\epsilon = 5.5 \times 10^{-10}$, $m = 3.2$.

EXERCISE 10.5

Figure 10.16 shows a wave elevation time record measured at a sea area in Africa. Table 10.1 shows the data of the first 60 elevation points in this measured wave elevation time record. Please develop a rain flow cycle count plot based on these data of the first 60 elevation points.

EXERCISE 10.6

We assume that the spectrum of a wide-band stress process is defined by a McCormick wave spectrum with $H_s = 5.0\ m$, $T_p = 11.0\ s$ and $T_z = 8.9573\ s$.

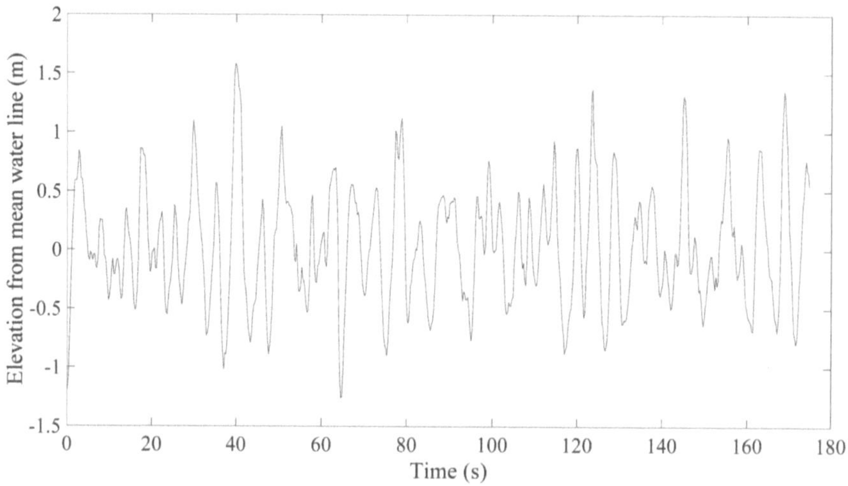

FIGURE 10.16 A wave elevation time record measured at a sea area in Africa.

TABLE 10.1
The Measured Wave Elevation Time Record at a Sea Area in Africa

Time Instant(*s*)	Wave Elevation (*m*)	Time Instant (*s*)	Wave Elevation (*m*)	Time Instant (*s*)	Wave Elevation (*m*)
0.050	−1.200494500	5.050	−0.070494540	10.050	−0.390494540
0.300	−1.090494500	5.300	−0.0904945400	10.30	−0.300494540
0.550	−0.7904945400	5.550	−0.0204945400	10.55	−0.160494540
0.800	−0.370494540	5.800	−0.0504945400	10.80	−0.080494540
1.050	−0.090494540	6.050	−0.0904945400	11.050	−0.210494540
1.300	0.2295054600	6.300	−0.0404945400	11.30	−0.210494540
1.550	0.399505460	6.550	−0.040494540	11.55	−0.140494540
1.80	0.589505460	6.80	−0.140494540	11.80	−0.080494540
2.050	0.589505460	7.050	−0.160494540	12.05	−0.090494540
2.300	0.589505460	7.30	−0.0704945400	12.30	−0.170494540
2.550	0.679505460	7.550	0.1695054600	12.55	−0.330494540
2.80	0.839505460	7.800	0.259505460	12.80	−0.420494540
3.050	0.799505460	8.050	0.2495054600	13.05	−0.400494540
3.300	0.609505460	8.300	0.2495054600	13.30	−0.190494540
3.550	0.599505460	8.550	0.139505460	13.55	0.079505460
3.800	0.519505460	8.800	−0.0504945400	13.80	0.319505460
4.050	0.349505460	9.050	−0.0704945400	14.05	0.349505460
4.300	0.299505460	9.30	−0.160494540	14.30	0.229505460
4.550	0.139505460	9.550	−0.300494540	14.55	0.139505460
4.800	0.0395054600	9.80	−0.430494540	14.80	0.089505460

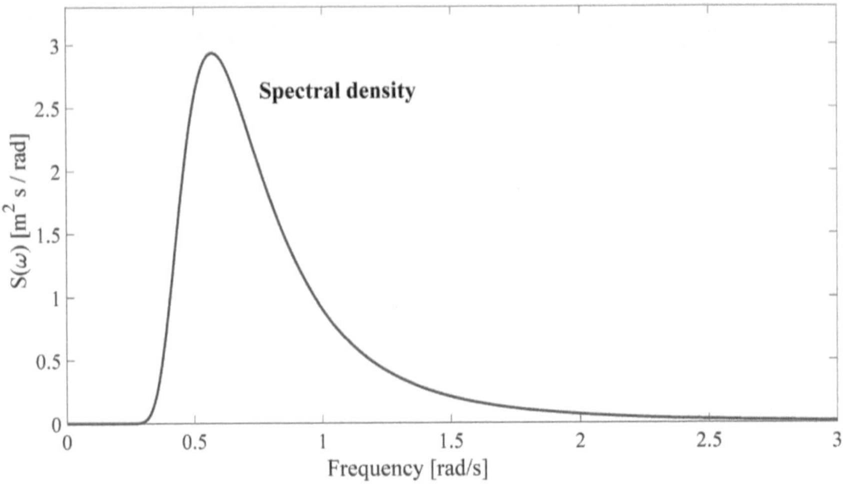

FIGURE 10.17 A McCormick spectrum with $H_s = 5.0\ m$, $T_p = 11.0\ s$ and $T_z = 8.9573\ s$.

The one-sided McCormick wave spectrum with $H_s = 5.0\ m$, $T_p = 11.0\ s$ and $T_z = 8.9573\ s$ is shown in Figure 10.17.

Please refer to EXERCISE 10.4 for the mathematical formulation of the McCormick wave spectrum. In fatigue damage calculation based on the McCormick spectrum shown in Figure 10.17, the slope of the $S-N$ curve is chosen to be $m = 3$, a typical value for welded joints. Please compare the rain flow correction factors of (a) Wirsching's wide band rain flow correction factor, (b) the Rice formula and (c) the Dirlik formula.

Appendix A: Derivation of the Parseval Theorem

In this Appendix, we will provide the detailed derivation of the following Parseval theorem:

$$\int_{-\infty}^{\infty} \{x(t)\}^2 \, dt = \frac{1}{2\pi} \int_{-\infty}^{\infty} |X(\omega)|^2 \, d\omega \tag{A.1}$$

First, we have the following relationship:

$$\begin{aligned} \int_{-\infty}^{\infty} \{x(t)\}^2 \, dt &= \int_{-\infty}^{\infty} x(t) \left\{ \frac{1}{2\pi} \int_{-\infty}^{\infty} X(\omega) e^{i\omega t} \, d\omega \right\} dt \\ &= \frac{1}{2\pi} \int_{-\infty}^{\infty} \left\{ \int_{-\infty}^{\infty} x(t) e^{i\omega t} \, dt \right\} X(\omega) \, d\omega \end{aligned} \tag{A.2}$$

According to the definition of Fourier transform, we have:

$$X(\omega) = \int_{-\infty}^{\infty} x(t) e^{-i\omega t} \, dt \tag{A.3}$$

Because the Fourier transform of the function $x(t)$ is a complex function of ω, we can write $X(\omega)$ as follows:

$$X(\omega) = G(\omega) + iH(\omega) = A(\omega) e^{i\theta(\omega)} \tag{A.4}$$

in which

$$A(\omega) = \sqrt{\{G(\omega)\}^2 + \{H(\omega)\}^2} \tag{A.5}$$

$$\theta(\omega) = \tan^{-1} \frac{H(\omega)}{G(\omega)} \tag{A.6}$$

Then, from equations (A.3) and (A.4), we have

$$G(\omega) = \int_{-\infty}^{\infty} x(t) \cos \omega t \, dt \tag{A.7}$$

$$H(\omega) = -\int_{-\infty}^{\infty} x(t)\sin\omega t\, dt \tag{A.8}$$

It is not difficult to prove that

$$G(-\omega) = \int_{-\infty}^{\infty} x(t)\cos(-\omega t)\, dt = \int_{-\infty}^{\infty} x(t)\cos(\omega t)\, dt = G(\omega) \tag{A.9}$$

$$H(-\omega) = -\int_{-\infty}^{\infty} x(t)\sin(-\omega t)\, dt = \int_{-\infty}^{\infty} x(t)\sin(\omega t)\, dt = -H(\omega) \tag{A.10}$$

As a result, we obtain

$$\begin{aligned} X(-\omega) &= G(-\omega) + iH(-\omega) = G(\omega) - iH(\omega) \\ &= \text{complex conjugate of } X(\omega) = X^*(\omega) \end{aligned} \tag{A.11}$$

From equation (A.3), we obtain

$$X(-\omega) = \int_{-\infty}^{\infty} x(t)e^{i\omega t}\, dt = X^*(\omega) \tag{A.12}$$

Combining equations (A.2) and (A.12) leads to

$$\begin{aligned} \int_{-\infty}^{\infty} \{x(t)\}^2\, dt &= \frac{1}{2\pi}\int_{-\infty}^{\infty} \left\{ \int_{-\infty}^{\infty} x(t)e^{i\omega t}\, dt \right\} X(\omega)\, d\omega \\ &= \frac{1}{2\pi}\int_{-\infty}^{\infty} X^*(\omega)X(\omega)\, d\omega \end{aligned} \tag{A.13}$$

Because we have

$$X^*(\omega)X(\omega) = |X(\omega)|^2 \tag{A.14}$$

Combining equations (A.13) and (A.14) leads to the Parseval theorem:

$$\int_{-\infty}^{\infty} \{x(t)\}^2\, dt = \frac{1}{2\pi}\int_{-\infty}^{\infty} |X(\omega)|^2\, d\omega \tag{A.15}$$

Appendix B: Derivation of the Wiener–Khintchine Theorem

In this Appendix, we will provide the detailed derivation of the Wiener–Khintchine theorem.

Clarification will be provided in this Appendix regarding the relationship that exists between the autocorrelation function of a random process that is defined in the time domain and its energy spectral density function that is defined in the frequency domain with regard to the random process.

We make the assumption, for the sake of simplicity, that the Fourier transform of a sample record $x(t)$ of a random process $\{X(t)\}$ exists across the entirety of the time domain. Because of this, the truncated function $x_T(t)$, which is defined in equation (4.63), will not be taken into consideration in the subsequent analysis. Keeping this in mind, let us take into consideration the Fourier transform of the autocorrelation function $R_{XX}(\tau)$ for the frequency ω. This is given by

$$\begin{aligned}\int_{-\infty}^{\infty} R_{XX}(\tau)e^{-i\omega\tau}d\tau &= \lim_{T\to\infty}\frac{1}{2T}\int_{-\infty}^{\infty}\int_{-\infty}^{\infty} x(t)x(t+\tau)e^{-i\omega\tau}\,dt\,d\tau \\ &= \lim_{T\to\infty}\frac{1}{2T}\int_{-\infty}^{\infty}\int_{-\infty}^{\infty} x(t)x(t+\tau)e^{-i\omega(t+\tau)}e^{i\omega t}\,dt\,d\tau\end{aligned} \tag{B.1}$$

According to the definition of Fourier transform, we have:

$$X(\omega) = \int_{-\infty}^{\infty} x(t)e^{-i\omega t}\,dt \tag{B.2}$$

Because the Fourier transform of the function $x(t)$ is a complex function of ω, we can write $X(\omega)$ as follows:

$$X(\omega) = G(\omega) + iH(\omega) \tag{B.3}$$

Then, from equations (B.3) and (B.2), we have

$$G(\omega) = \int_{-\infty}^{\infty} x(t)\cos\omega t\,dt \tag{B.4a}$$

$$H(\omega) = -\int_{-\infty}^{\infty} x(t)\sin\omega t\,dt \tag{B.4b}$$

It is easy to prove that

$$G(-\omega) = G(\omega) \tag{B.5}$$

$$H(-\omega) = -H(\omega) \tag{B.6}$$

Thus, we obtain

$$\begin{aligned} X(-\omega) &= G(-\omega) + iH(-\omega) = G(\omega) - iH(\omega) \\ &= \text{complex conjugate of } X(\omega) = X^*(\omega) \end{aligned} \tag{B.7}$$

From equations (B.2) and (B.7), we obtain

$$\int_{-\infty}^{\infty} x(t)e^{i\omega t}\, dt = X(-\omega) = X^*(\omega) \tag{B.8}$$

Obviously, we also have the following relationship:

$$\int_{-\infty}^{\infty} x(t+\tau)e^{-i\omega(t+\tau)}\, d\tau = \int_{-\infty}^{\infty} x(t+\tau)e^{-i\omega(t+\tau)} d(t+\tau) = X(\omega) \tag{B.9}$$

Combining equations (B.1), (B.8) and (B.9) leads to

$$\begin{aligned} \int_{-\infty}^{\infty} R_{XX}(\tau)e^{-i\omega\tau} d\tau &= \lim_{T\to\infty} \frac{1}{2T} \int_{-\infty}^{\infty}\int_{-\infty}^{\infty} x(t)x(t+\tau)e^{-i\omega\tau}\, dtd\tau \\ &= \lim_{T\to\infty} \frac{1}{2T} \int_{-\infty}^{\infty}\int_{-\infty}^{\infty} x(t)x(t+\tau)e^{-i\omega(t+\tau)}e^{i\omega t}\, dtd\tau \\ &= \lim_{T\to\infty} \frac{1}{2T} X(\omega)X^*(\omega) = \lim_{T\to\infty} \frac{1}{2T} \left|X(\omega)\right|^2 \end{aligned} \tag{B.10}$$

In accordance with the definition of the spectral density function that is given in equation (4.66), we obtain that

$$S_{XX}(\omega) = \frac{1}{\pi} \int_{-\infty}^{\infty} R_{XX}(\tau)e^{-i\omega\tau}\, d\tau \tag{B.11}$$

and thus

$$\int_{-\infty}^{\infty} R_{XX}(\tau)e^{-i\omega\tau}\, d\tau = \pi S_{XX}(\omega) \tag{B.12}$$

Through the application of the inverse Fourier transform, we obtain

$$R_{XX}(\tau)=\frac{1}{2}\int_{-\infty}^{\infty} S_{XX}(\omega)e^{i\omega\tau}\,d\omega \tag{B.13}$$

Upon examination of equations (B.11) and (B.13), it becomes evident that the autocorrelation function, denoted as $R_{XX}(\tau)$, and the spectral density function, denoted as $S_{XX}(\omega)$, constitute a pair of Fourier transforms. As a result, the following theorem can be derived:

Theorem. For a weakly stationary and ergodic random process $\{X(t)\}$, its autocorrelation function, denoted as $R_{XX}(\tau)$, and the spectral density function, denoted as $S_{XX}(\omega)$, are related through the Fourier transform. That is,

$$S_{XX}(\omega)=\frac{1}{\pi}\int_{-\infty}^{\infty} R_{XX}(\tau)e^{-i\omega\tau}\,d\tau \tag{B.14}$$

$$R_{XX}(\tau)=\frac{1}{2}\int_{-\infty}^{\infty} S_{XX}(\omega)e^{i\omega\tau}\,d\omega \tag{B.15}$$

in which

$$S_{XX}(\omega)=\lim_{T\to\infty}\frac{1}{2\pi T}\left|X(\omega)\right|^2 \tag{B.16}$$

and

$$R_{XX}(\tau)=\lim_{T\to\infty}\frac{1}{2T}\int_{-T}^{T} x(t)x(t+\tau)\,dt \tag{B.17}$$

This particular theorem is known as the ***Wiener–Khintchine Theorem***.

It should be noted that the definition of the spectral density function of a random process $\{X(t)\}$ can also be defined as follows:

$$S_{XX}(\omega)=\lim_{T\to\infty}\frac{1}{2T}\left|X(\omega)\right|^2 \tag{B.18}$$

In accordance with this definition of the spectral density function and taking equation (B.10) into consideration, we obtain that

$$S_{XX}(\omega)=\int_{-\infty}^{\infty} R_{XX}(\tau)e^{-i\omega\tau}\,d\tau \tag{B.19}$$

Through the application of the inverse Fourier transform, we obtain

$$R_{XX}(\tau)=\frac{1}{2\pi}\int_{-\infty}^{\infty} S_{XX}(\omega)e^{i\omega\tau}\,d\omega \tag{B.20}$$

Then we find that the Wiener–Khintchine Theorem can also be written as

$$S_{XX}(\omega)=\int_{-\infty}^{\infty} R_{XX}(\tau)e^{-i\omega\tau}\,d\tau \tag{B.21}$$

$$R_{XX}(\tau)=\frac{1}{2\pi}\int_{-\infty}^{\infty} S_{XX}(\omega)e^{i\omega\tau}\,d\omega \tag{B.22}$$

In a similar manner, we can also define the spectral density function of a random process $\{X(t)\}$ can also be defined as follows:

$$S_{XX}(\omega)=\lim_{T\to\infty}\frac{1}{4\pi T}|X(\omega)|^2 \tag{B.23}$$

In accordance with this definition of the spectral density function and taking equation (B.10) into consideration, we obtain that

$$S_{XX}(\omega)=\frac{1}{2\pi}\int_{-\infty}^{\infty} R_{XX}(\tau)e^{-i\omega\tau}\,d\tau \tag{B.24}$$

Through the application of the inverse Fourier transform, we obtain

$$R_{XX}(\tau)=\int_{-\infty}^{\infty} S_{XX}(\omega)e^{i\omega\tau}\,d\omega \tag{B.25}$$

Then we find that the Wiener–Khintchine Theorem can also be written as

$$S_{XX}(\omega)=\frac{1}{2\pi}\int_{-\infty}^{\infty} R_{XX}(\tau)e^{-i\omega\tau}\,d\tau \tag{B.26}$$

$$R_{XX}(\tau)=\int_{-\infty}^{\infty} S_{XX}(\omega)e^{i\omega\tau}\,d\omega \tag{B.27}$$

Appendix C: Damping

C.1 DIFFERENT TYPES OF DAMPING

Oscillation or vibration is the term used to describe any motion that repeats itself after an interval of time. The spring-mass-damper system as depicted in Figure C.1 is a typical example of a vibration system.

In general, a vibratory system consists of a means for storing potential energy (such as the blue spring as depicted in Figure C.1), a means for storing kinetic energy (such as the mass as depicted in Figure C.1) and a mechanism by which energy is gradually lost (such as the damper as depicted in Figure C.1).

The vibration of a system is characterized by the alternating transfer of its potential energy to kinetic energy and its kinetic energy to potential energy. In the event that the system is dampened, a certain amount of energy is dissipated throughout each cycle of vibration. This energy must be replenished by an external source in order to maintain a state of steady vibration.

The vibration is referred to be an undamped vibration if it does not experience any loss or dissipation of energy as a result of friction or other resistance throughout the oscillation process. On the other hand, if any energy is lost in this manner, it is referred to as a damped vibration. In a great number of physical systems, the quantity of damping is so negligible that it may be ignored for the majority of engineering objectives. The consideration of damping, on the other hand, becomes of utmost significance when assessing vibratory systems that are close to resonance.

In a great number of practical systems, the vibrational energy is gradually transformed into either heat or sound. The response, such as the displacement

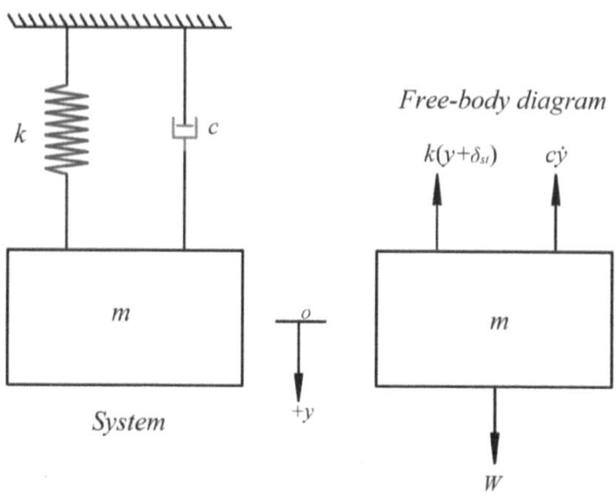

FIGURE C.1 A damped single degree of freedom system.

of the vibration system, gradually diminishes as a result of the reduction in the amount of energy. The term "damping" refers to the mechanism by which the vibrational energy is gradually transformed into either heat or sound.

In spite of the fact that the quantity of energy that is transformed into heat or sound is relatively small, the consideration of damping becomes important if one wishes to make an accurate prediction of the vibration response of a system. It is assumed that a damper does not possess either mass or elasticity, and damping force exists only when there is relative velocity between the two ends of the damper. The factors that lead to damping in practical systems are difficult to identify. As a result, damping is modeled as one or more of the types listed below.

C.1.1 Viscous Damping

Within the field of vibration analysis, viscous damping is the damping mechanism that is used the most commonly. When mechanical systems vibrate in a fluid medium such as the air, sea water or oil, the resistance that the fluid offers to the moving body causes energy to be dissipated. In this scenario, the quantity of energy that is dissipated because of vibration depends on a lot of factors. These include the size and shape of the vibrating body, the viscosity of the fluid, the frequency of the vibration and the velocity of the vibrating body. Through the process of viscous damping, the damping force is proportional to the velocity of the body that is vibrating. The following are some typical examples of viscous damping: (1) Fluid film between sliding surfaces; (2) fluid flow around a piston in a cylinder; (3) fluid "flow" around the submerged surface of a rolling ship; etc.

C.1.2 Coulomb Damping

Coulomb damping is also called the dry-friction damping. In this type of damping, the damping force is constant in size but acts in the opposite direction to the motion of the body that is vibrating. Friction between rubbing surfaces that are either dry or have insufficient lubrication is the root cause of this type of damping.

The phenomenon of Coulomb damping occurs when bodies slide over dry surfaces. In order for motion to start, there must be a force acting upon the body that is large enough to overcome the resistance to motion that is caused by friction. The dry friction force is parallel to the surface and is proportional to the force that is normal to the surface. In the case of the mass-spring system depicted in Figure C.2, the normal force is equal to the weight *W*.

The constant of proportionality is the static friction coefficient μ_s, which is a numerical value that ranges from 0 to about 1.2 and is dependent on the materials comprising the surface. After the initiation of the motion, the force decreases to $\mu_k W$, where μ_k represents the kinetic friction coefficient. In most cases, the value of μ_k is less than that of μ_s. The value of the kinetic friction coefficient (μ_k) depends on the materials in contact and the condition of the surfaces in contact. For example, $\mu_k \simeq 0.07$ for metal on metal (lubricated), 0.3 for steel on steel (unlubricated), 0.2 for wood on wood and nearly 1.0 for rubber on steel. The

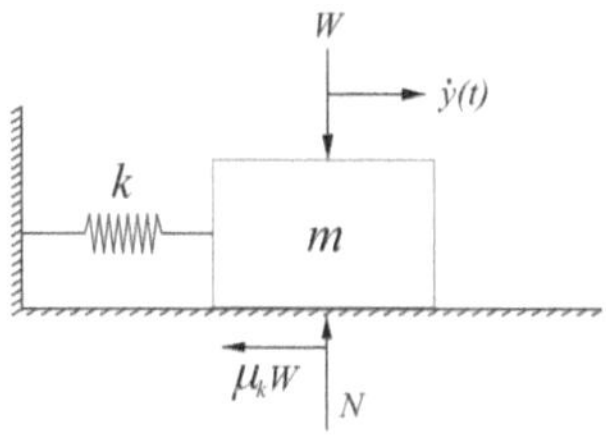

FIGURE C.2 A mass-spring system subjected to Coulomb damping

friction force acts in the opposite direction to the velocity and remains constant in magnitude as long as the forces acting on the mass m, specifically, the inertia force and the restoring force due to the spring, are sufficiently large to overcome the dry friction. When these forces are no longer sufficiently large, the motion will simply come to a halt.

Coulomb dampers, also known as dry-friction dampers, are utilized in a variety of mechanical systems due to the fact that they are mechanically straightforward and convenient. Additionally, dry-friction damping is a phenomenon that occurs internally in vibrating structures anytime the components slide relative to each other. The phenomenon of Coulomb damping occurs when bodies slide on dry surfaces. Coulomb's law of dry friction states that the force necessary to generate sliding between two bodies that are in contact is proportional to the normal force acting in the plane of contact. Therefore, the friction force F can be expressed as

$$F = \mu_k N = \mu_k W = \mu_k mg \tag{C.1}$$

in which N represents the normal force, which is equal to the mass's weight ($W = mg$), and μ_k represents the coefficient of sliding or kinetic friction. The force of friction acts in a direction that is opposite to the direction of velocity. Coulomb damping is also sometimes referred to as constant damping due to the fact that the damping force is independent of the velocity and displacement; rather, it is solely dependent on the normal force N between the sliding surfaces.

Take into consideration a system with dry friction that has a single degree of freedom, as seen in Figure C.3. Considering the fact that the friction force varies with the direction of velocity, we need to take into account two different scenarios, which are depicted in Figures C.2 and C.4.

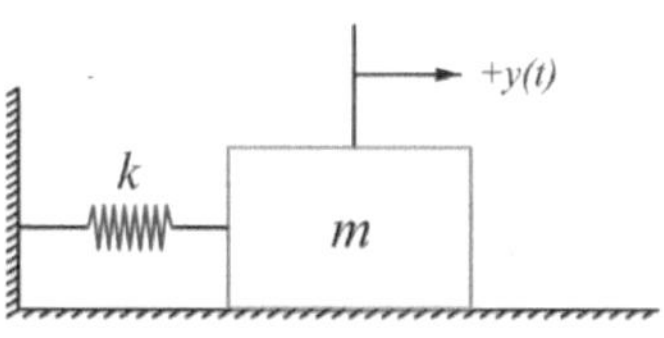

FIGURE C.3 A single degree of freedom with dry friction.

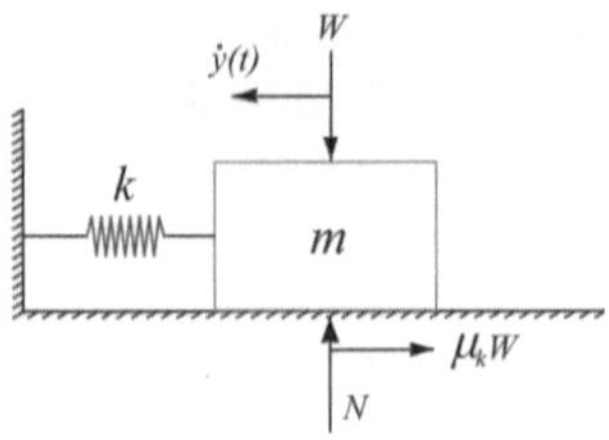

FIGURE C.4 A mass-spring system subjected to Coulomb damping.

C.1.3 Scenario 1

When y is positive and dy/dt is positive or when y is negative and dy/dt is positive (that is, for the half cycle during which the mass moves from left to right), the equation of motion of the vibration system can be found by applying Newton's second law (see Figure C.2):

$$m\ddot{y} = -ky - \mu_k mg \tag{C.2}$$

Equation (C.2) is a non-homogeneous differential equation of the second order. A verification of the solution of equation (C.2) can be accomplished by substituting equation (C.3) into equation (C.2):

$$y(t) = A_1 \cos\omega_n t + A_2 \sin\omega_n t - \frac{\mu_k mg}{k} \tag{C.3}$$

in which $\omega_n = \sqrt{k/m}$ is the natural frequency of the vibration system and A_1 and A_2 are constants whose values are dependent on the initial conditions of this half cycle.

C.1.4 Scenario 2

When y is positive and dy/dt is negative or when y is negative and dy/dt is negative (that is, for the half cycle during which the mass moves from right to left), the equation of motion of the vibration system can be found by applying Newton's second law (see Figure C.4):

$$m\ddot{y} = -ky + \mu_k mg \tag{C.4}$$

The solution of equation (C.4) is given as follows:

$$y(t) = A_3 \cos\omega_n t + A_4 \sin\omega_n t + \frac{\mu_k mg}{k} \tag{C.5}$$

in which $\omega_n = \sqrt{k/m}$ is the natural frequency of the vibration system and A_3 and A_4 are constants whose values are dependent on the initial conditions of this half cycle.

C.2 CONSTRUCTION OF A VISCOUS DAMPER

There are a few different ways that viscous dampers can be constructed. For example, a viscous damper can be obtained when a plate moves relative to another parallel plate while a viscous fluid is present in the space between the two plates. A specific example is provided below to show the method of the construction of a viscous damper that can be utilized in a specific application scenario.

Example C.1

▸ Take into consideration two plates that are parallel to one another and are separated by a distance h. In between the plates, there is a fluid with a viscosity of μ. Determine an expression for the damping constant in the case when one plate moves with a velocity v relative to the other plate, as depicted in Figure C.5.

SOLUTION

Let one of the plates remain in place while the other plate is moved in its own plane at a velocity denoted by v. The fluid layers that are in contact with the moving plate move at a velocity denoted by the letter v, whilst the fluid layers that are in contact with the fixed plate do not move. As can be seen in Figure C.5, it is assumed that the velocities of the intermediate fluid layers vary in a linear fashion between 0 and v. Newton's law of viscous flow states that the shear stress (τ) that is developed in the fluid layer at a distance z from the fixed plate can be calculated using the following equation:

$$\tau = \mu \frac{du}{dz} \tag{C.6}$$

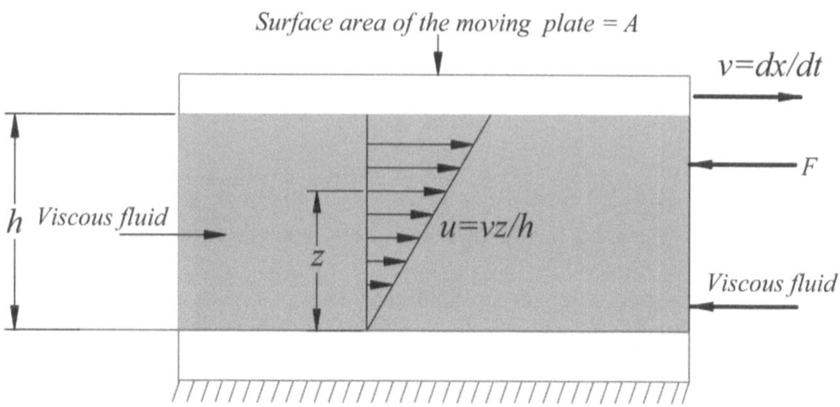

FIGURE C.5 Parallel plates containing a viscous fluid in the space between them.

in which $du/dz = v/h$ represents the gradient of the velocity. The shear force, also known as the resistive force (F), that is developed at the bottom surface of the moving plate is

$$F = \tau A = \frac{\mu A v}{h} \tag{C.7}$$

in which A represents the surface area of the plate that is moving. Through the expression of F as

$$F = cv \tag{C.8}$$

the damping constant c can be found to be

$$c = \frac{\mu A}{h} \tag{C.9}$$

◀

Therefore, according to the information presented in Example C.1, the viscous damping force F is linearly proportional to the velocity v and can be expressed as $F = cv$, in which c represents the damping constant or coefficient of viscous damping.

1. Linearization of a nonlinear damper
   In the event that the force (F)-velocity (v) relationship of a damper is nonlinear:

$$F = F(v) \tag{C.10}$$

A linearization process can be implemented with respect to the operating velocity (v^*). The equivalent damping constant can then be determined by the linearization process as to be:

$$c = \left.\frac{dF}{dv}\right|_{v^*} \tag{C.11}$$

We next provide the detailed derivation procedures for obtaining equation (C.11).

Some actual dampers are nonlinear. However, even though the force-velocity relation of a damper is nonlinear, as demonstrated in Figure C.6, we frequently use a linearization procedure to approximate it as if it were linear. In order to demonstrate the linearization process, let the static equilibrium load F of the damper corresponds to a velocity of v^*. When an incremental force ΔF is added to the force F, the damper experiences a

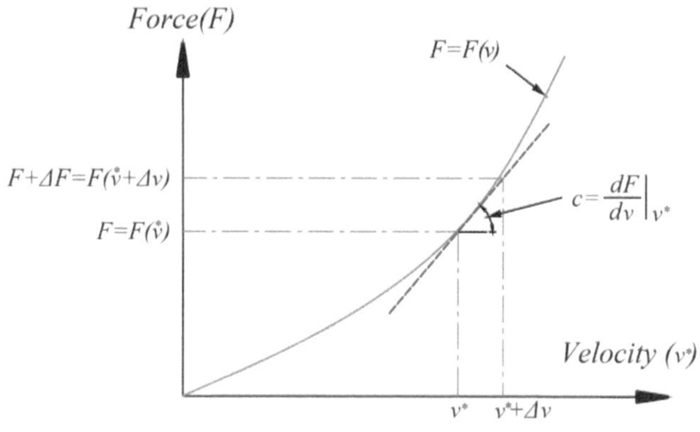

FIGURE C.6 The linearization process.

velocity of an additional quantity Δv. It is possible to represent the new damping force $F+\Delta F$ by employing Taylor's series expansion with regard to the position of static equilibrium $v*$ as:

$$\begin{aligned} F+\Delta F &= F\left(v^*+\Delta v\right) \\ &= F\left(v^*\right)+\left.\frac{dF}{dv}\right|_{v^*}(\Delta v)+\frac{1}{2!}\left.\frac{d^2F}{dv^2}\right|_{v^*}(\Delta v)^2+\cdots \end{aligned} \tag{C.12}$$

When the values of Δv are small, it is possible to disregard the higher-order derivative terms to obtain

$$F+\Delta F = F\left(v^*\right)+\left.\frac{dF}{dv}\right|_{v^*}(\Delta v) \tag{C.13}$$

Because $F = F\left(v^*\right)$, we can then express ΔF as follows:

$$\Delta F = c\Delta v \tag{C.14}$$

in which c is the linearized damping constant at v^* given by

$$c = \left.\frac{dF}{dv}\right|_{v^*} \tag{C.15}$$

2. Equivalent damping constant
   There are some dynamic systems that use multiple dampers. In such instances, a single equivalent damper can be used to replace all the dampers. We can employ procedures similar to those employed to determine the equivalent spring constant of multiple springs when dampers appear in combination to identify a single equivalent damper. For instance, when two

translational dampers, with damping constants, c_1 and c_2, appear in combination, the equivalent damping constant $\left(c_{eq}\right)$ can be determined as follows:

For dampers in parallel

$$c_{eq} = c_1 + c_2 \tag{C.16}$$

For dampers in series

$$\frac{1}{c_{eq}} = \frac{1}{c_1} + \frac{1}{c_2} \tag{C.17}$$

Appendix D: Residue Integration in Random Vibrational Analysis

D.1 INTRODUCTION

In Chapter 5, we have mentioned that it was necessary for us to evaluate the integral in the following equation:

$$E\left[y^2\right] = S_0 \int_{-\infty}^{\infty} \frac{1}{\left(k - m\omega^2\right)^2 + c^2\omega^2} d\omega \tag{5.108}$$

which can be observed to be independent of the magnitude of the mass m. We define

$$\beta = \omega / \omega_n = \omega\sqrt{m/k} \tag{5.109}$$

$$\xi = c/c_c = \frac{c}{2m\omega_n} = \frac{c}{2m\sqrt{k/m}} = \frac{c}{2\sqrt{km}} \tag{5.110}$$

$$r_1 = i\xi + \sqrt{1-\xi^2} \tag{5.111}$$

$$r_2 = i\xi - \sqrt{1-\xi^2} \tag{5.112}$$

Then we have

$$\begin{aligned}
&(\beta - r_1)(\beta - r_2)(\beta + r_1)(\beta + r_2) \\
&= (\omega/\omega_n - r_1)(\omega/\omega_n - r_2)(\omega/\omega_n + r_1)(\omega/\omega_n + r_2) \\
&= \left(\frac{\omega}{\omega_n} - i\xi - \sqrt{1-\xi^2}\right)\left(\frac{\omega}{\omega_n} - i\xi + \sqrt{1-\xi^2}\right)\left(\frac{\omega}{\omega_n} + i\xi + \sqrt{1-\xi^2}\right)\left(\frac{\omega}{\omega_n} + i\xi - \sqrt{1-\xi^2}\right) \\
&= \left(\left(\frac{\omega}{\omega_n} - i\xi\right)^2 - \left(1-\xi^2\right)\right)\left(\left(\frac{\omega}{\omega_n} + i\xi\right)^2 - \left(1-\xi^2\right)\right) \\
&= \left(\frac{\omega^2}{\omega_n^{\,2}} - 1\right)^2 - \left(2i\xi\frac{\omega}{\omega_n}\right)^2 = \left(\frac{\omega^2 m}{k} - 1\right)^2 - \left(2i\frac{c}{2\sqrt{km}}\omega\sqrt{\frac{m}{k}}\right)^2 \\
&= \frac{1}{k^2}\left[\omega^4 m^2 - 2\omega^2 km + k^2 + c^2\omega^2\right] = \frac{1}{k^2}\left[\left(k - m\omega^2\right)^2 + c^2\omega^2\right]
\end{aligned} \tag{5.113}$$

and we can thus obtain

$$k^2\left[(\beta - r_1)(\beta - r_2)(\beta + r_1)(\beta + r_2)\right] = \left(k - m\omega^2\right)^2 + c^2\omega^2 \tag{5.114}$$

Substituting equation (5.114) into equation (5.108) leads to

$$\begin{aligned} E\left[y^2\right] &= S_0 \int_{-\infty}^{\infty} \frac{1}{\left(k - m\omega^2\right)^2 + c^2\omega^2} d\omega \\ &= S_0 \int_{-\infty}^{\infty} \frac{1}{k^2\left[(\beta - r_1)(\beta - r_2)(\beta + r_1)(\beta + r_2)\right]} d\omega \\ &= \frac{\omega_n S_0}{k^2} \int_{-\infty}^{\infty} \frac{1}{\left[(\beta - r_1)(\beta - r_2)(\beta + r_1)(\beta + r_2)\right]} d\beta \end{aligned} \tag{5.115}$$

In Chapter 5, we have mentioned that equation (5.115) can be evaluated by utilizing the contour integration (residue integration) method. In the following, the detailed mathematical theories of the residue integration method will be elucidated.

D.2 COMPLEX FUNCTIONS AND COMPLEX DIFFERENTIATION

The field of complex analysis has a lot of applications, including those in the fields of engineering vibrations, fluid flow and other areas. The well-known "real calculus" is extended into the more advanced "complex calculus" through the introduction of complex numbers and complex functions. Despite the fact that many concepts are shared between calculus and complex analysis, there is a significant distinction between the two branches of mathematics. As an illustration, analytic functions, which are considered to be the "good functions" (differentiable in some domains) of complex analysis, possess derivatives of all orders. This is in contrast to calculus, in which real-valued functions of real variables may only have derivatives up to a specific order. Therefore, in certain respects, problems that are challenging to solve in a real calculus setting could be much simpler to solve in a complex analysis setting. For the following two primary reasons, complex analysis is an important part of applied mathematics:

1. The utilization of complex integration allows for the very elegant solution of a great number of difficult integrals, whether they are real or complex, that are encountered in applications.
2. In engineering mathematics, the majority of functions are analytic functions. The study of these functions as functions of a complex variable leads to a more in-depth comprehension of their properties and to interrelations in complex that do not have an analog in real calculus.

We now proceed to analytic functions. We prefer functions to be analytic because they are "useful" in the sense that they are differentiable in some domain and can be subjected to the operations of complex analysis. Due to the fact that they enable a test of the analyticity of such functions, the Cauchy–Riemann equations that are presented in the following are consequently the most important equations.

Equations that did not have real solutions, such as $x^2 = -4$ or $x^2 - 5x + 30 = 0$, were recognized at an early stage in history, which ultimately led to the development of complex numbers. According to the definition, a complex number z is an ordered pair (x, y) of real numbers x and y, and is written as

$$z = (x, y) = x + iy \tag{D.1}$$

Similar to how the study of calculus or real analysis necessitated the understanding of concepts like domain, neighborhood, function, limit, continuity, derivative and so on, the study of complex analysis also necessitates the understanding of these concepts. As a result of the fact that the functions are located in the complex plane, the concepts are somewhat more challenging or distinct from those that are found in real analysis.

Figure D.1 shows a unit circle $|z| = 1$ in the complex plane. The equation for z can be written as

$$z = \cos\theta + i\sin\theta = e^{i\theta} \tag{D.2}$$

Figure D.2 illustrates a general circle with a radius of ρ and center a. It is represented by the equation

$$|z| - a| = \rho \tag{D.3}$$

The equation for z in Figure D.2 can be also written as

$$z = a + \rho e^{i\theta} \tag{D.4}$$

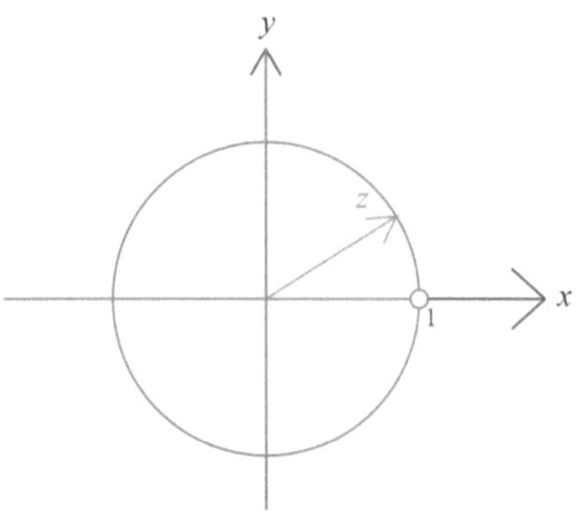

FIGURE D.1 A unit circle.

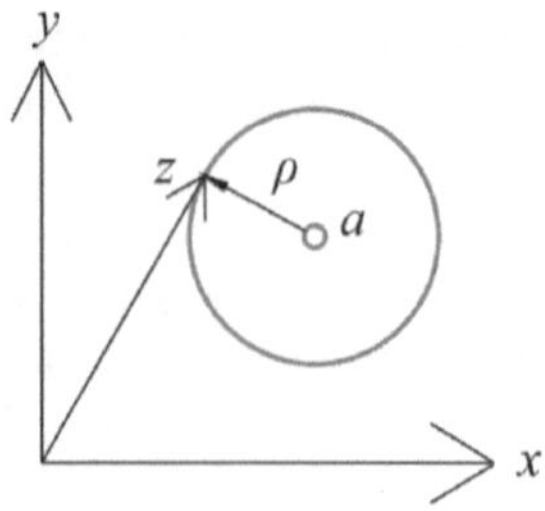

FIGURE D.2 A general circle.

In complex analysis, complex functions that are differentiable in some domains are the focus. Therefore, we need first define what a complex function is before going on to discuss the meaning of the terms derivative and limit in complex analysis. This discussion will be analogous to that in calculus. However, it requires careful scrutiny since it will reveal interesting fundamental distinctions between real and complex calculus.

It is important to remember from calculus that a real function f defined on a set S of real numbers is a rule that assigns a real number $f(x)$ to every x in the set S of real numbers (often an interval). This real number $f(x)$is referred to as the value of f at x. Now, in the context of complex analysis, S is a set of complex numbers in the complex domain. Furthermore, a function f that is defined on S is a rule that assigns a complex number w to each and every z in S. This complex number w is referred to as the value of f at z. We can write

$$w = f(z) = u(x, y) + iv(x, y) \tag{D.5}$$

This demonstrates that a complex function $f(z)$ is equivalent to a pair of real functions $u(x, y)$ and $v(x, y)$, each of which is dependent on the two real variables x and y.

The derivative of a complex function f at a point z_0 is denoted by the notation $f'(z_0)$ and is defined by the following equation:

$$f'(z_0) = \lim_{z \to z_0} \frac{f(z) - f(z_0)}{z - z_0} = \lim_{\Delta z \to 0} \frac{f(z_0 + \Delta z) - f(z_0)}{\Delta z} \tag{D.6}$$

This brings up a very important point. It is important to keep in mind that, according to the definition of limit, $f(z)$ is defined in a neighborhood of z_0, and z in equation (D.6) can approach z_0 from any direction in the complex plane. Therefore, differentiability at z_0 indicates that the quotient in equation (D.6) always approaches a certain value, and all of these values are equal. This holds true regardless of the path that z takes in order to get closer to z_0. Remembering this is essential, as it is of great significance.

When we talk about complex analysis, we are talking about the theory and application of "analytic functions", which are functions that are differentiable in some domains. This allows us to do "calculus in complex".

A function $f(z)$ is said to be analytic in a domain D if $f(z)$ can be defined and differentiated at every point in D. This is the definition of an ***analytic function***. The function $f(z)$ is said to be analytic at a point $z = z_0$ in D if $f(z)$ is analytic in a neighborhood of z_0. In addition, when we employ the term "analytic function", we are referring to a function that is analytic in some domain. Therefore, the analyticity of $f(z)$ at z_0 indicates that $f(z)$ has a derivative at each and every point in some neighborhood of z_0.

The Cauchy–Riemann equations are one of the foundational pillars which complex analysis is built upon. They offer a criterion, or a test, for determining whether or not a complex function is analytic. $f(z)$ in equation (D.5) is considered to be analytic in a domain D if and only if the first partial derivatives of u and v satisfy both of the following Cauchy–Riemann equations

$$\frac{\partial u}{\partial x} = \frac{\partial v}{\partial y}, \frac{\partial u}{\partial y} = -\frac{\partial v}{\partial x} \tag{D.7}$$

everywhere in D.

D.3 COMPLEX INTEGRATION

We will now go on to the next part of complex calculus, that is, we will discuss the first method of complex integration. It is centered on the very important Cauchy integral theorem. The significance of this theorem lies in the fact that it enables the assessment of integrals that have an analytic integrand by means of the Cauchy integral formula that will be presented in the following. In addition to this, the Cauchy integral formula demonstrates the surprising finding that derivatives of analytic functions can be found in all orders. Therefore, in this regard, complex analytic functions behave in a manner that is far more straightforward than that of real-valued functions of real variables, which may only have derivatives up to a particular order.

There are several reasons why complex integration is appealing. The proof of certain fundamental properties of analytic functions is challenging to accomplish using other approaches. This includes the existence of derivatives of all orders that were just discussed. A main practical reason for the importance of integration in the complex plane is that such integration allows for the evaluation of certain real integrals that are encountered in applications but are not accessible through the use of real integral calculus.

When performing complex analysis, we make a distinction between definite integrals and indefinite integrals, also known as antiderivatives, much as we do when performing calculus. Within the context of this discussion, an indefinite integral is a function whose derivative is equal to a particular analytic function in a region. In the process of inverting known differentiation formulas, we have the potential to find a wide variety of indefinite integrals.

Complex definite integrals are referred to as (complex) line integrals. They are written as follows:

$$\int_C f(z)dz \tag{D.8}$$

In this scenario, the integrand $f(z)$ is integrated over a given curve C or a piece of it. The path of integration is the name given to this curve C that is located in the complex plane. A parametric representation is one possible way for us to express C.

$$z(t) = x(t) + iy(t) \ (a \leqq t \leqq b) \tag{D.9}$$

We make the assumption that C is a smooth curve, which means that C has a derivative at each point that is both continuous and nonzero:

$$\dot{z}(t) = \frac{dz}{dt} = \dot{x}(t) + i\dot{y}(t) \tag{D.10}$$

From a geometric point of view (see Figure D.3), this indicates that C possesses everywhere a tangent that is continually turning, as follows directly from the definition

$$\dot{z}(t) = \lim_{\Delta t \to 0} \frac{z(t+\Delta t) - z(t)}{\Delta t} \tag{D.11}$$

Let's say that C is a smooth curve in the complex plane according to equation (D.9), and let's say that $f(z)$ is a continuous function that is given (at least) at each point on C. The interval $a \leqq t \leqq b$ in equation (D.9) is now subdivided (we "partition") by having points

$$t_0 (= a),\ t_1,\ \cdots,\ t_{n-1},\ t_n (= b) \tag{D.12}$$

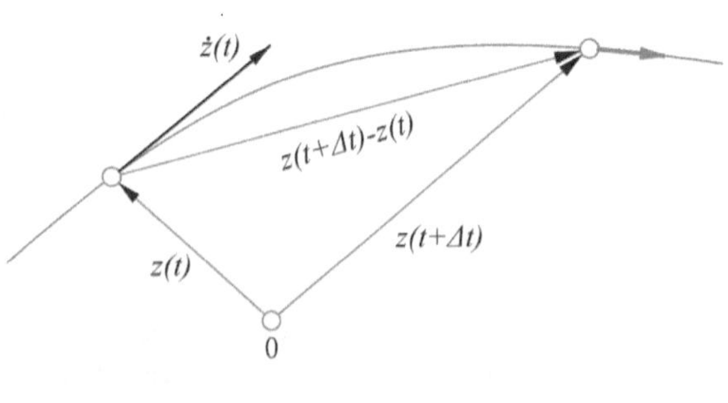

FIGURE D.3 Tangent vector $\dot{z}(t)$ of a curve C.

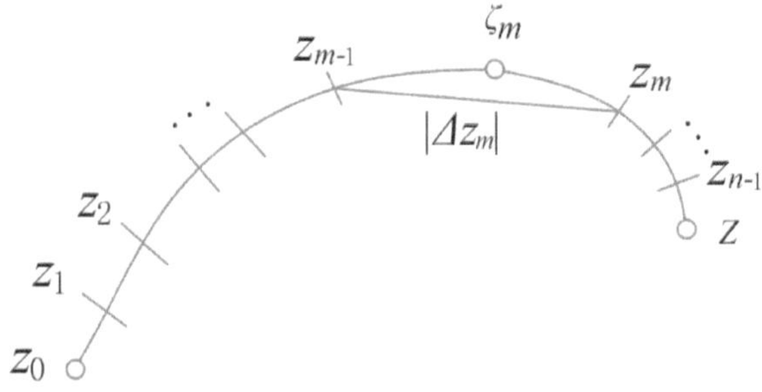

FIGURE D.4 Partition of a curve C in the complex line integral.

in which $t_0 < t_1 < \cdots < t_n$. With regard to this subdivision, there corresponds a subdivision of C by the following points (see Figure D.4)

$$z_0,\, z_1,\, \cdots,\, z_{n-1},\, z_n\,(=Z) \tag{D.13}$$

in which $z_j = z(t_j)$. An arbitrary point is selected for each portion of the subdivision of C. For example, we select a point ζ_1 between z_0 and z_1 (that is, $\zeta_1 = z(t)$ where t satisfies $t_0 \leqq t \leqq t_1$), a point ζ_2 between z_1 and z_2, and so on. The sum is then calculated by

$$S_n = \sum_{m=1}^{n} f(\zeta_m)\Delta z_m \text{ where } \Delta z_m = z_m - z_{m-1} \tag{D.14}$$

We carry out this process for every $n = 2, 3, \cdots$ in a manner that is fully independent, but in such a way that the maximum value of $|\Delta t_m| = |t_m - t_{m-1}|$ approaches zero as n approaches ∞. As a result, it may be deduced that the highest value of $|\Delta z_m|$ also approaches zero. As a matter of fact, it is not possible for it to exceed the length of the arc of C that extends from z_{m-1} to z_m, and this arc length goes to 0. This is because the arc length of the smooth curve C is a continuous function of t. The limit of the sequence of complex numbers $S_2, S_3, \cdots$ that is obtained in this manner is referred to as the line integral (or simply the integral) of $f(z)$ over the path of integration C with the orientation specified by equation (D.9) Notation for this line integral is as follows:

$$\int_C f(z)dz, \text{ or by } \oint_C f(z)dz \tag{D.15}$$

in the event that C is a closed path.

Let us assume that $f(z)$ is analytic in a domain D that is simply connected. Therefore, there is an indefinite integral of $f(z)$ in the domain D, which is an analytic function $F(z)$ such that $F'(z)=f(z)$ in D. Furthermore, for all paths in D that join two points z_0 and z_1 in D, we have the following:

$$\int_{z_0}^{z_1} f(z)dz = F(z_1)-F(z_0)\left[F'(z)=f(z)\right] \tag{D.16}$$

Let us say that C is a piecewise smooth path, represented by $z=z(t)$, in which $a \leqq t \leqq b$. Let us say that $f(z)$ is a continuous function (not restricted to an analytic function) on C. Then we have

$$\int_C f(z)dz = \int_a^b f[z(t)]\dot{z}(t)dt\left(\dot{z}=\frac{dz}{dt}\right) \tag{D.17}$$

Example D.1

We show that by integrating 1/z counterclockwise around the unit circle C (the circle with a radius of 1 and center 0), we obtain

$$\oint_C \frac{dz}{z} = 2\pi i \tag{D.18}$$

SOLUTION

We can represent the unit circle C by $z(t)=\cos t + i\sin t = e^{it}$. Then

$$\oint_C \frac{dz}{z} = \oint_C \frac{1}{z}dz = \oint_C \frac{1}{z(t)}\dot{z}(t)dt = \int_0^{2\pi} e^{-it}ie^{it}\,dt = \int_0^{2\pi} i\,dt = 2\pi i \tag{D.19}$$

◀

Example D.2

▶ Consider the equation $f(z)=(z-a)^m$, in which m is an integer and a is a constant. Using the red circle C with a radius of ρ and a center at a as shown in Figure (D.2), perform an integration of $f(z)$ in a counterclockwise direction around the circle C.

SOLUTION

The equation for the circle C in Figure D.2 can be written as

$$z = a + \rho e^{i\theta} \tag{D.20}$$

Therefore, we have

$$(z-a)^m = \rho^m e^{im\theta},\ dz = i\rho e^{i\theta}d\theta \tag{D.21}$$

We can represent the unit circle C by $z(t)=\cos t+i\sin t=e^{it}$. Then

$$\oint_C (z-a)^m\,dz=\int_0^{2\pi}\rho^m e^{im\theta}i\rho e^{i\theta}\,d\theta=i\rho^{m+1}\int_0^{2\pi}e^{i(m+1)\theta}\,d\theta$$
$$=i\rho^{m+1}\left[\int_0^{2\pi}\cos(m+1)\theta\,d\theta+i\int_0^{2\pi}\sin(m+1)\theta\,d\theta\right] \quad \text{(D.22)}$$

If $m=-1$, we have

$$\oint_C (z-a)^m\,dz=i\rho^{m+1}\left[\int_0^{2\pi}\cos(m+1)\theta\,d\theta+i\int_0^{2\pi}\sin(m+1)\theta\,d\theta\right]$$
$$=i\times1\times\left[\int_0^{2\pi}1\times d\theta+i\int_0^{2\pi}0\times d\theta\right]=2\pi i \quad \text{(D.23)}$$

If $m\neq-1$, we have

$$\oint_C (z-a)^m\,dz=i\rho^{m+1}\left[\int_0^{2\pi}\cos(m+1)\theta\,d\theta+i\int_0^{2\pi}\sin(m+1)\theta\,d\theta\right]$$
$$=i\rho^{m+1}[0+0]=0 \quad \text{(D.24)}$$

◀

The following is a highly important fact. It is generally the case that the integrals will have different values if we integrate a given function $f(z)$ from a point z_0 to a point z_1 along distinct paths. To phrase it another way, a complex line integral is dependent not only on the endpoints of the path but also, in general, on the path itself. Most of the time, this dependence makes things more difficult. Therefore, the circumstances that lead to the absence of this happening are of utmost significance. To be more specific, if $f(z)$ is analytic in a domain D and D is simply connected, then the integral will not be dependent on the choice of a path between the points that are given.

Let us provide the further discussion of simple connectedness. As can be seen in Figure D.5, a simple closed path is a closed path that does not have any intersections or touches with itself. As an illustration, the green path in Figure D.5 is a simple shape, however, a curve that is fashioned like an 8 is not simple.

A domain D that is simply connected is a domain in the complex plane in which every simple closed path encloses only points of D. A domain that is not simply connected is referred to as multiply connected. See Figure D.6 for references. To be more precise, a bounded domain D (i.e., a domain that is entirely

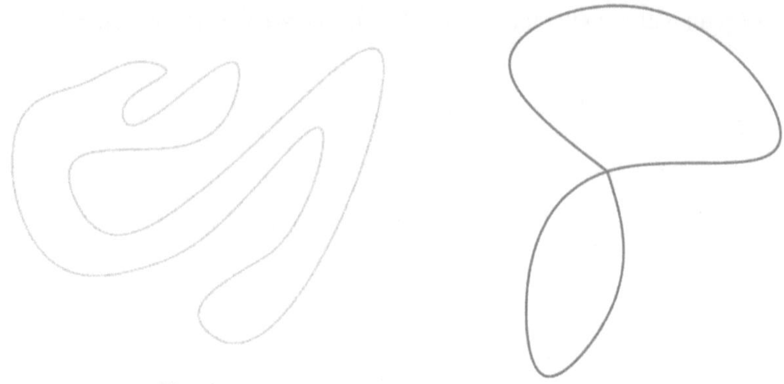

FIGURE D.5 Simple and non-simple closed paths.

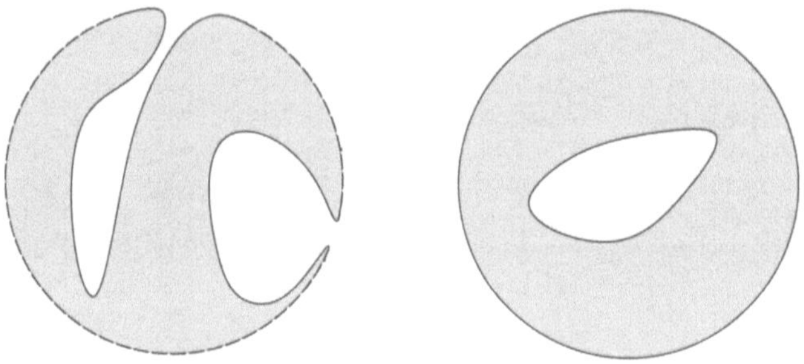

FIGURE D.6 Simply and non-simply connect domains.

contained within some circle centered on the origin) is referred to as p-fold connected if its boundary is composed of p closed connected sets that do not have common points. Therefore, an annulus is doubly connected ($p = 2$).

Cauchy's integral theorem states that if $f(z)$ is an analytic function in a simply connected domain D, then for every simple closed path C in D, we have

$$\oint_C f(z)dz = 0 \tag{D.25}$$

Please refer to Figure D.7 for an illustration of Cauchy's integral theorem. Cauchy's integral theorem also states that if $f(z)$ is an analytic function in a simply connected domain D, then the integral of $f(z)$ is independent of the path in D.

If $f(z)$ is analytic in a domain D that is simply connected, then there exists an indefinite integral $F(z)$ of $f(z)$ in D. Therefore, $F'(z) = f(z)$ -which is analytic

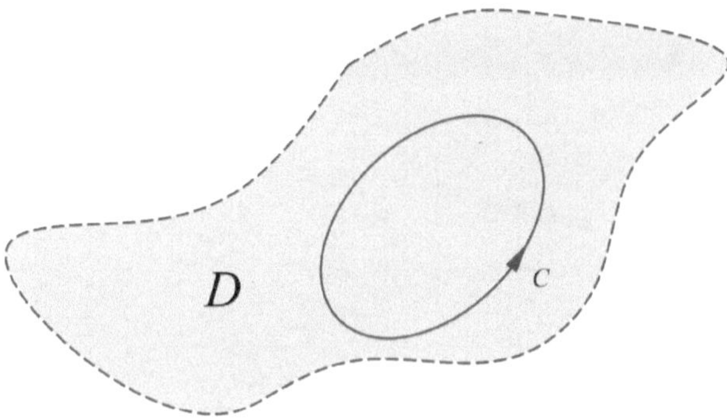

FIGURE D.7 Illustration of Cauchy's integral theorem.

in D. Furthermore, the integral of $f(z)$ from z_0 to z_1 may be evaluated using equation (D.16) for all paths in in D joining any two points z_0 and z_1 in D.

In the case of multiply connected domains, Cauchy's theorem is also applicable. In the first place, we will describe this for a domain D that is doubly connected and has an outer boundary curve C_1 and an inner boundary curve C_2 (Figure D.8). We assert that if a function $f(z)$ is analytic in any domain D^* that contains D and its boundary curves, then we can say that

$$\oint_{C_1} f(z)dz = \oint_{C_2} f(z)dz \tag{D.26}$$

Both integrals in the above equation are taken in the counterclockwise direction (or both in the clockwise direction and regardless of whether or not the entire interior of C_2 belongs to D^*).

We give the proof of equation (D.26) as follows. Through the utilization of two cuts, namely $\tilde{C}_1$ and $\tilde{C}_2$ (as depicted in Figure D.9), we have successfully partitioned D into two domains, D_1 and D_2, within which and on whose boundaries $f(z)$ is analytic. Since Cauchy's integral theorem states that the integral over the full boundary of D_1 (when interpreted in the sense of the arrows in Figure D.9)

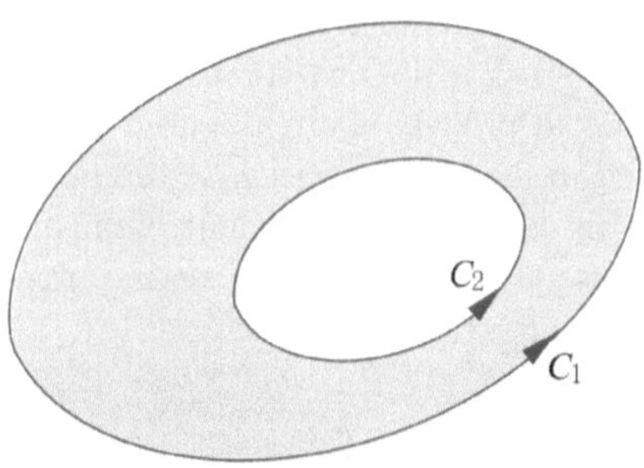

FIGURE D.8 A doubly connect domain.

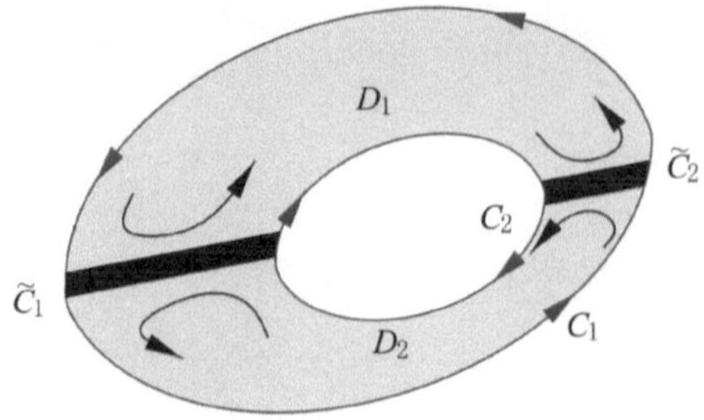

FIGURE D.9 A doubly connect domain and two cuts.

is 0, the integral over the boundary of D_2 is also 0, and the sum of these two integrals is therefore 0. Due to the fact that we integrate over the cuts $\tilde{C}_1$ and $\tilde{C}_2$ in both directions, the integrals over these cuts are canceled out in this sum. This is the key, and we are left with the integrals over C_1 (counterclockwise) and C_2 (clockwise; see Figure D.9); therefore, by reversing the integration over C_2 (to counterclockwise), we are able to obtain the following result:

$$\oint_{C_1} f\,dz - \oint_{C_2} f\,dz = 0 \tag{D.27}$$

and equation (D.26) follows.

We now take into consideration a function $f(z)$ that is analytic on a closed contour C and within the interior region that is bordered by C. The contour C is to be traversed in the counterclockwise direction. Our objective is to demonstrate the following result, which is commonly referred to as Cauchy's integral formula:

$$\oint_C \frac{f(z)}{z - z_0}\,dz = 2\pi i f(z_0) \tag{D.28}$$

where the value of z_0 can be any point within the interior region that is bordered by C. Note that because z is located on the contour C and z_0 is located in the interior, the equation $z - z_0 \neq 0$ holds. Therefore, the integral in equation (D.28) is well defined. In spite of the fact that $f(z)$ is presumed to be analytic, the integrand, which is $f(z)/(z - z_0)$, is not analytic at $z = z_0$ unless $f(z_0) = 0$. We will now deform the contour to produce a circle with a small radius r about $z = z_0$, which will be traversed in the counterclockwise direction, as was the case with the original contour. This does not alter the value of the integral, as demonstrated by equation (D.26). Consequently, we can write z as $z = z_0 + re^{i\theta}$, so $dz = ire^{i\theta}\,d\theta$, and the integration is from $\theta = 0$ to $\theta = 2\pi$, and

$$\oint_C \frac{f(z)}{z - z_0}\,dz = \int_0^{2\pi} \frac{f\left(z_0 + re^{i\theta}\right)}{re^{i\theta}}\,ire^{i\theta}\,d\theta \tag{D.29}$$

Taking the limit $r \to 0$ leads to

$$\oint_C \frac{f(z)}{z - z_0} dz = if(z_0)\int_0^{2\pi} d\theta = 2\pi i f(z_0) \tag{D.30}$$

One of the most remarkable things about complex analytic functions is that they have derivatives of all orders, as was discussed earlier. This is fundamentally different from real calculus. We cannot draw the conclusion that a real function is twice differentiable or that any of its higher derivatives exist, even if we are able to determine that the function is once differentiable. In this regard, the behavior of complex analytic functions is simpler than that of real functions. Cauchy's integral formula is employed to demonstrate this surprising fact.

The integral formula of Cauchy can be utilized in order to acquire an expression for the derivative of the function $f(z)$. Equation (D.28) is differentiated with regard to z_0 to obtain:

$$f'(z_0) = \frac{1}{2\pi i}\oint_C \frac{f(z)}{(z - z_0)^2} dz \tag{D.31}$$

Performing differentiation again,

$$f''(z_0) = \frac{2!}{2\pi i}\oint_C \frac{f(z)}{(z - z_0)^3} dz \tag{D.32}$$

Performing differentiation the third time, we obtain

$$f'''(z_0) = \frac{3!}{2\pi i}\oint_C \frac{f(z)}{(z - z_0)^4} dz \tag{D.33a}$$

In general, we have

$$f^{(n)}(z_0) = \frac{n!}{2\pi i}\oint_C \frac{f(z)}{(z - z_0)^{n+1}} dz \tag{D.33b}$$

in which C refers to any simple closed path in D that encloses z_0 and whose entire interior is a part of D; and we integrate in a counterclockwise direction over C.

D.4 TAYLOR AND LAURENT EXPANSIONS

Through the utilization of Cauchy's integral formula, which was derived from the well-known Cauchy integral theorem, we were able to directly evaluate complex integrals in the previous section. By now, we are going to move away from the method that Cauchy used to evaluate complex integrals and instead use a different method, which is to evaluate them by residue integration. A comprehensive understanding of Taylor series is required in order to proceed with this method, which will be covered in the next section. Please note that we continue to make use of Cauchy's integral theorem in order to come up with the theory of residue integration.

Within the scope of this section, we will be concentrating on the complex Taylor series. For those who are familiar with calculus, they are analogs of the real Taylor series.

The main result of this section is that every analytic function can be represented by a Taylor series.

The formula for the Cauchy integral that was shown in the part before this one paves the way for another derivation of Taylor's series, but this time it is for functions of a complex variable. Consider the following scenario: we are attempting to expand $f(z)$ about $z = z_0$, and we have $z = z_1$, representing the nearest point on the Argand diagram for which $f(z)$ is not analytic. As shown in Figure D.10, we construct a circle C centered at $z = z_0$ with a radius that is less than $|z_1 - z_0|$. Due to the fact that it was presumed that z_1 was the closest point at which $f(z)$ was not analytic, it follows that $f(z)$ is necessarily analytic on and within C.

In accordance with the Cauchy integral formula, equation (D.28),

$$\begin{aligned} f(z) &= \frac{1}{2\pi i} \oint_C \frac{f(z')}{z' - z} dz' = \frac{1}{2\pi i} \oint_C \frac{f(z')}{(z' - z_0) - (z - z_0)} dz' \\ &= \frac{1}{2\pi i} \oint_C \frac{f(z')}{(z' - z_0)\left[1 - (z - z_0)/(z' - z_0)\right]} dz' \end{aligned} \tag{D.34}$$

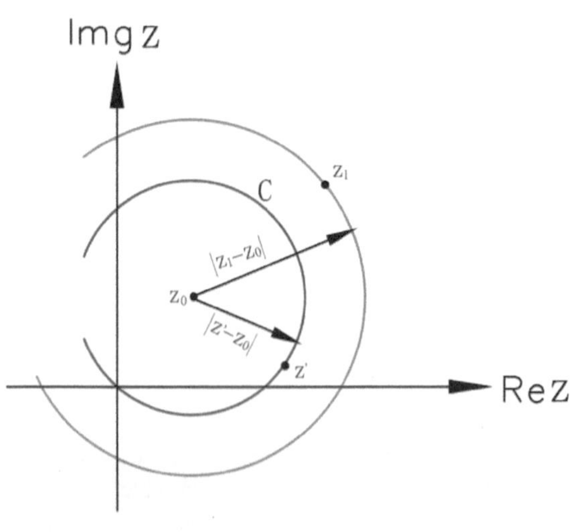

FIGURE D.10 Circular domains for Taylor series expansion.

in which z' is a point located on the contour C and z is any point located interior to C. We note the following identity

$$\frac{1}{1-t}=1+t+t^2+t^3+\cdots=\sum_{n=0}^{\infty}t^n, \tag{D.35}$$

that can easily be verified by multiplying both sides by $1-t$. The above infinite series is convergent for $|t|<1$. Now, for a point z that is located inside of C, $|z-z_0|<|z'-z_0|$, and by applying equation (D.35), equation (D.34) is transformed into

$$\begin{aligned} f(z)&=\frac{1}{2\pi i}\oint_C \frac{f(z')}{(z'-z_0)\left[1-(z-z_0)/(z'-z_0)\right]}dz' \\ &=\frac{1}{2\pi i}\oint_C \sum_{n=0}^{\infty}\frac{(z-z_0)^n f(z')}{(z'-z_0)^{n+1}}dz' \end{aligned} \tag{D.36}$$

By interchanging the order of integration and summation, which is valid due to the fact that equation (D.35) is uniformly convergent for $|t|<1-\varepsilon$, with $0<\varepsilon<1$, we are able to derive the following result:

$$\begin{aligned} f(z)&=\frac{1}{2\pi i}\oint_C \sum_{n=0}^{\infty}\frac{(z-z_0)^n f(z')}{(z'-z_0)^{n+1}}dz' \\ &=\frac{1}{2\pi i}\sum_{n=0}^{\infty}(z-z_0)^n\oint_C \frac{f(z')}{(z'-z_0)^{n+1}}dz' \end{aligned} \tag{D.37}$$

With reference to equation (D.33), we obtain

$$f(z)=\sum_{n=0}^{\infty}\frac{f^{(n)}(z_0)}{n!}(z-z_0)^n, \tag{D.38}$$

which is our desired Taylor series expansion. From the above Taylor series expansion for $f(z)$, the following binomial theorem can be derived:

$$(1+z)^m=1+mz+\frac{m(m-1)}{1\cdot 2}z^2+\cdots=\sum_{n=0}^{\infty}\binom{m}{n}z^n, \tag{D.39}$$

for m, any real number. This expansion is convergent for $|z|<1$.

As seen in Figure D.11, we often come across functions that are analytic in an annular region, say, between circles of inner radius r and outer radius R about a point z_0. We make the assumption that $f(z)$ is a function of this kind, with z

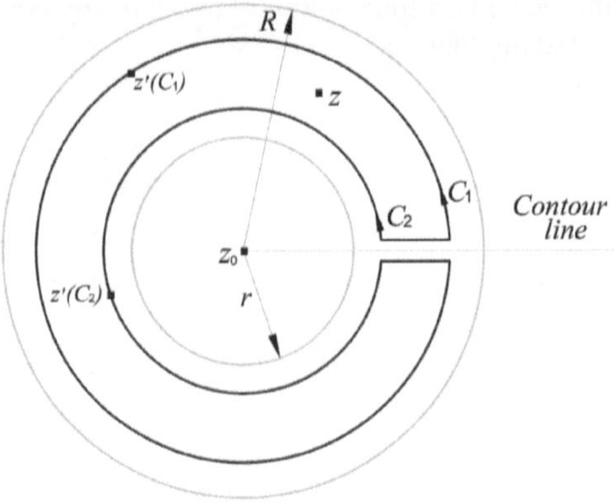

FIGURE D.11 An annular region for Laurent series.

being a typical point in the annular region. By constructing an imaginary barrier to transform our region into a simply connected region, we employ Cauchy's integral formula to evaluate $f(z)$ using the contour depicted in the figure. It should be noted that the contour is made up of two circles that are centered at z_0 and are labeled C_1 and C_2, plus segments on either side of the barrier whose contributions will cbe canceled The two circles can be regarded as closed because the barrier is fictitious. We assign the radii r_2 and r_1 to C_2 and C_1, respectively, where $r < r_2 < r_1 < R$. After that, using Cauchy's integral formula,

$$f(z) = \frac{1}{2\pi i} \oint_{C_1} \frac{f(z')}{z' - z} dz' - \frac{1}{2\pi i} \oint_{C_2} \frac{f(z')}{z' - z} dz' \tag{D.40}$$

It is important to take note that an explicit minus sign has been introduced into equation (D.40) in order to ensure that the contour C_2 (like C_1) is traversed in the positive direction, which is counterclockwise.

The treatment of equation (D.40) is now carried out in a manner that is identical to that of equation (D.34) in the development of the Taylor series. Each denominator in equation (D.40) is written as $(z' - z_0) - (z - z_0)$ and expanded by using equation (D.35).

Noting that for C_1, $|z' - z_0| > |z - z_0|$, we therefore have

$$\begin{aligned} \frac{1}{2\pi i} \oint_{C_1} \frac{f(z')}{z' - z} dz' &= \frac{1}{2\pi i} \oint_{C_1} \frac{f(z')}{(z' - z_0) - (z - z_0)} dz' \\ &= \frac{1}{2\pi i} \oint_{C_1} \frac{f(z')}{(z' - z_0)\left[1 - (z - z_0)/(z' - z_0)\right]} dz' \\ &= \frac{1}{2\pi i} \oint_{C_1} \sum_{n=0}^{\infty} \frac{(z - z_0)^n f(z')}{(z' - z_0)^{n+1}} dz' \end{aligned} \tag{D.41}$$

Noting that for C_2, $|z' - z_0| < |z - z_0|$, we therefore have

$$\frac{1}{2\pi i}\oint_{C_2}\frac{f(z')}{z'-z}dz' = \frac{1}{2\pi i}\oint_{C_2}\frac{f(z')}{(z'-z_0)-(z-z_0)}dz'$$
$$= \frac{-1}{2\pi i}\oint_{C_2}\frac{f(z')}{(z-z_0)\left[1-(z'-z_0)/(z-z_0)\right]}dz' \tag{D.42}$$
$$= \frac{-1}{2\pi i}\sum_{n=1}^{\infty}(z-z_0)^{-n}\oint_{C_2}(z'-z_0)^{n-1}f(z')dz'$$

Combining equations (D.40), (D.41) and (D.42) leads to

$$f(z) = \frac{1}{2\pi i}\oint_{C_1}\frac{f(z')}{z'-z}dz' - \frac{1}{2\pi i}\oint_{C_2}\frac{f(z')}{z'-z}dz'$$
$$= \frac{1}{2\pi i}\sum_{n=0}^{\infty}(z-z_0)^{n}\oint_{C_1}\frac{f(z')}{(z'-z_0)^{n+1}}dz' \tag{D.43}$$
$$+ \frac{1}{2\pi i}\sum_{n=1}^{\infty}(z-z_0)^{-n}\oint_{C_2}(z'-z_0)^{n-1}f(z')dz'$$

With the first series being labeled S_1, we obtain

$$S_1 = \frac{1}{2\pi i}\sum_{n=0}^{\infty}(z-z_0)^{n}\oint_{C_1}\frac{f(z')}{(z'-z_0)^{n+1}}dz' \tag{D.44}$$

The above expression has an identical form to the regular Taylor expansion, and it is convergent for $|z - z_0| < |z' - z_0| = r_1$. In other words, it converges for all z that lie within the larger circle, C_1.

With the second series being labeled S_2, we obtain

$$S_2 = \frac{1}{2\pi i}\sum_{n=1}^{\infty}(z-z_0)^{-n}\oint_{C_2}(z'-z_0)^{n-1}f(z')dz' \tag{D.45}$$

being convergent for $|z - z_0| > |z' - z_0| = r_2$, i.e., for all z that lie outside of the smaller circle, C_2. It is important to keep in mind that C_2 now goes counterclockwise.

These two series are combined into a single series, which is referred to as a Laurent series and has the following form:

$$f(z) = \sum_{n=-\infty}^{\infty}a_n(z-z_0)^{n}, \tag{D.46}$$

in which

$$a_n = \frac{1}{2\pi i}\oint_C \frac{f(z')}{(z'-z_0)^{n+1}}dz' \tag{D.47}$$

Laurent series is a generalization of the Taylor series. Indeed, a Laurent series is a series of positive and negative integer powers of $z - z_0$ that converges in an annulus (a circular ring) with z_0 as its center. This is in contrast to a Taylor series that has positive integer powers (plus a constant term) and converges in a disk. Therefore, we are able to express a given function $f(z)$ that is analytic in an annulus by using a Laurent series. This function $f(z)$, represented by the Laurent series, may have singularities both outside the ring and in the "hole" of the annulus.

If, in the context of an application, we wish to build a function $f(z)$ in powers of $z - z_0$ when $f(z)$ is singular at z_0, we cannot make use of a Taylor series. We might, as an alternative, make use of the new type of series known as the Laurent series.

D.5 SINGULARITIES

Generally speaking, a singular point of an analytic function $f(z)$ is a z_0 at which $f(z)$ stops being analytic. Precise definitions are provided in the following paragraphs. In this section, we demonstrate that the Laurent series may be utilized for the classification of singularities.

It is important to recall that, according to its definition, a function is a single-valued relation. A function $f(z)$ is said to be singular or to have a singularity at a point $z = z_0$ if $f(z)$ is not analytic (or possibly not even defined) at $z = z_0$, but every neighborhood of $z = z_0$ contains points at which $f(z)$ is analytic. We also state that the point $z = z_0$ is a singular point of the function $f(z)$. $z = z_0$ is considered to be an isolated singularity of $f(z)$ if $z = z_0$ is surrounded by a neighborhood that does not contain any additional singularities of $f(z)$.

Classification of isolated singularities of $f(z)$ at $z = z_0$ is possible through the use of the Laurent series

$$f(z) = \sum_{n=0}^{\infty} a_n (z - z_0)^n + \sum_{n=1}^{\infty} \frac{a_{-n}}{(z - z_0)^n} \tag{D.48}$$

being valid in the immediate vicinity of the singular point $z = z_0$, with the exception of being at z_0 itself, i.e., in a region of the form

$$0 < |z - z_0| < R. \tag{D.49}$$

According to what we learned in the previous sections, the sum of the first series at $z = z_0$ is analytic. The second series, which is comprised of the negative powers, is referred to as the principal part of equation (D.48). In the event that the second series possesses only a limited number of terms, it is of the form

$$\frac{a_{-1}}{z-z_0}+\cdots+\frac{a_{-m}}{(z-z_0)^m}\ (a_{-m}\neq 0) \tag{D.50}$$

Then the singularity of $f(z)$ at $z=z_0$ is referred to as a ***pole***, and the order of the pole is denoted by the letter m. Poles of the first order are also referred to as ***simple poles***.

D.6 RESIDUE METHOD FOR EVALUATING COMPLEX INTEGRALS

In this section, we will discuss a second approach to evaluating complex integrals. It is important to remember that in the previous sections, we solved complex integrals directly by using Cauchy's integral formula. Taylor series is a topic that was covered in the previous sections. The Taylor series was then generalized to the Laurent series. In the previous section, we also explored the singularities of a specific function. As a result of our laborious efforts, we are now able to observe how much of the theoretical underpinning is brought together in the process of evaluating complex integrals using the residue technique.

Cauchy's residue integration method is designed to be used for the purpose of evaluating the following integrals

$$\oint_C f(z)dz \tag{D.51}$$

taken around a simply closed path C. The idea is described as follows.

If the function $f(z)$ is analytic everywhere on the path C and within the region enclosed by C, then the integral in question is equal to 0 according to Cauchy's integral theorem, and everything is finished.

The situation is different if $f(z)$ has a singularity at a point $z=z_0$ inside C, but it is otherwise analytic on C and inside C as it was previously. Then the function $f(z)$ possesses a Laurent series

$$f(z)=\sum_{n=-\infty}^{\infty}a_n(z-z_0)^n, \tag{D.52}$$

In the event that the above Laurent expansion of the function $f(z)$ is integrated term by term through the utilization of a closed contour that encircles one isolated singular point z_0 once in a counterclockwise sense, we are able to obtain, through the application of equation (D.24),

$$a_n\oint(z-z_0)^n\,dz=0,\ n\neq -1. \tag{D.53}$$

However, when $n = -1$, equation (D.23) leads to

$$a_{-1}\oint(z - z_0)^{-1}\,dz = 2\pi i a_{-1}. \tag{D.54}$$

By combining equations (D.52), (D.53) and (D.54), we have the following:

$$\oint_C f(z)\,dz = 2\pi i a_{-1} \tag{D.55}$$

The constant a_{-1}, which represents the coefficient of $(z - z_0)^{-1}$ in the Laurent expansion, is referred to as the residue of $f(z)$ at the point $z = z_0$.

In light of the fact that we are able to obtain the Laurent series by a variety of techniques, without having to make use of the integral formulae for the coefficients, we are able to find b_1 through one of those techniques, and then we can use the formula in equation (D.55) to evaluate the complex integral of $f(z)$ over the path C.

The coefficient a_{-1} is defined as the residual of the function $f(z)$ at the point $z = z_0$, and we represent it by

$$a_{-1} = \text{Res}_{z=z_0} f(z). \tag{D.56}$$

In order to calculate a residue at a pole, we do not need to construct an entire Laurent series; rather, more economically, we can derive formulas for residues once and for all. For a simple pole at $z = z_0$, the Laurent series is

$$f(z) = \frac{a_{-1}}{z - z_0} + a_0 + a_1(z - z_0) + a_2(z - z_0)^2 + \cdots \; (0 < |z - z_0| < R). \tag{D.57}$$

By multiplying both sides of the above equation by $z - z_0$ and then letting $z \to z_0$, we obtain:

$$\lim_{z\to z_0}(z - z_0)f(z) = a_{-1} + \lim_{z\to z_0}(z - z_0)\left[a_0 + a_1(z - z_0) + \cdots\right] = a_{-1} \tag{D.58}$$

Thus, the following is a formula for the residue at a simple pole z_0:

$$\text{Res}_{z=z_0} f(z) = a_{-1} = \lim_{z\to z_0}(z - z_0)f(z) \tag{D.59}$$

There is a possibility of extending the scope of residue integration from the case of a single singularity to the case of multiple singularities that are contained inside the contour C. It is for this purpose that the residue theorem was developed. Quite surprisingly, the extension is not complicated.

Residue Theorem. Let $f(z)$ be analytic both inside a simple closed path C and on C, with the exception for a finite number of singular points $z_1, z_2, \cdots, z_k$ that are contained within C. When the integral of $f(z)$ is taken in a counterclockwise

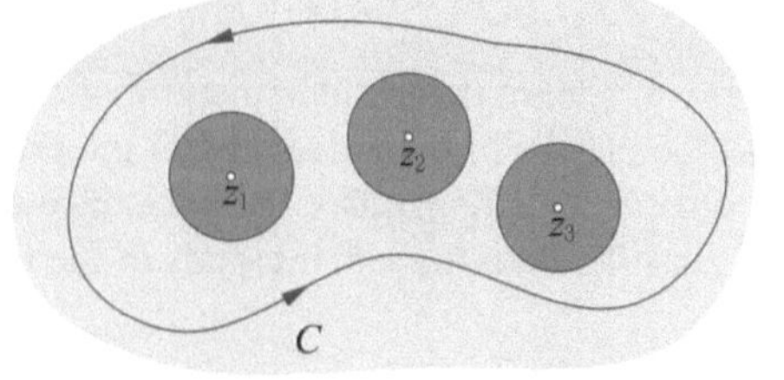

FIGURE D.12 The residue theorem.

direction around C, it is equal to $2\pi i$ times the sum of the residues of $f(z)$ at $z_1,\ z_2, \cdots, z_k$:

$$\oint_C f(z)dz = 2\pi i \sum_{j=1}^{k} \operatorname*{Res}_{z=z_j} f(z) \tag{D.60}$$

We now give the proof of the foregoing residue theorem as follows. We enclose each of the singular points z_j in a circle C_j that has a radius that is sufficiently small to ensure that those k circles and C are all kept apart from one another (see Figure D.12, where $k = 3$). Then the function $f(z)$ is analytic throughout the multiply connected domain D, which is bounded by C and $C_1,\cdots,C_k$, and $f(z)$ is also analytic on the entire boundary of D. According to Cauchy's integral theorem, we have

$$\oint_C f(z)dz + \oint_{C_1} f(z)dz + \oint_{C_2} f(z)dz + \cdots + \oint_{C_k} f(z)dz = 0, \tag{D.61}$$

the integral along C being taken in the counterclockwise direction, while the other integrals being taken in the clockwise direction. We next take the integrals over $C_1,\cdots,C_k$ to the right side of the equation and compensate the resulting minus sign by reversing the sense of integration (i.e., the integrals over $C_1,\cdots,C_k$ now being taken in the counterclockwise direction). Therefore,

$$\oint_C f(z)dz = \oint_{C_1} f(z)dz + \oint_{C_2} f(z)dz + \cdots + \oint_{C_k} f(z)dz \tag{D.62}$$

in which all the integrals are now being taken counterclockwise.

Note that we have the following relationship

$$\oint_{C_j} f(z)dz = 2\pi i \operatorname*{Res}_{z=z_j} f(z) \tag{D.63}$$

Combining equations (D.62) and (D.63) leads to equations (D.60) and the residue theorem is thus proved.

D.7 RESIDUE METHOD FOR EVALUATING REAL INTEGRALS

It may come as a surprise to learn that residue integration can also be utilized for the evaluation of specific classes of complicated real integrals. In comparison to real analysis or calculus, this demonstrates the advantage of complex analysis. First, let us take into consideration the real integrals of the form

$$\int_{-\infty}^{\infty} f(x)dx \tag{D.64}$$

An integral of this kind, in which the interval of integration is not finite, is referred to as an improper integral, and it has the characteristics described below:

$$\int_{-\infty}^{\infty} f(x)dx = \lim_{a\to-\infty}\int_{a}^{0} f(x)dx + \lim_{b\to\infty}\int_{0}^{b} f(x)dx. \tag{D.65}$$

In the event where both limits exist, it is possible to couple the two independent passages to $-\infty$ and ∞, and then write

$$\int_{-\infty}^{\infty} f(x)dx = \lim_{R\to\infty}\int_{-R}^{R} f(x)dx. \tag{D.66}$$

The limit that is found in equation (D.66) is referred to as the Cauchy principal value of the integral. It is written

$$\text{pr. v.} \int_{-\infty}^{\infty} f(x)dx \tag{D.67}$$

In equation (D.64), we make the assumption that the function $f(x)$ is a real rational function. More specifically, we assume that the denominator of this function $f(x)$ is not equal to 0 for any real x, and that its degree is at least two units higher than the degree of the numerator. If this is the case, then the limits in equation (D.65) exist, and we can begin with equation (D.66). We take into consideration the corresponding contour integral

$$\oint_{C} f(z)dz \tag{D.68}$$

around a path C in Figure D.13. As $f(x)$ is rational, $f(z)$ has a finite number of poles in the upper half-plane. If we select R to be sufficiently large, then C will

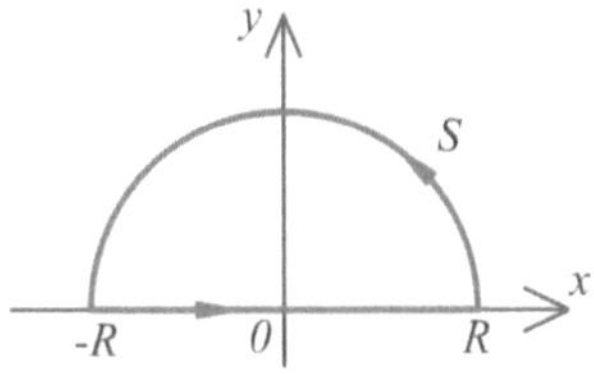

FIGURE D.13 Path C of the contour integral in equation (D.68).

enclose all of these poles. We subsequently obtain the following by applying the residue theorem:

$$\oint_C f(z)dz = \int_S f(z)dz + \int_{-R}^{R} f(x)dx = 2\pi i \sum \text{Res} f(z) \qquad \text{(D.69)}$$

in which the sum is composed of all the residues of $f(z)$ at the points in the upper half-plane where $f(z)$ has a pole. As a result of this, we have the following relationship:

$$\int_{-R}^{R} f(x)dx = 2\pi i \sum \text{Res} f(z) - \int_S f(z)dz \qquad \text{(D.70)}$$

We next demonstrate that the value of the integral $\int_S f(z)dz$ over the semicircle S approaches 0 when $R \to \infty$.

Let us suppose that $f(z) = P(z)/Q(z)$, and $P(z)$ and $Q(z)$ are polynomials of polynomial degree n and m with coefficients $b_n,\ldots,b_0$ and $c_m,\ldots,c_0$. Let us write $z \equiv Re^{i\theta}$. Because we have assumed that the denominator of the function $f(x)$ is not equal to 0 for any real x, and that its degree is at least two units higher than the degree of the numerator, we have $m \geq n+2$ and $n-m+1 \leq -1$. Therefore, we have

$$\begin{aligned}
\int_S f(z)dz &= \int_S [P(z)/Q(z)]dz = \lim_{R\to\infty} \int_0^{\pi} \frac{P(Re^{i\theta})}{Q(Re^{i\theta})} iRe^{i\theta} d\theta \\
&= \lim_{R\to\infty} \int_0^{\pi} \frac{b_n (Re^{i\theta})^n + b_{n-1}(Re^{i\theta})^{n-1} + \cdots + b_0}{c_m (Re^{i\theta})^m + c_{m-1}(Re^{i\theta})^{m-1} + \cdots + c_0} iRe^{i\theta} d\theta \\
&= \lim_{R\to\infty} \int_0^{\pi} \frac{b_n}{c_m} (Re^{i\theta})^{n-m} iRe^{i\theta} d\theta \\
&= \lim_{R\to\infty} \int_0^{\pi} \frac{b_n}{c_m} R^{n+1-m} i (e^{i\theta})^{n-m+1} d\theta = 0
\end{aligned} \qquad \text{(D.71)}$$

The result in the above equation has been obtained because $n - m + 1 \le -1$ and we have the following relationship:

$$\lim_{R \to \infty} R^{n+1-m} = 0 \tag{D.72}$$

Combining equations (D.70) and (D.71) leads to

$$\int_{-\infty}^{\infty} f(x)dx = 2\pi i \sum \text{Res}\, f(z) \tag{D.73}$$

in which we add up all of the residues of $f(z)$ that are located at the poles of $f(z)$ in the upper half-plane.

In equation (5.115), the integrand is an analytic function everywhere in the complex β plane, with the exception of the points $\beta = r_1$, $\beta = r_2$, $\beta = -r_1$ and $\beta = -r_2$, where poles of order 1 exist. Specifically, the points $\beta = r_1$ and $\beta = r_2$ are located in the upper half plane, whilst the points $\beta = -r_1$ and $\beta = -r_2$ are situated on the lower half plane. The process of contour integration is carried out in the upper half plane. Within the upper half complex plane, the integrand possesses two simple poles, which are as follows:

$$\beta_1 = r_1 = i\xi + \sqrt{1-\xi^2} \tag{5.116}$$

$$\beta_2 = r_2 = i\xi - \sqrt{1-\xi^2} \tag{5.117}$$

When Cauchy's residue theorem is utilized, the integral in equation (5.115) can be easily carried out, resulting in the relation

$$\begin{aligned} E\left[y^2\right] &= \frac{\omega_n S_0}{k^2} \int_{-\infty}^{\infty} \frac{1}{\left[(\beta - r_1)(\beta - r_2)(\beta + r_1)(\beta + r_2)\right]} d\beta = \frac{\omega_n S_0}{k^2} \int_{-\infty}^{\infty} f(\beta) d\beta \\ &= \frac{\omega_n S_0}{k^2} 2\pi i \left\{ \sum residues\ \ of\ \ f(\beta)\ \ in\ \ the\ \ upper\ \ half\ \ complex\ \ plane \right\} \\ &= \frac{\omega_n S_0}{k^2} 2\pi i \left\{ \sum \left[(\beta - r_1) f(\beta)\right]_{\beta = r_1} + \left[(\beta - r_2) f(\beta)\right]_{\beta = r_2} \right\} \\ &= \frac{\omega_n S_0}{k^2} 2\pi i \left\{ \sum \frac{1}{\left[(r_1 - r_2)(r_1 + r_1)(r_1 + r_2)\right]} + \frac{1}{\left[(r_2 - r_1)(r_2 + r_1)(r_2 + r_2)\right]} \right\} \\ &= \frac{\omega_n S_0}{k^2} 2\pi i \left\{ \frac{1}{2r_2 r_1 2i\xi} \right\} = \frac{\pi S_0}{kc} \end{aligned} \tag{5.118}$$

in a manner that is not reliant on the magnitude of the mass m.

Appendix E: Theory of Springs

E.1 INTRODUCTION

Oscillation, often known as vibration, is a term that is used to describe any motion that repeats itself after an interval of time. One example of a vibration system that is commonly used is the spring-mass-damper system that is shown in Figure E.1.

A vibratory system typically consists of a way to store potential energy (like the blue spring shown in Figure E.1). A spring is a form of mechanical link that, in the majority of applications, is presumed to have a negligible amount of mass and damping. Helical-coil springs are the most prevalent type of spring, and they are utilized in retractable pens and pencils, staplers and the suspensions of freight trucks and other vehicles. There are a number of other kinds of springs that can be found in engineering applications. As a matter of fact, a spring can be thought of as any body or member that is elastic or deformable, such as a cable, bar, beam, shaft or plate. As seen in Figure E.2, a spring is typically depicted in this manner.

If the free length of the spring is symbolized by the letter l, but there are no forces acting on it, then the spring will undergo a change in length when an axial force is applied to it. When a tensile force F is given to the free end 2 of a spring, for instance, the spring undergoes an elongation y, as depicted in Figure E.3. On the other hand, when a compressive force F is applied to the free end 2, the spring undergoes a decrease in length y, as depicted in Figure E.4.

A linear elastic spring obeys the Hooke's Law. Hooke's Law is a fundamental law of mechanics that states that the extension of a spring, when subjected to a force, is directly proportional to the force applied, within the elastic limit of the

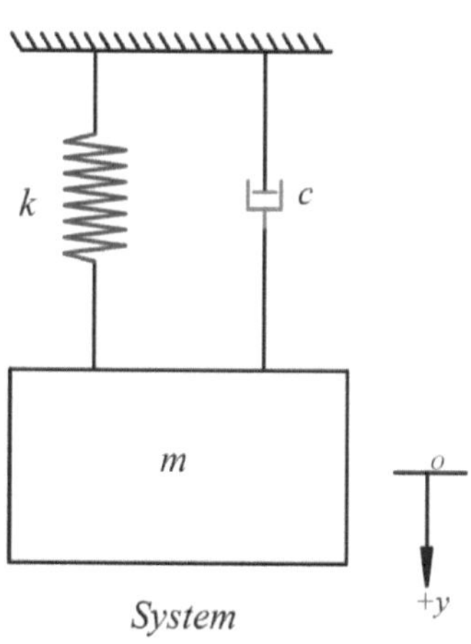

FIGURE E.1 A mass-spring-damper system.

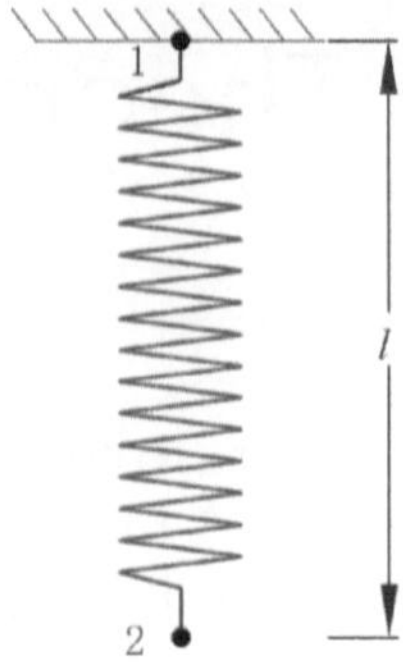

FIGURE E.2 A spring.

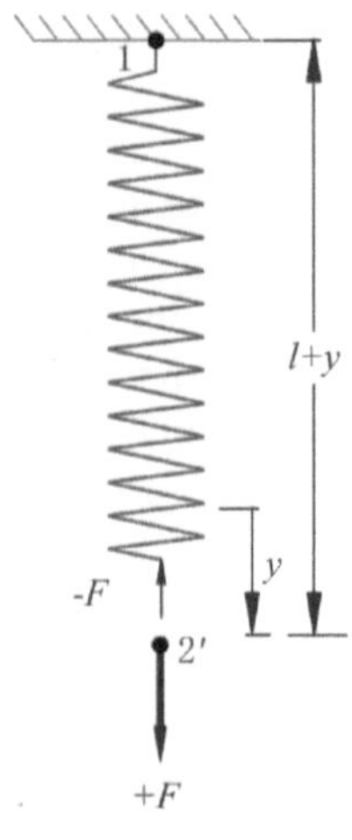

FIGURE E.3 The elongation of a spring.

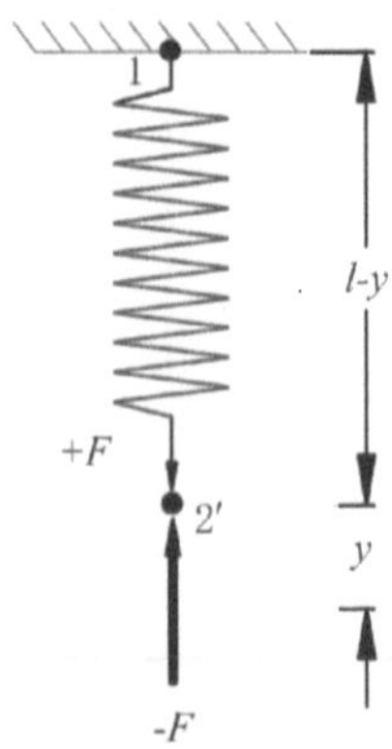

FIGURE E.4 The compression of a spring.

spring. In other words, the restoring force of a spring is directly proportional to its deformation or displacement.

A spring is considered to be linear if the elongation or reduction in length y is related to the force F that is applied to it in the following linear fashion:

$$F = ky \tag{E.1}$$

in which k is a constant, which is also referred to as the *spring constant*, *spring stiffness* or *spring rate*. The spring constant, denoted by the letter k, is always positive and represents the force, whether positive or negative, that is necessary to cause a unit deflection (elongation or reduction in length) in the spring. Newton's third law of motion states that when a spring is stretched (or compressed) under a tensile (or compressive) force F, a restoring force or reaction of magnitude $-F$ (or $+F$) is developed opposite to the force that is being applied to the spring. This restoring force makes an effort to bring the spring that has been stretched (or compressed) to its original unstretched or free length, as depicted in Figure E.3 (or Figure E.4). If we plot a graph between F and y, the outcome is a straight line in accordance with equation (E.1). The work that is done (U) in the process of deforming a spring is stored as strain or potential energy within the spring, and it is given by

$$U = \frac{1}{2}ky^2 \tag{E.2}$$

E.2 NONLINEAR SPRINGS

The majority of springs that are utilized in actual systems tend to display a force-deflection relation that is nonlinear, especially when the deflections are large. If a nonlinear spring undergoes small deflections, it can be substituted with a linear spring using the method explained in the next section. In the field of vibration analysis, nonlinear springs with force-deflection relationships are described by

$$F = ax + by^3;\ a > 0 \tag{E.3}$$

are commonly utilized. Within the equation (E.3), the letter a represents the constant that is associated with the linear component, and the letter b represents the constant that is associated with the (cubic) nonlinearity. The spring is considered to be hard if the value of b is greater than 0, linear if b is equal to zero, and soft if b is less than zero. As demonstrated in Figure E.5, the force-deflection relations for a number of different values of b are displayed.

It is possible for certain systems, which involve two or more springs, to display a force-displacement relationship that is nonlinear, despite the fact that the individual springs are linear. In Figure E.6, we see an example of a vibration system that falls within this category. The mass m is shown to move linearly through the clearances c and c that are present in the system in Figure E.6. Once the mass

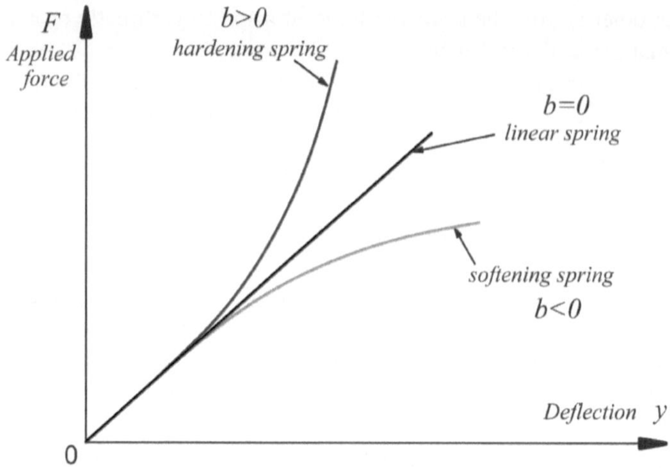

FIGURE E.5 Linear and nonlinear springs.

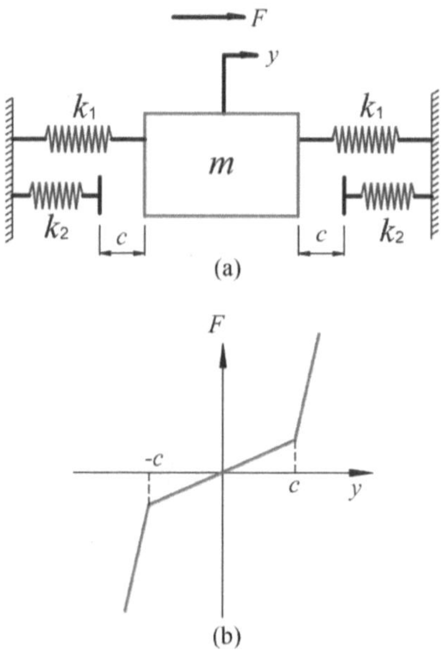

FIGURE E.6 Nonlinear force-displacement characteristic.

comes into contact with a particular spring k_2, after passing through the corresponding clearance, the spring force increases with the additional contribution from the spring constant of the particular spring k_2. It can be noted that the resulting force-deflection relation, although piecewise linear, represents a nonlinear relationship.

From the example shown in Figure E.6, we can find that mass-spring systems can be geometrically arranged so that the overall force-displacement is strongly nonlinear. Let us consider again the example shown in Figure E.6. Obviously, the linear springs with stiffness k_2 become activated only when the horizontal displacement, y, is such that $|y| > c$, where c is the gap between the oscillating body of mass m and the springs. The equation of motion of the vibration system can be written as follows:

$$m\ddot{y} + F(y) = 0 \tag{E.4}$$

in which

$$F(y) = 2k_1 y \quad |y| < c \tag{E.5}$$

and

$$F(y) = 2k_1 c + (y - c)(2k_1 + k_2) \quad |y| > c \tag{E.6}$$

The nonlinear (although piecewise linear) force-displacement characteristic for this system is shown in Figure E.6.

The type of nonlinearity depicted in Figure E.6 is encountered in a variety of applications where it is desired to restrict the amplitudes of vibrations by the implementation of high stiffness "buffers" or "barriers". Therefore, the flexural vibration of heat exchanger tubes, for instance, is often restricted by the incorporation of buffers, as shown in Figure E.7. As a similar example, fenders on

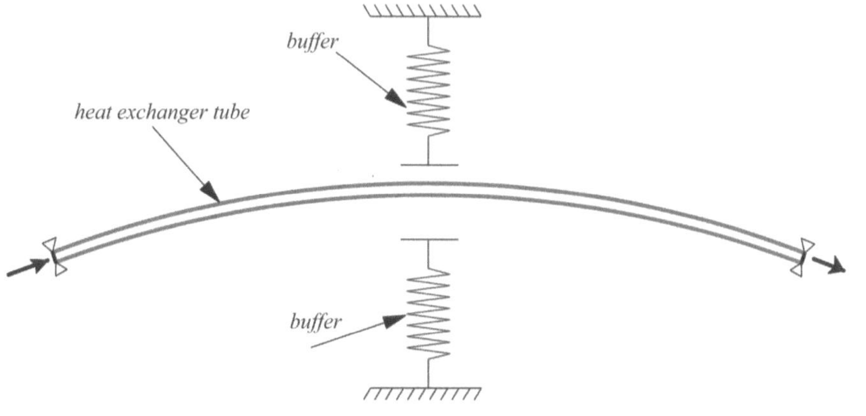

FIGURE E.7 A heat-exchanger tube with buffers.

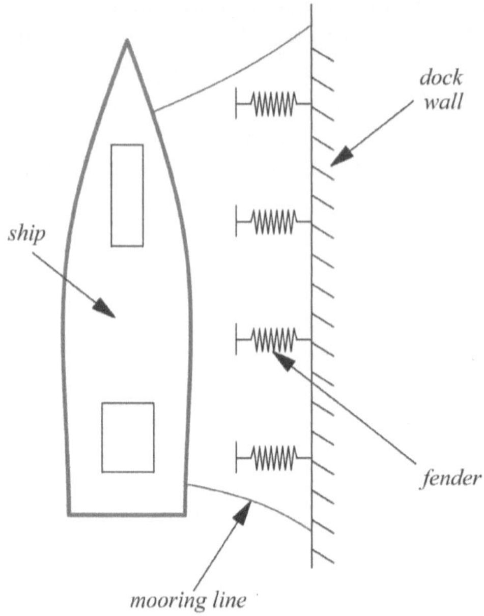

FIGURE E.8 A ship with mooring lines and fenders.

dock-side walls frequently act in parallel with ship mooring lines to produce an overall nonlinear restoring characteristic (see Figure E.8).

As another example, Figure E.9 shows a weightless rigid bar with springs. As shown in Figure E.9, the lengths of the two springs, which have stiffnesses k_1 and k_2, are distinct from one another. For the sake of simplicity, the spring with the

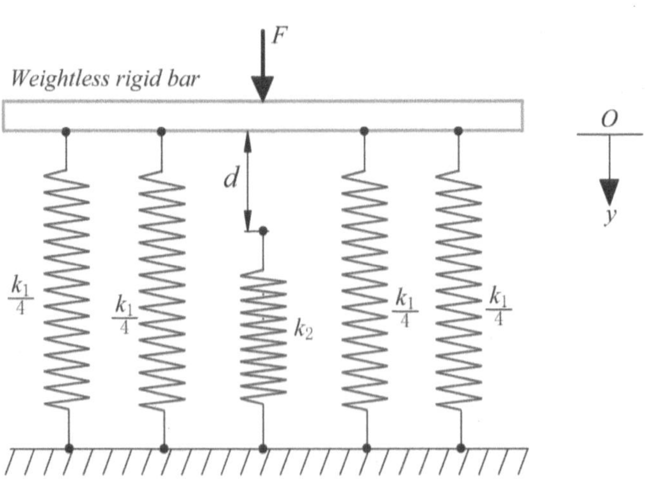

FIGURE E.9 A weightless rigid bar with springs.

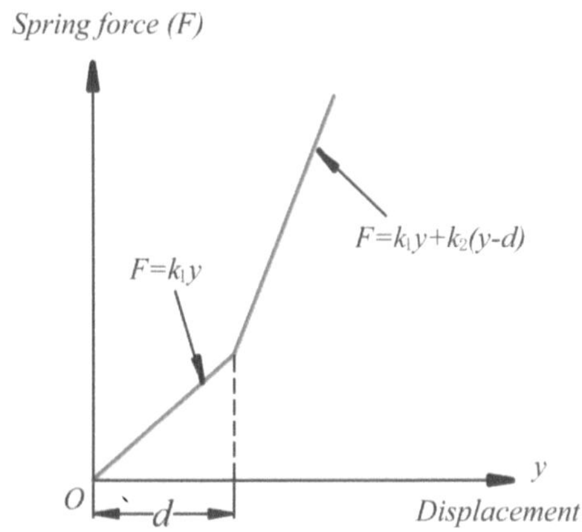

FIGURE E.10 Nonlinear force-displacement relationship.

stiffness k_1 is depicted as four parallel springs, each of which has a stiffness of $k_1/4$. Please note that $y = 0$ in Figure E.9 corresponds to the position of the bar with no force.

When the spring k_1 deflects by an amount $y = d$, the second spring starts to provide an additional stiffness k_2 to the vibration system. The resulting nonlinear force-displacement relationship is depicted in Figure E.10.

E.3 LINEARIZATION OF A NONLINEAR SPRING

The actual springs are nonlinear and only follow the equation (E.1) up to a specific point of deformation. When the spring reaches a specific magnitude of deformation (after point P in Figure E.11), the stress surpasses the material's proportional limit point and the relationship between force and deformation becomes

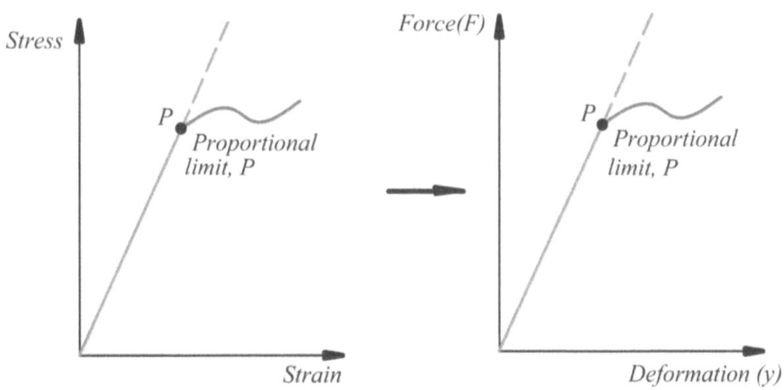

FIGURE E.11 Nonlinearity beyond the proportional limit.

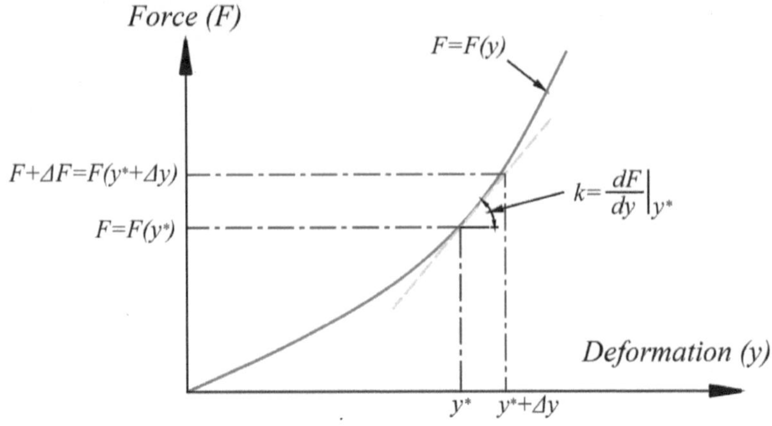

FIGURE E.12 The linearization process.

nonlinear. For a great number of practical applications, we adopt the assumption that the deflections are very small and employ the linear relationship found in equation (E.1).

Even if the force-deflection relation of a spring is nonlinear, as demonstrated in Figure E.12, we frequently use a linearization technique to approximate it as if it were linear rather than nonlinear. In order to demonstrate the linearization process, let the static equilibrium load F that is acting on the spring cause a deflection of y^*. When an incremental force ΔF is added to the force F, the spring experiences a deflection of an additional quantity Δy. It is possible to represent the new spring force $F+\Delta F$ by employing Taylor's series expansion with regard to the position of the static equilibrium y^* as

$$\begin{aligned} F+\Delta F &= F\left(y^*+\Delta y\right) \\ &= F\left(y^*\right)+\left.\frac{dF}{dy}\right|_{y^*}(\Delta y)+\frac{1}{2!}\left.\frac{d^2F}{dy^2}\right|_{y^*}(\Delta y)^2+\ldots \end{aligned} \tag{E.7}$$

In cases where the value of Δy is small, it is possible to disregard the higher-order derivative terms to obtain

$$F+\Delta F = F\left(y^*\right)+\left.\frac{dF}{dy}\right|_{y^*}(\Delta y) \tag{E.8}$$

Because $F = F\left(y^*\right)$, we can express ΔF as follows:

$$\Delta F = \left.\frac{dF}{dy}\right|_{y^*}(\Delta y) = k(\Delta y) \tag{E.9}$$

in which k is the linearized spring constant at y^* given by

$$k = \left.\frac{dF}{dy}\right|_{y^*} \tag{E.10}$$

Example E.1

This example shows the calculation procedures for obtaining an equivalent linearized spring constant.

A rubber mount is used to support a precision milling machine that weighs 6000 Newtons. The relationship between the force and the deflection of the rubber mount can be expressed as

$$F = 425y + 0.15y^3 \tag{E.11}$$

In the above equation, the force (F) and the deflection (y) are measured in newtons and millimeters, respectively. Find out what the equivalent linearized spring constant of the rubber mount is when it is in its position of static equilibrium.

SOLUTION

The static equilibrium position of the rubber mount, denoted as y^*, under the weight of the milling machine, can be calculated using equation (E.11) as follows:

$$0.15\left(y^*\right)^3 + 425y^* - 6000 = 0 \tag{E.12}$$

The roots of the cubic equation in equation (E.12) can be determined as

$$y^* = 13.2893; \quad y^* = -6.6447 + 54.4590i; \quad y^* = -6.6447 - 54.4590i$$

For the rubber mount, the position of static equilibrium can be determined by taking the real root of equation (E.11), which is $y^* = 13.2893$ mm. Equation (E.10) can be used to calculate the equivalent linear spring constant of the rubber mount when it is in its position of static equilibrium:

$$k_{eq} = \left.\frac{dF}{dy}\right|_{y^*} = 425 + 0.45\left(y^*\right)^2 = 425 + 0.45(13.29)^2 = 504.5\text{N/mm}$$

E.4 EQUIVALENT SPRING

There are some dynamic systems that use multiple springs. In such instances, a single equivalent spring can be used to replace all the springs.

E.4.1 Case 1: For Springs in Parallel

For the purpose of deriving an expression for the equivalent spring constant of springs that are connected in parallel, take into consideration the two springs that are depicted in Figure E.13(a). Following the application of a load F, the system experiences a static deflection denoted by the symbol δ_{st}, as depicted in Figure E.13(b). The equilibrium equation can then be obtained from the free-body diagram, which is depicted in Figure E.13(c).

$$F = k_1\delta_{st} + k_2\delta_{st} \tag{E.13}$$

If we consider the case where k_{eq} represents the equivalent spring constant of the combination of the two springs, then with the same static deflection δ_{st}, we should obtain the following:

$$F = k_{eq}\delta_{st} \tag{E.14}$$

Equations (E.13) and (E.13) give

$$k_{eq} = k_1 + k_2 \tag{E.15}$$

In general, if we have n springs that are parallel to one another and have spring constants $k_1, k_2, \ldots, k_n$, then we can obtain the equivalent spring constant k_{eq} as follows:

$$k_{eq} = k_1 + k_2 + \cdots + k_n \tag{E.16}$$

E.4.2 Case 2: For Springs in Series

In this case, we will develop a formula for the equivalent spring constant of springs that are connected in series by utilizing similar procedures as used in case 1. We merely summarize the developed formulas. When two springs that

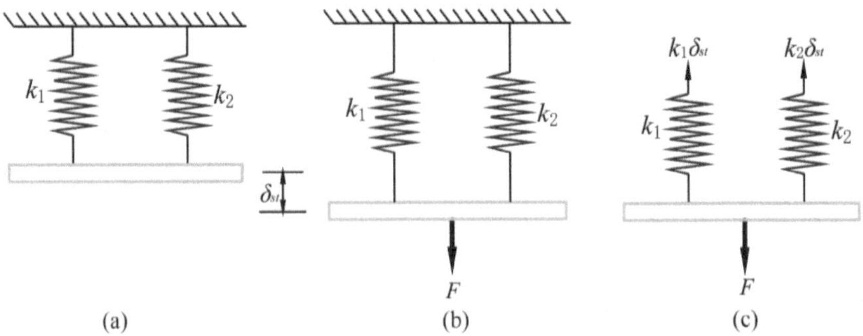

FIGURE E.13 Two springs in parallel.

have spring constants k_1 and k_2 are connected in series, the equivalent spring constant k_{eq} can be calculated as

$$\frac{1}{k_{eq}} = \frac{1}{k_1} + \frac{1}{k_2} \tag{E.17}$$

Equation (E.17) can be extended to the scenario where there are multiple springs arranged in a series configuration:

$$\frac{1}{k_{eq}} = \frac{1}{k_1} + \frac{1}{k_2} + \cdots + \frac{1}{k_n} \tag{E.18}$$

Bibliography

Abramowitz, M., Stegun, I. A., 1972. Handbook of Mathematical Functions with Formulas, Graphs, and Mathematical Tables, Dover Publications, New York.

Ang, A. H. S., Tang, W. H., 1975. Probability and Concepts in Engineering Planning and Design, Vol. 1, Wiley, New York.

Ang, A. H. S., Tang, W., 1984. Probability Concepts in Engineering Planning and Design, Vol. II: Decision, Risk, and Reliability, Wiley, New York.

ASTM, 1990. Cycle counting in fatigue analysis. Annual Book of ASTM Standards 03. 01.

Ayyub, B. M., 2014. Risk Analysis in Engineering and Economics. Second Edition, CRC Press, Boca Raton, FL.

Ayyub, B. M., McCuen, R. H., 2011. Probability, Statistics, and Reliability for Engineers and Scientists. Third Edition, CRC Press, Boca Raton, FL.

Benaroya, H., Han, S. M., 2005. Probability Models in Engineering and Science, Taylor & Francis, Boca Raton, FL.

Benaroya, H., Nagurka, M. L., 2010. Mechanical Vibration: Analysis, Uncertainties and Control. Third Edition, CRC Press, Boca Raton, FL.

Benasciutti, D., 2004. Fatigue analysis of random loadings. Ph.D. Dissertation, University of Ferrara.

Benson, H. T., 2002. Principles of Vibration, Oxford University Press, New York.

Cai, G. Q., Zhu, W. Q., 2016. Elements of Stochastic Dynamics, World Scientific, Singapore.

Caughey, T. K., 1963. Equivalent linearization techniques. Journal of the Acoustical Society of America. 35, 1706–1711.

Chopra, A. K., 2001. Dynamics of Structures: Theory and Applications to Earthquake Engineering. Second Edition, Prentice Hall, Inc., Upper Saddle River, NJ.

Clough, R. W., Penzien, J., 1993. Dynamics of Structures. Second Edition, McGraw-Hill, New York.

Cramér, H., 1946. Mathematical Methods of Statistics, Princeton University Press, Princeton.

Crandall, S. H., 1963. Perturbation techniques for random vibration of nonlinear systems. Journal of the Acoustical Society of America. 35, 1700–1705.

Crandall, S. H., 1983. Random vibration: A survey of recent developments. Journal of Applied Mechanics, ASME. 50, 953962.

Crandall, S. H., 2001. Is stochastic equivalent linearization a subtly flawed procedure? Probabilistic Engineering Mechanics. 16, 169–176.

Crandall, S. H., Mark, W. D., 1963. Random Vibration in Mechanical Systems, Academic Press, New York.

Crandall, S. H., Zhu, W. Q., 1983. Random vibration: A survey of recent developments. Journal of Applied Mechanics. 50 (4b), 953–962.

Crauel, H., Gundlach, M. (eds.), 1999. Stochastic Dynamics, Springer, New York.

Den Hartog, J. P., 1956. Mechanical Vibrations. Fourth Edition, McGraw Hill, New York.

Dimentberg, M. F., 1988. Statistical Dynamics of Nonlinear and Time-Varying Systems, Research Studies Press, Taunton, UK.

Dirlik, T., 1985. Application of computers in fatigue analysis. Ph. D. Dissertation, University of Warwick.

Doob, J. L., 1953. Stochastic Processes, Wiley, New York.

Dowling, N. E., 1999. Mechanical Behavior of Materials: Engineering Methods for Deformation, Fracture, and Fatigue, Prentice Hall, Upper Saddle River, New Jersey.

Duan, Y. Q., Luo, S. L., Wang, O. (eds.), 2010. Recent Developments in Stochastic Dynamics and Stochastic Analysis, World Scientific Singapore.

Duffy, D. G., 2024. Advanced Engineering Mathematics with MATLAB, CRC Press, Taylor & Francis Group, Boca Raton, FL.

Einstein, A., 1956. Investigations on the Theory of the Brownian Movement (edited by R. Furth). Dover Publications, Inc., New York.

Elishakoff, I., 1983. Probabilistic Methods in the Theory of Structures, John Wiley and Sons, New York.

Elishakoff, I., 1995. Random vibration of structures: A personal perspective. Applied Mechanics Reviews. 48, 809–825.

Elishakoff, I., 2000a. Stochastic linearization technique: A new interpretation and a selective review. The Shock and Vibration Digest. 32, 179–188.

Elishakoff, I., 2000b. Multiple combinations of the stochastic linearization criteria by the moment approach. Journal of Sound and Vibration. 237, 550–559.

Elishakoff, I., 2004. Safety Factors and Reliability. Friends or Foes? Kluwer Academic Publishers, Dordrecht, the Netherlands.

Elishakoff, I., 1999. Whys and Hows Uncertainty Modelling: Probability, Fuzziness and Anti-Optimization, Springer, Vienna.

Elishakoff, I., Cai, G. Q., 1993a. Approximate solution for nonlinear random vibration problems by partial stochastic linearization. Probabilistic Engineering Mechanics. 8, 233–237.

Elishakoff, I., Colombi, P., 1993b. Successful combination of the stochastic linearization and Monte Carlo methods. Journal of Sound and Vibration. 160, 554–558.

Elishakoff, I., Colojani, P., 1997. Stochastic linearization critically re-examine. Chaos, Solitons and Fractals. 8, 1957–1972.

Feller, W., 1968. An Introduction to Probability Theory and Its Applications. In Volume I. John Wiley and Sons, New York.

Feller, W., 1972. An Introduction to Probability Theory and Its Applications, Vol II, John Wiley and Sons, New York.

Ghanem, R. G., Spanos, P., 2003. Stochastic Finite Elements: A Spectral Approach. In Revised Edition. Dover Publications, Mineola, NY.

Grigoriu, M., 1998. Simulation of stationary non-Gaussian translation processes. Journal of Engineering Mechanics. 124 (2), 121–126.

Grigoriu, M., 2002. Stochastic Calculus: Applications in Science and Engineering, Springer, New York.

Grigoriu, M., 2012. Stochastic Systems: Uncertainty Quantification and Propagation, Springer, London.

Goda, Y., 2010. Random Seas and Design of Maritime Structures. Second Edition, World Scientific, Singapore.

Houbolt, J. C., 1950. A recurrence matrix solution for the dynamic response of elastic aircraft. Journal of Aeronautical Sciences. 17, 540–550, 594.

Hughes, T. J. R., 1976. A note on the stability of Newmark's algorithm in nonlinear structural dynamics. International Journal for Numerical Methods in Engineering. 11, 383–386.

Hurty, W. C., Rubinstein, M. F., 1964. Dynamics of Structures, Prentice Hall, Englewood Cliffs, NJ.

Ibrahim, R. A., 1985. Parametric Random Vibration, John Wiley and Sons, New York.

Inman, D. J., 1994. Engineering Vibration, Prentice-Hall, Englewood Cliffs, NJ.

Klyatskin, V. I., 2011. Lectures on Dynamics of Stochastic Systems, Elsevier, Amsterdam.

Kreyszig, E. 2010. Advanced Engineering Mathematics, Wiley Global Education.

Lin, Y. K., 1967. Probabilistic Theory of Structural Dynamics, McGraw-Hill, New York (Republished in 1976 by Krieger, Huntington, NY).

Lin, Y. K., 1970. First excursion failure of randomly excited structures. American Institute of Aeronautics and Astronautics Journal. 8 (4), 720–725.

Lin, Y. K., Cai, G. Q., 1995. Probabilistic Structural Dynamics-Advanced Theory and Applications, McGraw-Hill, New York.

Lindgren, G., Rootzen, H., Sandsten, M., 2013. Stationary Stochastic Processes for Scientists and Engineers-Chapman and Hall, CRC Press.

Lutes, L. D., Sarkani, S., 2004. Random Vibrations - Analysis of Structural and Mechanical Systems, Elsevier Butterworth-Heineman, Oxford, UK.

Matsuishi, M., Endo, T., 1968. Fatigue of Metals Subjected to Varying Stress. In Japan Society of Mechanical Engineering, pp. 37–40. Fukuoka, Japan.

Montgomery, D. C., Runger, G. C., 2018. Applied Statistics and Probability for Engineers, John Wiley & Sons, Inc, Hoboken, NJ.

Naess, A., Moe, V., 2000. Efficient path integration methods for nonlinear dynamic systems. Probabilistic Engineering Mechanics. 15, 221–231.

Naess, A., Moan, T., 2013. Stochastic Dynamics of Marine Structures, Cambridge University Press, Cambridge.

Newland, D. E., 1984. Random Vibrations and Spectral Analysis, Longman, New York.

Newland, D. E., 1991. An Introduction to Random Vibrations and Spectral Analysis. Second Edition, Longman, London.

Newland, D. E., 2005. An Introduction to Random Vibrations, Spectral & Wavelet Analysis. Third Edition, Dover Publications, Mineola, NY.

Newmark, N. M., 1959. A method of computation for structural dynamics. Journal of the Engineering Mechanics Division, ASCE. 85, 67–94.

Nigam, N. C., 1983. Introduction to Random Vibrations, The MIT Press, Cambridge, MA.

Nikolaidis, E., Ghiorel, D. M., Singhal, S. (eds.), 2004. Engineering Design Reliability Handbook, CRC Press, Boca Raton, FL.

Nowak, A. S., Collins, K. R., 2013. Reliability of Structures, Second Edition, CRC Press, Boca Raton, FL.

Ochi, M. K., 1992. Applied Probability and Stochastic Processes-In Engineering and Physical Sciences, John Wiley & Sons, New York.

Ochi, M. K., 1998. Ocean Waves - The Stochastic Approach, Cambridge University Press, Cambridge, UK.

Ochi, M. K., Ahn, K., 1994. Probability distribution applicable to non-Gaussian random processes. Probabilistic Engineering Mechanics. 9, 255–264.

Papoulis, A., 1962. The Fourier Integral and Its Applications, McGraw-Hill, New York.

Papoulis, A., Pillai, S. U., 2002. Probability, Random Variables, and Stochastic Processes. Fourth Edition, McGraw-Hill, Boston.

Parzen, E., 1962a. On estimation of a probability density function and mode. Annals of Mathematical Statistics. 33, 1065–1076.

Papadrakakos, M., Stefanou, G., Papadopoulos, V. (eds.), 2013. Computational Methods in Stochastic Dynamics, Springer, Dordrecht.

Parzen, E., 1962b. Stochastic Processes, Holden-Day, San Francisco.

Phoon, K. K., Ching, J. eds., 2015. Risk and Reliability in Geotechnical Engineering, CRC Press, Boca Raton, FL.

Podofillini, L., Sudret, B., Stojadinovic, B., Zio, E., Kröger, W. (eds.), 2015. Safety and Reliability of Complex Engineering Systems, CRC Press, Boca Raton, FL.

Pradlwarter, H. J., Schuëller, G. I., 1993. Equivalent linearization - a suitable tool for analyzing MDOF-systems. Probabilistic Engineering Mechanics. 8, 115–126.

Pradlwarter, H. J., 2001. Non-linear stochastic response distributions by local statistical linearization. International Journal of Non-Linear Mechanics. 36, 1135–1151.

Proppe, C., Pradlwarter, H. J., Schuëller, G. I., 2003. Equivalent linearization and Monte Carlo simulation in stochastic dynamics. Probabilistic Engineering Mechanics. 18, 1–15.

Rao, S. S., 2018. Mechanical Vibrations. In CM17 9NA. Pearson Education, Inc, United Kingdom.

Rice, S. O., 1944. Mathematical analysis of random noise. Bell System Technical Journal. 23, 282–332.

Rice, S. O., 1945. Mathematical analysis of random noise. Bell System Technical Journal. 24, 46–156.

Risken, H., 1984. The Fokker-Planck Equation-Methods of Solution and Applications, Springer-Verlag, New York.

Roberts, J. B., 1974. Probability of first passage failure for stationary random vibration. American Institute of Aeronautics and Astronautics Journal. 12, 1636–1643.

Roberts, J. B., 1981. Response of nonlinear mechanical systems to random excitation: Part II; equivalent linearization and other methods. Shock and Vibration Digest. 13, 15–29.

Roberts, J. B., 1991. Multiple solutions generated by statistical linearization and their physical significance. International Journal of Non-Linear Mechanics. 26, 945–959.

Roberts, J. B., Spanos, P. D., 1990. Random Vibration and Statistical Linearization, Wiley, New York.

Robson, J. D., 1963. An Introduction to Random Vibration, Edinburgh University Press, Edinburgh.

Rychlik, I., 1987. A new definition of the rainflow cycle counting method. International Journal of Fatigue. 9 (2), 119–121.

Rychlik, I., Johannesson, P., Leadbetter, M. R., 1997. Modelling and statistical analysis of ocean wave data using transformed Gaussian processes. Marine Structures. 10, 13–47.

Shinozuka, M., 1974. Digital Simulation of Random Processes in Engineering Mechanics with the Aid of FFT Techniques. In Stochastic Problems in Mechanics, pp. 277–286. University of Waterloo Press, Waterloo, Ontario, Canada.

Shinozuka, M., Deodatis, G., 1991. Simulation of stochastic processes by spectral representation. Applied Mechanics Reviews. 44 (4), 191–204.

Silverman, B. W., 1986. Density Estimation for Statistics and Data Analysis, Chapman and Hall, London.

Simiu, E., Scanlan, R. H., 1996. Wind Effects on Structures. Third Edition, John Wiley & Sons, New York, NY.

Sobczyk, K., 2013. Stochastic Differential Equations with Applications to Physics and Engineering, Springer, Berlin.

Sobczyk, K., Spencer, B. F., 1991. Random Fatigue; From Data to Theory, Academic, New York.

Socha, L., 2007. Linearization Methods for Stochastic Dynamic Systems, Springer, Berlin.

Society of Automotive Engineers (SAE), 1977. Fatigue Under Complex Loading, AE-6, SAE, Warrendale, PA.

Society of Automotive Engineers (SAE), 1988. Fatigue Design Handbook, AE-10, SAE, Warrendale, PA.

Soong, T. T., 1973. Random Differential Equations in Science and Engineering, Academic Press, New York.

Soong, T. T., 1981. Probabilistic Modeling and Analysis in Science and Engineering, Wiley, New York.

Soong, T. T., 2004. Fundamentals of Probability and Statistics for Engineers, John Wiley & Sons Ltd, the Atrium, Southern Gate, Chichester, West Sussex PO19 8SQ, England.

Soong, T. T., Grigoriu, M., 1993. Random Vibration of Mechanical and Structural Systems, Prentice Hall, Englewood Cliffs, New Jersey.

Spanos, P. D., 1981. Stochastic linearization in structural dynamics. Applied Mechanics Review. 34 (1), 1–8.

Spanos, P. D. ed., 1999. Computational Stochastic Mechanics, Balkema, Rotterdam.

St. Denis, M., Pierson, W. J., 1953. On the Motions of Ships in Confused Seas. Transactions of SNAME 61, 280–357.

Strutt, J. W., Rayleigh, L., 1877. The Theory of Sound, Macmillan, London (reprinted by Dover Publications, New York, in 1945).

Sun, J. Q., 1995. Random vibration analysis of a nonlinear system with dry friction damping by the short-time Gaussian cell mapping method. Journal of Sound and Vibration. 180 (5), 785–795.

Sun, J. Q., Hsu, C. S., 1990. The generalized cell mapping method in nonlinear random vibration based upon short-time Gaussian approximation. Journal of Applied Mechanics. 57, 1018–1025.

Svetlivsky, V. A., 2003. Statistical Dynamics and Reliability Theory for Mechanical Structures, Springer, Berlin.

Timoshenko, S. P., Young, D. H., Weaver, W. Jr., 1974. Vibration Problems in Engineering. Fourth Edition, Wiley, New York.

To, C. W. S., 2012. Nonlinear Random Vibration: Analytical Techniques and Applications. Second Edition, CRC Press, Boca Raton, FL.

To, C. W. S., 2010. Nonlinear Random Vibration: Computational Methods, Zip Publishing, Columbus, OH.

To, C. W. S., 2014. Stochastic Structural Dynamics: Applications to Finite Element Methods, Wiley, Chichester.

Vanmarcke, E., 1983. Random Fields: Analysis and Synthesis, The MIT Press, Cambridge, MA.

Wang, Y. G., 2010. Spectral fatigue analysis of a ship structural detail – A practical case study. International Journal of Fatigue. 32 (2), 310–317.

Wang, Y. G., 2010. Melnikov analysis of a ship's stability with water on deck. Journal of Marine Science and Technology. 18 (1), 106–111.

Wang, Y. G., 2014. A path integration algorithm for stochastic structural dynamic systems. Applied Mathematics and Computation. 228, 423–431.

Wang, Y. G., 2015. Robust frequency-domain identification of parametric radiation force models for a floating wind turbine. Ocean Engineering. 109, 580–594.

Wang, Y. G., 2017. Optimal threshold selection in the POT method for extreme value prediction of the dynamic responses of a Spar-type floating wind turbine. Ocean Engineering. 134, 119–128.

Wang, Y. G., 2018. A novel simulation method for predicting power outputs of wave energy converters. Applied Ocean Research. 80, 37–48.

Wang, Y. G., 2019. Efficient prediction of wave energy converters power output considering bottom effects. Ocean Engineering. 181, 89–97.

Wang, Y. G., 2020. Efficient computational method for the dynamic responses of a floating wind turbine. Ships and Offshore Structures. 15 (3), 269–279.

Wang, Y. G., 2020. Bottom effects on the tower base shear forces and bending moments of a shallow water offshore wind turbine. Marine Structures. 70, 102705.

Wang, Y. G., 2020. Predicting absorbed power of a wave energy converter in a nonlinear mixed sea. Renewable Energy. 153, 362–374.

Wang, Y. G., 2021. Predicting loads and dynamic responses of an offshore wind turbine in a nonlinear mixed sea. Ships and Offshore Structures. 16 (4), 373–385.

Wang, Y. G., 2021. A robust methodology for predicting extreme structural responses of offshore wind turbines. Ships and Offshore Structures. 16 (10), 1078–1086.

Wang, Y. G., 2022. An efficient method for predicting long term extreme design forces of wave energy converters. Applied Ocean Research. 121, 103094.

Wang, Y. G., 2023. Robust adaptive analysis of dynamic responses of offshore sustainable energy systems. Ocean Engineering. 273, 114022.

Wang, Y. G., Huang, Z. L., Tan, J. H., 2008. First passage probability of nonlinear ship rolling in random seas. Journal of Ship Mechanics. 12 (6), 870–879.

Wang, Y. G., Liu, X. J., Tan, J. H., 2007. Calculation of the level crossing rates of a random oscillation system by time domain simulation. Journal of Vibration and Shock. 26 (2), 37–38, 55.

Wang, Y. G., Tan, J. H., 2007. Global geometric analysis of ship rolling and capsizing in random waves. China Ocean Engineering. 21 (4), 577–586.

Wang, Y. G., Tan, J. H., 2007. A review of various methods of response analysis for nonlinear marine structures under random excitations. Ocean Engineering. 25 (4), 112–119.

Wang, Y. G., Tan, J. H., 2007. Path integral solution of a strongly nonlinear stochastic oscillation system. Journal of Vibration and Shock. 26 (11), 153–155.

Wang, Y. G., Tan, J. H., 2008. Markov modeling for slow drift oscillations of moored vessels in irregular waves. Journal of Ship Mechanics. 12 (3), 368–376.

Wang, Y. G., Tan, J. H., 2009. Nonlinear analysis of slow drift extreme responses of a compliant offshore structure. Acta Mechanica Sinica. 25 (5), 651–657.

Wang, Y. G., Tan, J. H., Xue, L. P., 2009. On the exceedance probabilities of extreme drift motions of an offshore structure. China Ocean Engineering. 23 (1), 27–35.

Wang, Y. G., Tan, J. H., 2009. Analyzing an ocean structure's dynamic stability by using Melnikov method. Ocean Engineering. 27 (1), 77–82.

Wang, Y. G., Tan, J. H., 2010. Research progress on ship stability and capsizing in random waves. Journal of Ship Mechanics. 14 (1), 201–211.

Wang, Y. G., Wang, L. F., 2018. Towards realistically predicting the power outputs of wave energy converters: Nonlinear simulation. Energy. 144, 120–128.

Wang, Y. G., Xia, Y. Q., 2012. Simulating mixed sea state waves for marine design. Applied Ocean Research. 37, 33–44.

Wang, Y. G., Xue, L. P., Tan, J. H., 2007. Surge response of a compliant offshore structure by time domain simulation. Journal of Shanghai Jiao Tong University (Science). E-12 (6), 838–844.

Wiener, N., 1933. The Fourier Integral and Certain of Its Applications, Cambridge University Press, New York.

Wilson, E. L., Farhoomand, I., Bathe, K. J., 1973. Nonlinear dynamic analysis of complex structures. International Journal of Earthquake Engineering and Structural Dynamics. 1, 241–252.

Winterstein, S. R., 1988. Nonlinear vibration models for extremes and fatigue. Journal of Engineering Mechanics. 114 (10), 1772–1790.

Wirsching, P. H., Chen, Y. N., 1988. Consideration of probability based fatigue design criteria for marine structures. Marine Structures. 1, 23–45.

Wirsching, P. H., Light, M. C., 1980. Fatigue under wide band random stresses. ASCE Journal of Structural Division. 106, 1593–1607.

Wirsching, P. H., Paez, T. L., Ortiz, H., 1995. Random Vibrations-Theory and Practice, John Wiley and Sons, New York.

Yang, C. Y., 1986. Random Vibration of Structures, John Wiley and Sons, New York.

Index